ENVIRONMENTAL INORGANIC CHEMISTRY FOR ENGINEERS

ENVIRONMENTAL INORGANIC CHEMISTRY FOR ENGINEERS

DR. JAMES G. SPEIGHT
CD & W inc., Laramie, Wyoming, United States

Butterworth-Heinemann is an imprint of Elsevier
The Boulevard, Langford Lane, Kidlington, Oxford OX5 1GB, United Kingdom
50 Hampshire Street, 5th Floor, Cambridge, MA 02139, United States

Library of Congress Cataloging-in-Publication Data
A catalog record for this book is available from the Library of Congress

British Library Cataloguing-in-Publication Data
A catalogue record for this book is available from the British Library

ISBN: 978-0-12-849891-0

For information on all Butterworth-Heinemann publications
visit our website at https://www.elsevier.com/books-and-journals

Publisher: Matthew Deans
Acquisition Editor: Ken McCombs
Editorial Project Manager: Peter Jardim
Production Project Manager: Kiruthika Govindaraju
Cover Designer: Vitoria Pearson

Typeset by SPi Global, India

CONTENTS

Author Biography *ix*
Preface *xi*

1. Inorganic Chemicals in the Environment **1**
1.1 Introduction 1
1.2 The Environment 8
1.3 Inorganic Chemistry and the Environment 30
1.4 Use and Misuse of Inorganic Chemicals 33
1.5 Inorganic Chemicals in the Environment 37
1.6 Chemistry and Engineering 44
References 46
Further Reading 49

2. Inorganic Chemistry **51**
2.1 Introduction 51
2.2 Nomenclature of Inorganic Compounds 56
2.3 Classification of Inorganic Chemicals 63
2.4 The Periodic Table 81
2.5 Bonding and Molecular Structure 83
2.6 Reactions and Stoichiometry 93
2.7 Acid-Base Chemistry 98
2.8 Minerals 108
References 109

3. Industrial Inorganic Chemistry **111**
3.1 Introduction 111
3.2 The Inorganic Chemicals Industry 115
3.3 The Production of Inorganic Chemicals 122
3.4 Inorganic Polymers 161
3.5 Inorganic Pigments 164
References 168

4. Properties of Inorganic Compounds **171**
4.1 Introduction 171
4.2 Bond Lengths and Bond Strengths 181
4.3 Physical Properties 187
4.4 Critical Properties 213

4.5 Reactive Chemicals 216
4.6 Catalysts 223
4.7 Corrosivity 227
References 228
Further Reading 229

5. Sources and Types of Inorganic Pollutants 231
5.1 Introduction 231
5.2 Sources 238
5.3 Fly Ash and Bottom Ash 267
5.4 Characterization of Inorganic Compounds 274
References 281
Further Reading 282

6. Introduction Into the Environment 283
6.1 Introduction 283
6.2 Minerals 294
6.3 Release Into the Environment 305
6.4 Types of Chemicals 319
6.5 Physical Properties and Distribution in the Environment 323
References 332

7. Transformation of Inorganic Chemicals in the Environment 333
7.1 Introduction 333
7.2 Chemical and Physical Transformation 338
7.3 Inorganic Reactions 345
7.4 Catalysts 355
7.5 Sorption and Dilution 357
7.6 Biodegradation 364
7.7 Chemistry in the Environment 366
References 382

8. Environmental Regulations 383
8.1 Introduction 383
8.2 Environmental Impact of Production Processes 389
8.3 Environmental Regulations in the United States 394
8.4 Outlook 421
References 426
Further Reading 426

9. Removal of Inorganic Compounds From the Environment **427**
- 9.1 Introduction 427
- 9.2 Cleanup 435
- 9.3 Bioremediation 453
- 9.4 Remediation of Heavy Metal-Contaminated Sites 455
- 9.5 Pollution Prevention 473
- References 477

Appendix *479*
Conversion Factors *527*
Glossary *533*
Index *567*

AUTHOR BIOGRAPHY

Dr. James G. Speight CChem., FRSC, FCIC, and FACS earned his BSc and PhD degrees from the University of Manchester, England; he also holds a DSc in geologic sciences and a PhD in petroleum engineering. Dr. Speight is the author of more than 70 books in fossil fuel and engineering and environmental sciences. Now, being an independent consultant, he was formerly the CEO of the Western Research and has served as adjunct professor in the Department of Chemical and Fuels Engineering at the University of Utah and in the Departments of Chemistry and Chemical and Petroleum Engineering at the University of Wyoming. In addition, he has also been a visiting professor in chemical engineering at the following universities: the University of Missouri-Columbia, the Technical University of Denmark, and the University of Trinidad and Tobago.

Dr. Speight was elected to the Russian Academy of Sciences in 1996 and awarded the gold Medal of Honor that same year for the outstanding contributions to the field of petroleum sciences. He has also received the Scientists without Borders Medal of Honor of the Russian Academy of Sciences. In 2001, the academy also awarded Dr. Speight the Einstein medal for the outstanding contributions and service in the field of geologic sciences.

PREFACE

The latter part of the 20th century saw the realization arise that all chemicals can act as environmental pollutants, and in addition, there came the realization that emissions of inorganic chemicals such as carbon dioxide (CO_2), methane (CH_4), and nitrous oxide (N_2O) to the atmosphere and a host of other inorganic chemicals to water systems and to land systems had either a direct or indirect impacts on the various floral and faunal systems. As a result, unprecedented efforts have been made to reduce all global emissions and chemicals disposal in order to maintain a *green perspective*. Furthermore, operations have been designed to reduce the direct emissions of inorganic chemical products (such as carbon dioxide, sulfur oxides, and nitrogen oxides) into the air (*the atmosphere*), inorganic chemicals into water systems (*the aquasphere*), and inorganic chemicals on to the soil (*the terrestrial biosphere*). This has been accompanied by efforts to recycle and to reuse as much of these chemicals and chemical waste as possible.

The primary purpose of this book is to focus on the various issues related to the production and use of inorganic chemical issues that are the focus of any environmental chemistry program. In the context of inorganic chemistry, the book also presents an understanding of information on environmentally important physiochemical properties of inorganic chemicals for use by engineers and managers as well as any nonchemists. This book is a companion text to a book (Environmental Organic Chemistry for Engineers, James G. Speight, Elsevier, 2017) previously prepared for environmentally important effects and properties of organic chemicals. Like the "organics" book, this "inorganics" book describes available information that relates to the properties of inorganic chemicals and the effects of these chemicals on the environment. This information supports the primary objective of the book, which is to assist environmental engineers and managers (many of whom may not have a detailed knowledge or understanding of inorganic environmental chemistry) in overcoming the common problem of property data gaps and developing timely responses to environmental problems. This "inorganics" book presents not only generic discussion of these properties but also a summary of environmentally important data for the most common elements and pollutants.

The topics covered in this book are the basis topics that serve to introduce the reader to not only inorganic chemistry but also the effect of inorganic chemicals on various ecosystems. Basic rules of nomenclature are presented. Understanding the mechanism of how a reaction takes place is particularly crucial in this and of necessity; the book brings a logic and simplicity to the reactions of the different functional groups. This in turn transforms a list of apparently unrelated facts into a sensible theme. Thus, this chapter will serve as an introduction to the physicochemical properties of inorganic chemicals and their effect on the floral and faunal environments.

The book will serve as an information source the engineers in presenting details of the various aspects of inorganic chemicals as they pertain to pollution of the environment. To accomplish this goal, the initial section (Chapters 1–4) presents an introduction to and a description of the nomenclature of inorganic compounds and the properties of these materials. The remaining part of the book (Chapters 5–9) presents information relevant to the sources of inorganic contaminants, the behavior of inorganic chemicals, the fate and consequences of chemicals in the environment, and cleanup of the environment.

From the book, the reader will gain an understanding of the fundamental inorganic chemistry and chemical processes that are central to a range of important environmental problems and to utilize this knowledge in making critical evaluations of these problems. Specific knowledge will be in the area of (i) an understanding of the chemistry of the stratospheric ozone layer and of the important ozone depletion processes; (ii) an understanding of the chemistry of important tropospheric processes, including photochemical smog and acid precipitation; (iii) understanding of the basic physics of the greenhouse effect and of the sources and sinks of the family of greenhouse gases; (iv) an understanding of the nature, reactivity, and environmental fates of toxic inorganic chemicals; and (v) an understanding of societal implications of some environmental problems. An appendix contains a selection of tables that contain data relating to the properties and characterization of the elements and inorganic compounds, and a comprehensive glossary will help the reader to understand the common terms that are employed in this field of science and engineering.

Dr. James G. Speight
Laramie, Wyoming
Jan. 2017

CHAPTER ONE

Inorganic Chemicals in the Environment

1.1 INTRODUCTION

Environmental chemistry focuses on environmental concerns about materials, energy, and production cycles and demonstrates how fundamental chemical principles and methodologies can protect the floral (plant) and faunal (animal, including human) species within the environment (Anastas and Kirchhoff, 2002). More specifically, the principles of chemistry can be used to develop how global sustainability can be supported and maintained. For this, future environmental chemists and environmental engineers must acquire the scientific and technical knowledge to design products and chemical processes. They must also acquire an increased awareness of the environmental impact of chemicals on the environment and develop an enhanced awareness of the importance of sustainable strategies in chemical research and the chemical industry, specifically in the context of this book the inorganic chemicals industry.

By way of introduction, although other classification systems have been published, a general classification of chemical pollutants is based on the chemical structure of the pollutant and includes (i) organic chemical pollutants and (ii) inorganic chemical pollutants. For the purposes of this text, organic chemical pollutants are those chemicals of organic origin or that could be produced by living organisms or are based of matter formed by living organisms (Speight, 2017a).

On the other hand, inorganic chemical pollutants are those chemicals of mineral origin in (not produced by living organisms). In general, substances of mineral origin (such as ceramics, metals, synthetic plastics, and water) as opposed to those of biological or botanical origin (such as crude oil, coal, wood, and food). In addition, minerals are the inorganic, crystalline solid

Environmental Inorganic Chemistry for Engineers
http://dx.doi.org/10.1016/B978-0-12-849891-0.00001-1

that makes up rocks. With certain exceptions, inorganic substances do not contain carbon or its compounds. In scientific terms, no clear line divides organic and inorganic chemistry.

Inorganic chemistry focuses on the classification of inorganic compounds based on the properties of the compound(s) (Weller et al., 2014). Partly, the classification focuses on the position in the periodic table (Fig. 1.1) of the heaviest element (the element with the highest atomic weight) in the compound, partly by grouping compounds on the basis of structural similarities. Also, inorganic compounds are generally structured by ionic bonds and do not contain carbon chemically bound to hydrogen (hydrocarbons) or any of their derivatives that contain elements such as nitrogen, oxygen, sulfur, and metals. Examples of inorganic compounds include sodium chloride (NaCl) and calcium carbonate ($CaCO_3$) and pure elements (Cox, 1995).

Thus, this text relates to an introduction to the planned and unplanned effects of inorganic chemicals on the various environmental systems. Inorganic chemicals are an essential component of life, but some chemicals are extremely toxic and can severely damage the floral (plant life) and faunal (animal life) environment (Table 1.1).

As with organic chemicals (Speight, 2017a), contamination of the environment by inorganic chemicals is a global issue, and toxic inorganic chemicals are found practically in all ecosystems because, at the end of the various inorganic chemical life cycles, inorganic chemicals have been either recycled for further use or sent for disposal as chemical waste (Bodek et al., 1988). Current regulations do not permit the unmanaged disposal of chemical waste but, in the past (particularly in the first decades of the 20th century), the inappropriate management of chemical waste (e.g., through haphazard disposal and unregulated burning) has led to a series of negative and lingering impacts on the floral and faunal species that are part of the environment.

Briefly, inorganic chemistry deals with the synthesis and behavior of inorganic and organometallic compounds. The exception to the subdiscipline is the multitude of chemical compounds that fall within the subdiscipline of organic chemistry that covers the multitude of organic compounds (carbon-based compounds, usually containing C—H bonds). The distinction between the two subdisciplines is far from absolute, as there is much overlap within the subdiscipline of organometallic chemistry. Nevertheless, the principles of inorganic chemistry have application in every aspect of the chemical industry, including materials science, catalysis, surfactants, pigments, coatings, medications, fuels, and agriculture.

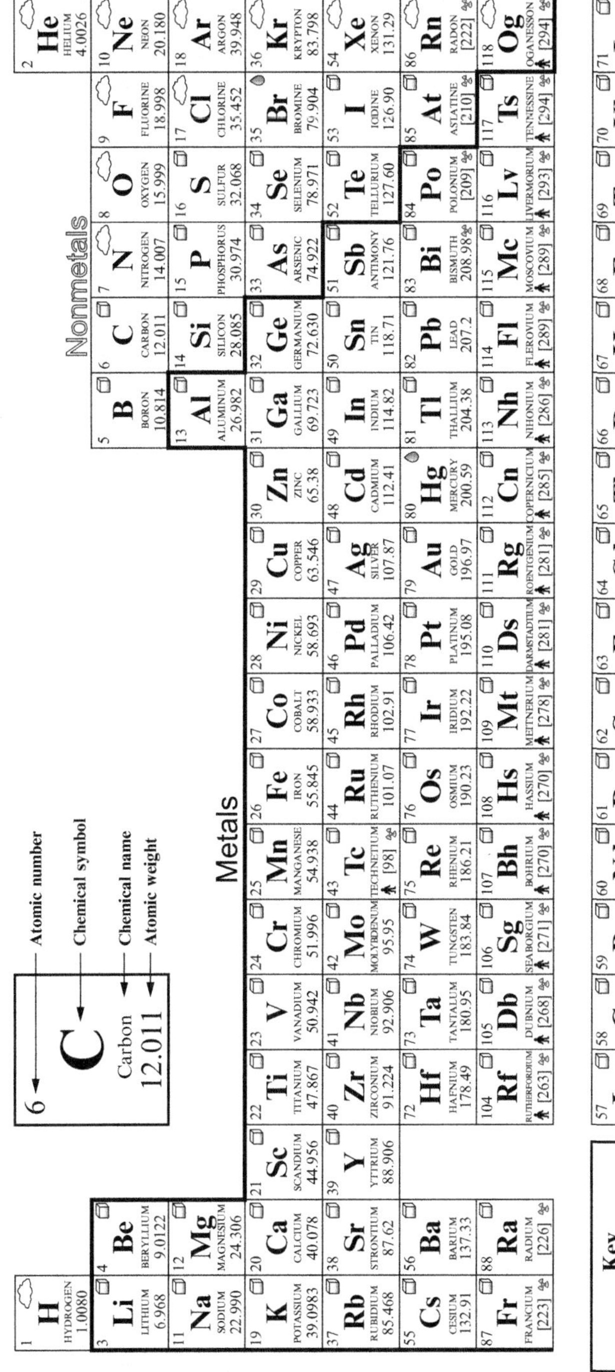

Fig. 1.1 The periodic table of the elements.

Table 1.1 Examples of the Classification of Elements According to Their Effects in the Biological Systems[a]

Essential
O, C, H, N, P, Na, K, Mg, Cl, Ca, S
Trace
I, Fe, Cu, Zn, Mn, Co, Mo
Nonessential
Al, Sr, Ba, Sn
Toxic
Cd, Pb, Hg

[a]Essential elements are necessary for life processes; trace elements are also necessary for life processes; nonessential elements are not essential. If they are absent, other elements may serve the same function; toxic elements disturb the natural functions of the biological system.

Many inorganic compounds are ionic compounds, consisting of cations and anions joined by ionic bonding (Chapter 2). An ionic compound is a chemical compound comprising ions (charged species) that are held together by electrostatic forces (ionic bonding). These can be simple ions such as sodium (Na^+) and chloride (Cl^-) in sodium chloride, sodium hydroxide that consists of sodium cations (Na^+) and hydroxide anions (OH^-), or polyatomic species such as the ammonium (NH_4^+) and carbonate (CO_3^{2-}) ions in ammonium carbonate. In any salt, the proportions of the ions are such that the electric charges cancel out, so that the bulk compound is electrically neutral. The ions are described by their oxidation state, and their ease of formation can be inferred from the ionization potential (for cations) or from the electron affinity (anions) of the parent elements (Chapter 2). Ionic compounds containing hydrogen ions (H^+) are classified as acids, and those containing basic ions, such as the hydroxide anion (OH^-) or oxide anion (O^{2-}), are classified as bases. Some ions are classed as amphoteric because of the ability of these ions to react with either an acid or a base (Davidson, 1955). This is also true of some compounds with ionic character, typically oxide derivatives or hydroxide derivatives of the less electropositive metals (which results in the compound having significant covalent character), such as zinc oxide (ZnO) aluminum hydroxide [$Al(OH)_3$], aluminum oxide (Al_2O_3), and lead(II) oxide (PbO).

Important classes of inorganic compounds are the oxide derivatives ($—O^{2-}$), the carbonate derivatives ($—CO_3^{2-}$), the sulfate derivatives

($—SO_4^{2-}$), and the halide derivatives (such as the chlorides, $—Cl^{-1}$)—all of these derivatives occur as minerals in the Earth (Chapter 2). Many inorganic compounds are characterized by high melting points (Chapter 4). Other important features include their high melting point and ease of crystallization, where some salts are very soluble in water (such as sodium chloride, NaCl) and crystallize form concentrated solutions of the salt others (such as silica—silicon dioxide, SiO_2). Thus,

Inorganic chemistry: oxygen and metals are the dominant elements.

Organic chemistry: carbon and hydrogen are the dominant elements.

By way of further definition and to alleviate any potential confusion, *bio-inorganic chemistry* (*bioinorganic chemistry* and *biological inorganic chemistry*) is a subcategory of chemistry that examines the role of metals in biology systems including the associated environmental issues (Bertini et al., 1994; Fraústo da Silva and Williams, 2001). Thus, bioinorganic chemistry includes the study of both natural phenomena such as the behavior of metalloproteins and artificially introduced metals, including those that are nonessential (Table 1.1) in medicine and toxicology. Many biological processes, such as respiration, depend upon the molecular species that fall within the realm of inorganic chemistry and, as a blend of biochemistry and inorganic chemistry, bioinorganic chemistry is important in elucidating the activity of proteins and the effect of the properties of inorganic compounds as they pertain to the activity and well-being of floral and faunal organisms and an understanding of the various elements inorganic compounds on floral and faunal systems.

As a side but relevant note, it must be assumed that all chemicals are toxic unless proved otherwise. In relation to the environmental effects of inorganic chemicals, consideration must also be given to the effect of the so-called *harmless chemicals* (the *indigenous chemicals* and the *natural products chemicals*) when they are added back to the environment in amounts that exceed the natural abundance. Within the local environment, these chemicals will be present in a measurable concentration, but the flora and fauna present in that ecosystem may be fatally susceptible to the effects of such chemicals when they are present in a concentration that is above the indigenous (natural) concentration of the chemicals. Using the human experience as an example, a sprinkling salt on a cooked vegetable or meat may add to the taste of the meal. However, it is extremely dangerous (the consequences can even be fatal), and therefore, unadvisable for a human to attempt to consume several ounces of salt with that same meal. Not only would the taste be ruined, but also the high concentration of salt could have a serious health effects (even death) on the consumer.

Thus, there is an increase in environmental issues and problems that can be partially explained because of the use of chemicals. In fact, many man-made chemicals are found in the most remote places in the environment. However, in order to successful manage the environment and protect the flora and fauna from such chemicals, knowledge of chemical, specifically inorganic chemicals (in the context of this book), is a decided advantage. Moreover, the chief reason for studying this subject is not only to understand the effects of inorganic chemicals on the environment, which may be caused by unforeseen side effects of a chemical substance during its production, transport, use, and disposal. These effects provide the motivation for the build of scientific knowledge on the effects of inorganic chemicals on the floral and faunal environments. Ideally, environmental scientists and environmental engineers should be able to predict the possible (if not, likely) effects of an inorganic chemical directly on the environment even before the chemical substance is released, enabling a more realistic appraisal to be made of any effects of the chemical on an ecosystem.

Thus, a first approximation to predicting a potentially harmful inorganic chemical may involve the following criteria: (i) whether the chemical is biologically essential; (ii) whether the chemical is biologically nonessential; (iii) whether the chemical is toxic in a variety of concentrations; (iv) whether the chemical is likely to undergo transformation in the environment and form highly stable inert compounds; (v) whether the chemical is unlikely to undergo transformation in the environment and form highly stable inert compounds; (vi) whether the chemical persists in the environment; (vii) whether the chemical interacts favorably or adversely with the flora and fauna of the ecosystem into which the chemical is released; (viii) whether the chemical is environmentally mobile in, or interferes with, any of the biological or biogeochemical cycles; and (ix) whether the chemical is environmentally immobile in, or interferes with, any of the biological or biogeochemical cycles.

Thus, as for any technical discipline, the nonchemist is faced with understanding many new terms that are related to the discipline of inorganic chemistry and that need to be understood to place inorganic chemicals in the correct environmental context. Some terms may seem familiar, but in chemistry, especially in inorganic chemistry where the terms may have meanings that are not quite the same as when used in popular commentaries, even at various technical meetings. For example, in the subdiscipline of inorganic chemistry and, indeed, in all subdisciplines of chemistry, the terms need to have definite and specific meanings to make the subject matter understandable, and this is the *raison d'être* for the acceptable systems of nomenclature (Chapter 2). One of the purposes of any system of chemical nomenclature is to provide definitions for many of these terms in a form and at a level that will make the meaning clear to technical persons in other discipline who need to deal with the various aspects of chemistry.

In this respect, the International Union of Pure and Applied Chemistry (IUPAC)—an international organization of chemists and national chemistry societies and the world authority on chemical nomenclature and terminology with a secretariat in Research Triangle Park, North Carolina, https://iupac.org/who-we-are/—makes the final determination of terminology and nomenclature in chemistry. Among other projects, this organization (IUPAC) authorizes and establishes systematic rules for naming compounds so that any inorganic (or organic) chemical structure can be defined uniquely and in a meaningful manner. Compounds are frequently called by common names or by trade names, often because the IUPAC names may be long and complex and difficult to understand for the nonchemist or the company need to protect the identity of the chemical; this is especially true in the pharmaceutical industry where the true (correct) chemical name could reveal the company secrets of drug synthesis. However, the IUPAC name permits the chemist and the nonchemist to know the structure of any chemical compound based on the rules of the IUPAC terminology, while the common name or trade name may be outside of the realms of the reality of chemical nomenclature.

Engineering students, on the other hand, undergo a much different form of training and education when compared with training to which the student of chemistry is subjected. While some (but not all) engineering disciplines (especially the chemical engineer) may require some background knowledge of chemistry, the practicing engineer is more concerned with practical applications of chemical synthesis (such as reactor construction, reactor operation, process parameters, and product yield), and there are differences in, for example, reactor novelty and reactor scale—areas that are not always pertinent to the chemist outside of laboratory chemistry who is accustomed to employing scientific glassware for his/her laboratory experiments.

Thus, a chemist is more likely to be engaged in developing new compounds and materials using novel or new synthetic routes, while a chemical engineer is more likely to be working with existing substances and improving the reaction parameters to suit industrial synthesis of the chemical. A chemist may be involved in laboratory synthesis program to produce several grams of a new compound, while a chemical engineer will focus on scale up of the synthetic process to produce the chemical (say, on a tonnage scale) at a profit. Thus, the chemical engineer will be more concerned with heating and cooling large reaction vessels, pumps and piping to transfer chemicals, plant design, plant operation, and process optimization, while a chemist will be more concerned with establishing the parameters of the reaction from which the engineer will design the plant. However, these differences are generalizations, and there is often much overlap. As a consequence, and throughout the pages of this book, the reader will be presented with the definitions and explanations of terms related to inorganic chemistry and the means by which inorganic chemistry can be understood and used (Weller et al., 2014).

An alternative quantitative approach to inorganic chemistry focuses on properties of inorganic chemicals and the energies of the various reactions (Chapter 4). This approach is highly traditional and empirical, but it is useful in terms of understanding the behavior of inorganic chemicals in the environment (Chapter 7). The general concepts of inorganic chemistry that fall under the umbrella of thermodynamics include (i) acidity or alkalinity, (ii) redox potential, and (iii) phase changes in addition to (iv) the Born-Haber cycle, which is used for assessing the energies of elementary processes such as electron affinity and hence theoretical reactivity of inorganic chemicals (Moore and Stanitski, 2014).

1.2 THE ENVIRONMENT

There is the need for two definitions to enhance the understanding of the concepts presented in this book, thus: (i) ecosystem and (ii) environment. There is a general tendency to use these terms interchangeably, but to be more correct, the definitions are (i) the term *ecosystem* relates to a community of organisms together within their physical environment, viewed as a system of interacting and interdependent relationships and including such processes as the flow of energy through trophic levels and the cycling of chemical elements and compounds through living and nonliving components of the system and (ii) the term *environment* relates to the conditions that surround someone or something; the conditions and influences that affect the growth, health, progress, etc., of someone or something; the total living and nonliving conditions (internal and external surroundings) that are an influence on the existence and complete life span of the organism.

Thus, the natural environment encompasses all living and nonliving flora and fauna that occur naturally and environment encompasses the interaction of all living species, climate, weather, and natural resources that affect survival and activity of the flora and fauna (Manahan, 2009). Furthermore, the concept of the *natural environment* can be distinguished by components: (i) complete ecological units that function as natural systems without massive civilized human intervention, including the atmosphere, all vegetation, microorganisms, soil, rocks, and natural phenomena that occur within the boundaries of the ecological unit and (ii) the natural resources and physical phenomena that lack clear-cut boundaries, such as air, water, and climate, as well as energy, radiation, and other effects that do not originate from anthropogenic (human) activity. In contrast to the natural environment is the anthropogenic environment in which human activity has fundamentally transformed landscapes such as urban settings and agricultural land conversion, the natural environment is greatly modified into a simplified human environment.

In fact, there has been the suggestion (from the International Geological Congress in Cape Town) that the name of the present geologic name of the Holocene (Fig. 1.2) should be changed to Anthropocene (Pöschl and Shiraiwa, 2015; Robinson, 2016). This is an epoch that is defined by human impact on the environment and is an indication that humans have changed the planet. However, the proponents of this change—while correct to a point—seem to have ignored the fact that the Earth is in an interglacial period when the climate is expected to change (as the Earth warms due to natural effects), and the human effects may be somewhat less than calculated (Speight and Islam, 2016). Defining an epoch normally relies on evidence laid down over thousands or millions of years and whether renaming the Holocene to the Anthropocene remains to be clarified and fully justified.

Renaming the Holocene aside, to understand the influence of any chemical on the environment, it is first necessary to understand the structure of the environment though an understanding of the and the various subdivisions of the air, water, and land. When an inorganic chemical or any chemical for that matter (Speight, 2017a) is introduced into the environment, it is eventually distributed among the four major environmental compartments, which are often referred to as separate but interrelated ecosystems: (i) air, (ii) water, (iii) soil, and (iv) biota (living organisms). Depending upon the situation, further subdivision of the categories can be made into (i) floral systems, i.e., plant systems and (ii) faunal systems, i.e., animals, including human systems.

The fraction of the chemical that will move into each environmental subdivision is governed by the physical properties and the chemical

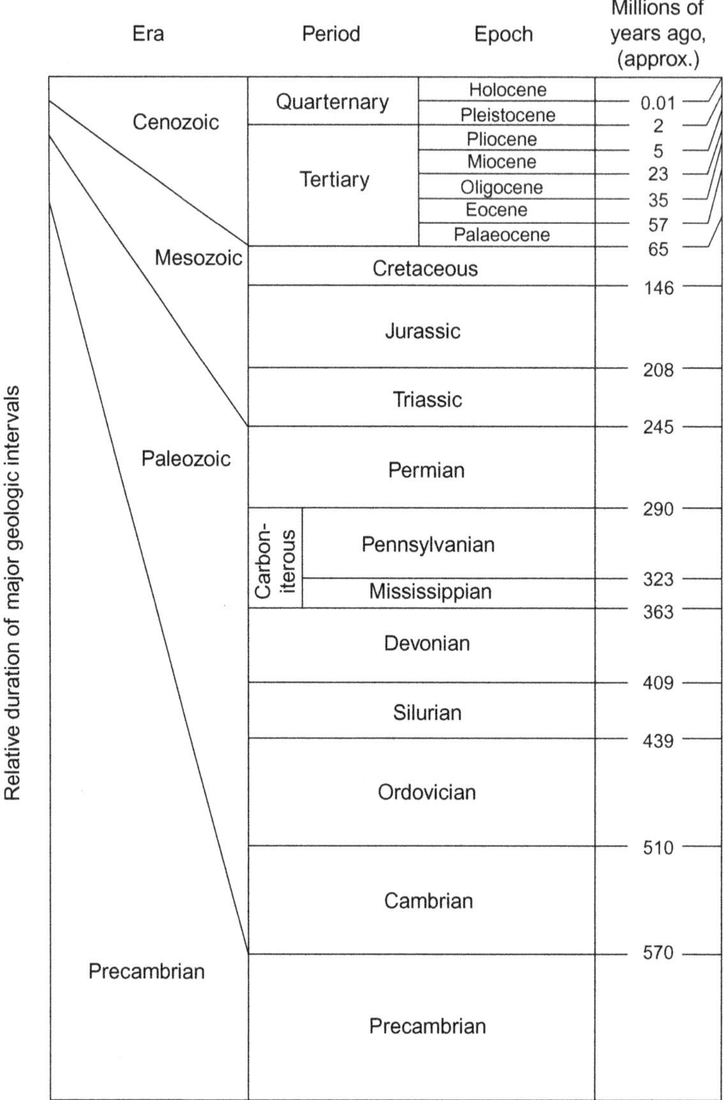

Fig. 1.2 Geologic era, epochs, and period.

properties of the chemical (Table 1.2) (Hickey, 1999; Yaron et al., 2011). In addition, the distribution of inorganic chemicals in the environment is governed by physical processes such as solubility in aqueous systems, sedimentation, adsorption, and volatilization evaporation, or sublimation and the mechanisms by which these chemicals can be degraded by chemical

Table 1.2 Important Common Physical Properties of Inorganic Compounds

State: Gas
Density
Critical temperature, critical pressure (for liquefaction)
Solubility in water, selected solvents
Odor threshold
Color
Diffusion coefficient
State: Liquid
Vapor pressure-temperature relationship
Density; specific gravity
Viscosity
Miscibility with water, selected solvents
Odor
Color
Coefficient of thermal expansion
Interfacial tension
State: Solid
Melting point
Density
Odor
Solubility in water, selected solvents
Coefficient of thermal expansion
Hardness/flexibility
Particle size distribution/physical form, such as fine powder, granules, pellets, lumps
Porosity

processes and/or biological processes. In contrast, the solubility of organic chemicals in aqueous systems is somewhat less than the solubility of inorganic chemicals, whereas organic chemicals (Speight, 2017a) are typically less prone to being soluble in aqueous systems but may be more subject to evaporation or vaporization than inorganic chemicals. In addition, chemical transformation processes (Chapter 7) generally occur in the atmosphere, in water, or in the soil and follow one of four reactions such as oxidation, reduction, hydrolysis, and photolysis.

Biological mechanisms in soil and living organisms typically utilize oxidation, reduction, and hydrolysis to degrade chemicals (Speight and Arjoon, 2012). The process of degradation will largely be governed by the compartment (water, soil, atmosphere, and biota) in which the inorganic chemical is distributed and the distribution is governed, in turn, by the physical processes already mentioned (i.e., solubilization, sedimentation, adsorption, and volatilization/evaporation/sublimation). The latter processes—i.e., volatilization, evaporation, and sublimation—are generally less obvious or effective in the field of inorganic chemistry as they are in the field of organic chemistry (Speight, 2017a).

The impact on the environment of the transformation gin the chemical state of inorganic chemicals is only partially elucidated but can be expected to be significant in many cases. Changes in the atmospheric abundance of various gases have been claimed (with some justification) to lead to observable changes in the climate of the Earth, including (i) changes in temperature, (ii) patterns of precipitation, and (iii) the frequency of occurrence of extreme natural events, such as hurricanes. Ecosystem damage also results from regional and global pollution. Moreover, acidic precipitation (acid rain) is claimed to be the reason for the suppression life in several lakes of North America and Europe and, together with enhanced ozone (O_3) levels, to have damaged forests in those same parts of the world.

In fact, when assessing the impact of inorganic chemicals on the environment, the most critical characteristics are (i) the types and properties of the discharged chemicals, which depend on the type of chemicals and processes used to produce the chemicals and (ii) the amount and concentration of the discharged chemicals. Solid waste (containing inorganic chemicals) and/or gaseous emissions generated from industrial sources also contribute to the amount and concentration (and effects) of inorganic chemicals in the environment.

This has led to the introduction of the concept of an *emissions factor* that is a representative value that attempts to relate the quantity of a pollutant released to the atmosphere with an activity associated with the release of that pollutant. The emission factor is usually expressed as the weight of pollutant divided by a unit weight, volume, distance, or duration of the activity emitting the pollutant (such as the kilogram of pollutant per kilogram of produced product). The factor facilitates an estimation of emissions from various sources of air pollution. In most cases, these factors may be averages of all available data of acceptable quality, which are generally (correctly or incorrectly) assumed to be representative of long-term averages for all facilities in the source category (i.e., a population average). Thus, the general equation for the estimation of emissions is

$$E = A \times \mathrm{EF} \times (1 - \mathrm{ER}/100)$$

In this equation, E = emissions, A = activity rate, EF = emission factor, and ER = overall emission reduction efficiency, %.

However, caution is advised when making such assumptions because the word *average* is not always indicative of the damage to the environment by that amount of chemical. The upper level data (used to compute the average) can be amounts of the chemical that lead to severe destruction of an ecosystem. On the other hand, the lower level data (used to compute the average) can be amounts of the chemical that can be tolerated by the ecosystem without causing serious disruption of that ecosystem. In either case, if there are data points that are *flyers* (data points that lie outside of the acceptable limits of variance—another assumption that may be erroneous), such data may skew the factor to the high end or low end. Or, if there are data points that could be classed as *flyers* and that have been included in the calculation of the factor, such data may skew the factor to the high end or low end.

1.2.1 Structure of the Atmosphere

The atmosphere is the gaseous envelope that surrounds the solid body of the Earth. Although the atmosphere has a thickness on the order of 600 miles, approximately half of the atmospheric mass is concentrated in the lower 3 miles. The atmosphere (i) is a protective blanket that nurtures life on the Earth and protects the planet from the hostile environment of outer space; (ii) is the source of carbon dioxide for plant photosynthesis and of oxygen for respiration; (iii) provides the nitrogen that nitrogen-fixing bacteria and ammonia-manufacturing industrial plants use to produce chemically bound nitrogen, an essential component of molecules of life; (iv) transports water from the oceans to land, thus acting as the condenser in a vast solar powered still; and (v) serves a vital protective function, absorbing harmful ultraviolet radiation from the sun and stabilizing the temperature of the Earth.

Physically, the atmosphere of the Earth is the thin and fragile envelope of air surrounding the Earth that is held in place around the Earth by gravitational attraction and that has a substantial effect on the environment. The total dry mass of the atmosphere (annual mean), three quarters of which is within approximately 36,000 ft of the surface, is estimated to be more than 5×10^{21} tons (Trenberth and Guillemot, 1994). In addition, the atmosphere is largely transparent and allows incoming sunlight to reach the surface of the Earth, provided the sunlight is not reflected or absorbed by clouds.

A fraction of the light that does reach the planet is absorbed, based on the degree of reflectivity of the surface, and most of the absorbed light is converted to energy, and heat is radiated back out at infrared wavelengths. Although the atmosphere allows most of the visible light through, many of the inorganic gases, such as water vapor (H_2O) and carbon dioxide (CO_2) absorb infrared radiation, thereby converting the infrared radiation to rotational and vibrational energy. This raises the energy content of the atmosphere and the average temperature. Thus, the higher the content of these gases in the atmosphere, the greater the chance of the infrared light being absorbed before it escapes into space.

If all other influences are maintained in a constant state, increased levels of greenhouse gases will necessarily produce increased atmospheric temperatures. However, the impact of greenhouse gases (any of the gases whose absorption of solar radiation is responsible for the greenhouse effect, including carbon dioxide, ozone, methane, and the fluorocarbons) differs based on the chemistry of the individual gases. Moreover, the impact of any gas (greenhouse or otherwise) is influenced by the lifetime of the gas in the atmosphere; for example, water vapor is removed from the atmosphere as precipitation, while organic gases (not the subject of this book but worthy of mention here) such as methane are typically oxidized to carbon dioxide and water within decades of its appearance in the atmosphere:

$$CH_4 + 2O_2 \rightarrow CO_2 + 2H_2O$$

In general, air pressure and air density decrease with altitude in the atmosphere, but the influence of the temperature (which has a more complicated profile with altitude) may remain relatively constant or even increase with altitude in some regions. Because the general pattern of the temperature-altitude profile is relatively constant and measurable (by means of instrumented balloon soundings), the temperature behavior provides a useful paradigm by which the atmospheric layers can be subcategorized.

Thus, the atmosphere can be divided (called atmospheric stratification) into five primary areas that differ in properties such as composition, temperature, and pressure and that include, from the lowest to the highest, the five main layers: (i) the troposphere, (ii) the stratosphere, (iii) the mesosphere, (iv) the thermosphere, and (v) the exosphere. The exosphere is the highest but thin, atmosphere-like volume surrounding the Earth that is gravitationally bound to the Earth, but the density is too low for any molecular collisions to occur on a regular basis. Although, in a more general sense, the atmosphere is defined by the homosphere and the heterosphere, in turn, they are defined by whether the atmospheric gases are well mixed.

By way of explanation, the surface-based *homosphere* includes the troposphere, stratosphere, mesosphere, and the lowest part of the thermosphere, where the chemical composition of the atmosphere is not fully dependent on the molecular weight of the gaseous constituents because the gases are mixed by turbulence. This relatively homogeneous layer ends at the *turbopause* found at approximately 62 miles (330,000 ft), which places it approximately 12 miles (66,000 ft) above the mesopause.

The *heterosphere* lies above the homosphere and includes the exosphere and most of the thermosphere. In this atmospheric layer, the chemical composition varies with altitude because the distance that particles can move without colliding with one another is large compared with the extent of the size of motion that causes mixing. This allows the gases to stratify (segregate) by molecular weight, with gases such as oxygen and nitrogen present only near the bottom of the heterosphere. The upper part of the heterosphere is composed almost completely of hydrogen, the lightest element.

The planetary boundary layer is the part of the troposphere that is closest to the surface of the Earth and is directly affected by it, mainly through turbulent diffusion. During the day, the planetary boundary layer usually is well mixed, whereas at night it becomes stably stratified with weak or intermittent mixing. The depth of the planetary boundary layer ranges from as little as approximately 300 ft on clear, calm nights to 10,000 ft or more during the afternoon in dry regions.

Approximately three quarters (75%, v/v) of the atmosphere's mass resides within the troposphere and is the layer within which the weather systems develop. The depth of this layer varies between 548,000 ft at the equator to 23,000 ft over the polar regions. The stratosphere, extends from the top of the troposphere to the bottom of the mesosphere, contains the ozone layer that ranges in altitude between 49,000 and 115,000 ft, and is where most of the ultraviolet radiation from the Sun is absorbed. The top of the mesosphere, ranges from 164,000 to 279,000 ft and is the layer wherein most meteors burn up. The thermosphere extends from 279,000 ft to the base of the exosphere at approximately 2,300,000 ft altitude and contains the ionosphere, a region where the atmosphere is ionized by incoming solar radiation.

1.2.1.1 Composition

Thus, the atmosphere is a mixture of chemical constituents—the most abundant of are nitrogen (N_2, (78%, v/v)) and oxygen (O_2, (21%, v/v)). These gases and the noble gases (argon, neon, helium, krypton, and xenon) possess very long lifetimes against chemical destruction and hence are relatively well mixed throughout the entire homosphere. Minor constituents, such as water vapor, carbon dioxide, ozone, and many others, also play an important role despite their lower concentration. These constituents influence the transmission of solar radiation and terrestrial radiation in the atmosphere and are therefore linked to the physical climate system. They are also key components ofbiological cycles and biogeochemical cycles and, in addition, they determine the *oxidizing capacity* of the atmosphere and hence the atmospheric lifetime of biogenic gases (gases produced by life processes) and anthropogenic gases (gases produced by human activities).

Water vapor accounts for approximately 0.25% (v/v) (~0.25%, w/w) of the atmosphere (Table 1.3), but the concentration of water vapor (a greenhouse gas) varies significantly from approximately 10 ppm (v/v) in the coldest portions of the atmosphere to as much as 5% (v/v) in hot, humid air masses. The concentrations of other atmospheric gases are typically quoted in terms of dry air (without water vapor) and are often referred to as trace gases, among which are the inorganic greenhouse gases, principally carbon dioxide (CO_2), nitrous oxide (N_2O), and ozone (O_3). The spatial and temporal distribution of chemical species in the atmosphere is determined by several processes that include (i) surface emissions, (ii) deposition, (iii) chemical reactions, (iv) photochemical reactions, and (v) transport. Surface emissions are associated with volcanic eruptions; floral and faunal activity on the continental land masses and in the oceanic activities; and anthropological activity such as biomass burning, agricultural practices, and industrial activity. Wet deposition results from precipitation of soluble species within the moisture, while the rate of dry deposition is affected by the nature of the surface (such as the type of soil, the type of vegetation, and the presence of a large water system, such as an ocean).

Table 1.3 Composition of the Atmosphere

Gas	Formula	Volume (ppm)	Volume (%)
Nitrogen	N_2	80,840	78.084
Oxygen	O_2	209,460	20.946
Argon	Ar	9340	0.9340
Carbon dioxide	CO_2	397	0.0397
Neon	Ne	18.18	0.001818
Helium	He	5.24	0.000524
Methane	CH_4	1.79	0.000179
Water vapor[a]	H_2O	10–50,000[(D)]	0.001–5[(D)]

[a]Water vapor is not included in above dry atmosphere and is approximately 0.25% (v/v) over the full atmosphere but does vary considerably.

Furthermore, knowledge of the inorganic components of biomass feedstock is important for process control and for handling coproducts and wastes resulting from energy and fuel utilization of biomass. Analytic survey of forestry thinnings (wood chips), agricultural residues (rice straw, wheat straw, and corn stover), and dedicated perennial grass crops (switchgrass, wheatgrass, and miscanthus) shows that, potentially, the whole periodic table may be present in biomass. The main effect of ashing is bonding of oxygen in the ash mainly as silicate, oxides, hydroxides, phosphates, and carbonate residual minerals. Carbon is partially retained as carbonates and graphite (char). Nitrogen is dominantly released to the flue gas, while sulfur is mostly retained in the ash as sulfates. Small losses (~19%, w/w) for both sulfur and chlorine were detected during ashing at 575°C (1070°F). The majority of the alkali metals (lithium, Li; sodium, Na; potassium, K; and rubidium, Rb) will substantially modify soil if applied as a fertilizer. Only magnesium (Mg), calcium (Ca), and strontium (Sr) of the alkali earth metals; manganese (Mn), copper (Cu), and zinc (Zn) of the period 4 transition metals; and molybdenum (Mo) and cadmium (Cd) of the period 5 transition metals may exceed regulatory limits if used as a fertilizer. The heavy elements occur in concentrations too low to cause concern except for selenium, Se. The high alkali content of some biomass ash thus makes them good candidates for use as fertilizers provided that they are applied in low proportions (<50%) to soil. Ash of wood material is a carrier of many of the alkali elements (Li, K, Rb, Mg, Ca, Sr, and Ba) and some transition elements (Mn, Cu, Zn, Mo, Ag, and Cd). In contrast, ash of herbaceous plant material is in addition to K only variably enriched: Li, Na, Se, and Mo in wheatgrass; Mg and Ca in switchgrass; Mg, Ca, and Cd in corn stover; Mn in rice straw; Mo in wheat straw; and Mn and Cd in miscanthus. Water leaching results in significant losses for anionic chlorine and sulfur and for most of the alkali metals, thus making resulting ash from such treated feedstock less attractive as potassium-containing fertilizers although fuel properties are enhanced for thermal conversion (Thy et al., 2013).

By volume, dry air contains (subject to minor rounding of the data) 78.09% nitrogen, 20.95% oxygen, 0.93% argon, 0.039% carbon dioxide, and small amounts of other gases (Table 1.3). Air also contains a variable amount of water vapor, approximately 1% (v/v) at sea level and 0.4% (v/v) throughout the entire atmosphere. Air content and atmospheric pressure vary at different layers, and air suitable for use in photosynthesis by terrestrial plants and breathing of terrestrial animals is found only in the troposphere.

1.2.1.2 Chemical Activity

Some examples of atmospheric pollutants include nitrogen dioxide (NO_2), sulfur dioxide (SO_2), and carbon monoxide (CO). Furthermore, nitrogen dioxide and sulfur dioxide combine with water to form acids:

$$NO_2 + H_2O \rightarrow \underset{\text{Nitric acid}}{HNO_3}$$

$$SO_2 + H_2O \rightarrow \underset{\text{Sulfurous acid}}{H_2SO_3}$$

These acids contribute to the long-term destruction of the environment due to the generation of acid rain. Carbon monoxide, generated by the incomplete combustion of hydrocarbons fuels and other carbonaceous materials, displaces and prevents oxygen from binding to hemoglobin and causes asphyxiation. Also, carbon monoxide binds with metallic pollutants and causes them to be more mobile in air and water. These reactions greatly reduce the protective effects of ozone against ultraviolet radiation.

Chemical compounds released at the surface by natural and anthropogenic processes are oxidized in the atmosphere before being removed by wet or dry deposition. Key chemical species of the troposphere include oxygenated inorganic species and carbon monoxide and nitrogen oxides (nitric oxide, NO; nitrous oxide, N_2O; nitrogen dioxide, NO_2; and dinitrogen tetroxide, N_2O_4, which forms an equilibrium mixture with nitrogen dioxide):

$$2NO_2 \rightleftharpoons N_2O_4$$

Nitrogen dioxide is also produced by lightning discharges in thunderstorms and nitric acid (HNO_3) and peroxyacetyl nitrate (PAN, an unstable secondary pollutant present in photochemical smog and that decomposes into peroxyethanoyl radicals and nitrogen dioxide).

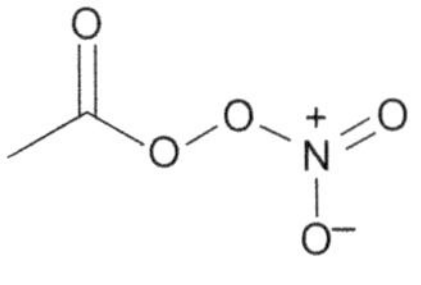

Peroxyacetyl nitrate

Other chemical species include hydrogen compounds—specifically the hydroxy radical, OH, and the hydroperoxy radical, HO_2, as well as hydrogen peroxide, H_2O_2; ozone, O_3; and sulfur compounds (dimethyl sulfide (DMS) CH_3SCH_3; sulfur dioxide, SO_2; and sulfuric acid, H_2SO_4). The hydroxyl radical (OH) has the capability of reacting with and efficiently destroying a large number of inorganic chemical compounds and hence has the potential to contribute directly to the oxidative capacity (reactivity) of the atmosphere. Ozone also plays an important role in the troposphere: together with water vapor ozone, it is the source of the hydroxy radical, and in addition, it contributes to climate forcing.

By way of explanation, climate forcing is any influence on climate that originates from outside the climate system itself and is a major cause of climate change. This includes the temperature rise of the Earth during an *interglacial period* that exists at the present. The presence of ozone in the troposphere results not only from the intrusion of ozone-rich stratospheric air masses through the tropopause but also from photochemical reactions involving hydrocarbon derivatives, nitrogen oxides (NO_x), and carbon monoxide (CO). One major question is to what extent the oxidizing capacity of the atmosphere has changed because of human activities.

Finally, the release of sulfur-containing chemicals at the surface of the Earth and the subsequent oxidation of the sulfur compounds in the atmosphere leads to the formation of small liquid or solid particles that remain in suspension in the atmosphere. The release to the atmosphere of sulfur compounds is a result of human activities, specifically coal combustion and the combustion of various petroleum products (Speight and Arjoon, 2012; Speight, 2013, 2014).

1.2.2 The Aquasphere

The aquasphere (also called the *hydrosphere* and, on occasion also referred to as the *aquatic biome*) is the layer of water that, in the form of the oceans, rivers, and lakes, covers approximately 71% of the surface of the Earth (some estimate put this at 75%); more than 97% (v/v) of this water exists in oceans (Charette and Smith, 2010). The oceans present a vast reservoir of saltwater, on land as surface water in lakes and rivers, underground as groundwater, in the atmosphere as water vapor, in the polar icecaps as solid ice, and in many segments of the anthrosphere such as in boilers or municipal water distribution systems.

Water is an essential part of all living systems (Jackson et al., 2001) and is the medium from which life evolved and in which life exists. Water also (i) carries energy and matter are through various spheres of the environment,

(ii) leaches soluble constituents from mineral matter and carries them to the ocean or leaves them as mineral deposits some distance from their sources, (iii) carries plant nutrients from soil into the bodies of plants by way of plant roots, and (iv) absorbs solar energy in oceans, and this energy is carried as latent heat and released inland when it evaporates from oceans; the accompanying release of latent heat provides a large fraction of the energy that is transported from equatorial regions of the Earth to the poles of the Earth and is the energy behind storms.

Fresh, clean, and drinkable water is a necessary but limited resource. Industrial, agricultural, and domestic wastes can contribute to the pollution of this valuable resource, and water pollutants can damage human and animal health. Three important classes of water pollutants are (i) inorganic pollutants, of which heavy metals are a part, and organic pollutants (Speight, 2017a). Heavy metals include transition metals such as cadmium, mercury, and lead, all of which can contribute to pollution and have serious effects on the floral species (plants) and faunal species (animals) in the environment. Inorganic pollutants such as hydrochloric acid (HCl), sodium chloride (NaCl), and sodium carbonate ($NaCO_3$) change the acidity or alkalinity (the pH) and the salinity of the water, making it undrinkable or unsuitable for the support of animal and plant life. These effects can result in dire consequences for higher mammals such as humans.

Due to overuse, pollution, and ecosystem degradation, the sources of most freshwater supplies—groundwater (water located below the soil surface), reservoirs, and rivers—are under severe and increasing environmental stress. The majority of the urban sewage in developing countries is discharged untreated into surface waters such as rivers and harbors. Approximately 65% (v/v) of the global freshwater supply is used in agriculture, and 25% (v/v) is used for industrial processes. Freshwater conservation therefore requires a reduction in wasteful practices such as inefficient irrigation, reforms in agriculture and industry, and strict pollution controls worldwide.

Aquatic regions house numerous species of plants and animals, both large and small. In fact, this is where life began billions of years ago when amino acids first started to come together to form the more complex proteins. Without water, most life forms would be unable to sustain themselves, and the Earth would be a barren, desert-like place. Although water temperatures can vary widely, aquatic areas tend to be more humid and the air temperature on the cooler side. The aquasphere can be broken down into two basic regions: (i) freshwater regions, such as ponds and rivers, and (ii) marine regions, i.e., the oceans.

1.2.2.1 Freshwater Regions

A freshwater region is an area where the water has a low salinity (a low salt concentration, usually on the order of <1%, w/w). Plants and animals in freshwater regions are adjusted to the low salt content and would not be able to survive in areas of high salt concentration (such as the ocean). There are different types of freshwater regions: (i) ponds and lakes, (ii) streams and rivers, and (iii) wetlands (Speight, 2017a).

Pond and lakes range in size from just a few square meters to thousands of square kilometers. Many ponds are seasonal, lasting just a couple of months (such as sessile pools), while lakes may exist for hundreds of years or more. Ponds and lakes may have limited species diversity since they are often isolated from one another and from other water sources like rivers and oceans. The temperature varies in ponds and lakes seasonally. During the summer, the temperature can range from 4°C (39°F) near the bottom to 22°C (72°F) at the top. During the winter, the temperature at the bottom can be 4°C (39 F) while the top is 0°C (32°F, ice). In between the two layers, there is a narrow zone called the thermocline where the temperature of the water changes rapidly. During the spring and fall seasons, there is a mixing of the top and bottom layers, usually due to winds, which results in a uniform water temperature of around 4°C (39°F). This mixing also circulates oxygen throughout the lake. There are many lakes and ponds that do not freeze during the winter, in which case the top layer would be a little warmer.

Streams and rivers are bodies of flowing water moving in one direction; they typically start at headwaters, which may be springs, snowmelt or even lakes, and then travel all the way to their mouths, usually another water channel or the ocean. The characteristics of a river or stream change during the journey from the source to the mouth. The temperature is cooler at the source than it is at the mouth. The water is also clearer and has higher oxygen levels, and freshwater fish such as trout and heterotrophs can be found there. Toward the middle part of the stream/river, the width increases, as does species diversity; numerous aquatic green plants and algae can be found. Toward the mouth of the river/stream, the water becomes murky from all the sediments that it has picked up upstream, decreasing the amount of light that can penetrate through the water. Since there is less light, there is less diversity of flora, and because of the lower oxygen levels, fish that require less oxygen, such as catfish and carp, can be found.

Wetlands are areas of standing water that support aquatic plants. Marshes, swamps, and bogs are all considered wetlands. Plant species adapted to the very moist and humid conditions are called hydrophytes. These include pond lilies, cattails, sedges, tamarack, and black spruce. Marsh flora also includes such species as cypress and gum. Wetlands have the highest species diversity of all ecosystems. Many species of amphibians, reptiles, birds (such as ducks and waders), and furbearers can be found in the wetlands. Wetlands are not considered freshwater ecosystems as there are some, such as salt marshes, that have high salt concentrations, which support different species of animals, such as shrimp, shellfish, and various grasses.

1.2.2.2 Marine Regions

Marine regions, more often referred to as *oceans*, cover approximately 71% of the surface of the Earth and that also (under the general term *marine regions*) includes coral reefs and estuaries; estuaries are areas where freshwater streams or rivers merge with the ocean (Charette and Smith, 2010). This mixing of waters with such different salt concentrations creates a very interesting and unique ecosystem. Microflora (such as algae) and macroflora (such as seaweeds, marsh grasses, and mangrove trees) can be found in these regions and, in fact, estuaries support a diverse fauna, including a variety of worms, oysters, crabs, and waterfowl. Marine algae supply much of the oxygen required by life on Earth and, at the same time, remove a substantial amount of carbon dioxide from the atmosphere. In addition, the evaporation of the seawater (as part of the water cycle) provides rainwater for the land.

The ocean can be divided into two general regions: (i) a warm, surface pool—typically 18°C (64°F)—that is approximately 3280 ft thick and (ii) the deep water—typically 3°C (37°F)—that outcrops to the surface at high latitudes and forms the bulk of the ocean volume. Unlike the atmosphere, heating of the ocean surface stabilizes the water and prevents rapid exchange (or mixing) between the surface water and the deeper water. In fact, contact between the surface water and deep water is limited to localized polar regions where the resulting thermohaline circulation (part of the large-scale ocean circulation that is driven by global density gradients that are created by surface heat and freshwater flow) is especially important over long timescales (such as glacial cycles).

In terms of the chemical effects of the ocean, the ocean influences the atmosphere through the exchanges of gases across the air-sea interface.

The transfer of carbon dioxide from the atmosphere to the ocean is controlled by the two competing factors of temperature: warming of surface waters, which releases carbon dioxide to the atmosphere, and biological productivity. Photosynthesis by marine phytoplankton converts dissolved carbon dioxide into organic carbon compounds, leading to a reduction in surface carbon dioxide values and a carbon dioxide flow into the ocean. The amount of carbon dioxide dissolved in seawater is quite large due to its high solubility and its reactivity with water to form carbonic acid and its dissociation products.

1.2.3 The Geosphere and Terrestrial Biosphere

The *geosphere* is that part of the Earth upon which humans live and from which food, minerals, and fuels are extracted. It is divided into layers, which include the solid, iron-rich inner core, molten outer core, and the lithosphere (the upper mantle and the crust). Environmental science (whether it is inorganic or organic science) is most concerned with the *lithosphere* that extends to depths on the order of 62 miles below the surface of the Earth and comprises two systems: (i) the *crust* and (ii) the *upper mantle*. The crust (the outer skin of the Earth) is the layer that is accessible to humans and is extremely thin compared with the diameter of the earth, ranging from 5 to 20 miles thick.

The *biosphere*, a subdivision of the geosphere, is the relatively thin zone of air, soil, and water that covers the Earth and is capable of supporting life, and ranges from approximately 6 miles from the deepest floor of the ocean and into the atmosphere. Life in the geosphere is dependent on the energy from the sun energy and on the circulation of heat and essential nutrients.

Thus, the biosphere (i) is virtually contained by the geosphere and hydrosphere in the very thin layer where these environmental spheres interface with the atmosphere, (ii) strongly influences, and in turn is strongly influenced by, the other parts of the environment, (iii) strongly influences the various parts of the aquasphere by producing the biomass required for life in the water and mediating oxidation-reduction reactions in the water, (iv) is involved with weathering processes that break down rocks in the geosphere and convert the rock matter to soil, and (v) is based upon plant photosynthesis, which fixes solar energy and carbon from atmospheric carbon dioxide in the form of high-energy biomass.

The *anthroposphere* is a name given to that part of the environment *made* or *modified* by humans and used for their activities. The anthroposphere

(i) is a strong interconnection to the biosphere, (ii) has strongly influenced the biosphere and change it drastically; for example, destruction of wild life habitats has resulted in the extinction of vast numbers of species, in some cases even before they are discovered, and (iii) it is the responsibility of humankind to make such changes intelligently and to protect and nurture the biosphere.

The terrestrial biosphere (*land*) is important to atmospheric chemistry as a source and sink for many compounds; a major activity within atmospheric chemistry has been (and remains) the determination of such flows. The structure of the biosphere is controlled by the interaction of the climate with the patterns of soils and topography, which result from geologic processes on a range of time scales and are modified further by the biogeographic distribution of organisms.

The structure of the biosphere is important for the understanding issues related to the source of trace gas sources and sinks for trace gases. However, the spatial structure of the biosphere cannot be conveniently described as the outcome of a series of physical calculations (unlike the atmosphere and oceans) but rather requires the use of large databases describing fine-scale structures within ecosystems. Not only does the terrestrial biosphere play an important role in the functioning of the global climate system, but also pollution can have major impacts on the biosphere itself. Climate change is projected to impact agricultural production, forestry, natural ecosystems, and biodiversity through changes to the soil (Gouin et al., 2013).

Technically, soil is a mixture of mineral constituents; the inorganic components of soil are principally produced by the weathering of rocks and minerals, plant materials, and animal materials, which forms during a long process that may take thousands of years, and it is an unconsolidated, or loose, combination of inorganic and organic materials. Soil is necessary for most plant growth and is essential for all agricultural production. The inorganic materials are composed of debris from plants and from the decomposition of animals and the many tiny (microscopic) life forms that inhabit the soil. The chemical composition and physical structure of soils is determined by a number of factors such as the kinds of rocks, minerals, and other geologic materials from which the soil is originally formed. The vegetation that grows in the soil is also important.

Food sources grown on soils are predominately composed of carbon, hydrogen, oxygen, phosphorous, nitrogen, potassium, sodium, and calcium. Plants take up these elements from the soil and configure them into the plants that are recognized as food plants. Each plant has unique nutritional requirements that are obtained through the roots from the soil. Nutrients are stored in soil on exchange sites of the inorganic and clay components. Calcium, magnesium, ammonium, potassium, and the vast majority of the micronutrients are present as cations in soils of varying acidity and alkalinity (varying under most soil pH).

The use of pesticides in agriculture—which are used to control the growth of insects, weeds, and fungi compete with humans in the consumption of crops—is an important issue to environmental scientists and, since there are many different pests that can damage the crops, each kind of pest requires different chemical compounds to kill it. The use of pesticides is not new, but in modern agriculture, most of the pesticides are synthetic (often inorganic) chemicals. On the basis of the definition of inorganic chemicals, inorganic pesticides are chemical compounds that do not include carbon within the molecule. Inorganic pesticides usually contain toxic elements, such as mercury and arsenic, which can remain in the soil and ecosystem long after they have been applied to crops and enter the food chain.

The pesticides are created in a laboratory and can be designed so that the products are specific to a particular pest. While this control seems adequate and necessary, it can result in a never-ending battle between to constantly develop new chemical formulas for the pesticides. Agricultural pests, like all other organisms, strive to survive and can adapt to chemicals over a period of several generations. When pests are exposed to a chemical compound that is designed to as a pesticide poison them, there will be some success, but the next generation of the pest is likely to be have some resistance to the deadly chemical that came for the patent pests that had an interest genetic mutation that allowed them to survive (survival of the fittest).

The use of pesticides in agriculture makes a significant contribution to environmental pollution. However, the spraying of crops and the water runoff from irrigation transports these harmful chemicals to the flora and fauna of the environment. Chemicals can build up in the tissues of the plants animals (including humans) leading to destruction of flora and fauna. The most significant pesticide of the 20th century was DDT, which was highly effective as an insecticide but did not break down in the environment and led to the death of birds, fish, and some humans. It was not until the appearance of the book *Silent Spring* (Carson, 1962), that the public, through the author of the book (Rachel Carson), awakened to the adverse effects of chemicals to the dangers to all natural floral and faunal ecosystems from the misuse of pesticides (specifically, chemicals such as DDT, dichlorodiphenyltrichloroethane) (Chapter 9).

1.2.3.1 Composition of Soil

Soil comprises a mixture of inorganic and organic components: minerals, air, water, and the decayed (or decaying) remnants of plant and animal material. Mineral and inorganic particles generally compose approximately 50% (v/v) of the soil. The other 50% (v/v) consists of open areas (pores) that are of various shapes and sizes. The networks of the pore systems hold water within the soil and also provide a means of the transport of water (including water-soluble chemicals) through the soil. Oxygen and other gases move through pore spaces in soil, and the pores also serve as passageways for small animals and provide space for the growth of plant roots.

The mineral component of soil consists of an arrangement of particles that are less than 2.0 mm in diameter. More specifically and technically, soil is composed of particles that fall into three main mineral groups, each of which is determined by particle size: (i) sand, 0.05–2.00 mm in size; (ii) silt, 0.002–0.05 mm in size; and (iii) clay, less than 0.002 mm in size. Depending upon the parent rock materials from which these minerals were derived, the assorted mineral particles ultimately release the chemicals on which plants depend for survival, such as potassium, calcium, magnesium, phosphorus, sulfur, iron, and manganese. Clearly, the insertion of nonindigenous inorganic chemicals into this cycle can interfere with plant survival leading to interference with faunal cycles in which the various faunal groups depend upon the natural cycle for survival.

Other sources of the naturally occurring inorganic materials provide more of the essential components of soils. Some of the inorganic material arises from the residue of plants, such as the remains of the roots of plants deep within the soil, or materials that fall on the ground, such as leaves on a forest floor or even from a dead animal. These materials become part of a cycle of decomposition and decay, a cycle that provides important nutrients to the soil. In general, soil fertility depends on a high content of inorganic materials.

Soils are also characterized according to how effectively the soil retains water and transports water. Once water enters the soil from rain or irrigation, gravity can be the dominant force by causing water to trickle downward. Soil differs in the capacity to retain moisture against the pull exerted by gravity and by plant roots. Coarse soil, such as soil consisting of mostly of sand, tend to hold less water than do soils with finer textures, such as those with a greater proportion of clays.

Water also moves through soil pores by capillary action (the ability of a liquid to flow in narrow spaces without the assistance of, or even in opposition to, external forces such as gravity), and that is the type of movement in which the water molecules move because they are more attracted to the pore walls (adhesion) than to one another (cohesion). Such movement tends to occur from wetter to drier areas of the soil. The attraction of water molecules to each other is an example of cohesion in which the water molecules are attracted to each other, usually though the formation of hydrogen bonds (Chapter 2).

1.2.3.2 Soil Pollution

Soils may become contaminated by the accumulation of heavy metals and metalloids through emissions from the rapidly expanding industrial areas, mine tailings, disposal of high metal wastes, leaded gasoline and paints, land application of fertilizers, animal manures, sewage sludge, pesticides, wastewater irrigation, coal combustion residues, spillage of petrochemicals, and atmospheric deposition (Khan et al., 2008; Zhang et al., 2010). Heavy metals constitute an ill-defined group of inorganic chemical hazards, and those most commonly found at contaminated sites are lead (Pb), chromium (Cr), arsenic (As), zinc (Zn), cadmium (Cd), copper (Cu), mercury (Hg), and nickel (Ni). Soils are the major sink for heavy metals released into the environment by anthropogenic activities, and unlike organic contaminants that are oxidized to carbon (IV) oxide by microbial action, most metals do not undergo microbial or chemical degradation (Kirpichtchikova et al., 2006), and their total concentration in soils persists for a long time after their introduction (Adriano, 2003). Changes in their chemical forms (speciation) and bioavailability are, however, possible. The presence of toxic metals in soil can severely inhibit the biodegradation of organic contaminants (Maslin and Maier, 2000). Heavy metal contamination of soil may pose risks and hazards to humans and the ecosystem through direct ingestion or contact with contaminated soil, the food chain (soil-plant-human or soil-plant-animal-human), drinking of contaminated ground water, reduction in food quality (safety and marketability) via phytotoxicity, reduction in land usability for agricultural production causing food insecurity, and land tenure problems (McLaughlin et al., 2000a,b; Ling et al., 2007).

Some of the most common toxic soil pollutants include not only inorganic chemicals but also asbestos and other hazardous inorganic materials. These substances commonly arise from (i) the percolation of contaminated surface water to subsurface strata, (ii) leaching of mobile or soluble chemical from landfill waste, (iii) direct discharge of industrial waste materials into the soil, (iv) way of atmospheric deposition, (v) direct spreading onto land, (vi) contamination by wastewater, and (vii) the various forms of waste disposal.

Inorganic pollutants that are directly applied into soils or deposited from the atmosphere may be taken up by plants or leached into water bodies. Ultimately, they affect floral and faunal well-being when taken up into the species of flora or fauna in the form of a food source or water source. Some of these chemicals can mimic one or more of the essential chemicals required by a plant or animal for survival (Table 1.2) and thereby interfere, with the reproductive and developmental functions of the plant or animal; the substances are generally known as *endocrine disrupters*. Although, soil might be affected less by pollution compared to water or air, cleaning polluted soil is more difficult, complex, and expensive than cramming water and air.

Soil pollution (soil contamination) as part of degradation of the land is caused by the presence of anthropogenic (xenobiotic, human-made) chemicals or by any other alteration in the natural soil environment. Typically, soil pollution has been caused by industrial activity, by agricultural chemicals, or by improper disposal of waste. The most common inorganic chemicals involved have been (and continue to be) lead and other heavy metals. In fact, soil contamination in a region can be correlated with the degree of industrialization and intensity of chemical usage.

A heavy metal is generally classed as a metal with a relatively high density, high atomic weight, or a high atomic number as evidence by the position of the metal in the periodic table of the elements (Table 1.4) (Chapter 2). The criteria used, and whether or not a metalloid is included in the definition, varies depending on the context and must be used with caution. In metallurgy, for example, a heavy metal may be classed (defined) based on density, whereas in physics the distinguishing criterion might be atomic number, while in the chemistry field the metal may be classed on the basis of chemical behavior. Generally, the definitions have not been widely accepted.

Some heavy metals are either essential nutrients (typically iron, cobalt, and zinc) or are considered to be relatively harmless, but remembering that dosage or concentration in an ecosystem can be the deciding factor, can be toxic in large amounts or in certain forms. In summary, the physical characterization, chemical characterizations, and general classification of a metal as a heavy metal needs to be treated with caution as the metals involved are not always consistently defined. As well as being relatively dense, heavy metals tend to be less reactive than lighter metals and have much less soluble sulfides and hydroxides. While it is relatively easy to distinguish a heavy

Table 1.4 General Properties of Heavy Metals and Light Metals

Physical Properties	Light Metals	Heavy Metals
Density	Usually lower	Usually higher
Hardness	Tend to be soft, easily cut	Most are quite hard
Thermal expansivity	Mostly higher	Mostly lower
Melting point	Mostly low	Low to very high
Tensile strength	Mostly lower	Mostly higher
Chemical properties	Light metals	Heavy metals
Periodic table location	Generally: groups 1 and 2	Generally, groups 3–16
Abundance in Earth's crust	More abundant	Less abundant
Main occurrence (or source)	Lithophiles	Lithophiles or chalcophiles[a]
Reactivity	More reactive	Less reactive
Sulfides	Soluble to insoluble	Extremely insoluble
Hydroxides	Soluble to insoluble	Generally insoluble
Salts	Generally form colorless solutions in water	Generally form colored solutions in water
Complexes	Mostly colorless	Mostly colored
Biological role	Include macronutrients (Na, Mg, K, Ca)	Include micronutrients (V, Cr, Mn, Fe, Co, Ni, Cu, Zn, Mo)

[a]Lithophile—an element that forms silicates or oxides and is concentrated in the minerals of the crust of the Earth. Chalcophile—an element that preferentially sulfide minerals if sufficient sulfur is available. Siderophile—an element that tends to bond with metallic iron.

metal from a lighter metal, some of the so-called heavy metals (such as zinc, mercury, and lead) may have some of the characteristics of lighter metals and heavier metals, while some of the lighter metals may have the characteristics of heavier metals. Second guessing is not sufficient to classify a metal as a heavy metal or as a light metal. Caution is advised, and a study of properties of behavior of the metal is a necessity.

As part of the biosphere, soil (when polluted) cannot support floral growth such as the proliferation of forests. Forests are very important for maintaining ecological balance and provide many environmental benefits. In addition to timber and paper products, forests provide wildlife habitat, prevent flooding and soil erosion, help provide clean air and water, and contain tremendous biodiversity. Forests are also an important defense against global climate change though the production of life-giving oxygen and consumption of carbon dioxide, the compound that is claimed to be the most responsible for global warming through photosynthesis, thereby reducing the effects of global warming.

1.3 INORGANIC CHEMISTRY AND THE ENVIRONMENT

The 20th century came into being in much the same manner as the 19th century ended insofar as there was a continuation of the less-than-desirable disposal methods for chemicals and the various forms of chemical waste, which included inorganic chemical waste. However, as the 20th century evolved, especially from the 1970s onward, there came the realization that all chemicals (whether in large concentrations or in small concentrations) were toxic and the unabated and unmanaged disposal of chemicals had to change.

Inorganic chemistry is the study of inorganic compounds, many of which are ionic compounds (salts) consisting of cations (positively charged ions) and anions (negatively charged ions) held together by ionic bonding. Examples of salts are magnesium chloride ($MgCl_2$) that consists of magnesium cations (Mg^{2+}) and chloride anions (Cl^-). In any salt, the proportions of the ions are such that the electric charges cancel each other (Mg^{2+} plus $2Cl^-$) and so that the bulk compound ($MgCl_2$) is electrically neutral.

The inorganic chemicals industry (which, within the context of this book, also includes the fossil fuels industry because of the production of inorganic gases from the use of fossil fuels) and its products provides many real and potential benefits, but at the same time that benefits accrue, the production and use of inorganic chemicals create risks to the environment at all stages of the production cycle. The generation and intentional and unintentional release of the various products (and the process by-products, such as ash and slag for coal-based processes and inorganic residues from petroleum-based processes as well as sludge from natural-gas processing) have contributed to environmental contamination and degradation at multiple levels—local, regional, and global—and even though steps have been taken to mitigate such pollution, in many instances, the impact will, more than likely, continue to be felt for generations.

Thus, it is now (some observers would use the word *finally* instead of the word *now*) recognized, and the necessary legislation passed that any process waste (whether the waste is hazardous or nonhazardous wastes) should never be discarded without proper guidance, management, and authorization. Following from this and with the establishment of environmental protection agencies or departments by various levels (e.g., local, state, and federal governments) in the United States and many other countries, serious consideration was given to the need for investigation of the methods by which chemicals were disposed and affected the environment. As a result, methods were devised for handling chemical wastes with minimal effect on the environment (Carson and Mumford, 1988, 1995). In fact, during the last five decades, it has become increasingly clear that the inorganic chemicals can cause serious environmental problems if methods of disposal remain unchecked. Thus, the disposal of chemical inorganic wastes in the various forms (gases, liquids, and solids) has been subjected to strict legislation that requires minimization or, preferably, elimination of the various waste streams by process modifications (Sheldon, 2010).

Furthermore, because of the importance of chemical contamination of the environment that involves a study of the effects of chemicals on the environment, the recent subdiscipline of *environmental chemistry* with the subcategories of *environmental organic chemistry* and *environmental inorganic chemistry* has arisen and evolved (Speight, 2017a). Both environmental inorganic chemistry and environmental inorganic chemistry are now a component (optional or even a necessity) of many chemistry degree courses in universities and are included in environmental science courses and environmental engineering courses as elements of increasing substance. These relatively new disciplines focus on the various environmental factors that govern the processes that determine the fate of organic chemicals and inorganic chemicals in ecosystems. The information discovered is then combined with the properties of the chemical and applied to a quantitative assessment of the environmental behavior of a wide variety of chemicals (Mackay et al., 2006; Speight, 2017a).

Relating to the current theme of this book, inorganic chemistry (a subdiscipline of *chemistry*) is generally the chemistry of noncarbon compounds, with the exception of the chemistry of carbon monoxide (CO) and carbon dioxide (CO_2), and involves the scientific study of the structure, properties, and reactions of inorganic compounds (Table 1.5). Also in the context of this book, environmental inorganic chemistry addresses the influence of inorganic chemicals on the environment, which includes (i) the study of the structure of inorganic compounds, (ii) the physical properties of inorganic compounds, (iii) the chemical properties of inorganic compounds, and (iv) the reactivity of inorganic compounds. This latter category has the goal of understanding the behavior of inorganic compounds not only in the pure form (when possible) but also in gaseous, liquid (solution), and solid forms and the chemistry of complex mixtures. Such studies are necessary to reflect the way such chemicals exist and react in the environment.

Table 1.5 Examples of Inorganic Compounds[a]

Acids
All inorganic acids, such as the following:
Hydrochloric acid (HCl)
Hydrofluoric acid (HF)
Nitric acid (HNO_3)
Bases
All inorganic bases, such as the following:
Ammonium hydroxide (ammonia water)
Calcium hydroxide (lime water)
Magnesium hydroxide
Sodium bicarbonate (baking soda)
Salts
All inorganic salts, such as the following:
Calcium chloride
Potassium dichromate
Sodium chloride
Ionic components
Alkali metals (Li, Na, K, etc.)
Alkali earth metals (Be, Mg, Ca, etc.)
Metalloids (B, Si, Ge, etc.)
Nonmetals (C, N, O, P, S)
Transition metals (Ti, Fe, Ni, etc.)
Halogenated compounds
All halogens, such as the following:
Bromine derivatives
Chlorine derivatives
Fluorine derivatives
Iodine derivatives
Other
Bioinorganic compounds
Organometallic compounds

[a]As used in the text of this book.

For example, the presence of a specific functional group (such as nitrate, NO_3, or sulfate, SO_4) consisting of atoms or bonds within inorganic molecules that are responsible for the characteristic chemical reactions and behavior of those inorganic molecules as, for example, (i) the properties of the chemical, (ii) the reactivity of the chemical, (iii) the manner by which the chemical reacts to external influences, (iv) the manner by which the chemical can dissipate into the atmosphere, (v) the solubility of the chemical in aqueous systems, and (vi) the manner by which the compound can adhere to and remain in the soil.

These properties can assist in developing an understanding of the manner in which the chemical will behave under various environmental conditions. In concert with the chemical properties of inorganic chemicals, the physical properties that are typically of interest to the environmental scientist and environmental engineer include both quantitative and qualitative properties (Table 1.2). In addition, solubility of inorganic compounds in water is also an important property that must be acknowledged. Inorganic compounds tend to dissolve in aqueous systems compared with the relative insolubility of many organic chemicals (Speight, 2017a,b), whereas the solubility of inorganic chemicals in organic solvents such as ether (diethyl ether, $C_2H_5OC_2H_5$) and in paraffinic solvents such as the various types of petroleum-derived naphtha and kerosene as well as in a variety of aromatic solvents (aromatic naphtha and aromatic kerosene) is typically minimal. Finally, modern inorganic chemistry is a dynamic discipline, and it is evolving rapidly, and the concepts are applicable to all aspects of inorganic chemistry, especially the inorganic chemistry of the environment. Thus, inorganic compounds can be described in terms of simple noncarbon systems of molecular structures in which atoms are held together by chemical bonds, often ionic bonds rather than covalent (nonionic) bonds. Indeed, the concepts upon which inorganic chemistry is based have persisted for more almost 200 years and seem unlikely to be superseded, no matter how much the discipline is refined and modified (Smith, 2013).

1.4 USE AND MISUSE OF INORGANIC CHEMICALS

Inorganic chemicals, and their various derivatives, are widely used in many sectors of the modern world including the chemicals industry, the fossil fuels industry, agriculture, mining, water purification, and public health. However, not only the dedicated use of inorganic chemicals but also the production, storage, transportation, and disposal of these chemicals can pose risks to the environment if safe handling protocols are not followed. Developing an effective management system for inorganic chemicals has been and continues to be a challenge for the inorganic chemicals industry and requires addressing the specific challenges that arise because of the individual chemicals and chemical mixtures as well as the past irregular management of obsolete inorganic chemicals and chemical mixtures, stockpiles, and waste that continue to present serious threats to the environment. As the use of inorganic chemicals and production increases, chemical management, which already has limited resources and capacity, will be further constrained and overburdened and may fail if not regulated and managed responsibly. Measures and systems need to be developed to reduce exposure to negative impacts and to reduce vulnerability of the environment.

The initial moves in the development of an efficient management system is to ensure that there are education programs that prepare professionals to enter the field of environmental technology and education programs that prepare individuals to meet the challenges of environmental management in the forthcoming decades (Speight and Singh, 2014). There is no single discipline by which these challenges can be met; young professionals should be skilled in the sciences, the engineering technologies, and the relevant subdisciplines that enable them to crossover from one discipline to another as the occasion demands.

The use of inorganic chemicals for domestic and commercial purposes increased phenomenally during the 19th century and during the 20th century, but although bringing benefits to communities, the use of these chemicals also had negative impacts on the integrity of terrestrial and marine ecosystems and on air and water quality as well as on human health and safety. Whatever the chemical, there are risks—known and unknown—to the use of the chemical and some chemicals, including heavy metals, and any persistent inorganic pollutants (PIPs) present risks that have been known for decades. On the other hand, there has been the release of chemicals into the environment, many of which are long lived and transform into by-products whose behavior, synergies, and impacts are not well known (Jones and De Voogt, 1999).

By way of recall, a heavy metal is generally classed as a metal with a relatively high density, high atomic weight, or a high atomic number as evidence by the position of the metal in the periodic table of the elements (Table 1.4) (Chapter 2). The criteria used for the definition varies depending on the context and must be used with caution.

A PIP is an inorganic chemical that is stable in the environment and (i) is liable to long-range transport, (ii) may be capable of bioaccumulation in floral and faunal issue, and (iii) may have significant impacts on human health and the environment. Persistence is usually described as the half-life ($T\frac{1}{2}$) of a chemical in water, soil, sediment, or air. The half-life is the amount of time necessary for a given amount of chemical released into the environment to decrease to one-half of its initial value.

Examples of persistent chemicals are arsenide derivatives (such as sodium arsenide, Na_3As), fluoride derivatives, cadmium salts, and lead salts. Some inorganic chemicals, such as asbestos, are persistent in almost all ecosystems and environments, while other chemicals, such as metal sulfides, are persistent only in unreactive environments. However, sulfides can undergo chemical transformation (Chapter 7) and generate hydrogen sulfide (H_2S) in a reducing environment or sulfate species (SO_4) and sulfuric acid (H_2SO_4) by oxidation, which also exert a detrimental influence on the environment. As with organic substances, persistence is often a function of environmental properties. Thus, sound chemical management across the lifecycle of a chemical—from extraction or production to disposal—is possible essential (under current legislative guidelines) to avoid risks to the floral and faunal ecosystems.

Nonetheless, the modern world has adapted to the use of inorganic chemicals, and there are continuing debates that crude oil the largest source of inorganic chemicals—while in good supply at the present time—may be in short supply in the next 50–100 years and there will be the need to rely on alternate sources of energy, which are not immune from causing damage to the environment (Speight, 2011a, 2014; Lee et al., 2014; Speight and Islam, 2016). Thus, to develop an effective management system that protects the environment from inorganic chemicals, there is the need to recognize that the modern world relies on both natural and synthetic chemicals that can be tailored to serve specific purposes. In fact, the gasification of coal (or, for that matter, other carbonaceous material such as biomass) to produce synthesis gas (a mixture of carbon monoxide, CO, and hydrogen, H_2) is an established process from which a variety of organic chemicals can be synthesized (Chadeesingh, 2011; Speight, 2011b, 2013).

As the knowledge and understanding of inorganic chemicals increases, as well as the variety of routes for the synthesis of these chemicals, so does the ease of industrial production of such compounds. This makes understating the different aspects of the use of these chemicals a greater task than ever, and the legislated regulation of multiuse chemicals become an essential part of the inorganic chemicals industry. However, as an understanding of inorganic chemistry increases, so must a sense of responsibility related to the use of inorganic chemicals and the effects of these chemicals on the floral and faunal ecosystems and the various protocols for the disposal of inorganic chemicals. Laws relating to the regulation and use of such chemicals must be vigorously policed and updated, especially since there is the continual search for new synthetic routes to the chemicals (accompanied by the efforts of the would-be multiuse chemical abusers) to avoid being outstripped by the development of science and suffer the resulting environmental consequences of this negligence.

When an event occurs that is detrimental to any floral or faunal ecosystem, allocation of chemical responsibility to the company or persons disposing of the chemicals is often a difficult process. The issues relating to the responsibility for the development and dispersal of inorganic chemicals continue to be debated. However, given that the use and disposal of chemicals is a global problem, the responsibility to deal with the problem must fall to policy makers in the various levels of government (local, state, and federal) who are involved in the creation of regulatory laws not only in the United States but also in the industrialized nations of the world; it is important to create codes of conduct to guide behavior and actions about this complex problem. For example, signatory countries to the Kyoto Protocol—an international treaty that extends the 1992 United Nations Framework Convention on Climate Change (UNFCCC) that commits the signatories to reduce greenhouse gas emissions, which was adopted in Kyoto, Japan, on Dec. 11, 1997 and entered into force on Feb. 16, 2005—cannot assume that immunity to other forms of pollution is afforded by the protocol. In addition, governments that fail to create responsible regularity laws must also share some of the blame for the misuse of chemicals. The politicians cannot consider themselves immune from blame when the necessary laws are not passed or especially when such laws are passed but are not policed.

There are two types of codes related to the use/misuse of inorganic chemicals and their subsequent disposal: (i) enforceable codes of conduct and (ii) aspirational codes of conduct. An *enforceable code* of conduct deals with the necessary protocols for regulation and enforcement of the code while an *aspirational code* of conduct presents the ideals of performance so that those bound to the code may be reminded of their obligations to perform ethically and responsibly. Nevertheless, there are many observers who are in serious doubt about the practical effectiveness of such codes, which may even prescribe ambiguous (and often unattainable) ideals that can be circumvented if the producers and/or the users of the chemicals wish to do so.

Typically, the value of a code of conduct is usually evident to the creators and writers of the code. Those who must consider every word and phrase included in the code must also explain the importance of expressing the meaning of the code in an unambiguous, straightforward, understandable, and effective manner. Furthermore, it is also essential to involve the various groups with different interests and perspectives at the time when the code is being formulated to inform the various groups of the issues addressed in the code and to remind all participants and the users of the responsible use of inorganic chemicals. In doing so, a code of conduct can be written to be highly effective, which should assist the scientists, the engineers, and the public of the issues at hand. From this understanding, there should come the responsibilities and the guidelines for each party to act in a responsible and ethical manner.

Thus, effective management of inorganic chemicals to protect all types of flora and fauna from all chemicals should carry with it the reminder that to ensure the proper use of chemistry and chemicals there is the need to develop and hold to strict codes of conduct that establish guidelines for ethical scientific development and protection of the environment.

1.5 INORGANIC CHEMICALS IN THE ENVIRONMENT

Inorganic chemicals and inorganic chemical waste (such as inorganic hazardous waste) pose substantial or potential threats to the floral and faunal ecosystems. In the United States, the treatment, storage, and disposal of any type of waste (but for the purposes of this text, specifically inorganic waste and inorganic hazardous waste) are regulated under the Resource Conservation and Recovery Act (RCRA). In this act, hazardous wastes are defined 40 CFR 261 (CFR, 2016) and are also divided into two major categories: (i) characteristic wastes and (ii) listed wastes.

By way of explanation, the Code of Federal Regulations (CFR) is the codification of the general and permanent rules published in the Federal Register by the departments and agencies of the Federal Government produced by the Office of the Federal Register (OFR) and the Government Publishing Office.

Characteristic hazardous wastes are materials that are known or tested to exhibit one or more of the following four hazardous traits: (i) ignitability, (ii) reactivity, (iii) corrosivity, and (iv) toxicity. Thus, chemicals in the environment can be designated as hazardous or nonhazardous (Carson and Mumford, 2002), generally as a category of wastes. Listed hazardous wastes are materials specifically listed by regulatory authorities as hazardous wastes that are from nonspecific sources, specific sources, or discarded chemical products. These wastes may be found in different physical states such as gaseous, liquids, or solids. A hazardous waste is a special type of waste because it cannot be disposed of by common means like other by-products of our everyday lives. Depending on the physical state of the waste, treatment and solidification processes might be required.

In relation to the environmental effects of inorganic chemicals, consideration must also be given to the effect of the so-called *harmless chemicals* (*indigenous chemicals and natural products chemicals*); these are harmless but nonindigenous chemicals on the environment. Within the local environment, these chemicals will be present in a measurable concentration, but the flora and fauna present in that ecosystem may be fatally susceptible to the concentrations of such chemicals when they are present in a concentration that is above the indigenous concentration of the chemicals. For example, a sprinkling salt on a meal may add to the taste of the meal, but it is inadvisable for a human to attempt to consume several ounces of salt with that same meal. Not only would the taste be ruined, but also the high concentration of salt could have a serious health effects (even death) on the consumer.

1.5.1 Indigenous Chemicals

Naturally occurring inorganic chemicals are chemicals that occur naturally in the Earth and do not have an anthropogenic origin and are produced by living organisms (Chapter 2) (Wenk and Bulakh, 2004; Mills et al., 2009). A mineral is an element or chemical compound that is normally crystalline and that has been formed by geologic processes. On the other hand, biogenic substances are chemical compounds produced entirely by biological processes without a geologic component (e.g., urinary calculi, oxalate crystals in plant tissues, and shells of marine mollusks) and are not regarded as minerals. However, if geologic processes were involved in the genesis of the compound, then the product can be accepted as a mineral. Thus, the general definition of a mineral is based on the following criteria: (i) is naturally occurring; (ii) is stable at room temperature or natural temperature; (iii) can be represented by a chemical formula; (iv) is usually abiogenic, i.e., does not result from the activity of living organisms; and (v) has an ordered atomic arrangement.

The most common indigenous chemicals are minerals that (excluding crude oil, coal, and natural gas that are often classed as minerals) are naturally occurring inorganic chemical compounds (Table 1.6). Most often, they are crystalline and abiogenic in origin. A mineral is different from a rock, which can be an aggregate of minerals or nonminerals and does not have one specific chemical composition, as a mineral does. The exact definition of a mineral is under debate, especially with respect to the requirement that a valid species be abiogenic, and to a lesser extent with regard to the mineral having an ordered atomic structure. Thus, minerals are distinguished by various chemical and physical properties and differences in chemical composition and crystal

Table 1.6 Some Common Minerals[a]

Aggregates
Natural aggregates include sand, gravel, and crushed stone. Aggregates are composed of rock fragments that may be used in their natural state or after mechanical processing, such as crushing, washing, or sizing. Recycled aggregates consist mainly of crushed concrete and crushed asphalt pavement
Basalt
Basalt is an extrusive igneous rock. Crushed basalt is used for railroad ballast, aggregate in highway construction, and is a major component of asphalt
Clays
There are many different clay minerals that are used for industrial applications. Clays are used in the manufacturing of paper, refractories, rubber, ball clay, dinnerware and pottery, floor and wall tile, sanitary wear, fire clay, firebricks, foundry sands, drilling mud, iron-ore pelletizing, absorbent and filtering materials, construction materials, and cosmetics
Diamond
Industrial diamonds are those that cannot be used as gems. Large diamonds are used in tools and drilling bits to cut rock and small stone. Small diamonds, also known as dust or grit, are used for cutting and polishing stone and ceramic products
Diatomite
Diatomite is a rock composed of the skeletons of diatoms, single-celled organisms with skeletons made of silica, which are found in fresh and salt water. Diatomite is primarily used for filtration of drinks, such as juices and wines, but it is also being used as filler in paints and pharmaceuticals and environmental cleanup technologies
Dolomite
Dolomite is the near twin-sister rock to limestone. Like limestone, it typically forms in a marine environment but also as has a primary magnesium component. Dolomite is used in agriculture, chemical and industrial applications, cement construction, refractories, and environmental industries
Feldspar
Feldspar is a rock-forming mineral. It is used in glass and ceramic industries, pottery, porcelain and enamelware, soaps, bond for abrasive wheels, cement, glues, fertilizer, and tarred roofing materials and as a sizing, or filler, in textiles and paper applications
Fluorite
Fluorite is used in production of hydrofluoric acid, which is used in the pottery, ceramics, optical, electroplating, and plastics industries. It is also used in the metallurgical treatment of bauxite, as a flux in open-hearth steel furnaces, and in metal smelting, as well as in carbon electrodes, emery wheels, electric arc welders, and toothpaste as a source of fluorine

Continued

Table 1.6 Some Common Minerals—cont'd

Garnet
Garnet is used in water filtration, electronic components, ceramics, glass, jewelry, and abrasives used in wood furniture and transport manufacturing. Garnet is a common metamorphic mineral that becomes abundant enough to mine in a few rocks
Gold
Gold is used in dentistry and medicine, jewelry and arts, medallions and coins, and in ingots. It is also used for scientific and electronic instruments, computer circuitry, as an electrolyte in the electroplating industry, and in many applications for the aerospace industry
Granite
Granite can be cut into large blocks and used as a building stone. When polished, it is used for monuments, headstones, countertops, statues, and facing on buildings. It is also suitable for railroad ballast and for road aggregate in highway construction
Graphite
Graphite is the crystal form of carbon. Graphite is used as a dry lubricant and steel hardener and for brake linings and the production of "lead" in pencils. Most graphite production comes from Korea, India, and Mexico
Gypsum
Processed gypsum is used in industrial or building plaster, prefabricated wallboard, cement manufacture, and for agriculture
Halite
Halite (salt) is used in the human and animal diet, primarily as food seasoning and as a food preservation. It is also used to prepare sodium hydroxide, soda ash, caustic soda, hydrochloric acid, chlorine, and metallic sodium, and it is used in ceramic glazes, metallurgy, curing of hides, mineral waters, soap manufacture, home water softeners, highway deicing, photography, and scientific equipment for optical parts
Iron ore
Iron ore is used to manufacture steels of various types and other metallurgical products, such as magnets, auto parts, and catalysts. Most US production is from Minnesota and Michigan. The Earth's crust contains about 5% iron, the fourth most abundant element in the crust.

Table 1.6 Some Common Minerals—cont'd

Limestone
"A sedimentary rock consisting largely of the minerals calcite and aragonite, which have the same composition $CaCO_3$." Limestone, along with dolomite, is one of the basic building blocks of the construction industry. Limestone is used as aggregate, building stone, cement, and lime and in fluxes, glass, refractories, fillers, abrasives, soil conditioners, and a host of chemical processes
Mica
Mica minerals commonly occur as flakes, scales, or shreds. Sheet muscovite (white) mica is used in electronic insulators, paints, as joint cement, as a dusting agent, in drilling mud and lubricants, and in plastics, roofing, rubber, and welding rods
Phosphate rock
Primarily, a sedimentary rock used to produce phosphoric acid and ammoniated phosphate fertilizers, feed additives for livestock, elemental phosphorus, and a variety of phosphate chemicals for industrial and home consumers. The majority of production in the United States comes from Florida, North Carolina, Idaho, and Utah
Potash
Potash is an industry term that refers to a group of water-soluble salts containing the element potassium and to ores containing these salts. Potash is used in fertilizer, medicine, the chemical industry and to produce decorative color effects on brass, bronze, and nickel
Pyrite
Pyrite (fool's gold) is used in the manufacture of sulfur, sulfuric acid, and sulfur dioxide; pellets of pressed pyrite dust are used to recover iron, gold, copper, cobalt, and nickel
Quartz
Quartz crystals are popular as a semiprecious gemstone; crystalline varieties include amethyst, citrine, rose quartz, and smoky quartz. Because of its piezoelectric properties (the ability to generate electricity under mechanical stress), quartz is used for pressure gauges, oscillators, resonators, and wave stabilizers. Quartz is also used in the manufacture of glass, paints, abrasives, refractories, and precision instruments
Sandstone
Sandstone is used as a building stone, road bases and coverings, construction fill, concrete, railroad ballast, and snow and ice control

Continued

Table 1.6 Some Common Minerals—cont'd

Silica
Silica is used in the manufacture of computer chips, glass and refractory materials, ceramics, abrasives, and water filtration and is a component of hydraulic cements, a filler in cosmetics, pharmaceuticals, paper, and insecticides; as an anticaking agent in foods; a flatting agent in paint, and as a thermal insulator
Sulfur
Sulfur is of importance to every sector of the world's manufacturing processes, drugs, and fertilizer complexes. Sulfur is used as an industrial raw material through its major derivative, sulfuric acid. Sulfuric acid production is the major end use for sulfur. Most sulfur goes into fertilizer; oil refining is another major use and a source of sulfur
Talc
The primary use for talc is in the production of paper. Ground talc is used as filler in ceramics, paint, paper, roofing, plastics, cosmetics, and agriculture. Talc is found in many common household products, such as baby (talcum) powder, deodorant, and makeup. Very pure talc is used in fine arts and is called soapstone. It is often used to carve figurines
Trona
Trona is used in glass container manufacture, fiberglass, specialty glass, flat glass, liquid detergents, medicine, food additives, photography, cleaning and boiler compounds, and control of water pH. Trona is mined mainly in Wyoming
Zeolites
Some of the uses of zeolite minerals include aquaculture (for removing ammonia from the water in fish hatcheries), water softener, catalysts, cat litter, odor control, and removing radioactive ions from nuclear plant effluent

[a]Listed alphabetically and not by order of abundance.

structure distinguish the various species, which were determined by the mineral's geologic environment when formed. Changes in the temperature, pressure, or bulk composition of a rock mass cause changes in its minerals.

Minerals can be described by their various physical properties, which are related to the chemical structure and composition. Common distinguishing characteristics include crystal structure, hardness, luster, color, and specific gravity. More specific tests for describing minerals include magnetism, radioactivity, and reaction to acid. The silicate class of minerals composes over 90% of the crust of the Earth, and since silicon and oxygen constitute approximately 75% of the Earth's crust, this translates into the predominance of silicate minerals. Other important mineral groups include the native elements, sulfides, oxides, halides, carbonates, sulfates, and phosphates.

1.5.2 Nonindigenous Chemicals

Common nonindigenous chemicals are atmospheric aerosols that originate from the condensation of gases and from the action of the wind on the surface of the Earth (Jacobson, 1998). *Fine* aerosol particles (less than 1 mm in radius) originate almost exclusively from condensation of precursor gases. A key precursor gas is sulfuric acid (H_2SO_4), which is produced in the atmosphere by oxidation of sulfur dioxide (SO_2) emitted from fossil fuel combustion, volcanoes, and other sources. Sulfuric acid has a low vapor pressure over sulfuric acid-water solutions and condenses under all atmospheric conditions to form aqueous sulfate particles. The composition of these sulfate particles can then be modified by condensation of other gases with low vapor pressure including ammonia (NH_3) and nitric acid (HNO_3). Organic carbon is contributed mainly by condensation of large hydrocarbons of biogenic and anthropogenic origin. Another important component of the fine aerosol is soot produced by condensation of gases during combustion. Soot as commonly defined includes both elemental carbon and black organic aggregates.

Household hazardous waste (HHW) (also referred to as domestic hazardous waste or home generated special materials) is waste that is generated from residential households. HHW only applies to wastes that are the result of the use of materials that are labeled for and sold for *home use*. Wastes generated by a company or at an industrial setting are not HHW. The following list includes categories often applied to HHW, and it is important to note that many of these inorganic chemical categories overlap and that many household wastes can fall into one or more categories: (i) paints and solvents, (ii) automotive wastes, such as used motor oil and glycol-based antifreeze, and (iii) pesticides, which include insecticides, herbicides, and fungicides.

More specific to the present text are the inorganic chemicals designated as hazardous waste (US EPA, 2015). Proper management of chemicals and chemical waste is an essential part of maintain a sustainable environment. The RCRA, passed in 1976, created the framework for hazardous and nonhazardous chemical solid waste management programs, and only materials that meet the definition of solid waste under RCRA can be classified as hazardous wastes, which are subject to additional regulation. EPA developed detailed regulations that define what materials qualify as solid wastes and hazardous wastes.

If an inorganic chemical has a listing as a hazardous waste, there will be a narrative description of a specific type of waste that United States Environmental Protection Agency (US EPA) considers to be sufficiently dangerous to warrant regulation. Hazardous waste listings describe (i) wastes from specific processes, (ii) wastes from very specific sectors of industry, or (iii) wastes in the form of very specific chemical formulations. Before developing a hazardous waste listing, the US EPA thoroughly studies a waste stream and the threat it can pose to human health and the environment. If the waste poses enough of a threat, the US EPA includes a precise description of that waste on one of the hazardous waste lists in the regulations. Thereafter, any waste fitting that narrative listing description is considered hazardous, regardless of its chemical composition or any other potential variable.

The RCRA requires a *cradle to grave* system of accounting for hazardous waste; the Department of Transportation requires compliance with Federal Motor Carrier Safety Regulations during transportation of hazardous waste. Furthermore, it is often required that state department of environmental quality license, inspect, and regulate generators, haulers, and disposal facilities handling hazardous waste. These modern safety guidelines, regulations, and procedures presented are intended to help generators comply with governmental rules and regulations designed to protect human health and the environment. Strict compliance with these regulations ensures the waste is managed, transported, and disposed of safely and properly while reducing potential liability to waste generator.

Finally, a chemical is an active ingredient in a formulation if that chemical serves the function of the formulation. For instance, a pesticide made for killing insects may contain a poison such as heptachlor and various solvent ingredients that act as carriers or lend other desirable properties to the poison. Although all of these chemicals may be capable of killing insects, only the heptachlor serves the primary purpose of the insecticide product. The other chemicals involved are present for other reasons, not because they are poisonous. Therefore, heptachlor is the *active* ingredient in such a formulation even though it may be present in low concentrations; this formulation would carry the P059 waste code.

1.6 CHEMISTRY AND ENGINEERING

There is often the question: Why teach engineers chemistry and why teach chemists engineering? Both questions can be answered by understanding the need to establish process knowledge (reactor construction and reactor parameters) for the chemist and to establish chemical knowledge (reaction parameters, feedstock properties, and product properties) for engineers. This will help to establish a link between the disciplines that can then (in the context of this book) be applied to the development of pathways for a sustainable environment. This cross fertilization of chemistry with the various engineering disciplines is especially useful when many technical issues cannot be dealt with successfully by a chemist of by an engineer working individually. Moreover, need is for teamwork in which professionals from both the chemical and engineering disciplines (and the related subdisciplines) work together for a better environment, sometimes referred to as a *green environment*, which in turn is brought about by the application of chemistry and engineering to solving environmental issues. Furthermore, when applied to the development of a sustainable environment (and to add some confusion to the terminology), chemistry and engineering are referred to not only as *environmental chemistry* and *environmental engineering* but also as *green chemistry* and *green engineering*. As an historical aside, environmental engineering (formerly known as *sanitary engineering*) originally developed as a subdiscipline of civil engineering.

The term *green chemistry* is often used in the context of environmental science to which *green engineering*, in the current context, can be added. By way of explanation, *green chemistry* and *green engineering* focus on the environmental concerns related to the use of materials, the generation of energy, and the various production cycles and can also be used to demonstrate how the fundamental chemical principles, engineering principles, and the various chemical and engineering methodologies can be applied to the protection of the environment (Anastas and Kirchhoff, 2002). The principles of both disciplines (chemistry and engineering) are central to chemical education and to engineering education because professionals entering the environmental field need to develop the tools and skills to support the concept of global sustainability. Thus, future chemists and engineers will acquire the technical knowledge to synthesize products and design the necessary industrial processes though an increased awareness of environmental impact and understand the importance of sustainable strategies to protect the environment.

Furthermore, both environmental chemistry and environmental engineering course (which should be taught as crossover interrelated courses) have the potential to add considerable enhancement to chemistry learning and to engineering leading to an improved and understanding of the chemical and engineering concepts. Incorporation of the principles of both chemistry and engineering into course material can be coupled with specific inserts that will complement the chemistry curriculum and complement the engineering curriculum and that will serve as a reminder that the practice of chemistry and the practice of engineering can lead to important developments in environmental technology (Braun et al., 2006).

The implementation of the principles of environmental chemistry for engineers and environmental engineering for chemists in any university curriculum will contribute not only to the general aims of science and engineering education but also to important elements in the development of scientific and engineering literacy and knowledge (Van Eijck and Roth, 2007). Crossover studies will help the fledgling chemist (or the industrially inexperience chemist) and the fledgling engineer (or the industrially inexperience chemist) make the necessary connections among the disciplines of chemistry and engineering, which, in turn, will contribute to the education of chemical and engineering professionals and bring about practices related to protection of the environment (Karpudewan et al., 2012).

In addition, crossover studies will provide the required knowledge and awareness that lead to development of technologies that are necessary to achieve the ultimate goal of environmental protection. Teaching environmental chemistry and environmental engineering at different levels of the university chemistry and engineering degree programs education has received significant attention recently (Andraos and Dicks, 2012; Eilks and Rauch, 2012; Burmeister and Eilks, 2012; Burmeister et al., 2012; Mandler et al., 2012; Karpudewan et al., 2012). The importance of this type of education, beyond the basics of chemical and engineering learning, relates to the ability of the chemist and engineer to participate as a meaningful and knowledgeable team member in the development of sustainable environmental practices (Eilks and Rauch, 2012).

In summary, an understanding of the chemical types that contribute to pollution can lead to an understanding of the chemical and physical methods (and the related process parameters) for mitigating pollution. Mitigation of such effects is not only a matter of knowing the elemental composition of the pollutant but also a matter of understanding the bulk properties as they relate to the chemical or physical composition of the material relating to the behavior of (in the context of this book) the inorganic chemical in the environment.

REFERENCES

Adriano, D.C., 2003. Trace Elements in Terrestrial Environments: Biogeochemistry, Bioavailability, and Risks of Metals, second ed. Springer Science, New York, NY.

Anastas, P.T., Kirchhoff, M.M., 2002. Origins, current status, and future challenges of green chemistry. Acc. Chem. Res. 35 (9), 686–694.

Andraos, J., Dicks, A.P., 2012. Green chemistry teaching in higher education: a review of effective practices. Chem. Educ. Res. Practice 13, 69–79.
Bertini, I., Gray, H.B., Lippard, S.J., Valentine, J.S., 1994. Bioinorganic Chemistry. University Science Books, Mill Valley, CA.
Bodek, I., Lyman, W.J., Reehl, W.F., Rosenblatt, D.H., 1988. Environmental Inorganic Chemistry: Properties, Processes, and Estimation Methods. Elsevier, Amsterdam.
Braun, B., Charney, R., Clarens, A., Farrugia, J., Kitchens, C., Lisowski, C., Naistat, D., O'Neil, A., 2006. Completing our education. Green chemistry in the curriculum. J. Chem. Educ. 83 (8), 1126–1128.
Burmeister, M., Eilks, I., 2012. An example of learning about plastics and their evaluation as a contribution to education for sustainable development in secondary school chemistry teaching. Chem. Educ. Res. Practice 13 (2), 93–102.
Burmeister, M., Rauch, F., Eilks, I., 2012. Education for sustainable development (ESD) and chemistry education. Chem. Educ. Res. Practice 13 (2), 59–68.
Carson, R., 1962. Silent Spring. Houghton Mifflin Company, Houghton Mifflin Harcourt International, Geneva, IL.
Carson, P., Mumford, C., 1988. The Safe Handling of Chemicals in Industry, vols. 1 and 2. John Wiley & Sons Inc., New York, NY.
Carson, P., Mumford, C., 1995. The Safe Handling of Chemicals in Industry, vol. 3. John Wiley & Sons Inc., New York.
Carson, P., Mumford, R., 2002. Hazardous Chemicals Handbook, second ed. Butterworth-Heinemann, Oxford.
CFR, 2016. Title 40: Protection of Environment. Code of Federal Regulations. United States Government Publications Office, Washington DC. http://www.ecfr.gov/cgi-bin/text-idx?tpl=/ecfrbrowse/Title40/40cfr261_main_02.tpl.
Chadeesingh, R., 2011. The fischer-tropsch process. In: Speight, J.G. (Ed.), The Biofuels Handbook. The Royal Society of Chemistry, London, pp. 476–517 (Part 3, Chapter 5).
Charette, M., Smith, W.H.F., 2010. The volume of earth's ocean. Oceanography 23 (2), 112–114.
Cox, P.A., 1995. The Elements on Earth: Inorganic Chemistry in the Environment. Oxford University Press, Oxford.
Davidson, D., 1955. Amphoteric molecules, ions and salts. J. Chem. Educ. 32 (11), 550.
Eilks, I., Rauch, F., 2012. Sustainable development and green chemistry in chemistry education. Chem. Educ. Res. Practice 13 (2), 57–58.
Fraústo da Silva, J.J.R., Williams, R.J.P., 2001. The Biological Chemistry of the Elements: The Inorganic Chemistry of Life, second ed. Oxford University Press, Oxford.
Gouin, T., James, Y., Armitage, M., Cousins, I.T., Muir, D.C.G., Ng, C.A., Reid, L., Tao, S., 2013. Influence of global climate change on chemical fate and bioaccumulation: the role of multimedia models. Environ. Toxicol. Chem. 32 (1), 20–31.
Hickey, J.P., 1999. Estimating the Environmental Behavior of Inorganic and Organometal Contaminants: Solubilities, Bioaccumulation, and Acute Aquatic Toxicities. Contribution No. 1066. USGS Great Lakes Science Center, Ann Arbor, MI. http://toxics.usgs.gov/pubs/wri99-4018/Volume2/sectionD/2512_Hickey/pdf/2512_Hickey.pdf.
Jackson, R.B., Carpenter, S.R., Dahm, C.N., McKnight, D.M., Naiman, R.J., Postel, S.L., Running, S.W., 2001. Water in a changing world. Ecol. Appl. 11 (4), 1027–1045.
Jacobson, M.Z., 1998. Fundamentals of Atmospheric Modeling. Cambridge University Press, Cambridge.
Jones, K.C., De Voogt, P., 1999. Persistent inorganic pollutants (POPs): state of the science. Environ. Pollut. 100, 209–221.
Karpudewan, M., Ismail, Z., Roth, W.M., 2012. Ensuring sustainability of tomorrow through green chemistry integrated with sustainable development concepts (SDCs). Chem. Educ. Res. Practice 13 (2), 120–127.

Khan, S., Cao, Q., Zheng, Y.M., Huang, Y.Z., Zhu, Y.G., 2008. Health risks of heavy metals in contaminated soils and food crops irrigated with wastewater in Beijing, China. Environ. Pollut. 152 (3), 686–692.

Kirpichtchikova, T.A., Manceau, A., Spadini, L., Panfili, F., Marcus, M.A., Jacquet, T., 2006. Speciation and solubility of heavy metals in contaminated soil using X-ray microfluorescence, EXAFS spectroscopy, chemical extraction, and thermodynamic modeling. Geochim. Cosmochim. Acta 70 (9), 2163–2190.

Lee, S., Speight, J.G., Loyalka, S., 2014. Handbook of Alternative Fuel Technologies, second ed. CRC Press/Taylor & Francis Group, Boca Raton, FL.

Ling, W., Shen, Q., Gao, Y., Gu, X., Yang, Z., 2007. Use of bentonite to control the release of copper from contaminated soils. Aust. J. Soil Res. 45 (8), 618–623.

Mackay, D., Shiu, W., Ma, K., Lee, S., 2006. Handbook of Physical-Chemical Properties and Environmental Fate for Inorganic Chemicals, second ed. CRC Press/Taylor & Francis Group, Boca Raton, FL.

Manahan, S.E., 2009. Environmetnal Chemistry, ninth ed. CRC Press/Taylor and Francis Group, Boca Raton, FL.

Mandler, D., Mamlok-Naaman, R., Blonder, R., Yayon, M., Hofstein, A., 2012. High school chemistry teaching through environmentally oriented curricula. Chem. Educ. Res. Practice 13 (2), 80–92.

Maslin, P., Maier, R.M., 2000. Rhamnolipid-enhanced mineralization of phenanthrene in organic-metal Co-contaminated soils. Bioremediat. J. 4 (4), 295–308.

McLaughlin, M.J., Zarcinas, B.A., Stevens, D.P., Cook, N., 2000a. Soil testing for heavy metals. Commun. Soil Sci. Plant Anal. 31 (11–14), 1661–1700.

McLaughlin, M.J., Hamon, R.E., McLaren, R.G., Speir, T.W., Rogers, S.L., 2000b. A bioavailability-based rationale for controlling metal and metalloid contamination of agricultural land in Australia and New Zealand. Aust. J. Soil Res. 38 (6), 1037–1086.

Mills, S.J., Hatert, F., Nickel, E.H., Ferraris, G., 2009. The standardization of mineral group hierarchies: application to recent nomenclature proposals. Eur. J. Mineral. 21, 1073–1080.

Moore, J.W., Stanitski, C.L., 2014. Chemistry: The Molecular Science, fifth ed. Cengage Learning, Independence, KY.

Pöschl, U., Shiraiwa, M., 2015. Multiphase chemistry at the atmosphere—biosphere interface influencing climate and public health in the anthropocene. Chem. Rev. 115, 4440–4475.

Robinson, P., 2016. Our Eponymous Epoch. Chemistry World, vol. 13. The Royal Society of Chemistry, London. no. 10, p. 3.

Sheldon, R., 2010. Introduction to green chemistry, inorganic synthesis and pharmaceuticals. In: Dunn, P.J., Wells, A.S., Williams, M.T. (Eds.), Green Chemistry in the Pharmaceutical Industry. Wiley-VCH Verlag GmbH & Co. KGaA, Weinheim, Germany.

Smith, M.B., 2013. March's Advanced Inorganic Chemistry: Reactions, Mechanisms, and Structure, seventh ed. John Wiley & Sons Inc., Hoboken, NJ.

Speight, J.G., 2011a. An Introduction to Petroleum Technology, Economics, and Politics. Scrivener Publishing, Salem, MA.

Speight, J.G. (Ed.), 2011b. The Biofuels Handbook. Royal Society of Chemistry, London.

Speight, J.G., 2013. The Chemistry and Technology of Coal, third ed. CRC Press/Taylor and Francis Group, Boca Raton, FL.

Speight, J.G., 2014. The Chemistry and Technology of Petroleum, fifth ed. CRC Press/Taylor and Francis Group, Boca Raton, FL.

Speight, J.G., 2017a. Environmental Organic Chemistry for Engineers. Butterworth-Heinemann/Elsevier, Cambridge, MA.

Speight, J.G. (Ed.), 2017b. Lange's Handbook of Chemistry, 17th ed. McGraw-Hill Education, New York, NY.

Speight, J.G., Arjoon, K.K., 2012. Bioremediation of Petroleum and Petroleum Products. Scrivener Publishing, Beverly, MA.
Speight, J.G., Islam, M.R., 2016. Peak Energy—Myth or Reality. Scrivener Publishing, Salem, MA.
Speight, J.G., Singh, K., 2014. Environmental Management of Energy from Biofuels and Biofeedstocks. Scrivener Publishing, Salem, MA.
Thy, P., Yu, C., Jenkins, B.M., Lesher, C.E., 2013. Inorganic composition and environmental impact of biomass feedstock. Energy Fuels 27 (7), 3969–3987.
Trenberth, K.E., Guillemot, C.J., 1994. The total mass of the atmosphere. J. Geophys. Res. 99, 23079–23088.
US EPA, 2015. List of Lists: Consolidated List of Chemicals Subject to the Emergency Planning and Community Right-to-Know Act (EPCRA), Comprehensive Environmental Response, Compensation and Liability Act (CERCLA) and Section 112(r) of the Clean Air Act. Report No. EPA 550-B-15-001, Office of Solid Waste and Emergency Response, United States Environmental Protection Agency, Washington, DC. March.
Van Eijck, M., Roth, W.M., 2007. Improving science education for sustainable development. PLoS Biol. 5, 2763–2769.
Weller, M., Overton, T., Rourke, J., Fraser, A., 2014. Inorganic Chemistry, sixth ed. Oxford University Press, Oxford.
Wenk, H.-R., Bulakh, A., 2004. Minerals: Their Constitution and Origin. Cambridge University Press, Cambridge.
Yaron, B., Dror, I., Berkowitz, B., 2011. Properties and behavior of selected inorganic and organometallic contaminants. Soil Subsurf. Chang, 39–74.
Zhang, M.K., Liu, Z.Y., Wang, H., 2010. Use of single extraction methods to predict bioavailability of heavy metals in polluted soils to rice. Commun. Soil Sci. Plant Anal. 41 (7), 820–831.

FURTHER READING

Swaddle, J.W., 1997. Inorganic Chemistry: An Industrial and Environmental Perspective. Academic Press Inc., New York, NY.

CHAPTER TWO

Inorganic Chemistry

2.1 INTRODUCTION

If organic chemistry is defined as the chemistry of hydrocarbon compounds and their derivatives, *inorganic chemistry* can be described very generally as the chemistry of noncarbon compounds or as the chemistry of *everything else*. This includes all the remaining elements in the periodic table (Figs. 2.1 and 2.2) and some compounds of carbon (such as carbon monoxide (CO) and carbon dioxide (CO_2)), which plays a major role in many inorganic compounds. Thus, inorganic chemistry is the subcategory of chemistry concerned with the properties and reactions of inorganic compounds, which includes all chemical compounds without the chains or rings of carbon atoms that fall into the subcategory of organic compounds.

A common differentiation to help distinguish between inorganic compounds and organic compounds is that inorganic compounds are either the result of natural processes unrelated to any life form or the result of human experimentation in the laboratory, whereas organic compounds result from the activity of living beings. However, caution is advised when using such a definition because organic compounds can be artificially created in the laboratory. Another definition pertains to the salt-making property of inorganic compounds, which is absent in organic compound, but even then, this definition too is not truly correct since organic acids (RCO_2H) sacrosanct can also form salts. There is also the argument that inorganic compounds do not have carbon-hydrogen bonds—a characteristic of organic compounds—but this, also, is not strictly true since perfluorocarbons (carbon-fluorine compounds where all of the hydrogen atoms have been replaced by fluorine atoms) do not have any carbon-hydrogen bonds but are still organic compounds. Another often-quoted difference is that inorganic compounds contain metal atoms, whereas organic compounds do not. Again, this is not true since organometallic compounds contain metal atoms. Thus, caution is advised when accepting any definition that purports to define the differences between inorganic compounds and organic compounds.

Environmental Inorganic Chemistry for Engineers
http://dx.doi.org/10.1016/B978-0-12-849891-0.00002-3

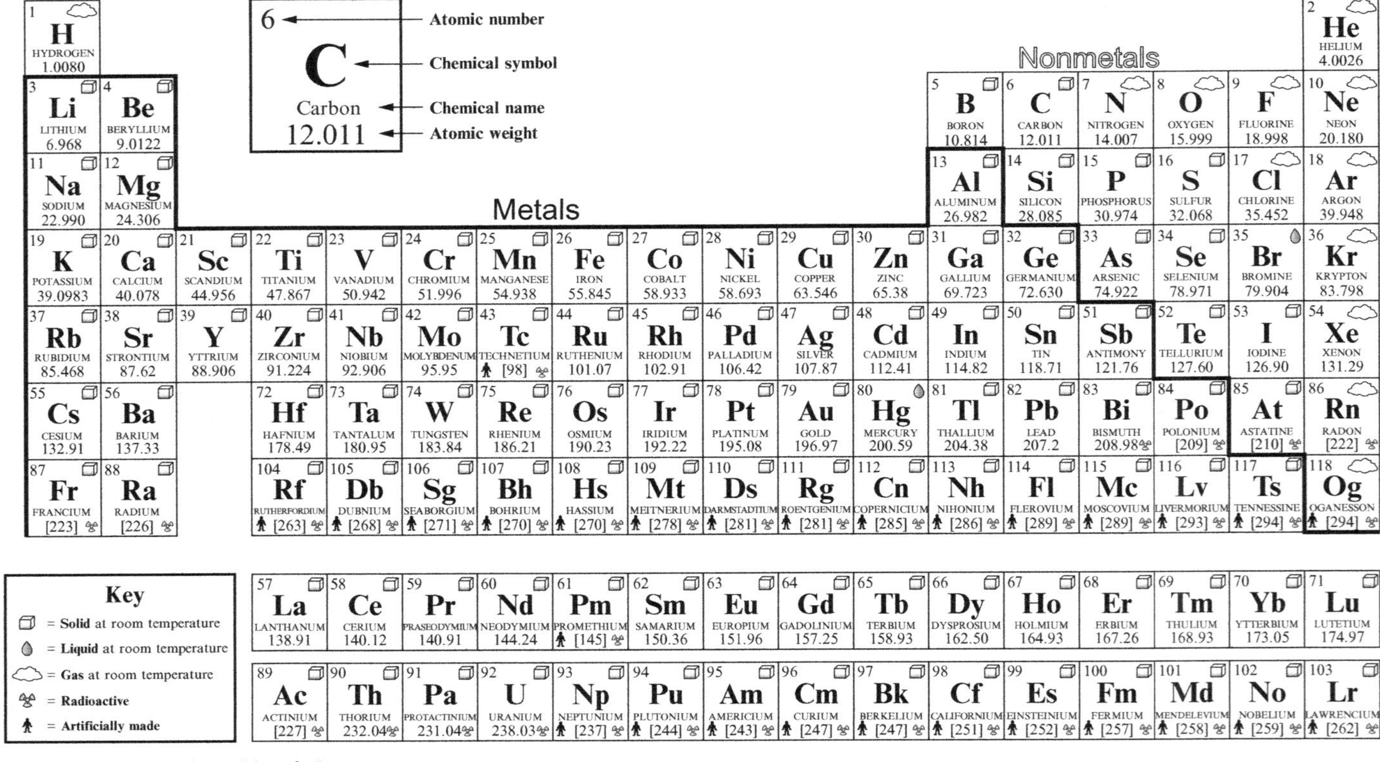

Fig. 2.1 The periodic table of elements.

Group→ ↓Period	1	2		3	4	5	6	7	8	9	10	11	12	13	14	15	16	17	18
1	1 H																		2 He
2	3 Li	4 Be												5 B	6 C	7 N	8 O	9 F	10 Ne
3	11 Na	12 Mg												13 Al	14 Si	15 P	16 S	17 Cl	18 Ar
4	19 K	20 Ca		21 Sc	22 Ti	23 V	24 Cr	25 Mn	26 Fe	27 Co	28 Ni	29 Cu	30 Zn	31 Ga	32 Ge	33 As	34 Se	35 Br	36 Kr
5	37 Rb	38 Sr		39 Y	40 Zr	41 Nb	42 Mo	43 Tc	44 Ru	45 Rh	46 Pd	47 Ag	48 Cd	49 In	50 Sn	51 Sb	52 Te	53 I	54 Xe
6	55 Cs	56 Ba	*	71 Lu	72 Hf	73 Ta	74 W	75 Re	76 Os	77 Ir	78 Pt	79 Au	80 Hg	81 Tl	82 Pb	83 Bi	84 Po	85 At	86 Rn
7	87 Fr	88 Ra	**	103 Lr	104 Rf	105 Db	106 Sg	107 Bh	108 Hs	109 Mt	110 Ds	111 Rg	112 Cn	113 Nh	114 Fl	115 Mc	116 Lv	117 Ts	118 Og

*	57 La	58 Ce	59 Pr	60 Nd	61 Pm	62 Sm	63 Eu	64 Gd	65 Tb	66 Dy	67 Ho	68 Er	69 Tm	70 Yb
**	89 Ac	90 Th	91 Pa	92 U	93 Np	94 Pu	95 Am	96 Cm	97 Bk	98 Cf	99 Es	100 Fm	101 Md	102 No

Fig. 2.2 Periodic table of elements showing the groups and periods including the lanthanide elements and the actinide elements.

Organometallic chemistry, a very large and rapidly growing field, bridges both areas by considering compounds containing direct metal-carbon bonds and includes catalysis of many chemical reactions. Organometallic compounds contain at least one bond between a metal atom and a carbon atom. They are named as coordination compounds, using the additive nomenclature system. The name for an organic ligand binding through one carbon atom may be derived either by treating the ligand as an anion or as a neutral substituent group. In addition, *bioinorganic chemistry* bridges biochemistry and inorganic chemistry, and because environmental chemistry includes the study of both inorganic and organic compounds, studies of these various subdivisions of chemistry are essential fields of knowledge. As can be imagined, the realm of inorganic chemistry is extremely broad, providing essentially limitless areas for investigation.

In the very broadest (or general) sense, inorganic chemicals and compounds are defined by what they are not: (i) they are not organic in nature and (ii) anything beyond biological, hydrocarbon, and other similar carbon-based chemicals may be considered to be inorganic. From a more practical sense, inorganic chemicals are substances of mineral origin that do not contain carbon in their molecular structure and are typically based on the most abundant chemicals on earth: oxygen, silicon, aluminum, iron, calcium, sodium, potassium, and magnesium. The exceptions are CO and CO_2, as well as the mineral carbonate ($—CO_3$) derivatives and bicarbonate ($—HCO_3$) derivatives.

In terms of environmental issues related to spills or disposal of inorganic chemicals, a good deal is already known about the influence of molecular structure on the toxicity to human beings of chemicals, much less is known about the influence of molecular structure on the environmental persistence of a chemical. For ecosystems in which floral and faunal species exist, the persistence of any chemicals (inorganic and organic) is an extremely important criterion for predicting potential harm because there are, inevitably, some species that are sensitive to any chemical and any persistent chemical. Although some chemicals may be harmless to a limited number of organisms, they will eventually be delivered through the course of biogeochemical cycles to a sensitive species in an ecosystem. Thus, highly toxic, readily biodegradable substances may pose much less of an environmental problem than a relatively harmless persistent chemical that may well damage a critical floral or faunal species.

Thus, the study of chemical effects in the environment can be resolved into two study areas: (i) a study of the levels of a substance accumulating in air, water, soils including sediments and biota and (ii) a study of the effects of chemicals when the threshold action level has been reached, particularly the effects produced in biota, which constitute a significant adverse response (i.e., environmental dose-response curve). In order to predict trends in levels of a chemical, much more information is needed about the rates of injection; flow and partitioning between air, water, soils, and biota; and loss through degradation, which gives rise to the concept of an environmental balance sheet for an ecosystem. These dynamic phenomena are governed by the physical properties and by the chemical properties of the molecular contaminant.

Fluid mechanics and meteorology may in future provide the conceptual and technical tools for producing predictive models of such systems. Most of the knowledge of effects derives from acute toxicology and medical studies on various (but not all) organism (including humans), but since environmental effects are usually associated with chronic exposure, studies are being increasingly made of long-term continuous exposure to minute amounts of a chemical. The well-known difficulty of recognizing such effects when they occur in an ecosystem is aggravated by the fact that many of the effects are nonspecific and can often be masked by similar effects deriving from exposure to natural phenomena such as famines, droughts, and any one (or more) of several meteorological or catastrophic phenomena. Even when a genuine effect is recognized, a candidate causal agent must be found and correlated with the effect. This process must be accompanied by experimental studies (laboratory studies and/or field work), which link beyond reasonable doubt the causal inorganic chemical(s) and any adverse effect(s) on the floral and/or faunal ecosystem. This can only be achieved by the assiduous collection and assimilation of the technical knowledge of inorganic chemistry as it relates to the properties and behavior of inorganic chemicals.

However, it not expected that the engineer will accumulate as much chemical knowledge as the professional inorganic chemist—just as the chemist might become squeamish at having to be proficient in one or more of the engineering disciplines. But, the accumulation of sufficient knowledge to (i) understand the behavior of inorganic chemicals in the environment followed by (ii) the ability to make a reasonable prediction (based on properties) of the behavior of inorganic chemicals in the environment. Failure to recognize the mutually interactive roles of the chemist and engineer will hinder and despoil the development of a united environmental management policy that will apply to the sustainability of any ecosystem or broad environmental area (Chapter 1).

Thus, inorganic chemicals have application in every aspect of the chemical industry including catalysts, pigments, surfactants, coatings, medicine, fuel, and agriculture. The products of inorganic chemical processes are used as (i) basic chemicals for industrial processes, which include acids, bases, salts, oxidizing agents, gases, and halogens; (ii) chemical additives, which include pigments, alkali metals, and colors; and (iii) finished products, which include fertilizers, glass, and construction materials.

From an industry perspective, there are two main classes of inorganic chemicals: (i) alkali chemicals, including soda ash, which is predominantly sodium carbonate ($NaCO_3$), caustic soda (NaOH), and liquid chlorine (Cl_2) and (ii) basic inorganic compounds such as aluminum fluoride (AlF_3), calcium carbide (CaC_2), potassium chlorate ($KClO_3$), and titanium dioxide (TiO_2). Furthermore, the chlor-alkali industry is a major component of the world chemical economy. A prime reaction of the industry is a reaction in which saltwater (brine—water which contains sodium chloride, NaCl) is decomposed by an electrolysis process to generate NaOH (sodium hydroxide, NaOH), chlorine gas (Cl_2), and hydrogen (H_2) gas:

$$2NaCl + 2H_2O \rightarrow Cl_2 + H_2 + 2NaOH$$

Chlorine is produced at the positive electrode (anode), while hydrogen (H_2) and sodium hydroxide are produced at the negative electrode (cathode). These three materials are feedstocks for the production of bleach (sodium hypochlorite, NaOCl), and a variety of other products, including soda ash (Na_2CO_3).

Finally, inorganic chemistry is a subject that should not be approached with any degree of dismay or hesitancy insofar as the subject becomes easier as the individual researcher works delves into it. The topics covered in this book are the basic topics that serve to introduce the reader not only to inorganic chemistry but also to an understanding of the effect(s) of inorganic chemicals on various ecosystems. Furthermore, understanding the mechanism by means of which a reaction occurs is particularly crucial, and of necessity, the book brings a logical and simplistic approach to the reactions of the different inorganic functional groups. This in turn transforms a list of apparently unrelated facts into a sensible and coherent theme. This chapter will serve as an introduction to the nature of and production of inorganic chemicals and the structure of inorganic chemical, which will serve as an introduction and the production of inorganic chemicals (Chapter 3), and the ensuing chapter (Chapter 4) presents an introduction to the chemical properties and the physical properties of inorganic chemicals (Chapter 4) from which an understanding of the effects of inorganic chemicals on floral and faunal environments can be estimated.

2.2 NOMENCLATURE OF INORGANIC COMPOUNDS

The universal adoption of an agreed chemical nomenclature is key for communication in any of the chemical sciences and for regulatory purposes, such as those associated with any commercial activity and for environmental health and safety or commercial activity. Names and formulas (sometimes written in the Latin version as formulae) have only served half their role when they are created and used to describe or identify compounds, for example, in publications. Achieving their full role requires that the reader of a name or formula can interpret that formula successfully in order to, for example, produce a structural diagram.

Thus, throughout the field of chemistry, nomenclature is the process of naming chemical compounds with different names so that they can be easily identified as separate chemicals. In the subdiscipline of inorganic chemistry, there are two general types of inorganic compounds that can be formed: (i) ionic compounds and (ii) molecular compounds (Connelly et al., 2005).

The system of nomenclature for inorganic chemicals is recommended by the International Union of Pure and Applied Chemistry (IUPAC); ionic compounds are named according to the composition of the compound and not the structure of the compound. This allows international recognition of a chemical to be recognized by the name of the compound, no matter what the language of the chemist. In the simplest case of a binary ionic compound with no possible ambiguity about the charges and the stoichiometry, the common name is written using two words. The name of the cation (the unmodified element name for monatomic cations) comes first, followed by the name of the anion. For example, $MgCl_2$ is formally named magnesium chloride, and Na_2SO_4 is formally named sodium sulfate. In addition, the sulfate ion (SO_4^{2-}) is an example of a polyatomic ion (i.e., a multiatom ion). To obtain the empirical formula from these names, the stoichiometry can be deduced from the charges on the ions and the requirement of molecular neutrality in terms of the overall charge.

If there are multiple (different) cations and/or anions, multiplicative prefixes (*di-*, *tri-*, and *tetra-*) are often required to indicate the relative compositions after which the cations and then the anions are listed in alphabetical order. For example, $KMgCl_3$ (which can also be written as $KCl \cdot MgCl_2$) is named correctly identified as magnesium potassium trichloride to distinguish it from K_2MgCl_4 magnesium dipotassium tetrachloride ($MgCl_2 \cdot 2KCl$). In both of the empirical formula and the written name, the cations appear in alphabetical order, but the order varies between them because the symbol for potassium is K. When one of the ions already has a multiplicative prefix within its name, the alternate multiplicative prefixes (*bis-*, *tris-*, and *tetrakis-*) are used. For example, $Ba(BrF_4)_2$ is written in full as barium bis(tetrafluoridobromate).

Compounds containing one or more elements that can exist in a variety of charged states (oxidation states) will have a stoichiometry that depends on the oxidation state in order to ensure overall neutrality of the molecule. This can be indicated in the name by specifying either the oxidation state of the elements present or the charge on the ions. Because of the risk of ambiguity in allocating oxidation states, the preference of the IUPAC is an indication of the ionic charge numbers. These are written as an Arabic (not Roman) integer followed by the sign (2^-, 1^-, 1^+, and 2^+) in parentheses directly after the name of the cation (without a space separating them). For example, ferrous sulfate ($FeSO_4$) is named iron(2^+) sulfate (with the 2^+ charge on the iron ion (Fe^{2+}) thereby balancing the 2^- charge on the sulfate ion), whereas ferric sulfate [$Fe_2(SO_4)_3$] is written as iron(3^+) sulfate (because each of the two iron ions in the molecular formula have a charge of 3^+ to balance the 2^- on each of the three sulfate ions). However, there is the option, which remains a common use (and is used throughout this text) to write the oxidation number in Roman numerals (I, II, III, and IV), and using the example presented above, the names would be written as iron(II) sulfate and iron(III) sulfate, respectively. For simple ions, the ionic charge and the oxidation number are identical, but for polyatomic ions, they often differ. An even older naming system (often referred to as the *classical system*) still exists for metal cations and remains in wide use involves placement of the suffixes *-ous* and *-ic* to the Latin root of the name to give special names for the low and high oxidation states. Again, using the above compounds as the example, this form of nomenclature uses *ferrous* and ferric (from the Latin for iron, *ferrum*), for iron(II) and iron(III), respectively, and as a result, the examples given above would be (and are still often written) as ferrous sulfate and ferric sulfate. Although the system recommended by the IUPAC is used throughout the world, there is still the tendency for chemists to use the Arabic and Roman systems of nomenclature.

Thus, compounds consisting of a metal and nonmetal are commonly known as ionic compounds, where the compound name has an ending of *-ide*. Cations have positive charges, while anions have negative charges (Tables 2.1 and 2.2) with the name derived from the parent metal (Table 2.3). The transition metals may form more than one ion, and it must be specified which ion is indicated by assigning a Roman numeral after the metal, which denotes the charge and the oxidation state of the metal (Table 2.4). For example, iron can form two common ions, the ferrous

Table 2.1 Cations and Anions

+1 Charge	+2 Charge	−1 Charge	−2 Charge	−3 Charge	−4 Charge
Hydrogen: H^+	Beryllium: Be^{2+}	Hydride: H^-	Oxide: O^{2-}	Nitride: N^{3-}	Carbide: C^{4-}
Lithium: Li^+	Magnesium: Mg^{2+}	Fluoride: F^-	Sulfide: S^{2-}	Phosphide: P^{3-}	
Sodium: Na^+	Calcium: Ca^{2+}	Chloride: Cl^-			
Potassium: K^+	Strontium: Sr^{2+}	Bromide: Br^-			
Rubidium: Rb^+	Barium: Ba^{2+}	Iodide: I^-			
Cesium: Cs^+					

Table 2.2 Transition Metal and Metal Cations

+1 Charge	+2 Charge	+3 Charge	+4 Charge
Copper(I): Cu^{+}	Copper(II): Cu^{2+}	Aluminum: Al^{3+}	Lead(IV): Pb^{4+}
Silver: Ag^{+}	Iron(II): Fe^{2+}	Iron(III): Fe^{3+}	Tin(IV): Sn^{4+}
	Cobalt(II): Co^{2+}	Cobalt(III): Co^{3+}	
	Tin(II): Sn^{2+}		
	Lead(II): Pb^{2+}		
	Nickel: Ni^{2+}		
	Zinc: Zn^{2+}		

Table 2.3 Examples of the Nomenclature of Metals Ions

Metal Ion[a]	Latin Name
Copper (I): Cu^{+}	Cuprous
Copper (II): Cu^{2+}	Cupric
Iron (II): Fe^{2+}	Ferrous
Iron (III): Fe^{3+}	Ferric
Lead (II): Pb^{2+}	Plumbous
Lead (IV): Pb^{4+}	Plumbic
Mercury (I): $Hg_2{}^{2+}$	Mercurous
Mercury (II): Hg^{2+}	Mercuric
Tin (II): Sn^{2+}	Stannous
Tin (IV): Sn^{4+}	Stannic

[a]The Roman numeral indicates the valency of the metal.

Table 2.4 Oxidation State Guidelines and Rules

1. All pure elements have an oxidation sate of zero
2. Simple, monatomic ions have an oxidation state equal to the net ion charge
3. The charge of a polyatomic molecule or ion is equal to the sum of the oxidation states of all the constituent atoms, which allows unknown values to be determined if the oxidation states of all other species are known
4. Hydrogen has oxidation state of +1; oxygen has oxidation state of −2 in most compounds; the exceptions include some metal hydrides, where hydrogen has oxidation state of −1 due to the electropositivity of the metal, and peroxides, where oxygen has oxidation state of −1
5. As fluorine is the most electronegative element, its oxidation state does not change from −1; for the other halogens, they also have oxidation state of −1, except when bonded to oxygen, nitrogen, or another halogen with greater electronegativity
6. Alkali metals generally have oxidation state of +1, while alkaline earth metals generally have an oxidation state of +2

ion (Fe^{2+}) and the ferric ion (Fe^{3+}), and to distinguish between the two, the ferrous ion is named iron (II), and ferric ion is named iron (III).

The net charge of any ionic compound must be zero, which also means it must be electrically neutral. For example, one sodium ion (Na^{+}) is paired with one chloride ion (Cl^{-}), and one calcium ion (Ca^{2+}) is paired with two bromide ions (Br^{-}). The rules of nomenclature state that: (i) the cation (metal) is always named first with its name unchanged and (ii) the anion (nonmetal) is written after the cation, modified to end in *-ide*. In summary, Ionic compounds may be produced from a combination of an ionic component and an anionic component where stoichiometric and other rules are met.

Some of the charges on transition metals have specific Latin names. Just like the other nomenclature rules, the ion of the transition metal that has the lower charge has the Latin name ending with *-ous* and the one with the higher charge has a Latin name ending with *-ic*. However, several exceptions apply to the Roman numeral assignment: aluminum, zinc, and silver. Although they belong to the transition metal category, these metals do not have Roman numerals written after their names because these metals only exist in one ion. Instead of using Roman numerals, the different ions can also be presented in plain words. The metal is changed to end in *-ous* or *-ic*.

Although HF can be named hydrogen fluoride, it is given a different name for emphasis that it is an acid—a substance that dissociates into hydrogen ions (H^{+}) and anions in water. A quick way to identify acids is to see if there is an H (denoting hydrogen) in front of the molecular formula of the compound. To name acids, the prefix *hydro-* is placed in front of the nonmetal modified to end with *-ic*. The state of acids is aqueous (aq) because acids are found in water. Some common binary acids include

$$\text{HF (g) (hydrogen fluoride)} \rightarrow \text{HF (aq) (hydrofluoric acid)}$$
$$\text{HBr (g) (hydrogen bromide)} \rightarrow \text{HBr (aq) (hydrobromic acid)}$$
$$\text{HCl (g) (hydrogen chloride)} \rightarrow \text{HCl (aq) (hydrochloric acid)}$$
$$\text{H}_2\text{S (g) (hydrogen sulfide)} \rightarrow \text{H}_2\text{S (aq) (hydrosulfuric acid)}$$

Polyatomic ions (meaning two or more atoms) (Table 2.5) are joined together by covalent bonds. Thus, a polyatomic ion, as a molecular ion, is a charged chemical species composed of two or more atoms that are covalently bonded or composed of a metal complex that can be considered to act as a single unit (e.g., a functional group). An example of a polyatomic ion is the hydroxide ion (OH—), which consists of one oxygen atom and one

Table 2.5 Common Polyatomic Ions

Name: Cation Anion	**Formula**
Ammonium ion	NH_4^+
Hydronium ion	H_3O^+
Acetate ion	$C_2H_3O_2^-$
Arsenate ion	AsO_4^{3-}
Carbonate ion	CO_3^{2-}
Hypochlorite ion	ClO^-
Chlorite ion	ClO_2^-
Chlorate ion	ClO_3^-
Perchlorate ion	ClO_4^-
Chromate ion	CrO_4^{2-}
Dichromate ion	$Cr_2O_7^{2-}$
Cyanide ion	CN^-
Hydroxide ion	OH^-
Nitrite ion	NO_2^-
Nitrate ion	NO_3^-
Oxalate ion	$C_2O_4^{2-}$
Permanganate ion	MnO_4^-
Phosphate ion	PO_4^{3-}
Sulfite ion	SO_3^{2-}
Sulfate ion	SO_4^{2-}
Thiocyanate ion	SCN^-
Thiosulfate ion	$S_2O_3^{2-}$

hydrogen atom and hiving a -1 charge. Similarly, an ammonium ion (NH_4^+) is composed of one nitrogen atom and four hydrogen atoms and has a resulting charge of $+1$.

Polyatomic ions are often useful in the context of acid-base chemistry or in the formation of salts. A polyatomic ion can often be considered as the conjugate acid/base of a neutral molecule. For example, the conjugate base of sulfuric acid (H_2SO_4) is the polyatomic hydrogen sulfate anion (HSO_4^-). Although there may be an element with positive charge, such as the hydrogen ion (like H^+), it is not joined with another element with an ionic bond. This occurs because if the atoms formed an ionic bond, then it would have already become a compound, thus not needing to gain or lose any electrons. Polyatomic anions have negative charges, while polyatomic cations have positive charges. To correctly specify how many oxygen atoms are in the ion, prefixes and suffixes are used.

A later class of *inorganic compounds* that has emerged is ionic liquids, which are salts in the liquid state or salts with melting points lower than 100°C

Table 2.6 General Properties of Ionic Liquids

Property	**Comment**
A salt	Cation is usually large
A salt	Anion is usually small
Melting point	Preferably below 100°C
Freezing point	Preferably below 100°C
Liquid range	>200°C
Thermal stability	High
Viscosity	<100 cP
Dielectric constant	<30
Polarity	Moderate
Solvent properties	Good
Catalytic properties	Good
Vapor pressure	Low to negligible

(212°F) (Table 2.6). A typical liquid is predominantly electrically neutral while ionic liquids are composed predominantly of ions and short-lived ion pairs. These substances are variously called liquid electrolytes, ionic melts, ionic fluids, fused salts, liquid salts, or ionic glasses. Ionic liquids are powerful solvents and electrically conducting fluids (electrolytes). Any salt that melts without decomposing or vaporizing usually yields an ionic liquid. Conversely, when an ionic liquid is cooled, it often forms an ionic solid that may be either crystalline or glass-like. Examples include compounds based on the 1-ethyl-3-methylimidazolium (EMIM) cation and include $(C_2H_5)(CH_3)C_3H_3N^+$ $N(CN)^-$ that melts at −21°C (−6°F) and 1-butyl-3,5-dimethylpyridinium bromide, which becomes a glass below −24°C (−11°F).

Low-temperature ionic liquids can be compared with ionic solutions, liquids that contain both ions and neutral molecules, and, in particular, the so-called deep eutectic solvents in which mixtures of ionic and nonionic solid substances have much lower melting points than the pure compounds. Certain mixtures of nitrate salts can have melting points below 100°C (212°F). By their very low vapor pressure, temperature stability, nonflammability, and noncorrosivity, ionic liquids are ideal candidates to replace conventional organic solvents also are volatile and toxic for membranes technology.

Finally, an *isotope* is a variant of a chemical element that differs in the number of neutrons in the atom of the element. All isotopes of a given element have the same number of protons in each atom, and different isotopes of a single element occupy the same position on the periodic table of the elements. The number of protons within the nucleus of an atom (the *atomic number*) is equal to the number of electrons in the neutral (nonionized) atom. Each atomic number identifies a specific element but not the isotope; an atom of a given element may have a wide range in its number of neutrons, and the number of *nucleons* (both protons and neutrons) in the nucleus is the *mass number* of the atom, and each isotope of a given element has a different mass number. For example, carbon-12, carbon-13, and carbon-14 are three isotopes of the element carbon with mass numbers 12, 13, and 14, respectively. The atomic number of carbon is 6, which means that every carbon atom has 6 protons, so that the neutron numbers of these isotopes are 6, 7, and 8, respectively.

The term *nuclide* refers to a nucleus rather than to an atom. *Isotope* (the older term) is better known than the term *nuclide* and is still sometimes used in contexts where the use of the term *nuclide* might be more appropriate. Identical nuclei belong to one nuclide, for example, each nucleus of the carbon-13 nuclide is composed of six protons and seven neutrons. The *nuclide* concept (referring to individual nuclear species) emphasizes nuclear properties over chemical properties, whereas the *isotope* concept (grouping all atoms of each element) emphasizes chemical properties over nuclear properties. The neutron number has large effects on nuclear properties, but its effect on chemical properties is negligible for most elements. Even in the case of the lightest elements where the ratio of neutron number to atomic number varies the most between isotopes, it usually has only a small effect, although it does matter in some circumstances (for hydrogen, the lightest element, the isotope effect is large enough to strongly affect the chemistry).

2.3 CLASSIFICATION OF INORGANIC CHEMICALS

Based on general molecular composition, there are two classes of inorganic chemicals: (i) simple inorganic chemicals and (ii) complex inorganic chemicals. Simple inorganic chemicals are, for example, molecules that consist of one-type atoms (atoms of one element), which, in chemical reactions, cannot be decomposed to form other chemicals. On the other hand, complex inorganic chemicals are molecules that consist of different types of atoms (atoms of different chemical elements), which, in chemical reactions, are decomposed with the formation of several other chemicals. Thus,

Simple inorganic chemicals	Metals
	Nonmetals
Complex inorganic chemicals	Acids
	Bases
	Oxides
	Salts

However, a sharp line of demarcation between metals and nonmetals does not exist, since simple substances show dual properties. On the other hand, inorganic chemicals have also been classified on the basis that there are two types of inorganic compounds that can be formed: (i) ionic compounds and (ii) molecular compounds. Nevertheless, nomenclature and classification is the process of identifying chemical compounds with different names so that they can be easily classified as separate chemicals.

In the very broadest sense, inorganic chemicals are classified in a manner that excludes anything beyond biological molecules except for biinorganic compounds, hydrocarbon derivatives, and other similar carbon-based chemicals. Inorganic compounds, due to their lack of carbon-based chemical bonds, are usually very simple. Nevertheless, simplicity or not, from a practical standpoint, inorganic chemicals are based on chemical elements that are classified by the periodic table (Fig. 2.1, Tables 2.7 and 2.8) and compounds of mineral origin (Tables 2.9 and 2.10) that do not contain carbon in their molecular structure and are typically based on the most abundant chemicals on earth: oxygen, silicon, aluminum, iron, calcium, sodium, potassium, and magnesium. Inorganic synthesis, the process of synthesizing inorganic chemical compounds, is used to produce many basic inorganic chemical compounds. For example, an inorganic pigment is a natural or synthetic metallic oxide, sulfide, or other salt that is calcined during processing at 1200–2100°F (650–1150°C). Inorganic pigments have outstanding heat stability, light stability, weather resistance, and migration resistance. These compounds are widely used in many applications, including electroplating, dye and precision casting, alcohol distillation, and papermaking. Sulfur dioxide (SO_2) and sulfite derivatives ($—SO_3$) are inorganic chemicals used as preservatives. Inorganic compounds are also used as feed additives, insecticides, wood preservatives, and antiseptics.

According to the shared physical and chemical properties, the elements can be classified into the major categories of (i) metals, (ii) metalloids, and (iii) nonmetals. Metals are generally shiny, highly conducting solid elements that form alloys with one another and saltlike ionic compounds with

Table 2.7 Tabular Form of the Periodic Table Showing the Ionic Components

Group	Description	Examples
Alkali metals	Group 1 of the periodic table	Lithium (Li) Sodium (Na) Potassium (K) Rubidium (Rb) Cesium (Cs) Francium (Fr)
Alkali earths	Group 2 of the periodic table	Beryllium (Be) Magnesium (Mg) Calcium (Ca) Strontium (Sr) Barium (Ba) Radium (Ra)
Metalloids	Semimetals that cannot be clearly defined as either a metal or a nonmetal	Boron (B) Silicon (Si) Germanium (Ge) Arsenic (As) Antimony (Sb) Tellurium (Te) Polonium (Po)
Nonmetals	Elements that are poor conductors with low density and melting point	Hydrogen (H) Carbon (C) Nitrogen (N) Phosphorus (P) Oxygen (O) Sulfur (S) Selenium (Se)
Transition metals	Groups 3–10 of the periodic table	Silver (Ag) Gold (Au) Platinum (Pt) Iron (Fe) Titanium (Ti)
Other metals	Groups 13, 14, and 15 of the periodic table	Aluminum (Al) Gallium (Ga) Indium (In) Tin (Sn) Thallium (Tl) Lead (Pb) Bismuth (Bi)
Rare earths	A collection of 17 elements	Scandium Yttrium Lanthanides

Table 2.8 Tabular Form of the Periodic Table Showing Anionic Components

Group	Description	Examples
Halogens	Group 17 of the periodic table	Fluorine (F) Chlorine (Cl) Bromine (Br) Iodine (I) Astatine (At)
Nonmetals	Elements that are poor conductors with low density and melting point	Hydrogen (H) Carbon (C) Nitrogen (N) Phosphorus (P) Oxygen (O) Sulfur (S) Selenium (Se)
Metalloids	Semimetals that cannot be clearly defined as either a metal or a nonmetal	Boron (B) Silicon (Si) Germanium (Ge) Arsenic (As) Antimony (Sb) Tellurium (Te) Polonium (Po)
Oxoanions	Compounds that contain a simple polyatomic anion incorporating oxygen	Borates ($BO_{3,4}$) Bromates (BrO_3) Carbonates (CO_3) Chlorates (ClO_3) Cyanates (NCO) Nitrates (NO_3) Phosphates (PO_4) Silicates (SiO_4) Sulfates (SO_4)
Other		Cyanides (CN) Hydroxides (OH)

nonmetals (other than the noble gases—helium, neon, argon, krypton, xenon, and radon). The majority of the nonmetals are colored or colorless insulating gases and nonmetals that form compounds with other nonmetals feature covalent (shared electron) bonding. In between the metals and the nonmetals are the metalloids, which have intermediate properties or mixed properties (Silberberg, 2006; Emsley, 2011).

Metals and nonmetals can be further divided into subcategories that show a transition from metallic properties (on the left-hand side of the table) to nonmetallic properties (on the right-hand side of the table). The metals

Table 2.9 Some Common Elements and Minerals

Aluminum	The most abundant metal element in Earth's crust. Aluminum originates as an oxide called alumina. Bauxite ore is the main source of aluminum and must be imported from Jamaica, Guinea, Brazil, and Guyana. Used in transportation (automobiles), packaging, building/construction, electric, machinery, and other uses
Antimony	A native element; antimony metal is extracted from stibnite ore and other minerals. Used as a hardening alloy for lead, especially storage batteries and cable sheaths; also used in bearing metal, type metal, solder, collapsible tubes and foil, sheet and pipes, and semiconductor technology. Antimony is used as a flame retardant, in fireworks, and antimony salts are used in the rubber, chemical and textile industries, and medicine and glassmaking
Barium	A heavy metal contained in barite. Used as a heavy additive in oil well-drilling; in the paper and rubber industries; as a filler or extender in cloth, ink, and plastic products; in radiography (often know as a *barium milk shake*); as a deoxidizer for copper; a sparkplug in alloys; and in making expensive white pigments
Bauxite	Rock composed of hydrated aluminum oxides. In the United States, it is primarily converted to alumina (Al_2O_3)
Beryllium	Used in the nuclear industry and to make light, very strong alloys used in the aircraft industry. Beryllium salts are used in fluorescent lamps, in X-ray tubes and as a deoxidizer in bronze metallurgy. Beryl is the gem stones emerald and aquamarine. It is used in computers, telecommunication products, aerospace and defense applications, appliances and automotive, and consumer electronics. Also used in medical equipment
Chromite	The United States consumes about 6% of the world chromite ore production in various forms of imported materials, such as chromite ore, chromite chemicals, chromium ferroalloys, chromium metal, and stainless steel. Used as an alloy and in stainless and heat-resisting steel products. Used in chemical and metallurgical industries (chrome fixtures, etc.)

Continued

Table 2.9 Some Common Elements and Minerals—cont'd

Clay	Used in floor and wall tile as an absorbent, in sanitation, mud drilling, foundry sand bond, iron pelletizing, brick, light weight aggregate, and cement. It is produced in 40 states. Ball clay is used in floor and wall tile. Bentonite is used for drilling mud, pet waste absorbent, iron ore pelletizing, and foundry sand bond. Kaolin is used for paper coating and filling, refractory products, fiberglass, paint, rubber, and catalyst manufacture. Common clay is used in brick, light aggregate, and cement
Cobalt	Used primarily in super alloys for aircraft gas turbine engines, in cemented carbides for cutting tools and wear-resistant applications, chemicals (paint dryers, catalysts, and magnetic coatings), and permanent magnets. The United States has cobalt resources in Minnesota, Alaska, California, Idaho, Missouri, Montana, and Oregon. Cobalt production comes principally from Congo, China, Canada, Russia, Australia, and Zambia
Copper	Used in building construction, electric, and electronic products (cables and wires, switches, plumbing, and heating); transportation equipment; roofing; chemical and pharmaceutical machinery; alloys (brass, bronze, and beryllium alloyed with copper are particularly vibration resistant); alloy castings; electroplated protective coatings and undercoats for nickel, chromium, and zinc. More recently, copper is being used in medical equipment due to its antimicrobial properties. The United States has mines in Arizona, Utah, New Mexico, Nevada, and Montana. Leading producers are Chile, Peru, China, the United States, and Australia
Feldspar	A rock-forming mineral; industrially important in glass and ceramic industries; patter and enamelware; soaps; bond for abrasive wheels; cements; insulating compositions; fertilizer; tarred roofing materials; and as a sizing, or filler, in textiles and paper. In pottery and glass, feldspar functions as a flux. End uses for feldspar in the United States include glass (70%) and pottery and other uses (30%)
Fluorite (fluorspar)	Used in production of hydrofluoric acid, which is used in the pottery, ceramics, optical, electroplating, and plastics industries; in the metallurgical treatment of bauxite; as a flux in open hearth steel furnaces and in metal smelting; in carbon electrodes; emery wheels; electric arc welders; toothpaste; and paint pigment. It is a key ingredient in the processing of aluminum and uranium

Gallium	Gallium is used in integrated circuits, light-emitting diodes (LEDs), photodetectors, and solar cells. It has a new use in chemotherapy for some types of cancer. Integrated circuits are used in defense applications, high-performance computers and telecommunications. Optoelectronic devices were used in areas such as aerospace, consumer goods, industrial equipment, medical equipment, and telecommunications. Leading sources are Germany, the United Kingdom, China, and Canada
Gold	Used in jewelry and arts; dentistry and medicine; in medallions and coins; in ingots as a store of value; for scientific and electronic instruments; as an electrolyte in the electroplating industry. Mined in Alaska and several western states. Leading producers are China, Australia, the United States, Russia, and Canada
Gypsum	Processed and used as prefabricated wallboard or an industrial or building plaster; used in cement manufacturing, agriculture, and other uses
Halite (sodium chloride—salt)	Used in human and animal diet, food seasoning, and food preservation; used to prepare sodium hydroxide, soda ash, caustic soda, hydrochloric acid, chlorine, metallic sodium; used in ceramic glazes; metallurgy, curing of hides; mineral waters; soap manufacturing; home water softeners; highway deicing; photography; in scientific equipment for optical parts. Single crystals used for spectroscopy, ultraviolet, and infrared transmission
Indium	Indium tin oxide is used for electric conductivity purposes in flat panel devices—most commonly in liquid crystal displays (LCDs). It is also used in solders, alloys, compounds, electric components, semiconductors, and research. Indium ore is not recovered from ores in the United States; China is the leading producer. It is also produced in Canada, Japan, and Belgium
Iron ore	Used to manufacture steels of various types. Powdered iron, used in metallurgy products, magnets, high-frequency cores, auto parts, catalyst. Radioactive iron (Fe^{59}), in medicine and tracer element in biochemical and metallurgical research. Iron blue, in paints, printing inks, plastics, cosmetics, and paper dyeing. Black iron oxide, as pigment, in polishing compounds, metallurgy, medicine, and magnetic inks. Most US production is from Michigan and Minnesota. China, Australia, Brazil, and Russia are the major producers

Continued

Table 2.9 Some Common Elements and Minerals—cont'd

Lead	Used in lead-acid batteries, gasoline additives (now being eliminated), and tanks and solders, seals, or bearing; electric and electronic applications; TV tubes and glass, construction, communications and protective coatings; ballast or weights; ceramics or crystal glass; X-ray and gamma radiation shielding; soundproofing material in construction industry; and ammunition. Industrial-type batteries are used as a source of uninterruptible power equipment for computer and telecommunications networks and mobile power. US mines lead mainly in Missouri but also in Alaska and Idaho
Lithium	Compounds are used in ceramics and glass, batteries, lubricating greases, air treatment, in primary aluminum production, in the manufacture of lubricants and greases, rocket propellants, vitamin A synthesis, silver solder, batteries, and medicine. Lithium-ion batteries have become a substitute for nickel-cadmium batteries in handheld/portable electronic devices. There is one brine operation in Nevada. Australia, Chile, and China are major producers
Manganese	Ore is essential to iron and steel production. Also used in the making of manganese ferroalloys. Construction, machinery, and transportation end uses account for most United States consumption of manganese. Manganese ore has not been produced in the United States since 1970. Major producers are South Africa, Australia, China, Gabon, and Brazil
Mica	Micas commonly occur as flakes, scales, or shreds. Ground mica is used in paints, as joint cement, as a dusting agent, in oil well-drilling muds, and in plastics, roofing, rubber, and welding rods. Sheet mica is fabricated into parts for electronic and electronic equipment. China and Russia are leading producers
Molybdenum	Used in alloy steels to make automotive parts, construction equipment, and gas transmission pipes; stainless steels; tool steels; cast irons; super alloys; and chemicals and lubricants. As a pure metal, molybdenum is used because of its high melting temperatures (2610°C, 4730°F) as filament supports in light bulbs, metalworking dies and furnace parts. Major producers are China, the United States, Chile, and Peru
Nickel	Vital as an alloy to stainless steel; plays key role in the chemical and aerospace industries. End uses were transportation, fabricated metal products, electric equipment, petroleum and chemical industries, household appliances, and industrial machinery. Major producers are the Philippines, Indonesia, Russia, Australia, and Canada

Perlite	Expanded perlite is used in building construction products like roof insulation boards; as fillers, for horticulture aggregate, and filter aids. It is produced in New Mexico and other western states and is processed in over 20 states. Leading producers are the United States, Greece, and Turkey
Platinum group metals (PGM)	Includes platinum, palladium, rhodium, iridium, osmium, and ruthenium. Commonly occur together in nature and are among the scarcest of the metallic elements. Platinum is used principally in catalysts for the control of automobile and industrial plant emissions; in jewelry; in catalysts to produce acids, organic chemicals and pharmaceuticals. PGMs used in bushings for making glass fibers are used in fiber-reinforced plastic and other advanced materials, in electric contacts, in capacitors, in conductive and resistive films used in electronic circuits, and in dental alloys used for making crowns and bridges. South Africa, Russia, the United States, and Canada are major producers
Phosphate rock	Used to produce phosphoric acid for ammoniated phosphate fertilizers, feed additives for livestock, elemental phosphorus, and a variety of phosphate chemicals for industrial and home consumers. US production occurs in Florida, North Carolina, Idaho, and Utah
Potash	A carbonate of potassium; used as a fertilizer, in medicine, in the chemical industry, and to produce decorative color effects on brass, bronze, and nickel. The leading producers are Canada, Russia, and Belarus
Pyrite	Used in the manufacture of sulfur, sulfuric acid, and sulfur dioxide; pellets of pressed pyrite dust are used to recover iron, gold, copper, cobalt, and nickel; used to make inexpensive jewelry
Quartz (silica)	As a crystal, quartz is used as a semiprecious gemstone. Crystalline varieties include amethyst, citrine, rose quartz, smoky quartz, etc. Cryptocrystalline forms include agate, jasper, and onyx. Because of its piezoelectric properties, quartz is used for pressure gauges, oscillators, resonators, and wave stabilizes; because of its ability to rotate the plane of polarization of light and its transparency in ultraviolet rays, it is used in heat-ray lamps, prism, and spectrographic lenses. Also used in manufacturing glass, paints, abrasives, refractory materials, and precision instruments

Continued

Table 2.9 Some Common Elements and Minerals—cont'd

Rare-earth elements (lanthanum, cerium, praseodymium, neodymium, promethium, samarium, europium, gadolinium, terbium, dysprosium, holmium, erbium, thulium ytterbium, and lutetium)	Used mainly in petroleum fluid cracking catalysts, metallurgical additives and alloys, glass polishing and ceramics, permanent magnets and phosphors. It is estimated that 40 pounds of rare earths are used in a hybrid car for rechargeable battery, permanent magnet motor, and the regenerative braking system. The United States now has one rare-earth (bastnasite) mine in California. More than 85% of global production is in China
Silica	Aluminum and aluminum alloy producers and the chemical industry are major users of silicon metal. Silica is also used in the manufacture of computer chips, glass, and refractory materials; ceramics; abrasives; water filtration; component of hydraulic cements; filler in cosmetics, pharmaceutical, paper, insecticides; anticaking agent in foods; flatting agent in paints; thermal insulator; and photovoltaic cells. China is the leading producer
Silver	Used in coins and medals, electric and electronic devices, industrial applications, jewelry, silverware, and photography. The physical properties of silver include ductility, electronics conductivity, malleability, and reflectivity. Used in lining vats and other equipment for chemical reaction vessels, and water distillation; a catalyst in the manufacture of ethylene; mirrors; silver plating; table cutlery; dental, medical, and scientific equipment; bearing metal; magnet windings; brazing alloys and solder. Also used in catalytic converters, cell phone covers, electronics, circuit boards, bandages for wound care, and batteries. Silver is produced in the United States at over 30 bases and precious metal mines primarily in Alaska and Nevada. The leading global producers include Mexico, China, Peru, Chile, Australia, Bolivia, and the United States
Sodium carbonate (soda ash or trona)	Used in glass container manufacture; in fiberglass and specialty glass; also, used in production of flat glass; in liquid detergents; in medicine; as a food additive; photography; cleaning and boiler compounds; pH control of water. Most of the trona production in the United States comes from Wyoming

Sulfur	Used in the manufacture of sulfuric acid, fertilizers, petroleum refining; and metal mining. Elemental sulfur and by-product sulfuric acid were produced in over 100 operations in 26 states and the Virgin Islands. The United States, Canada, China, and Germany are major producers
Tantalum	A refractory metal with unique electric, chemical, and physical properties used to produce electronic components and tantalum capacitors (in auto electronics, pagers, personal computers, and portable telephones) and for high-purity tantalum metals in products ranging from weapon systems to superconductors; high-speed tools; catalyst; sutures and body implants; electronic circuitry; and thin-film components. Used in optical glass and electroplating devices. Leading producers are Mozambique, Brazil, and Congo
Titanium	Titanium mineral concentrates are used primarily by titanium dioxide pigment producers. A small amount is used in welding rod coatings and for manufacturing carbides, chemicals, and metals. It is produced in Florida and Virginia. Leading producing countries are South Africa, Australia, Canada, and China
	Titanium and titanium dioxide are used in aerospace applications (in jet engines, airframes, and space and missile applications). It is also used in armor, chemical processing, marine, medical, power generation, sporting goods, and other nonaerospace applications. Titanium sponge metal was produced in three operations in Nevada and Utah. The leading global producers are China, Japan, Russia, and Kazakhstan
Tungsten	More than half of the tungsten consumed in the United States was used in cemented carbide parts for cutting and wear-resistant materials, primarily in the construction, metalworking, mining, and oil- and gas-drilling industries. The remaining tungsten was consumed to make tungsten heavy alloys for applications requiring high density; electrodes, filaments, wires, and other components for electric, electronic, heating, lighting, and welding applications; steels, super alloys, and wear-resistant alloys; and chemicals for various applications. China is by far the leading producer. Russia, Canada, Austria, and Bolivia also produce tungsten
Uranium	Nearly 20% of America's electricity is produced using uranium in nuclear generation. It is also used for nuclear medicine, atomic dating, powering nuclear submarines, and other uses in the US defense system

Continued

Table 2.9 Some Common Elements and Minerals—cont'd

Vanadium	Metallurgical use, primarily as an alloying agent for iron and steel, accounted for about 93% of the domestic vanadium consumption. Of the other uses for vanadium, the major nonmetallurgical use was in catalysts for the production of maleic anhydride and sulfuric acid. China, South Africa, and Russia are largest producers
Zeolites	Used in animal feed, cat litter, cement, aquaculture (fish hatcheries for removing ammonia from the water); water softener and purification; in catalysts; odor control; and for removing radioactive ions from nuclear plant effluent
Zinc	Of the total zinc consumed in the United States, about 55% is used in galvanizing, 21% in zinc-based alloys, 16% in brass and bronze, and 8% in other uses. Zinc compounds and dust were used principally by the agriculture, chemical, paint, and rubber industries. Major coproducts of zinc mining and smelting, in order of decreasing tonnage, were lead, sulfuric acid, cadmium, silver, gold, and germanium. Zinc is used as protective coating on steel, as die casting, as an alloying metal with copper to make brass, and as chemical compounds in rubber and paints; sheet zinc and for galvanizing iron; electroplating; metal spraying; automotive parts; electric fuses; anodes; dry cell batteries; nutrition; chemicals; roof gutter; engravers' plates; cable wrappings; organ pipes; and pennies. Zinc oxide used in medicine, paints, vulcanizing rubber, and sun block. Zinc dust used for primers, paints, and precipitation of noble metals and removal of impurities from solution in zinc electrowinning. The US production is in three states and 13 mines. Leading producers are China, Australia, Peru, and the United States

The United States Geological Survey, Facts About Minerals (National Mining Association); Mineral Information Institute; Energy Information Administration.

Table 2.10 Common Minerals and Uses

Bauxite	Aluminum, foil, and airplane parts
Borax	Antiseptic soaps, welding flux, or cleaner (found in dry lake beds)
Calcite	Medicine, toothpaste, building. Materials (hard water deposit ancient seabeds)
Copper	Tubing, electric wires, and sculptures
Diamond	Cutting tools/blades/saws
Feldspar	Ceramics and porcelain, colors in granites (not black)
Galena	Source of lead
Graphite	Pencils, lubricant in machinery
Gypsum	Wall board, plaster of Paris
Halite	Salt
Hematite	Source of iron
Jade	Jewelry, figurines
Limonote/taconite	Source of iron (around Cedar City, Utah)
Muscovite (mica)	White, gray material in electric insulators
Quartz (massive type), quartz crystal	Glass manufacturing, radios, computers, and electronic equipment
Silver	Jewelry, photography, and electric equip
Sulfur	Fungicides, kills bacteria, vulcanizes rubber, in coal and fuels, fertilizer
Talc	Baby powder, soapstone, gymnastics to grasp bars

are subdivided into the highly reactive alkali metals to the less reactive alkaline earth metals, the lanthanide elements (elements 59–70, inclusive) and the actinide elements (elements 89–102, inclusive) and ending in the physically and chemically weak posttransition metals. The nonmetals are simply subdivided into the polyatomic nonmetals, which, being nearest to the metalloids, show some metallic character. Then, follows the diatomic nonmetals and the nonmetallic monatomic noble gases that are almost completely inert. Other classification schemes are possible such as the division of the elements into mineralogical occurrence categories or crystalline structure.

Also in addition to the elements, there are several important types (or groups) of inorganic compounds, including (i) bioinorganic compounds, which include natural and synthetic compounds that contain metallic elements bonded to proteins and other biological chemistries; (ii) cluster compounds, which are ensembles of bound atoms and are intermediate in molecular size, typically larger than a molecule; (iii) coordination compounds where the central ion, typically the ion of a transition metal, is surrounded by a group of anions or charged molecules; (iv) organometallic compounds that include carbon atoms directly bonded to a metal ion; and (v) solid state, which constitute a diverse class of compounds that are solid at standard temperature and pressure and exhibit unique properties as semiconductors.

Inorganic chemicals exist as gases, liquids (including colloids, emulsions, or dispersions), or solids (powder, bulk solid such as pellets, flakes, or granules) but are not typically classified on the basis of the physical state of the compound (gas, liquid, or solid) but more on data that is based on the chemical and physical properties of the chemicals (Table 2.11). Typically, at ambient temperature and pressure, gases and liquids take on the shape of the container in the bulk phase in which they are contained, while solids have definite shapes and volume and are held together by strong

Table 2.11 Important Common Physical Properties of Inorganic Compounds

State: Gas
Density
Critical temperature, critical pressure (for liquefaction)
Solubility in water, selected solvents
Odor threshold
Color
Diffusion coefficient
State: Liquid
Vapor pressure-temperature relationship
Density; specific gravity
Viscosity
Miscibility with water, selected solvents
Odor
Color
Coefficient of thermal expansion
Interfacial tension
State: Solid
Melting point
Density
Odor
Solubility in water, selected solvents
Coefficient of thermal expansion
Hardness/flexibility
Particle size distribution/physical form, such as fine powder, granules, pellets, lumps
Porosity

intermolecular and interatomic forces. For many solid inorganic chemicals, these forces are strong enough to maintain the atoms in a definite ordered crystal state, while some solids have little or no crystal structure and remain disordered (amorphous).

Inorganic gases have weaker attractive forces between individual molecules and therefore diffuse rapidly and assume the shape of the container but the volume of the gas is affected by temperature and pressure (often defined by the size of the container). On the other hand, the molecules of inorganic liquids are separated by relatively small distances (compared with gases), and the attractive forces between molecules tend to hold firm within a definite volume (the fixed size of the container) at a controlled temperature. Thus, inorganic solvents are compounds in which molecules are much closer together than in a gas and the intermolecular forces are therefore relatively strong.

Inorganic compounds are classified according to the presence of functional groups in the molecule (such as a nitrate group, NO_3, or a sulfate group, SO_4) and the functional group typically dictates the behavior of the inorganic compound in the environment. A functional group is a molecular moiety, and the reactivity of that functional group is assumed to be the same in a variety of molecules, within some limits and based on the assumption that steric effects do not interfere. Although not usually addressed in the inorganic chemical field to the same extent as functional groups in the organic chemistry field (Patrick, 2004; Smith, 2013; Speight, 2017a), the concept of functional groups in inorganic chemistry is important and is means of classifying the structure of inorganic compounds and for predicting physical and chemical properties.

In fact, a functional group is any atom or collection of atoms that is capable or reacting with a reactive species to produce a product. A functional group is also capable of affecting the properties of the original nonfunctional molecule in which the group occurs. Moreover, a functional group is the part of an inorganic molecule—either the cation (the positively charged ion) or the anion (the negatively charged ion) that contains an atom or a group of atoms that form a functional entity in which all of the atoms in that entity contribute collectedly to the properties (e.g., Table 2.12).

From a strict chemical sense, inorganic chemicals may be classified according to several different criteria, and one widely used method is based on the specific elements present (which is related to the functional group present but not always addressed as a functional group classification). For example, oxides contain one or more oxygen atoms, hydrides contain one or more hydrogen atoms, and halides contain one or more

Table 2.12 Examples of the Varying Reactivity of Metal Cations in Inorganic Chemistry

Metal	Periodic Table Group	Periodic Table Period	Reaction With *Air*	Reaction With *Water*	Reaction With *Steam*	Reaction With *Dilute Acid*	Solubility in *Water*
Sodium	1	3	Burns readily to form oxide	Violent reaction	Violent reaction	Violent reaction	Readily soluble
Potassium	1	4	Burns readily to form oxide	Violent reaction	Violent reaction	Violent reaction	Readily soluble
Magnesium	2	3	Burns readily to form oxide	No reaction	Strong reaction	Ready reaction	Sparingly soluble
Calcium	2	4	Burns readily to form oxide	Slow reaction	Violent reaction	Violent reaction	Slightly soluble
Iron	8	4	Forms oxide when heated	No reaction	Reversible reaction	Reaction	Insoluble
Copper	11	4	Forms oxide when heated	No reaction	No reaction	No reaction	Insoluble
Silver	11	5	No reaction	No reaction	No reaction	No reaction	Insoluble
Gold	11	6	No reaction	No reaction	No reaction	No reaction	Insoluble
Zinc	12	4	Forms oxide when heated	No reaction	Reaction	Reaction	Insoluble
Mercury	12	6	Reversible reaction	No reaction	No reaction	No reaction	Insoluble
Aluminum	13	3	Burns readily to form oxide	No reaction	Reaction	Reaction	Insoluble
Tin	14	5	Forms oxide when heated	No reaction	No reaction	Weak reaction	Insoluble
Lead	14	6	Forms oxide when heated	No reaction	No reaction	Weak reaction	Insoluble

See Fig. 2.2.

halogen atoms. As the name suggests, organometallic compounds are organic compounds bonded to metal atoms.

Another classification scheme for chemical compounds is based on the types of bonds that the compound contains. Ionic chemicals contain ions and are held together by the attractive forces among the oppositely charged ions; sodium chloride (common salt) is one of the best-known ionic compounds. Molecular compounds contain discrete molecules, which are held together by sharing electrons (covalent bonding). Examples are water (which contains H_2O molecules) and hydrogen fluoride (which contains HF molecules).

A third classification scheme is based on reactivity—specifically, the types of chemical reactions in which the chemicals participate. For example, acids are compounds that produce protons (H^+ ions) when dissolved in water to produce aqueous solutions and are defined as proton donors. The most common acids are hydrochloric acid (aqueous solutions of hydrogen chloride, HCl), sulfuric acid (H_2SO_4), nitric acid (HNO_3), and phosphoric acid (H_3PO_4). On the other hand, bases are proton acceptors; the most common base is the hydroxide ion (OH^-), which reacts with a proton (H^+) to form a molecule of water:

$$H^+ + OH^- \rightarrow HOH(\text{usually written as } H_2O)$$

Yet, another form of classification includes those inorganic compounds that participate in oxidation-reduction reactions; oxidation involves a loss of electrons, whereas reduction involves a gain of electrons. For example, in the reaction between sodium metal and chlorine gas to form sodium chloride, electrons (e^-) are transferred from sodium atoms to chlorine atoms to form Na^+ and Cl^- ions in the reaction product, sodium chloride.

$$2Na + Cl_2 \rightarrow 2NaCl$$

In this reaction, each sodium atom loses an electron and is thus oxidized, and each chlorine atom gains an electron and is thus reduced, and in this reaction, sodium is the reducing agent (which provides the electrons), and chlorine is the oxidizing agent (which accepts the electrons). The most common reducing agents are metals, for they tend to lose electrons in their reactions with nonmetals, while the most common oxidizing agents are the halogens: fluorine (F_2), chlorine (Cl_2), and bromine (Br_2), as well as certain oxy anions, such as the permanganate ion (MnO_4^-) and the dichromate ion ($Cr_2O_7^{2-}$).

Finally, an important aspect of chemistry is the concept of *valence*, which is used to determine the validity of a chemical reaction. The valence numbers (oxidation numbers or states) are an important number for all elements; the valence of a free atom is zero, but in a chemical compound, the value has some positive or negative value. For example, when carbon burns, the carbon atom has a zero valence, and the valence of CO_2 is +4:

$$C + O_2 \rightarrow CO_2$$

The valence of each oxygen atom is −2, so that the net valence of CO_2 is zero, which is the requirement for the chemical reaction to proceed. The increase in valence number from 0 to +4 (for the carbon atom in the CO_2) is referred to as an oxidation reaction. Another example is the reduction of copper oxide to metallic copper:

$$CuO + H_2 \rightarrow Cu + H_2O$$

In this reaction, where the valence of copper changes from +2 (in CuO) to zero, i.e., there is a decrease in the valence of the copper Cu in this reaction. Valence and stoichiometry (Section 2.6, below) are related by the balancing of both sides of the reaction. Furthermore, the guidance rule for classifying inorganic reactions is an element is oxidized when its valence number (oxidation number) increases.

Finally, for a chemical reaction, the *equilibrium constant* is the value of the reaction quotient for a system at equilibrium. The reaction quotient is the ratio of molar concentrations of the reactants to those of the products, each concentration being raised to the power equal to the coefficient in the equation. For the hypothetical chemical reaction,

$A + B \leftrightarrow C + D$, the equilibrium constant, K, is

$$K = [C][D]/[A][B]$$

The notation [A] signifies the molar concentration of species A. An alternative expression for the equilibrium constant can involve the use of partial pressures. The equilibrium constant can be determined by allowing a reaction to reach equilibrium, measuring the concentrations of the various solution-phase or gas-phase reactants and products, and substituting these values into the relevant equation.

In terms of acid-base reactivity, the *dissociation constant* of an acid K_a may conveniently be expressed in terms of the pK_a value where $pK_a = -\log 10$ $(K_a/\text{mol dm}^{-3})$. A similar can be derived for the dissociation constant of a base.

2.4 THE PERIODIC TABLE

The periodic table is a tabular arrangement of the chemical elements, ordered by the atomic number (i.e., number of protons in the nucleus), the electron configuration, and the recurring chemical properties (Figs. 2.1 and 2.2). This ordering of the elements shows trends within the elements (referred to as *periodic trends*) such as elements with similar behavior in the same column. Furthermore, the periodic table also shows rectangular blocks of elements with some approximately similar chemical properties. In general, within one row (period), the elements are metals on the left, and nonmetals are on the right.

In terms of elemental classification, the periodic table is the most important chemistry reference there is insofar as all of the elements are arranged in an informative manner. Thus, the elements are arranged left to right and top to bottom in order of increasing atomic number, which generally coincides with increasing atomic mass. The trends in the periodic table can help to predict the properties of various elements and the relations between properties. Thus, the periodic table is a useful framework for analyzing chemical behavior and as such is widely used in chemistry and in other sciences (Fig. 2.1, Tables 2.7 and 2.8).

An elemental group (or elemental family) in the table is a vertical column in the periodic table. Elements in the same group show patterns in atomic radius, ionization energy, and electronegativity. From top to bottom in a group, the atomic radii of the elements increase: Since there are more filled energy levels, valence electrons are found farther from the nucleus. From top to bottom, each successive element has a lower ionization energy because it is easier to remove an electron since the atoms are less tightly bound. Similarly, from top to bottom, elements decrease in electronegativity due to an increasing distance between the valence electrons and the nucleus. However, there are exceptions to these trends. For example, in group 11, the electronegativity increases down the group. A period is a horizontal row in the periodic table. Although groups generally show stronger trends, in some cases, horizontal trends can be more significant.

For example, in the so-called f-block, the lanthanide elements (elements 59–70, inclusive) and actinide elements (elements 89–102, inclusive) are two horizontal series of elements that follow significant trends. The term lanthanides was adopted, originating from the first element of the series, lanthanum. These elements were first classified as *rare-earth elements* due to the fact that they were obtained by minerals that were not readily available and were classed as *rare minerals*. However, this is can be misleading since the lanthanide elements have a significantly high abundance in the Earth and, like any other series of elements in the periodic table, the lanthanide elements share many similar characteristics. The characteristics of these elements include the following: (i) similarity in physical properties throughout the series; (ii) the adoption mainly of the +3 oxidation state that is usually found in crystalline compounds; (iii) these elements can also have an oxidation state of +2 or +4, though some lanthanides are most stable in the +3 oxidation state; (iv) the adoption of coordination numbers greater than 6—usually in the 8–9 range—in compounds; (v) the tendency to decreasing coordination number across the series; and (vi) a preference for more electronegative elements, such as bonding to oxygen or fluorine. Similarly, the lanthanide elements have similarities in their electron configuration, which explains most of the physical similarities. These elements are different from the main group elements in the fact that they have electrons in the f orbital—hence the name the *f-block elements*. After lanthanum, the energy of the 4f subshell falls below that of the 5d subshell. This means that the electron starts to fill the 4f subshell before the 5d subshell.

Elements in the same period show trends in atomic radius, ionization energy, electron affinity, and electronegativity. Moving left to right across a period, from the alkali metals to the noble gases, atomic radius usually decreases. This is because each successive element has an additional proton and electron, which causes the electrons to be drawn closer to the nucleus. This decrease in atomic radius also causes ionization energy to increase from left to right across a period; the more tightly bound an element is, the more energy is required to remove an electron. Electronegativity increases in the same manner as ionization energy because of the pull exerted on the electrons by the nucleus. Electron affinity also shows a slight trend across a period: metals (the left side of a period) generally have a lower electron affinity than nonmetals (the right side of a period), except for the noble gases.

Because of the importance of the outermost electron shell, the different regions of the periodic table are sometimes referred to as blocks, named according to the subshell in which the "last" electron resides. Specific regions of the periodic table are often referred to as *blocks* in recognition of the sequence in which the electron shells of the elements are filled. Each block is named according to the subshell in which the "last" electron notionally resides. For example, the s-block comprises the first two groups (alkali metals and alkaline earth metals) and hydrogen and helium. The p-block comprises the last six groups, which are groups 13–18 in IUPAC group numbering (3A–8A in American group numbering) and contains, among other elements, all of the metalloids. The d-block comprises groups 3–12 (or 3B–2B in American group numbering) and contains all of the transition metals. The f-block, often offset below the rest of the periodic table (Fig. 2.1), has no group numbers and comprises the lanthanide series of elements and the actinide series of elements.

Generally, the lanthanide metals are soft; their hardness increases across the series. The resistivity of the lanthanide metals is relatively high and falls into the range from 29 to 134 μ-ohm × cm—these resistivity values can be compared with the resistivity of a good conductor such as aluminum, which has a resistivity of 2.655 μ-ohm × cm. With the exceptions of lanthanum (La), ytterbium (Yb), and lutetium (Lu) (which have no unpaired f electrons), the lanthanide elements are highly paramagnetic, which is reflected in their magnetic susceptibilities. Gadolinium (Gd) becomes ferromagnetic at temperatures below 16°C (60.8°F, the Curie point). The remaining heavier lanthanide elements—terbium, dysprosium, holmium, erbium, thulium, and ytterbium—become ferromagnetic at much lower temperatures.

The actinide elements are also typical metals and all of these elements, although soft, have a silvery color with the tendency to tarnish in air. These elements have a relatively high density and plasticity. While some of the actinide elements can be cut with a knife, the hardness of thorium is similar to that of soft steel, so heated pure thorium can be rolled in sheets and pulled into wire. Thorium is nearly half as dense as uranium and plutonium but is harder than either of them. The electric resistivity of the actinide elements varies between 15 and 150 μ-ohm × cm, and all are radioactive and paramagnetic.

2.5 BONDING AND MOLECULAR STRUCTURE

Bonding occurs when two particles can exchange or combine their outer electrons in such a way that is *energetically favorable*. When two atoms are close to each other and their electrons are of the correct type, it is more energetically favorable for them to come together and share electrons (become *bonded*) than it is for them to exist as individual, separate atoms. When the bond is formed, the individual atoms have combined to form a compound.

Thus, a chemical bond is a lasting attraction between atoms and which contributes to the formation (in the current context) of inorganic chemical compounds. The bond may result by the sharing of electrons as in the formation of covalent bonds (such as the typical carbon-carbon bond or the typical carbon-hydrogen bond or from the electrostatic force of attraction between atoms with opposite charges). The strength of a chemical bond varies considerably—there are the so-called relatively strong bonds such as covalent or ionic bonds and the relatively *weak bonds* such as (i) hydrogen bonds and (ii) bonds formed by van der Waals interactions (Table 2.13).

Table 2.13 Different Types of Bond Arrangements

Covalent bond	A bond in which one or more electrons (often a pair of electrons) are drawn into the space between the two atomic nuclei. These bonds exist between two particular identifiable atoms and have a direction in space, allowing them to be shown as single connecting lines between atoms in drawings, or modeled as sticks between spheres in models
Ionic bond	Occurs between ionized functional groups such as carboxylic acids and amines
Hydrogen bond	Occurs between alcohol derivatives, carboxylic acid derivatives, amide derivatives, amine derivatives, and phenol derivatives. Hydrogen bonding involves the interaction of the partially positive hydrogen on one molecule and the partially negative heteroatom on another molecule. Hydrogen bonding is also possible with elements other than nitrogen or oxygen and can occur intermolecularly or intramolecularly
Dipole-dipole interaction	Possible between molecules having polarizable bonds, in particular the carbonyl group (C=O) that have a dipole moment and molecules can align themselves such that their dipole moments are parallel and in opposite directions. Ketones and aldehydes are capable of interacting through dipole-dipole interactions
Van der Waals interaction	Weak intermolecular bonds between regions of different molecules bearing transient positive and negative charges that are caused by the movement of electrons. Alkanes, alkenes, alkynes, and aromatic rings interact through van der Waals interactions
Intermolecular bond	Occurs between different molecules and can take the form of ionic bonding, hydrogen bonding, dipole-dipole interactions, and van der Waals interactions
Intramolecular bond	Occurs within a molecule and can take the form of ionic bonding, hydrogen bonding, dipole-dipole interactions, and van der Waals interactions

2.5.1 Bonding Types

Intermolecular bonding takes place between different molecules. This can take the form of ionic bonding, hydrogen bonding, dipole-dipole interactions, and van der Waals interactions. The type of bonding involved in inorganic compounds depends on the functional groups that are present in the molecule. On the other hand, *intramolecular bonding* occurs within the same molecule and also depends on the functional groups that are present in the molecule.

Ionic bonds are possible between ionized functional groups such as carboxylic acids and amines. Ionic bonding takes place between molecules having opposite charges and involves an electrostatic interaction between the two opposite charges. The functional groups that most easily ionize are amines and carboxylic acids, such as the reaction of ammonia (NH_3) with a carboxylic acid (carboxylate ion).

The *molecular structure* of simple molecules can be crucial in understanding the reactivity of the compound and the potential products from a given reaction. Electronic structure is necessary because it refers to the modes and orientation of bonds, which will give rise to the shape and properties of each molecule (Speight, 2017a,b).

In inorganic chemistry, as in organic chemistry, there are many atoms in even the simplest molecules, and the molecular structure theory focuses on considering the bonds formed by one or two central atoms, which are themselves bonded to two or more atoms; the geometry conferred to these selected atoms provides the shape of the overall molecule. Various structures can be used to determine the number of electron pairs in the valence shell of a selected atom; this can be extended to show electron lone pairs and projecting bonds.

Since each atom is considered to be a sphere, the bonds coming from the atom can orient themselves anywhere over the surface, producing a full three-dimensional geometry for that atom. An easy example to begin with is CO_2, which has two oxygen atoms bonded to one carbon atom, via two separate bonds; each c=o bond experiences the lowest repulsion from the other when they are diametrically positioned and the molecule exhibits linear molecular geometry.

Extension to three electron pairs sees the atoms bonded to the central atom and adopts the vertices of an equilateral triangle around the selected atom, as in a trigonal pyramidal geometry.

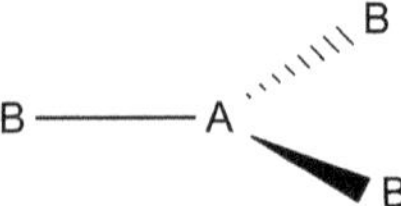

On the other hand, the four electron pairs move the arrangement from two dimensions, and the geometry becomes tetrahedral, with each of the connecting atoms adopting the vertices of a trigonal pyramid (an equilateral triangle, with all three points meeting at a fourth point in the space above) around the selected atom and, thus, exhibiting tetrahedral geometry.

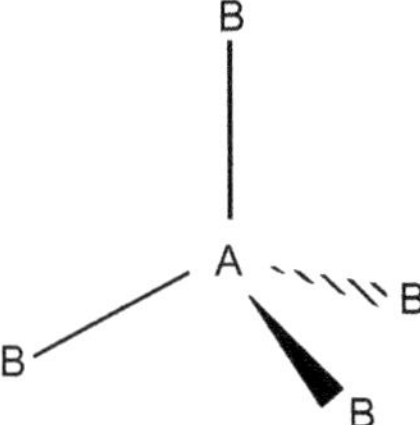

The geometry can be affected by whether the electron pair is shared with another atom in a bond or if it is a lone pair, which will be held more strongly to the central atom. The lone pairs, therefore, exert a greater repulsive force than bonding pairs, reducing the angle between bonding pairs.

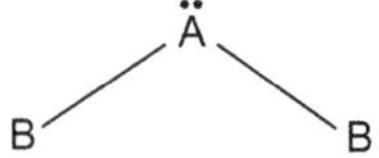

Lone pairs will most strongly repel other lone pairs, while bonding pair-bonding pair repulsions are weakest, with mixed repulsions somewhere in-between, hence, the lone pairs repelled by other lone pairs will tend to occupy the sites of least repulsive force. An example of the importance of inequivalent sites in structure is the five electron pair arrangement of trigonal pyramidal (imagine two trigonal pyramids glued together on one face). Here, the three sites in the same plane have the same repulsion, which is less than that felt by the two sites out of the plane (i.e., trigonal bipyramidal geometry).

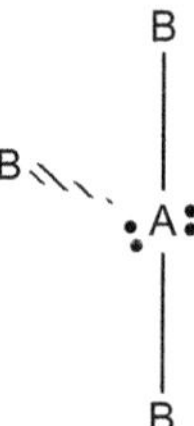

The existence of lone pairs, occupying given sites in a molecular geometry, can also explain the differences in shapes of molecules with equal numbers of atoms; for example, CO_2 is linear (O=C=O), as all the electrons are involved in two bonding pairs; however, sulfur dioxide is angular because there are two bonding pairs and one lone pair on the central sulfur atom; the three bonds adopt a trigonal arrangement, but there are only connecting atoms on two bonds, ideally the bonds should be 120 degrees apart but the lone pair repulsion to the bonding pairs will cause a contraction in the angle between the two oxygen atom bonds. It also explains the structure of ammonia (NH_3) which should be trigonal, but the lone pair of electrons from the nitrogen forces the geometry to be tetrahedral with the lone pair making the fourth point and rendering the molecule tetrahedral in structure:

For water, there are four pairs; hence, the structure should be trigonal pyramidal; however, there are two lone pairs and two bonding pairs; hence, the geometry adopted will again be bent, and the bond will be smaller than the 109.5 degrees expected for the trigonal pyramid, at only 104.5 degrees, as the lone pairs exert greater repulsion on the bonding pairs. Hence, the description of an angular molecule is based purely on the orientation of the atoms in the molecule, not the overall electronic arrangement within the molecule; thus, two lone pairs can cause the molecule to exhibit tetrahedral geometry.

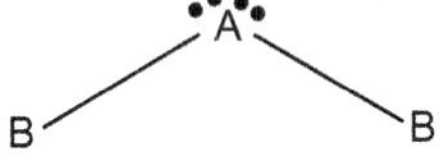

Thus, the steric number of an atom, A, in conjunction with the information on the number of bonding (B) and lone (L) pairs in a molecule allow the structure to be predicted, can be used to predict the geometry (and relative reactivity) that is to be expected for the simple steric arrangements most often encountered.

2.5.2 Ionization Energy

The ionization energy is the energy required to remove an electron completely from its atom, molecule, or radical (Cotton et al., 1999). Thus,

$$X + \text{energy} \rightarrow X^+ + e^-$$

X is any atom or molecule capable of being ionized; X^+ is that atom or molecule with an electron removed, and e^- is the removed electron. This is an endothermic process and, generally, the closer the electrons are to the nucleus of the atom, the higher the ionization energy of the atom. In chemistry, the units are the amount of energy it takes for all the atoms in a mole of substance to lose one electron each. The molar ionization energy or enthalpy is expressed as kilocalories per mole (kcal/mol) or as kilojoules per mole (kJ/mol). In physics, the unit is the amount of energy required to remove a single electron from a single atom or molecule (or radical) is expressed as an electron volt.

Comparison of the ionization energy of the elements in the periodic table reveals two patterns (Table 2.14 and Fig. 2.2). Thus, moving left to right within a period or upward within a group, the first ionization energy generally increases, with some exceptions such as aluminum and sulfur. As the nuclear charge of the nucleus increases across the period, the atomic radius decreases, and the electron cloud becomes closer toward the nucleus. The decrease in the ionization of the elements within a group from top to bottom is due to the outer electron shell being progressively further away from the nucleus with the addition of one inner shell per row as one moves down the column.

The *n*th ionization energy refers to the amount of energy required to remove an electron from the species with a charge of $(n - 1)$. For example, the first three ionization energies are defined as follows:

First ionization energy: $X \rightarrow X^+ + e^-$

Second ionization energy: $X^+ \rightarrow X^{2+} + e^-$

Third ionization energy: $X^{2+} \rightarrow X^{3+} + e^-$

The term *ionization potential* is an older name for ionization energy, because the oldest method of measuring ionization energies was based on ionizing a sample and accelerating the electron removed using an electrostatic

Table 2.14 Elements of the Periodic Table of Elements Organized by the First Ionization Potential (eV)

Ionization Potential	Name	Symbol	Atomic Number
3.83	Francium	Fr	87
3.894	Cesium	Cs	55
4.177	Rubidium	Rb	37
4.341	Potassium	K	19
5.139	Sodium	Na	11
5.17	Actinium	Ac	89
5.212	Barium	Ba	56
5.279	Radium	Ra	88
5.392	Lithium	Li	3
5.4259	Lutetium	Lu	71
5.46	Praseodymium	Pr	59
5.53	Neodymium	Nd	60
5.54	Cerium	Ce	58
5.554	Promethium	Pm	61
5.58	Lanthanum	La	57
5.64	Samarium	Sm	62
5.67	Europium	Eu	63
5.695	Strontium	Sr	38
5.786	Indium	In	49
5.86	Terbium	Tb	65
5.89	Protactinium	Pa	91
5.94	Dysprosium	Dy	66
5.986	Aluminum	Al	13
5.993	Americium	Am	95
5.999	Gallium	Ga	31
6.018	Holmium	Ho	67
6.02	Curium	Cm	96
6.05	Uranium	U	92
6.06	Plutonium	Pu	94
6.08	Thorium	Th	90
6.101	Erbium	Er	68
6.108	Thallium	Tl	81
6.113	Calcium	Ca	20
6.15	Gadolinium	Gd	64
6.184	Thulium	Tm	69
6.19	Neptunium	Np	93
6.23	Berkelium	Bk	97
6.254	Ytterbium	Yb	70
6.3	Californium	Cf	98
6.38	Yttrium	Y	39
6.42	Einsteinium	Es	99

Continued

Table 2.14 Elements of the Periodic Table of Elements Organized by the First Ionization Potential (eV)—cont'd

Ionization Potential	Name	Symbol	Atomic Number
6.5	Fermium	Fm	100
6.54	Scandium	Sc	21
6.58	Mendelevium	Md	101
6.65	Hafnium	Hf	72
6.65	Nobelium	No	102
6.74	Vanadium	V	23
6.766	Chromium	Cr	24
6.82	Titanium	Ti	22
6.84	Zirconium	Zr	40
6.88	Niobium	Nb	41
7.099	Molybdenum	Mo	42
7.28	Technetium	Tc	43
7.289	Bismuth	Bi	83
7.344	Tin	Sn	50
7.37	Ruthenium	Ru	44
7.416	Lead	Pb	82
7.435	Manganese	Mn	25
7.46	Rhodium	Rh	45
7.576	Silver	Ag	47
7.635	Nickel	Ni	28
7.646	Magnesium	Mg	12
7.726	Copper	Cu	29
7.86	Cobalt	Co	27
7.87	Iron	Fe	26
7.88	Rhenium	Re	75
7.89	Tantalum	Ta	73
7.899	Germanium	Ge	32
7.98	Tungsten	W	74
8.151	Silicon	Si	14
8.298	Boron	B	5
8.34	Palladium	Pd	46
8.42	Polonium	Po	84
8.641	Antimony	Sb	51
8.7	Osmium	Os	76
8.993	Cadmium	Cd	48
9	Platinum	Pt	78
9.009	Tellurium	Te	52
9.1	Iridium	Ir	77
9.225	Gold	Au	79
9.322	Beryllium	Be	4

Table 2.14 Elements of the Periodic Table of Elements Organized by the First Ionization Potential (eV)—cont'd

Ionization Potential	Name	Symbol	Atomic Number
9.394	Zinc	Zn	30
9.65	Astatine	At	85
9.752	Selenium	Se	34
9.81	Arsenic	As	33
10.36	Sulfur	S	16
10.437	Mercury	Hg	80
10.451	Iodine	I	53
10.486	Phosphorus	P	15
10.748	Radon	Rn	86
11.26	Carbon	C	6
11.814	Bromine	Br	35
12.13	Xenon	Xe	54
12.967	Chlorine	Cl	17
13.598	Hydrogen	H	1
13.618	Oxygen	O	8
13.999	Krypton	Kr	36
14.534	Nitrogen	N	7
15.759	Argon	Ar	18
17.422	Fluorine	F	9
21.564	Neon	Ne	10
24.587	Helium	He	2

potential. However, this term is now considered obsolete. Some factors affecting the ionization energy include (i) the nuclear charge as the greater the magnitude of nuclear charge, the more tightly the electrons are held by the nucleus and hence more will be ionization energy; (ii) the number of electron shells as the greater the size of the atom, the less tightly the electrons are held by the nucleus and ionization energy will be less; (iii) a screening effect as the greater the magnitude of screening effect, the less tightly the electrons are held by the nucleus and hence less will be the ionization energy; (iv) the type of orbital ionized as the atom having stable electronic configuration has less tendency to lose electrons and consequently has high ionization energy; and (v) the occupancy of the orbital matters as if the orbital is half or completely filled and then it is harder to remove electrons.

Generally, the $(n+1)$th ionization energy is larger than the nth ionization energy. When the next ionization energy involves removing an electron from the same electron shell, the increase in ionization energy is primarily due to the increased net charge of the ion from which the electron is being removed. Electrons removed from more highly charged ions of an element experience greater forces of electrostatic attraction; thus, their removal requires more energy. In addition, when the next ionization energy involves removing an electron from a lower electron shell, the greatly decreased distance between the nucleus and the electron also increases both the electrostatic force and the distance over which that force must be overcome to remove the electron. Both factors contribute to an increase in the ionization energy.

2.5.3 Electron Affinity

The electron affinity of an atom or molecule is the amount of energy *released* or *spent* when an electron is added to a neutral atom or molecule in the gaseous state to form a negative ion (Anslyn and Dougherty, 2006):

$$X + e^{-} \rightarrow X^{-} + \text{energy}$$

This property is measured for atoms and molecules in the gaseous state since in a solid or liquid state, the energy levels would be changed by contact with other atoms or molecules. Other theoretical concepts that use electron affinity include electronic chemical potential and chemical hardness. Another example, a molecule or atom that has a more positive value of electron affinity than another is often called an electron acceptor and the less positive an electron donor which, together, may undergo charge-transfer reactions.

In the use of electron affinity data, it is essential to keep track of the sign. For any reaction that *releases* energy, the *change* ΔE in the total energy has a negative value (exothermic process). Electron capture for almost all non-noble gas atoms involves the release of energy and thus are exothermic reactions. The positive values that are listed in tables of E_{ea} are amounts or magnitudes. However, if the value assigned to E_{ea} is negative, this implies a reversal of direction, and energy is *required* to attach an electron (endothermic process) and the relationship: $E_{ea} = -\Delta E_{attach}$ is valid. Negative values typically arise for the capture of a second electron, but also for the nitrogen atom. The expression for calculating E_{ea} when an electron is attached is

$$E_{ea} = \left(E_{initial} - E_{final}\right)_{attach} = -\Delta E_{attach}$$

Similarly, electron affinity is also the amount of energy *required* to detach an electron from a singly charged negative ion:

$$X^{-} \rightarrow X + e^{-}$$

Although E_{ea} varies greatly across the periodic table, some patterns emerge. Typically, (i) nonmetals have more positive E_{ea} than metals, (ii) atoms where anions are more stable than neutral atoms have a greater E_{ea}, (iii) chlorine most strongly attracts extra electrons, (iv) mercury most weakly attracts an extra electron, (iv) E_{ea} generally increases across a period (row) in the periodic table that is caused by the filling of the valence shell of the atom, and (v) group 17 atom releases more energy than a group 1 atom on gaining an electron because it obtains a filled valence shell and therefore is more stable. In addition, a trend of decreasing E_{ea} going down the groups in the periodic table might be expected since the additional electron will be entering an orbital farther away from the nucleus. Since this electron is farther from the nucleus, it is less attracted to the nucleus and would release less energy when added. However, a clear counterexample to this trend can be found in group 2, and inspecting the entire periodic table, it turns out that the proposed trend only applies to group 1 atoms. Thus, electron affinity follows the left-right trend of electronegativity but not the up-down trend.

2.6 REACTIONS AND STOICHIOMETRY

By way of introduction, most inorganic chemical reactions fall into four broad categories: (i) combination reactions, (ii) decomposition reactions, (iii) single-displacement reactions, and (iv) double-displacement reactions.

Combination reactions are reactions where two substances combine to form a third substance:

$$A + B \rightarrow AB$$

$$2Na(s) + Cl_2(g) \rightarrow 2NaCl(s)$$

$$8Fe + S8 \rightarrow 8FeS$$

$$S + O_2 \rightarrow SO_2$$

Decomposition reactions occur when a single compound reacts to give two or more substances. An example of a decomposition reaction is the decomposition of mercury (II) oxide into mercury and oxygen when the compound is heated:

$$2HgO \rightarrow 2Hg + O_2$$

A compound can also decompose into a compound and an element or two compounds.

A *single-displacement reaction* occurs when one element trades places with another element in a compound:

$$A + BC \rightarrow AC + B$$

For example, when magnesium replacing hydrogen in water to produce magnesium hydroxide and hydrogen gas or the production of silver crystals when a copper metal strip is dipped into silver nitrate:

$$Mg + 2H_2O \rightarrow Mg(OH)_2 + H_2$$
$$Cu(s) + 2AgNO_3(aq) \rightarrow 2Ag(s) + Cu(NO_3)_2(aq)$$
$$Zn(s) + CuSO_4(aq) \rightarrow Cu(s) + ZnSO_4(aq)$$

Double-displacement reactions occur when the anions and cations of two different molecules switch places to form two entirely different compounds:

$$AB + CD \rightarrow AD + CB$$

An example is the reaction of lead (II) nitrate with potassium iodide to form lead (II) iodide and potassium nitrate or when solutions of calcium chloride and silver nitrate are reacted to form insoluble silver chloride in a solution of calcium nitrate:

$$Pb(NO_3)_2 + 2KI \rightarrow PbI_2 + 2KNO_3$$
$$CaCl_2(aq) + 2AgNO_3(aq) \rightarrow Ca(NO_3)_2(aq) + 2AgCl(s)$$

Another type of double-displacement reaction takes place when an acid and base react with each other. The hydrogen ion in the acid reacts with the hydroxyl ion in the base causing the formation of water or the reaction of hydrochloric acid and sodium hydroxide to form sodium chloride and water:

$$HA + BOH \rightarrow H_2O + BA$$
$$HBr(aq) + NaOH(aq) \rightarrow NaBr(aq) + H_2O(l)$$
$$HCl(aq) + NaOH(aq) \rightarrow NaCl(aq) + H_2O(l)$$

More generally, the majority of the reactions of inorganic compounds take place at the site of the functional groups and are characteristic of that functional group. However, the reactivity of the functional group is affected by stereoelectronic effects.

The mole is a useful quantity in determining the amount of material that is present in a system or that might be required in a reaction. When considering chemical reactions, the ratio of each species required to take part in the reaction can be one of the most important pieces of information. It allows a chemical engineer to determine how much material is required or, more importantly, the minimum amount of a cheaper reagent that they can supply to maximize the conversion of a more expensive substance. The ratios required for each element or molecule are known as the stoichiometry of the reaction (Table 2.15).

Stoichiometry is the theory of the proportions in which chemical species combine with one another. A stoichiometric or compositional name provides information only on the composition of an ion, molecule, or compound, and may be related to either the empirical or molecular formula for that entity. It does not provide any structural information.

The use of the concept of the mole is vital when dealing with chemical reactions. Briefly, chemical reactions (shown as chemical equations) must be balanced by stoichiometric coefficients in front of each reactant and each

Table 2.15 Reaction Terminology

Limiting reactant	The reactant that is present in the smallest stoichiometric amount and which determines the maximum extent to which a reaction can proceed; if the reaction is 100% complete, then all of the limiting reactant is consumed and the reaction can proceed no further
Excess reactant	All other reactants within the process with exception of the limiting reactant; the term may refer to more than one reactant
Percentage excess	Based on the quantity of excess reactant above the amount required to react with the total quantity of limiting reactant
Percent conversion	The percentage of any reactant that has been converted to products
Degree of completion	The percentage or fraction of the limiting reactant that has been converted to products
Yield	The mass (or moles) of a chosen final product divided by the mass (or moles) of one of the initial reactants
Extent of reaction	The extent to which a reaction proceeds (i.e., the material actually reacting can be expressed by) the extent of reaction (a) in moles. A conventionally relates the feed quantities to the amount of each component present in the product stream, after the reaction has proceeded to equilibrium, through the stoichiometry of the reaction. To a term that appears in all reactions

product so there is an equal number of atoms or molecule (moles) on each side of the equation. Thus,

$$2HgO \rightarrow 2Hg + O_2$$

In this equation, there are two stoichiometric coefficients of the reactant (mercuric oxide, HgO) and two stoichiometric coefficients of one of the products (mercury, Hg). This is required so that a molecule of oxygen (O_2) can be produced. To write the equation in any other way (e.g., $HgO \rightarrow Hg + \frac{1}{2}O_2$) is stoichiometrically incorrect since oxygen does not exist as half a molecule in this reaction.

Thus, there is a molecular balance insofar as there is the same number of each atom type of the left-hand side and right-hand side of the reaction ($2HgO \rightarrow 2Hg + O_2$, i.e., 4 atoms $\rightarrow$ 2 atoms plus 2 atoms). This equation shows the relative number of atoms or molecules (or moles) of reactants and products involved in this particular chemical reaction, indicating the stoichiometry of the balanced reaction. The numbers written before the species in the chemical reaction are called the *stoichiometric coefficients*.

When performing stoichiometric calculations, remember that these are only valid on a molar basis, so convert all mass quantities provided to moles; this allows the other quantities to be determined by the use of ratios and multiples. It is always possible to convert the number of moles back to a mass at the end, if this is desired. When working with molar quantities, ensure that any molar units are consistent throughout the calculation, e.g., lb-mol, kmol, and mol; any can be used but must be maintained in working. In such calculations, total mass must be conserved, i.e., the total masses in and out of the system must tally, and it must be borne in mind that a balanced reaction is no proof of a feasible density flow reaction, which must be determined thermodynamically. Similarly, there is no information regarding the rate at which the reaction will proceed, the degree of completion that will be achieved, or the possibility of unwanted side reactions.

In many processes, there will not be a static system but a flowing process stream, so the molecular weight is used to relate the mass flow rate of a continuous stream to the molar flow rate:

$$\text{molar flow rate} = (\text{mass flow rate})/(\text{molecular weight})$$

This provides an easy route to convert volumetric flow rates into molar quantities, but using the methods outlived in mass and volume.

In spite of the focus stated above, the exact reaction stoichiometric amounts are rarely used in industrial processes and companies often increase the conversion of an expensive reactant, by adding other reactants in excessive quantities. This means that the product streams will also commonly contain these excess reactants. Such methods can be used to prevent unexpected side reactions that can also occur, limiting conversion to the desired product. The conversion can also be increased by recycling the necessary feedstocks since such a procedure can affect the equilibrium of a system and drive the reaction toward further production of desirable products.

The *equivalent mass* of an element is the number of parts by mass of the element that combines with or displaces 1.008 parts by mass of hydrogen or 35.45 parts by mass of chlorine or one equivalent mass of any other element. Elements combine among themselves in a definite ratio by mass to form compounds. The term equivalent mass expresses the combining capacity of an element or a compound in terms of mass with reference to a specific standard. Many metals do not either combine with hydrogen or displace hydrogen from acids. But almost all elements combine with oxygen and chlorine. Hence, elements such as oxygen (8 parts by mass of oxygen) and chlorine (35.45 parts by mass of chlorine) may be chosen as standards.

Finally, the rate of chemical reaction (sometimes generally but erroneously referred to as the *reactivity*) is generally a function of reactant concentration and temperature. For many homogeneous reactions, therefore, if they are exothermic,

$$\text{Rate of generation } \alpha e^{\mathrm{RT_r}}$$

R is the gas constant, and T_r is the absolute temperature. If the heat is removed by forced convection to a coolant in a jacket or coil, then

$$\text{Rate of removal } \alpha T_r - T_a$$

T_a is the coolant temperature.

Thus, since the generation rate is exponential, whereas the removal rate is linear, for any exothermic reaction in a specific reactor configuration, a critical condition may exist, i.e., a value of T_r beyond which "runaway" occurs. Also, a reaction that is immeasurably slow at ambient temperature may become rapid if the temperature is raised. With some reactions, which have a significant rate at ambient temperature, e.g., catalyzed reactions, such as oxidation, severe hazards may be associated with an elevation in temperature.

Exothermic reactions require control strategies that may involve temperature control, dilution of reagents, controlled addition of one reagent, containment/venting, and provision for emergencies. Many liquid phase or heterogeneous solid-liquid or gas-liquid reactions result in gaseous products or by-products. These products may be toxic or flammable or result in overpressurization of any sealed container or vessel. Unless pressure relief is provided, relatively small volumes of reactants—the presence of which may not be expected—may generate sufficient gas pressure to rupture a container.

2.7 ACID-BASE CHEMISTRY

In addition to the method of classification of inorganic chemicals as (i) simple organic chemicals and (2) complex inorganic chemicals, there is another form of classification that uses the concept (alphabetically) of (i) acids, (ii) bases, (iii) oxides, and (iv) salts.

Acids are complex chemicals that consist of hydrogen atoms and an acid radical which, when dissociating, form the H+ cation. Inorganic acids can be identified (or even classified) as those acids without oxygen in the molecule and those with oxygen in the molecule. In addition, the number of hydrogen atoms in the acid molecule is an indicator of the atoms that can be substituted on to form metal salts.

Nonoxygen-containing acids		***Salt***
HCl; Hydrogen chloride (hydrochloric acid)	Monobasic	Chloride
HBr; Hydrogen bromide (hydrobromic acid)	Monobasic	Bromide
HI; Hydrogen iodide (hydriodic acid)	Monobasic	Iodide
HF; Hydrogen fluorine (hydrofluoric)	Monobasic	Fluoride
Containing oxygen		
HNO_3; Nitric acid	Monobasic	Nitrate
H_2SO_3; Sulfurous acid	Dibasic	Sulfite
H_2SO_4; Sulfuric acid	Dibasic	Sulfate
H_2CO_3; Carbonic acid	Dibasic	Carbonate
H_2SiO_3; Silicic acid	Dibasic	Silicate
H_3PO_4; Orthophosphoric acid	Tribasic	Orthophosphate

The measure of acidity (other than the pH of the aqueous solution) is given by the acid dissociation constant, K_a, which is a quantitative measure of the strength of an acid in solution:

$$K_a = [A^-][H^+][HA]$$

In the above reaction, HA is the generic acid, A^- is the conjugate base of the acid, and H^+ is the hydrogen ion or proton that are in equilibrium when their concentrations do not change over time. As with all equilibrium constants, the value of K_a is determined by the concentrations (in mol/L) of each aqueous species at equilibrium. Due to the many orders of magnitude of the K_a values, a logarithmic measure of the acid dissociation constant is more commonly used in practice 0 the logarithmic constant (pK_a) is equal to $-\log_{10}(K_a)$. The larger the value of pK_a, the smaller the extent of dissociation. A weak acid has a pK_a value in the approximate range of -2 to 12 in water, and the pK_a values of strong acids is typically less than -2.

Acid dissociation constants are most often associated with weak acids, or acids that do not completely dissociate in solution because strong acids are presumed to ionize completely in solution and therefore their K_a values are large (Table 2.16). The larger the value of pK_a, the smaller the extent of dissociation. A strong acid is almost completely dissociated in aqueous solution to the extent that the concentration of the undissociated acid in negligible and becomes undetectable.

Table 2.16 Example of Acid Dissociation Constants K_a

Acid	Formula	K_{a1}	K_{a2}	K_{a3}
Arsenic	H_3AsO_4	5.8×10^{-3}	1.1×10^{-7}	3.2×10^{-12}
Carbonic	H_2CO_3	4.45×10^{-7}	4.69×10^{-11}	
Hydrogen cyanide	HCN	6.2×10^{-10}		
Hydrofluoric	HF	6.8×10^{-4}		
Hydrogen peroxide	H_2O_2	2.2×10^{-12}		
Hydrogen sulfide	H_2S	9.6×10^{-8}	1.3×10^{-14}	
Hydrochloric	HCl	Strong		
Hypochlorous	HOCl	3.0×10^{-8}		
Iodic	HIO_3	1.7×10^{-1}		
Nitric	HNO_3	Strong		
Nitrous	HNO_2	7.1×10^{-4}		
Perchloric	$HClO_4$	Strong		
Periodic	H_5IO_6	2×10^{-2}	5×10^{-9}	
Phosphoric	H_3PO_4	7.11×10^{-3}	6.32×10^{-8}	4.5×10^{-13}
Phosphorous	H_3PO_3	3×10^{-2}	1.62×10^{-7}	
Sulfamic	H_2NSO_3H	1.03×10^{-1}		
Sulfuric	H_2SO_4	Strong	1.02×10^{-2}	
Sulfurous	H_2SO_3	1.23×10^{-2}	6.6×10^{-8}	
Thiosulfuric	$H_2S_2O_3$	0.3	2.5×10^{-2}	

pK_{a1}, pK_{a2}, and pK_{a3} are the constants for the release of one proton at a time.

Acids are prepared by the interaction of acid oxides with water (for oxy-acids):

$$SO_3 + H_2O \rightarrow H_2SO_4$$
$$P_2O_5 + 3H_2O \rightarrow 2H_3PO_4$$

Another method of acid preparation by the interaction of hydrogen with nonmetals and following dissolution product in water (for acids without oxygen in the molecule):

$$H_2 + Cl_2 \rightarrow 2HCl$$
$$H_2 + S \rightarrow H_2S$$

The exchange reaction between a salt and an acid also produces a new acid:

$$Ba(NO_3)_2 + H_2SO_4 \rightarrow BaSO_4 + 2HNO_3$$

This method also includes the displacement of a weak or slightly soluble acid from a salt by means of a stronger acid:

$$Na_2SiO_3 + 2HCl \rightarrow H_2SiO_3 + 2NaCl$$
$$2NaCl + H_2SO_4(conc.) \rightarrow Na_2SO_4 + 2HCl$$

The chemical properties of acids include (i) interaction with bases, i.e., neutralization; (ii) interaction with basic oxides; (iii) interaction with metals; and (iv) interaction with salts, i.e.,

$$H_2SO_4 + 2KOH \rightarrow K_2SO_4 + 2H_2O$$
$$2HNO_3 + Ca(OH)_2 \rightarrow Ca(NO_3)_2 + 2H_2O$$
$$CuO + 2HNO_3 \rightarrow Cu(NO_3)_2 + H_2O$$
$$Zn + 2HCl \rightarrow ZnCl_2 + H_2$$
$$2Al + 6HCl \rightarrow 2AlCl_3 + 3H_2$$
$$BaCl_2 + H_2SO_4 \rightarrow BaSO_4 + 2HCl$$
$$K_2CO_3 + 2HCl \rightarrow 2KCl + H_2O + CO_2$$

Bases are also complex chemicals in which an atom of a metal is bonded with one or several hydroxyls groups which, when dissociating in water, form a metal cation (or the ammonium cation, NH_4^+) and a hydroxide anion (OH^-).

Bases are prepared by the interaction of active metals (alkaline and alkaline earth metals) with water:

$$2Na + 2H_2O \rightarrow 2NaOH + H_2$$
$$Ca + 2H_2O \rightarrow Ca(OH)_2 + H_2$$

Another method of preparation involves the interaction of the oxides of active metals with water:

$$BaO + H_2O \rightarrow Ba(OH)_2$$

The electrolysis of an aqueous solution of a salt can also produce a base:

$$2NaCl + 2H_2O \rightarrow 2NaOH + H_2 + Cl_2$$

The chemical reactions of bases, like the acids, are variable and include (i) interaction with acid oxides, (ii) interaction with acids—neutralization, and (iii) exchange reactions with salts:

$$KOH + CO_2 \rightarrow KHCO_3$$
$$2KOH + CO_2 \rightarrow K_2CO_3 + H_2O$$
$$NaOH + HNO_3 \rightarrow NaNO_3 + H_2O$$
$$Ba(OH)_2 + K_2SO_4 \rightarrow 2KOH + BaSO_4$$

Acids, such as hydrochloric acid, and bases, such as potassium hydroxide, that have a great tendency to dissociate in water are completely ionized in solution and are called strong acids or strong bases. On the other hand, acids, such as acetic acid, and bases, such as ammonia, that are reluctant to dissociate in water are only partially ionized in solution; they are called weak acids or weak bases. Strong acids in solution produce a high concentration of hydrogen ions, and strong bases in solution produce a high concentration of hydroxide ions and a correspondingly low concentration of hydrogen ions. The hydrogen ion concentration is often expressed in terms of its negative logarithm, or pH. Strong acids and strong bases make very good electrolytes insofar as solutions of storing acids and strong bases readily conduct electricity. Conversely, weak acids and weak bases make poor electrolytes.

The chemical interaction between an acid and a base (*acid-base reaction*) varies depending on the species involved (Table 2.17). All inorganic acids elevate the hydrogen concentration in an aqueous solution. All inorganic bases elevate the hydroxide concentration in an aqueous solution. Inorganic salts are neutral, ionically bound molecules and do not affect the concentration of hydrogen in an aqueous solution. Inorganic compounds, in a broader sense, consist of an ionic component and an anionic component. A very

Table 2.17 Examples of Inorganic Acids, Inorganic Bases, and Inorganic Salts

Inorganic acids

- Carbonic acid (H_2CO_3): a weak inorganic acid
- Hydrochloric acid (HCl): a highly corrosive, strong inorganic acid with many uses
- Hydrofluoric acid (HF): a weak inorganic acid that is highly reactive with silicate, glass, metals, and semimetals
- Nitric acid (HNO_3): a highly corrosive and toxic strong inorganic acid
- Phosphoric acid (H_3PO_4): not considered to be a strong inorganic acid. It is found in solid form as a mineral and has many industrial uses
- Sulfuric acid (H_2SO_4): a highly corrosive inorganic acid. It is soluble in water and widely used

Inorganic bases

- Ammonium hydroxide (ammonia water, NH_4OH): a solution of ammonia in water
- Calcium hydroxide (lime water, $Ca(OH)_2$): a weak base with many industrial uses
- Magnesium hydroxide [$Mg(OH)_2$]: referred to as brucite when found in its solid mineral form
- Sodium bicarbonate (baking soda, $NaHCO_3$): a mild alkali
- Sodium hydroxide (caustic soda. NaOH): a strong inorganic base. It is widely used in industrial and laboratory environments

Inorganic salts

- Calcium chloride ($CaCl_2$): many industrial uses
- Potassium dichromate ($K_2Cr_2O_7$): commonly used as an oxidizing agent
- Sodium chloride (NaCl): common table salt used in the food industry

large number of compounds occur naturally, while others may be synthesized. In all cases, charge neutrality of the compound is key to the structure and properties of the compound.

Historically, the theory of acid-base chemistry was thought to be related to the composition of the compound, specifically that it should contain oxygen; this was the theory of Antoine Lavoisier but was disproved some years later by Sir Humphry Davy when he proved nonoxygen-containing species, such as hydrogen sulfide and the hydrohalic acids (such as hydrochloric acid) existed. This work was extended by Liebig, who proposed that acids are substances that contain hydrogen that can be displaced by a metal species before being formulated into the currently accepted *Arrhenius definition* of an acid as a substance that dissociates in water to produce hydrogen cations (H^+), while a base dissociates in water to form hydroxide anions (OH^-). It should be appreciated that hydrogen ions (H^+) do not exist as separate species in a solution but rather that the hydronium cation (H_3O^+) is actually formed.

This forms the basis of the statement that the reaction of an acid and a base will produce a salt and water in a neutralization reaction, as the hydrogen cations form a compound (water) with the anion from the base (i.e., hydroxide anion) and the cation from the base, often a metal, will react with the anion from the acid to form a salt. For example, the reaction of potassium hydroxide and nitric acid:

$$\underset{\text{Acid}}{HNO_3} + \underset{\text{base}}{2KOH} \rightarrow \underset{\text{salt}}{K_2NO_3} + \underset{\text{water}}{2H_2O}$$

Note that the Arrhenius definition only holds for aqueous systems and that the dissolution of acids in other solvents would not necessarily be acidic. Similarly, molten metal hydroxides are not basic. The issue of the solvent used was resolved with the generalization offered by the *solvent system theory*, devised by Albert Germann who noted that in several solvents, there is a presence of solvonium cations and solvate anions in equilibrium (see chapter on equilibriums for more information) with nondissociated solvent molecules. The simplest example is water itself, which exists as water with some hydronium cations and hydroxide anions. The addition of another substance, i.e., a solute, to the solvent causes a shift in this equilibrium to form either more solvonium cations or solvate anions, depending on its nature. An acid increases the solvonium cations present, while a base increases the concentration of solvent anions. For example, water exists as

$$2xH_2O \leftrightarrow xH_3O^+ + xOH^-$$

Addition of the acid (HNO_3) will form more hydronium cations:

$$2xH_2O + yH_2NO_3 \leftrightarrow (x+2y)H_3O^+ + yNO_3^- + xOH^-$$

While the presence of the base (KOH) will increase the presence of hydroxide anions

$$2H_2O + KOH \rightarrow H_3O^+ + K^+ + OH^-$$

The strength of this approach, in terms of solvent rather than only water, is demonstrated by consideration of aprotic solvents, such as liquid dinitrogen tetroxide (N_2O_4):

$$\underset{\text{Base}}{AgNO_3} + \underset{\text{acid}}{NOCl} \sim \underset{\text{salt}}{AgCl} + \underset{\text{solvent}}{N_2O_4}$$

As a result of the interaction between the basic/acidic species and the solvent, the nature of a substance can be altered by changing the solvent,

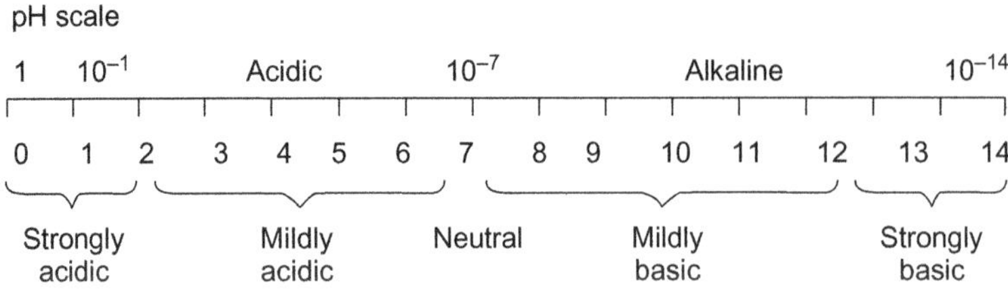

Fig. 2.3 The pH scale.

for example, perchloric acid, which is a strong acid in water and is a weak acid in ethanoic acid yet a weak base in fluorosulfonic acid. It is important to appreciate the changing nature of compounds with regard to their environment, which can have consequences for reaction and materials choice. The pH scale (Fig. 2.3) is a measure of acidity and alkalinity of an acid or base or a solution of either.

2.7.1 Reactions of Acids and Alkalis

This is a particular acid-base reaction, where the base is also an alkali species, meaning it is a basic, ionic salt of either an alkali or alkaline earth metal. In this instance, the reaction produces a metal salt and water in a neutralization reaction. The acid either contains hydrogen cations or causes them to be produced in solution, while the alkali is a soluble base either containing hydroxide ions or causing them to be produced in solution. For example, nitric acid and potassium hydroxide:

$$HNO_3 \rightarrow H^+(aq) + NO_3^-(aq)$$
$$NaOH \rightarrow Na^+(aq) + OH^-(aq)$$

As a result of the neutralization of either the acid or base, by the other species, into water, such reactions have a number of applications including within antacid formulations, where excess hydrochloric acid in the stomach is neutralized, most commonly, by species including $NaCO_3$ and aluminum hydroxide. Neutralization can also be used for the regulation of pH in agriculture via the application of fertilizers, or the abatement of acid formation by the removal of sulfur from fuel before combustion to avoid the formation of sulfur dioxide, and, subsequently, acid rain.

2.7.2 Brønsted-Lowry Acids and Bases

Working independently, Johannes Nicolaus Brønsted and Martin Lowry developed a definition, underpinned by the concept of base protonation and acid deprotonation, where acids have the ability to *donate* hydrogen cations (H^+) to bases, hence the acids *accept* these H^+ ions. This definition removes the reliance on a full system description, i.e., the solvent or salt formed, but rather focuses on the acid and base themselves. The species involved in the transfer of a proton are known as *conjugates* (i.e., a pair), with a conjugate acid and conjugate base produced.

By this definition, i.e., acids donate protons and bases accept them, an acid-base reaction is, therefore, the transfer of a hydrogen cation from an acid to a base. Thus, the acid and base react, not to given a salt and solvent, but rather a new (conjugate) acid, the base with accepted H^+, and new (conjugate) base, the acid with H+ removed. Consequently, eliminating the idea of neutralization. For example, the addition of hydrogen cations (H^+) to hydroxide anions (OH^-), which are basic, produces water (H_2O); hence, water is the conjugate acid of hydroxide ions:

$$H^+ + OH^- \rightarrow H_2O$$

As mentioned above, the solvent is no longer considered; however, the solvent system can coexist within the Brønsted-Lowry definition, which can also provide an explanation for the low concentrations of hydronium and hydroxide ions produced by the dissociation of water:

$$2H_2O \leftrightarrow H_3O^+ + OH^-$$

In this case, water, which is amphoteric, acts as both acid and base; as one molecule of water donates a H^+ ion, forming the conjugate base, OH^-, with a second molecule of water accepting the H^+ ion, forming the conjugate acid, H_3O^+. In general, for acid-base reactions, the Brønsted-Lowry definition is

$$AH + B^- + BW + A^-$$

AH is the acid, B is the base, BH^+ is the conjugate acid of B, and A^- is the conjugate base of AH.

2.7.3 Lewis Acids and Bases

In contrast to the Breasted-Lowry definition of acids and bases, Lewis defined the species in terms of electron transfer rather than hydrogen cations. Hence, a *Lewis base* is a species that donates an electron pair to a *Lewis acid*, which accepts the donated electron pair. This concept is most easily demonstrated by considering the established aqueous acid-base reaction:

$$HCl + NaOH^- + H_2O + NaCl$$

In this case, the acid combines with the base, rather than exchanging atoms with it and the acid is H^+ and the base OH^-, which has an unshared electron pair that is able to donate to form a bond with the electron-deficient proton. Hence, the acid- base reaction is not the transfer of H^+ but the donation of an electron pair from OH^-, forming a covalent bond to produce water. The electronic interactions involved are the donation of electrons from the highest occupied molecular orbital (HOMO) of the subsequently known base to the lowest unoccupied molecular orbital (LUMO) of the second species, then known as the acid.

The strength of a Lewis acid and base interaction is determined by the concentrations of all solution species, for a simple system:

$$\mathrm{A} + \mathrm{B} \rightarrow \mathrm{AB}$$

The stability constant is given by

$$K = \frac{[\mathrm{AB}]}{[\mathrm{A}][\mathrm{B}]}$$

A high value of K indicates a strong interaction as [AB] is large compared with the product of [A] [B], and a low value of K indicates a weak interaction. The interaction is typically governed by the relative strengths of the acid and/or the base.

Hard acids and bases are usually small species that are difficult to polarize, e.g., hard acids, H^+, Na, and AP^+ and hard bases, OH^-, F^-, and NH_3. In contrast, *soft acids and bases* are usually large species that are easily polarized, e.g., soft acids, Ag^+, Cd^{2+}, and Cu^+ and soft bases, H^-, 1^-, and CO. This is an important distinction as hard acids tend to bind to hard bases, as they both exhibit high ionic character so K is high, and soft acids tend to bind to soft bases, which both exhibit significant covalent character and, again, K is high. Any soft-hard interaction will have a low K as the interaction will be poor.

2.7.4 Oxides and Salts

Oxides consisting from two elements, one of which is oxygen, are produced by the interaction of simple and complex chemicals substances with oxygen:

$$2\mathrm{Mg} + \mathrm{O_2} \rightarrow 2\mathrm{MgO}$$

$$4\mathrm{P} + 5\mathrm{O_2} \rightarrow 2\mathrm{P_2O_5}$$

$$\mathrm{S} + \mathrm{O_2} \rightarrow \mathrm{SO_2}$$

$$2CO + O_2 \rightarrow 2CO_2$$
$$2CuS + 3O_2 \rightarrow 2CuO + 2SO_2$$
$$CH_4 + 2O_2 \rightarrow CO_2 + 2H_2O$$

Oxides are also produced by the thermal decomposition of inorganic chemicals (bases, acids, and salts) that contain oxygen within the molecular structure:

$$Cu(OH)_2 \rightarrow CuO + H_2O$$
$$2Pb(NO_3)_2 \rightarrow 2PbO + 4NO_2 + O_2$$

The reactions of oxides are varied and can produce a variety of products depending upon the electronegativity of the oxide. Thus,

Basic oxides	**Acid oxides**
Interaction with water	
A base is formed $Na_2O + H_2O \rightarrow 2NaOH$ $CaO + H_2O \rightarrow Ca(OH)_2$	An acid is formed $SO_3 + H_2O \rightarrow H_2SO_4$ $P_2O_5 + 3H_2O \rightarrow 2H_3PO_4$
Interaction with acid or base	
Reactions with an acid produce a salt $MgO + H_2SO_4 \rightarrow MgSO_4 + H_2O$ $CuO + 2HCl \rightarrow CuCl_2 + H_2O$	Reactions with base produce a salt $CO_2 + Ba(OH)_2 \rightarrow BaCO_3 + H_2O$ $SO_2 + 2NaOH \rightarrow Na_2SO_3 + H_2O$
An amphoteric oxide interacts	
With an acid as a base $ZnO + H_2SO_4 \rightarrow ZnSO_4 + H_2O$	With a base as an acid $ZnO + 2NaOH \rightarrow Na_2ZnO_2 + H_2O$
Interaction of basic oxide produces a salt	
$Na_2O + CO_2 \rightarrow Na_2CO_3$	
Oxides can also undergo reduction	
$3CuO + 2NH_3 \rightarrow 3Cu + N_2 + 3H_2O$ $P_2O_5 + 5C \rightarrow P + 5CO$	

Salts are also complex inorganic chemicals that consist from atoms of metal and acid moieties. They are generally classified into (i) medium salts, (ii) acidic salts, (iiii) basic salts, (iv) double salts, (v) mixed salts, and (vi) complex salts.

Medium salts, by dissociation, produce metal cations (or the ammonium cation, NH_4^+) and anions of the acid radical:

$$Na_2SO_4 \rightarrow 2Na^+ + SO_4^{2-}$$
$$CaCl_2 \rightarrow Ca_2^+ + 2Cl^-$$

On the other hand, acid salts produce only metal cations (or the ammonium cation, NH_4^+), hydrogen anions, and anions of the acid radical:

$$NaHCO_3 \rightarrow Na^+ + HCO_3^- \rightarrow Na^+ + H^+ + CO_3^{2-}$$

Basic salts, by dissociation, produce metal cations, hydroxyl anions, and anions of the acid radical:

$$Zn(OH)Cl \rightarrow [Zn(OH)]^+ + Cl^- \rightarrow Zn^{2+} + OH^- + Cl^-$$

Double salts, by dissociation, produce two cations and one anion:

$$KAl(SO_4)2 \rightarrow K^+ + Al^{3+} + 2SO_4^{2-}$$

Mixed salts dissociate as follows:

$$CaOCl_2 \rightarrow Ca^{2+} + Cl^- + OCl^-$$

whereas complex salts dissociate into complex cations and anions:

$$[Ag(NH_3)_2]Br \rightarrow [Ag(NH_3)_2]^+ + Br^-$$
$$Na[Ag(CN)2] \rightarrow Na^+ + [Ag(CN)2]^-$$

2.8 MINERALS

A mineral is a naturally occurring inorganic chemical compound, usually having a crystalline form and is abiogenic in origin (i.e., it does not result from the activity of living organisms) (Pellant, 2002; Speight, 2015). A mineral has one specific chemical composition, whereas a rock can be an aggregate of different minerals or mineraloids (a mineral-like substance that does not demonstrate crystallinity). Minerals are distinguished by various chemical and physical properties. Differences in chemical composition and crystal structure distinguish the various species, which were determined by the mineral's geological environment when formed. Changes in the temperature, pressure, or bulk composition of a rock mass cause changes in its minerals. In some states, the fossil fuel resources (coal, natural gas, and crude oil) are listed as components of the mineral resources of the state. However, for the most part, fossil fuel resources (perhaps with the exception) of oil shale are not classed as true minerals. Furthermore, the general definition of a mineral can be defined by the following criteria: (i) it is naturally

Table 2.18 Elements in the Crust of the Earth

Element Name	Symbol	% w/w of the Earth's Crust
Oxygen	O	47
Silicon	Si	28
Aluminum	Al	8
Iron	Fe	5
Calcium	Ca	3.5
Sodium	Na	3
Potassium	K	2.5
Magnesium	Mg	2
All other elements		1

occurring, (ii) it is stable at room temperature, (iii) it can be represented by a chemical formula, and (iv) it has an ordered atomic arrangement.

Approximately 99% of the minerals that occur in the crust of the Earth are made up of just eight elements (Table 2.18). Most of these elements are found combined with other elements as compounds. Minerals have a definite chemical composition and structure. There are over 3000 minerals known. Some are rare and precious such as gold and diamond, while others are more ordinary, such as quartz.

Minerals are chemical compounds, sometimes specified by crystalline structure and by composition, which are found in rocks (or pulverized rocks, known as sand). On the other hand, rocks consist of one or more minerals and fall into three main types depending on their origin and maturation history: (i) igneous rocks, which are rocks that have solidified directly from a molten state, such as volcanic lava; (ii) sedimentary rocks, which are rocks that have been remanufactured from previously existing rocks, usually from the products of chemical weathering or mechanical erosion, without melting; and (iii) metamorphic rocks, which are rocks that have resulted from processing, by heat and pressure (but not melting), of previously existing sedimentary or igneous rocks.

REFERENCES

Anslyn, E.V., Dougherty, D.A., 2006. Modern Physical Organic Chemistry. University Science Books, Herndon, VA.

Connelly, N.G., Damhus, T., Hartshorn, R.M., Hutton, A.T. (Eds.), 2005. Nomenclature of Inorganic Chemistry: IUPAC Recommendations 2005. The Royal Society of Chemistry, London.

Cotton, F.A., Wilkinson, G.A., Murillo, C.A., Bochmann, M., 1999. Advanced Inorganic Chemistry, sixth ed. John Wiley & Sons, Hoboken, NJ.

Emsley, J., 2011. Nature's Building Blocks: An A-Z Guide to the Elements—New Edition. Oxford University Press, Oxford.

Patrick, G.L., 2004. Instant Notes: Organic Chemistry: Instant Notes. Garland Science/BIOS Scientific Publishers, Abingdon, Oxfordshire.
Pellant, C., 2002. Rocks and Minerals. DK Books, New York.
Silberberg, M.S., 2006. Chemistry: The Molecular Nature of Matter and Change, fourth ed. McGraw-Hill, New York.
Smith, M.B., 2013. March's Advanced Organic Chemistry: Reactions, Mechanisms, and Structure, seventh ed. John Wiley & Sons, Hoboken, NJ.
Speight, J.G., 2015. Asphalt Materials Science and Technology. Butterworth-Heinemann, Elsevier, Oxford.
Speight, J.G., 2017a. Environmental Organic Chemistry for Engineers. Butterworth-Heinemann, Elsevier, Cambridge, MA.
Speight, J.G. (Ed.), 2017b. Lange's Handbook of Chemistry, 17th ed. McGraw-Hill Education, New York.

CHAPTER THREE

Industrial Inorganic Chemistry

3.1 INTRODUCTION

The previous chapter (Chapter 2) has introduced the reader to the varied, but fundamental, aspects of the subdiscipline of inorganic chemistry. The inorganic chemical industry manufactures chemicals that are often of a mineral origin, but not of a basic carbon molecular (Buchel et al., 2008). In addition, organic chemicals may be destined for use at during the manufacture of a great variety of other inorganic and organic chemical products. The products of the inorganic chemicals industry are used as (i) basic chemicals for industrial processes, such as acids, alkalis, salts, oxidizing agents, industrial gases, and halogens; (ii) chemical products to be used in manufacturing products, such as pigments and dry colors; and (iii) alkali metals and finished products for ultimate consumption, such as mineral fertilizers, glass, and construction materials. The largest use of inorganic chemicals is as processing aids in the manufacture of chemical and non-chemical products. Consequently, inorganic chemicals often do not appear in the final products.

Thus, industrial inorganic chemistry includes subdivisions of the chemical industry that manufacture inorganic products on a large scale such as the heavy inorganics (chlor-alkalis, sulfuric acid, and sulfate derivatives) and fertilizers (potassium, nitrogen, and phosphorus products) as well as segments of fine chemicals that are used to produce high-purity inorganics on a much smaller scale. Among these are reagents and raw materials used in high-tech industries, the pharmaceutical industry, and the electronics industry, for example, as well as in the preparation of inorganic specialties such as catalysts, pigments, and propellants. Metals find a variety of uses without being incorporated into salts. They are manufactured from ores and purified by many of the same processes as those used in the manufacture of inorganics. However, if they are commercialized as alloys or in their pure form such as iron, lead, copper, or tungsten, they are often considered products of the metallurgical rather than products of the inorganic chemical industry.

Environmental Inorganic Chemistry for Engineers
http://dx.doi.org/10.1016/B978-0-12-849891-0.00003-5

Moreover, inorganic chemistry as practiced on the industrial stage is not so simple and/or straightforward (Heaton, 1994, 1996; Budavari, 1996; Brock, 2000). Industrial inorganic chemistry is an extremely comprehensive and practical discipline and, although there are benefits from understanding the basic inorganic chemical science, there is still the need to gain a valuable insight into chemical technology. Basic inorganic chemistry does provide valuable insight into industrial inorganic chemistry and the various products, but other chemicals that are used and produced in industrial processes offer a considerable (but valuable) challenge to (1) an understanding of the processes, (2) the necessary process parameters, (3) the properties of the feedstocks, (4) the properties of the products, (5) the properties of the by-products, (6) the adaptability of the process, and (7) the influence of these various chemicals on the environment. These effects are difficult to understand on the basis of laboratory chemical studies alone, and it is for this reason that this chapter is included in the book.

Industrial inorganic chemistry includes subdivisions of the chemical industry that manufacture inorganic products on a large scale such as the heavy inorganics (chlor-alkali chemicals, sulfuric acid, and sulfate chemicals) and fertilizers (potassium, nitrogen, and phosphorus products) as well as segments of fine chemicals that are used to produce high-purity inorganic chemicals on a much smaller scale. Among these chemicals are reagents and raw materials used in the pharmaceutical industry, the high-tech industry, and the electronic industry as well as in the preparation of inorganic specialty chemicals such as catalysts, pigments, and propellants. In addition, metals are fall within the classification of industrial inorganic chemicals and are manufactured from ores and purified by many of the same processes as those used in the manufacture of inorganic fine chemicals. However, if metals are commercialized as alloys or in the pure form such as iron, lead, copper, or tungsten, they are considered products of the metallurgical industry rather than products of the inorganic chemical industry.

By way of explanation, the term *fine chemicals* is used in this text as a distinction to from other chemicals (often referred to as *heavy chemicals, crude chemicals*, or *commodity chemicals*) that are produced in large lots and are often in the crude state as starting materials for other chemical products (Table 3.1). Commodity chemicals are a subsector of the chemical industry that is differentiated by primarily their bulk of their manufacture (other subsectors are fine chemicals and specialty chemicals). Thus, fine chemicals are complex, single, and pure chemical substances, produced in limited quantities in multipurpose plants by multistep batch chemical or biotechnological processes. They are manufactured to precise specifications and may also be

Table 3.1 General Differentiation of Fine Inorganic Chemicals for Heavy Inorganic Chemicals

Inorganic Chemicals	
Heavy chemicals[a]	Bulk chemicals Used as intermediate chemicals Multipurpose chemicals Not always manufactured to specification
Fine chemicals[b]	Single pure chemicals Formulated chemicals Differentiated chemicals Performance chemicals Produced to specifications

[a]Also called *crude chemicals*.
[b]Also called *specification chemicals*.

used for further processing within the inorganic chemical industry. The class of fine chemicals is subdivided based on the added value of the chemicals when used as either building block chemicals—the basic components for chemical synthesis, advanced intermediate chemicals, or active ingredient chemicals. In addition, fine chemicals are produced in limited volumes (mainly referred to as *limited tonnage chemicals*) by traditional organic synthesis in multipurpose chemical plants although a significant amount of the total production of fine chemicals may also be manufactured in house by large users.

Thus, the umbrella term *the industrial inorganic chemicals industry* includes the myriad of chemicals (numbered in the thousands) produced by a variety of different processes. To accomplish this production, a set of basic chemicals (feedstocks) is combined in a series of reaction steps to produce both intermediate chemicals and end products. Relatively few inorganic chemical manufacturing facilities are a single-product (or a single-process) facility, and many process units are designed to be flexible (with process options) so that production levels of related chemical products (or chemical by-products) can be adapted over a wide range. This flexibility allows the chemicals producer to accommodate variations in feedstock quality (purity and physical state—gas, liquid, or solid) that can change the production rate and processes used, even on a short-term (less than a year) basis. Furthermore, the process by-products are may also valuable saleable products (or feedstocks for an alternate process) and the value of these by-products can change the economics of the process.

The typical process for the production of organic chemicals involves combining multiple feedstocks in a series of unit operations; the petroleum refining industry is the best example of the integration of a series of unit processes to produce the desired. This is not always the case in the inorganic chemicals industry where the first unit operation is a typically a unit to accomplish a chemical reaction to produce the product or an intermediate product that leads to the desired chemical. In addition, there is some differentiation between the chemicals to be produced; commodity chemicals (i.e., bulk commodities or bulk chemicals) are a group of chemicals that are made on a very large scale to satisfy the chemical market) and tend to be synthesized in a continuous reactor, while specialty chemicals usually are typically produced in a batch reactor. The yield of the inorganic chemical may determine the quantity and type of by-products, including gaseous emissions.

The production of many inorganic specialty chemicals often requires a series of two or more reaction steps and once the reaction is complete, the desired product must be separated from the by-products by a second unit operation. In the separation stage, several separation techniques such as settling, distillation, or refrigeration may be used and the final product may require further processing (e.g., by spray drying or pelletizing) to produce the saleable item. The separation technology employed depends on many factors including (1) the phases of the substances being separated—gas, liquid, or solid; (2) the number of components in the mixture; and (3) whether recovery of by-products is important. Numerous techniques such as solvent extraction, filtration, and settling can be used singly or in combination to accomplish the separation.

Regulatory laws regarding emissions of inorganic chemicals (especially hazardous emissions) generated during the production of inorganic chemicals are important dynamics that shape the industry (Chapter 8). To minimize the detrimental effects of chemical industry pollutants, multiple local, state, and federal laws govern the producers of these chemicals. For example, the federal Emergency Planning and Community Right-to-Know Act requires many manufacturers to submit details of any emission data to the United States Environmental Protection Agency (US EPA). Similarly, the Pollution Prevention Act requires those same companies to report their waste management and pollution reduction activities. Other federal regulations impacting producers include the Safe Drinking Water Act, the Clean Air Act and Amendments, and other laws that restrict hazardous wastes. In addition to legal restrictions, both the US EPA and the Chemical Manufacturers Association (CMA) sponsor successful voluntary pollution reduction programs that encourage environmental sensitivity. The US EPA has continued to monitor the industry, and in the light of current strong emphasis on chemical safety and pollution controls, it is likely that regulations will continue to be added and modified.

There are also voluntary programs where the member companies work with the public to address such issues as chemical safety and environmental protection. The mechanism for these programs involves a combination of soliciting information from the public about the various concerns that are addressed, and the progress is reported back to the public. While increasing federal and state regulations pose an ongoing challenge to chemical industry participants, positive signs indicated that the inorganic chemicals industry has been successful in clearing these hurdles. The overall chemical industry has managed to reduce the various emissions of waste that appear on the Toxics Release Inventory.

The purpose of this chapter is to give an outline of the background information regarding the development of the inorganic chemicals including the production methods of the common inorganic chemicals. Thus, this chapter deals with the fundamental aspects of industrial inorganic chemistry and presents the various processes (that are used regular basis) on which the industrial inorganic chemicals industry is. This will give the reader the ability to understand the necessary links between laboratory inorganic chemistry and industrial process chemistry that is a necessary within the industrial chemistry and process engineering communities. These chemical and engineering sectors of industry have long held strong ties, and the knowledge of the chemistry involved in the various processes will help to point the way to synthetic pathways by which these chemicals are produced on a commercial scale.

3.2 THE INORGANIC CHEMICALS INDUSTRY

The inorganic chemicals industry, like the organic chemicals industry and the petroleum refining industry (Speight, 2017a,b) adds value to raw materials by transforming them into the chemicals required for the manufacture of consumer products. Basic chemicals represent the starting point for the manufacture of inorganic industrial chemicals that are produced on a very large scale employing continuous processes. The unit price of these products is relatively low, and producing them cheaply and efficiently is a major concern for the companies that manufacture them. Sulfur, nitrogen, phosphorus, and chlor-alkali industries are the main producers of basic inorganic chemicals, and they will often sell them to other industries as well as using them in the manufacture of their own end products. The basic principles for their production and major uses are indicated here for each of these industries.

Thus, the inorganic chemical industry manufactures a range of chemicals, many of which arise from the use of naturally occurring (mineral) feedstocks, but there are also disadvantages insofar as most process produce effluents (Table 3.2) that must be managed to meet environmental regulations. The products of these manufacturing processes may also be used for

Table 3.2 Examples of Feedstocks and Effluents for Inorganic Chemicals Production

Process Input	Process Output
Ammonia Synthesis	
Natural gas Air Catalyst Additives	Anhydrous ammonia Carbon dioxide Recycle gas streams Spent catalyst Wastewater
Ammonium Nitrate Synthesis	
Ammonia Nitric acid Additives	Ammonium nitrate Ammonia Nitric acid Particulate matter
Ammonium Phosphate Synthesis	
Ammonia Phosphoric acid Sulfuric acid Phosphate rock Gypsum	Ammonium phosphate Process water Phosphate rock dust Gypsum slurry Gaseous fluorides (e.g., SiF4 and HF)
Ammonium Sulfate Synthesis	
Ammonia Sulfuric acid	Ammonium sulfate Scrubber products Gas
Brine Purification	
Brine Additives	Brine Brine mud Wash water
Caustic Soda Processing	
Caustic soda	Caustic soda Sodium sulfate
Chlor-Alkali Compounds	
Sodium chloride	Sodium hydroxide Sodium carbonate Chlorine
Chlorine Processing	
Chlorine	Chlorine Wastewater

Table 3.2 Examples of Feedstocks and Effluents for Inorganic Chemicals Production—cont'd

Process Input	**Process Output**
Hydrochloric Acid Manufacture	
Chlorine	Hydrochloric acid
Hydrogen	Variable (depending on process)
	Hydrogen chloride
Hydrofluoric Acid Manufacture	
Calcium fluoride	Hydrofluoric acid
Sulfuric acid	Sulfuric acid
	Particulate matter and dust
Hydrogen Production	
Water	Variable (depending on process)
Activated carbon	Sodium hydroxide
	Spent activated carbon
Nitric Acid and Nitrogen Compounds	
Ammonia	Nitric acid
Sulfuric acid	Particulate matter
Catalyst	Nitric acid
Air	Oxides of nitrogen
	Spent catalyst
Phosphoric Acid and Phosphorus Compounds	
Phosphate rock	Phosphoric acid
Sulfuric acid	Fluorosilicate salts
Water	Uranium oxide
	Silicon fluoride
	Gypsum slurry
	Process water
Sodium Carbonate	
Trona	Particulate matter
	Carbon oxides
	Nitrogen oxides
	Sulfur oxides
	Sulfuric acid
Sulfuric Acid Synthesis	
Elemental sulfur	Sulfuric acid
Air	Sulfur dioxide

Continued

Table 3.2 Examples of Feedstocks and Effluents for Inorganic Chemicals Production—cont'd

Process Input	Process Output
Catalyst	Process water
Water	Spent catalyst
Sulfur Recovery	
Hydrogen sulfide	Sulfur
Air	Sulfur oxides
Catalyst	Particulate matter
Titanium Dioxide	
Ilmenite	Titanium dioxide
	Mineral dust
	Ferrous sulfate

the manufacture of a multitude of inorganic chemical products. Generally, the products arising from the inorganic chemical industry are used as basic chemicals (such as acid derivatives, alkali derivatives, various salts, oxidizing agents, halogen derivatives, and industrial gases). Other inorganic chemical products include those to be used in the manufacture of products such as pigments, dry colors, and alkali metals as well as finished products for ultimate consumption, for example, mineral fertilizers, glass, and construction materials. Typically, the largest use of manufactured inorganic chemicals is as processing aids in the production of chemical products and nonchemical products, and thus, inorganic chemicals often do not always appear in the final products of the inorganic chemical industry.

3.2.1 Historical Aspects

The origins of the inorganic chemical industry (even the chemical industry as a whole) can be traced to the *industrial revolution* that commenced about 1760 and lasted until sometime between 1820 and 1840 after which the modern chemical industry evolved (Brock, 2000). In the early stages of the revolution, sulfuric acid (often called *oil of vitriol* in the early days of its use) and sodium carbonate were among the first industrial chemicals. Sulfuric acid played an important role in the manipulation of metals, but the production of this acid on an industrial scale required the development of materials that would resist attack. On the other hand, sodium carbonate was obtained in the anhydrous form as *soda ash* from vegetable material until the quantities produced could no longer meet the rapidly expanding needs of manufacturers of glass, soap, and textiles. This led the Royal Academy of

Sciences of Paris, in 1775, to establish a contest for the discovery of a process based on an abundant raw material, sodium chloride (NaCl) and to the Leblanc process for the preparation of soda by converting salt into sulfate followed by conversion of the sulfate to the carbonate using charcoal and chalk ($CaCO_3$):

$$2NaCl + H_2SO_4 \rightarrow Na_2SO_4 + 2HCl$$
$$Na_2SO_4 + 2C + CaCO_3 \rightarrow Na_2CO_3 + CaS + 2CO_2$$

This process is often referenced as or associated with the beginnings of the industrial chemicals industry.

Sulfuric acid was also an essential chemical for the manufacture of dyes, bleaches, and alkalis and the production of the acid on a large scale required the development of lead-lined chambers that could resist the acid (and corrosive) gases that were formed when sulfur was burned with nitrates:

$$SO_2 + NO_2 + H_2O \rightarrow H_2SO_4 + NO$$
$$2NO + O_2 \rightarrow 2NO_2$$

This process was improved in the mid-19th century when towers to recycle the product gases were finally introduced. However, the transportation of sulfuric acid was dangerous, and alkali manufacturers tended to produce the acid on-site.

Sulfuric acid was also used in the manufacture of *superphosphates*, which were produced as fertilizers on a large scale by the mid-19th century. By that time, a solution was found for the complex engineering problems that had hampered the use of the alternative process to produce soda:

$$NH_3 + H_2O + CO_2 \rightarrow NH_4HCO_3$$
$$NaCl + NH_4HCO_3 \rightarrow NaHCO_3 + NH_4Cl$$
$$2NaHCO_3 \rightarrow Na_2CO_3 + H_2O + CO_2$$

In the meantime, the Solvay process using a tower in which carbon dioxide reacted efficiently with solid salts came on line. The Solvay process did not generate as much waste and pollution as the Leblanc process and had additional advantages, such as the following: (1) The raw materials for the Solvay process—brine and ammonia—were readily available, the latter from coal-based gasworks; (2) the process required less fuel; and (3) sulfur-containing feedstocks or nitrates were not required.

An early process used in the manufacture of inorganic chemicals involved the catalytic conversion of nitrogen and hydrogen to ammonia (the Haber process of the high-pressure variant (the Haber-Bosch) process).

$$N_2 + 3H_2 \rightarrow 2NH_3$$

The process uses an iron-based catalyst and is still the primary method for ammonia synthesis used in a modern plant. The process does require hydrogen, which can be produced from a variety of hydrocarbon sources, and nitrogen, which is supplied from air. The production of ammonia from coal-derived synthesis gas (carbon monoxide plus hydrogen) is another option. However, since the 1930s, the primary source of hydrogen for ammonia production. Approximately 2% (v/v) of the hydrogen required for the Haber process is obtained from electrolysis of brine at chlorine plants. The late 19th century and the early 20th century saw an explosion in both the quantity of production and the variety of chemicals that were manufactured, and as a result, large chemical industries also took shape in Germany and later in the United States.

A typical process configuration for production of ammonia involves the use of natural gas that is mixed with steam and charged to a primary reformer, where it is passed over a nickel catalyst. In the primary reformer, which operates at temperatures on the order of 700–815°C (1300–1500°F), most of the gas is converted to hydrogen, carbon monoxide, and carbon dioxide. The exiting gas is mixed with air and charged to a secondary reformer operating at higher temperatures, 900–925°C (1650–1700°F), where the remaining natural gas is converted. The gas leaving the secondary reformer contains nitrogen, hydrogen, carbon monoxide, and carbon dioxide. The reformed gas is cooled in a waste heat boiler where high-pressure, superheated steam is generated. The cooled gas is then charged to high- and low-temperature shift converters containing different catalysts to convert the carbon monoxide into carbon dioxide to obtain additional hydrogen.

$$CO + H_2O \rightarrow CO_2 + H_2$$

The mixture of gases is then charged to a carbon dioxide removal plant. Methods most commonly used for this purpose include absorption or wet scrubbing (e.g., with hot potassium carbonated or methyl diethanolamine). Outlet gas from the recovery plant is further purified through methanation and drying. The resulting pure synthesis gas is compressed and fed through heat exchangers to ammonia converters containing iron oxide catalysts (the Haber process). The gas stream is refrigerated to condense ammonia, and unreacted gases are recycled. The resulting product is anhydrous ammonia.

3.2.2 The Modern Industry

As the basis for the inorganic manufacture of inorganic chemicals, there are many different sources of raw materials available to the production of industrial chemicals. Very few of these feedstocks are available naturally in found in the elemental form; sulfur is a notable exception that occurs in underground deposits and can be brought to the surface by an in situ recovery process involving the use of compressed air after the sulfur has been melted by superheated steam. In addition, substantial quantities of sulfur are recovered from crude oil processing and natural gas processing where the sulfur occurs in the hydrocarbon feedstocks as an impurity and is removed during the gas and oil refining processes (Mokhatab et al., 2006; Speight, 2007, 2014, 2017a; Kidnay et al., 2011; Bahadori, 2014).

In terms of other naturally occurring feedstocks, the air is also a major source. Air contains molecular nitrogen and oxygen, and these two gases may be separated by liquefaction followed by fractional distillation; the inert gases such as argon are also separated at this point in the process. Salt or brine can be used as sources of chlorine and sometimes bromine, sodium hydroxide, and sodium carbonate (Na_2CO_3), whereas metals such as iron, aluminum, copper, or titanium as well as phosphors, potassium, calcium, and fluorine are obtained from mineral ores. Saltpeter (potassium nitrate, KNO_3) was once an important source of nitrogen compounds, but in the modern inorganic chemical industry, most of the ammonia and nitrate derivatives are produced synthetically from the nitrogen gas obtained from the air (above). Recovery and recycling provide increasing amounts of some metal feedstock, and as environmental concerns increase, these operations will probably become an important source of materials used in the manufacture of many inorganic chemicals.

The inorganic chemicals industry adds value to raw materials by transforming them into the chemicals required for the manufacture of consumer products. Since there are usually several different processes that can be used for this purpose, the chemical industry is associated with intense competition for new markets. It is made up of companies of different sizes, including several giants that are engaged in the transformation of some very basic raw materials into final products and medium-sized or small-sized companies that concentrate on very few of these steps. The closer to the raw material, the larger the scale of operations; such heavy inorganic chemicals are usually manufactured by continuous processes. At the other extreme in terms of scale are the firms that manufacture specialty chemicals, mostly in batch processes, from intermediate chemicals that correspond to chemicals that have already gone through several steps of synthesis and purification.

Basic chemicals represent the starting point for the manufacture of inorganic industrial chemicals. They are usually one step away from the raw materials and are produced on a very large scale employing continuous processes. The unit price of these products is relatively low, and producing them cheaply and efficiently is a major concern for the companies that manufacture them. Sulfur, nitrogen, phosphorus, and chlor-alkali industries are the main producers of basic inorganic chemicals, and they will often sell them to other industries as well as using them in the manufacture of their own end products. The basic principles for their production and major uses are indicated here for each of these industries.

It is the purpose of this section to present the major industrial processes that are used in the inorganic chemical industry, including the materials and equipment used and brief description of the processes. The section is designed for those interested in gaining a general understanding of the industry. This section does not attempt to replicate published engineering information that is available for this industry but specifically contains a description of commonly used production processes, associated raw materials, the by-products produced or released, and the materials either recycled or transferred off-site.

3.3 THE PRODUCTION OF INORGANIC CHEMICALS

The production of inorganic chemicals for a large (but not the only) part focuses on the production of a large quantity of inorganic products like chlor-alkali chemicals, sulfuric acid, sulfate derivatives, and fertilizers (such as potassium, nitrogen, and phosphorus products). The origins of the modern industry can be traced to the production of sulfuric acid and sodium carbonate that played an important role in the development of processes to produce the various inorganic chemicals that are in modern use.

3.3.1 Ammonia

Ammonia is a compound of nitrogen and hydrogen (NH_3) and exists as a colorless gas with a characteristic pungent smell that boils at −33.3°C (−28°F) and freezes to from white crystals at −77.7°C (−107.86°F). Ammonium hydroxide (NH_4OH) for general sales and use is a solution of ammonia in water.

$$NH_3 + H_2O \rightarrow NH_4OH$$

The concentration of such solutions is measured in units of the Baumé density scale, with 26 degrees Baumé approximately 30%, w/w ammonia at 15.5°C (59.9°F)) being the typical high-concentration commercial product.

Synthetic ammonia (NH_3) refers to ammonia that has been synthesized from hydrogen (produced from natural gas), which has been purified and reacted with nitrogen to produce ammonia.

$$N_2 + 3H_2 \rightarrow 2NH_3$$

In the process, anhydrous ammonia is synthesized by reacting hydrogen with nitrogen at a molar ratio of 3:1, then compressing the gas and cooling it to −33°C (−27°F). Nitrogen is obtained from the air, while hydrogen is obtained from either the catalytic steam reforming of natural gas (methane) or naphtha, or the electrolysis of brine at chlorine plants.

Steam reforming of natural gas—often referred to as steam methane reforming—is the most common method of producing hydrogen on a commercial scale. Thus, at high temperatures (700–1100°C, 1290–2010°F) and in the presence of a metal-based catalyst (such as nickel), steam reacts with methane to yield synthesis gas (syngas)—a mixture of carbon monoxide and hydrogen:

$$CH_4 + H_2O \rightleftharpoons CO + 3H_2$$

Additional hydrogen can be recovered by a lower-temperature gas-shift reaction (in the presence of a copper or iron catalyst) in which carbon monoxide is produced:

$$CO + H_2O \rightleftharpoons CO_2 + H_2$$

Although represented by these simple equations, six process steps are required to produce synthetic ammonia using the catalytic steam reforming method: (1) natural gas desulfurization, (2) catalytic steam reforming of natural gas, (3) carbon monoxide shift, (4) carbon dioxide removal, (5) methanation, and (6) ammonia synthesis. The first, third, fourth, and fifth steps remove impurities such as sulfur, carbon monoxide, carbon dioxide, and water from the feedstock, hydrogen, and synthesis gas streams (Mokhatab et al., 2006; Speight, 2007, 2014, 2017a; Kidnay et al., 2011; Bahadori, 2014). In the second step, hydrogen is manufactured, and nitrogen (as air) is introduced into this two-stage process.

In the synthesis step, the synthesis gas from the methanation unit is compressed at pressures ranging from 2000 to 5000 psi, mixed with recycled synthesis gas, and cooled to 0°C (32°F). Condensed ammonia is separated from the unconverted synthesis gas in a liquid-vapor separator and sent to a let-down separator. The unconverted synthesis is compressed and preheated to

180°C (356°F) before entering the synthesis converter, which contains an iron oxide catalyst. Ammonia from the exit gas is condensed and separated and then sent to the let-down separator. A small portion of the overhead gas is purged to prevent the buildup of inert gases (such as argon) in the circulating gas system. Ammonia in the let-down separator is flashed to 14.7 psi at −33°C (−27°F) to remove impurities from the liquid. The flash vapor is condensed in the let-down chiller where anhydrous ammonia is drawn off and stored at low temperature.

Approximately 75% (v/v) of the ammonia produced by this process is used as a feedstock for the fertilizer industry, either directly as ammonia or indirectly after synthesis as urea (H_2NCONH_2), ammonium nitrate (NH_4NO_3), and monoammonium phosphate ($NH_4H_2PO_4$) or diammonium phosphate ($(NH_4)_2HPO_4$). The remaining ammonia is used as raw material in the manufacture of polymeric resins, explosives, nitric acid, and other products.

Pollutants from ammonia manufacture are emitted from regeneration of the desulfurization bed, heating of the catalytic steam, regeneration of carbon dioxide scrubbing solution, and steam stripping of the process condensate. Many ammonia plants use activated carbon fortified with metallic oxide additives to desulfurize feedstocks. These beds are regenerated about once a month. The vented regeneration steam contains sulfur oxides and hydrogen sulfide, some hydrocarbons, and carbon monoxide.

Carbon dioxide is a by-product of the reaction and is removed from the synthesis gas by scrubbing with hot potassium carbonate or similar compounds. Regeneration of this scrubbing solution liberates water, ammonia, carbon monoxide, and volatile scrubbing solution compounds. Stripping of process condensate yields steam, which is vented to the atmosphere and contains ammonia, carbon dioxide, and methanol.

Wastewaters from manufacture of ammonia and agricultural chemicals consist mostly of wash water, scrubber water, boiler and vaporizer blow down, or stripper water. These may contain phosphorus, fluorides, ammonia, carbon dioxide, or weak acids. Many of these waters are treated and recycled to the process. Valuable components (e.g., ammonia) may also be recovered. Water scrubbing of the purge gases in ammonia production, for example, creates an ammonia water solution that can be used in another process (e.g., urea production).

Solid wastes from ammonia production, for example, include spent catalysts and any other additives that are removed and sent off-site for removal of valuable precious metals. Sulfur may be recovered in plants that use partial oxidation.

3.3.2 Ammonium Nitrate

Ammonium nitrate is the nitrate salt of the ammonium cation (NH_4NO_3, sometimes written as $N_2H_4O_3$) that is a white crystal solid and is highly soluble in water. It is predominantly used in agriculture as a high-nitrogen fertilizer and is also used as a component of explosive mixtures in mining, quarrying, and civil construction.

Ammonium nitrate (NH_4NO_3) is produced by neutralizing nitric acid (HNO_3) with ammonia (NH_3). All ammonium nitrate plants produce an aqueous ammonium nitrate solution through the reaction of ammonia and nitric acid in a neutralizer according to the following equation:

$$NH_3 + HNO_3 \rightarrow NH_4NO_3$$

The process involves several unit process operations including (1) solution formation and concentration, (2) solids formation, (3) finishing, (4) screening and coating, and (5) product bagging and/or bulk shipping. In some cases, solutions may be blended for marketing as liquid fertilizers. The number of operating steps employed depends on the specification of the product. For example, plants producing ammonium nitrate solutions alone use only the solution formation, solution blending and bulk shipping operations. Plants producing a solid ammonium nitrate product may employ all of the operations. Approximately 15%–20% (v/v) of the ammonium nitrate prepared in this manner is used for explosives and the balance for fertilizer.

Additives such as magnesium nitrate or magnesium oxide may be introduced into the melt prior to solidification to raise the crystalline transition temperature, act as a desiccant (removing water) or lower the temperature of solidification. Products are sometimes coated with clays or diatomaceous earth to prevent agglomeration during storage and shipment, although additives may eliminate the need for coatings. The final solid products are screened and sized, and off-size particles are dissolved and recycled through the process.

Ammonium nitrate is marketed in several forms, depending upon its use. For example, liquid ammonium nitrate may be sold as a fertilizer, generally in combination with urea or the liquid ammonium nitrate may be concentrated to form an ammonium nitrate *melt* for use in solids formation processes. Solid ammonium nitrate may be produced in the form of prills, grains, granules, or crystals. Ammonium nitrate prills can be produced in either high- or low-density form, depending on the concentration of the melt. High-density prills, granules, and crystals are used as fertilizer, while ammonium nitrate grains are used solely in explosives, and low-density prills that are small aggregates or globules of the material—most often a dry sphere—formed from a melted liquid. The term *prill* is also used in manufacturing to refer to a product that has been pelletized.

The manufacture of ammonium nitrate produces particulate matter, ammonia, and nitric acid emissions. Emissions from ammonia and nitric acid occur primarily when they form solutions (neutralizers and concentrators), and when they are used in granulators. Particulate matter is the largest source and is emitted throughout the process during the formation of solids. Prill towers and granulators are the largest sources of particulates. Microprills can form and clog orifices, increasing fine dust loading and emissions.

Emissions occur from screening operations by the banging of ammonium nitrate solids against each other and the screens. Most of these screening operations are enclosed or have partial covers to reduce emissions. The coating of products may also create some particulate emissions during mixing in the rotary drums. This dust is usually captured and recycled to coating storage. Another source of dust is bagging and bulk loading, mostly during final filling when dust-laden air is displaced from bags.

Plants producing nitric acid and ammonium nitrate produce wastewaters containing these compounds and ammonia. Wastewater containing ammonia and nitric acid must be neutralized to produce ammonium nitrate.

3.3.3 Ammonium Phosphate

Ammonium phosphate is the salt of ammonia and phosphoric acid ($(NH_4)_3PO_4$), which is difficult to produce because of the inherent instability:

$$(NH_4)_3PO_4 \rightarrow (NH_4)_2HPO_4 + NH_3$$

In contrast to the unstable nature of the triammonium salt, diammonium phosphate ($(NH_4)_2HPO_4$) is a valuable material that finds major use in the fertilizer industry (Schrödter et al., 2008).

The phosphate fertilizer industry is divided into three segments: (i) phosphoric acid and super phosphoric acid, (ii) normal superphosphate and triple superphosphate, and (iii) granular ammonium phosphate. The focus of this subsection is on normal superphosphate, triple superphosphate, and ammonium phosphate. The term *normal superphosphate* (*normal superphosphate*) refers to fertilizer material containing 15%–21% (w/w) phosphorous as phosphorous pentoxide (P_2O_5). By further definition, superphosphate contains a large percentage of phosphate, but by definition, normal superphosphate contains not more than 22% (w/w) of available phosphorous pentoxide.

Ammonium phosphate ($NH_4H_2PO_4$) is produced by reacting phosphoric acid (H_3PO_4) with anhydrous ammonia (NH_3):

$$NH_3 + H_3PO_4 \rightarrow NH_4H_2PO_4$$

On the other hand, ammoniated superphosphates are produced by adding normal superphosphate or triple superphosphate to the mixture. The production of liquid ammonium phosphate and ammoniated superphosphates in fertilizer mixing plants is considered a separate process.

Normal superphosphates are prepared by reacting ground phosphate rock with 65%–75% sulfuric acid; an important factor in the production of normal superphosphate derivatives is the amount of iron and aluminum in the phosphate rock. Aluminum (as aluminum oxide or alumina, Al_2O_3) and iron (as ferric oxide, Fe_2O_3) amounts in excess of 5% (w/w) impart an extreme stickiness to the superphosphate, which makes the superphosphate difficult to handle. The two general types of sulfuric acid used in superphosphate manufacture are unused acid (often referred to as virgin acid) and spent acid. Virgin acid is produced from elemental sulfur, pyrites, and industrial gases and is relatively pure, while spent acid is a recycled waste product from various industries that use large quantities of sulfuric acid. Problems encountered with using spent acid include unusual color, unfamiliar odor, and toxicity.

In the process, ground phosphate rock and acid are mixed in a reaction vessel, held in an enclosed area for approximately 30 min until the reaction is partially completed, and then transferred, using an enclosed conveyer known as *the den*, to a storage pile for curing (the completion of the reaction). Following curing, the product is most often used as a high-phosphate additive for the production of granular fertilizers. It can also be granulated for sale as granulated superphosphate or granular mixed fertilizer.

To produce granulated normal superphosphate, cured superphosphate is fed through a clod breaker (lump breaker and lump crusher) and sent to a rotary drum granulator where steam, water, and acid may be added to aid in the granulation process. Rotating drum granulators are open-ended, slightly inclined rotary cylinders with a cutter mounted inside. A bed of dry material is maintained in the unit, while slurry is introduced through pipes under the bed. The product is then processed through a rotary drum granulator, a rotary dryer, a rotary cooler, and is then screened to specification and, finally, stored in bagged or bulk form prior to being sold. Two processes have been used to produce triple superphosphate: run-of-the-pile triple superphosphate (ROP-TSP) and granular (GTSP).

The ROP-TSP material is produced in a manner similar to normal superphosphate and appears as a pulverized product of variable particle size produced. In the process, wet-process phosphoric acid (50%–55% (w/w) phosphorus pentoxide) is reacted with ground phosphate rock in a cone mixer after which the resultant slurry begins to solidify on a slow-moving conveyer en route to the curing area. At the point of discharge from the den, the material passes through a rotary mechanical cutter that breaks up the solid mass. The coarse ROP-TSP product is sent to a storage pile and cured for 3–5 weeks. The product is then mined from the storage pile to be crushed, screened, and shipped in bulk. Granular triple superphosphate yields larger, more uniform particles with improved storage and handling properties. Most of Dorr-Oliver slurry granulation process material is manufactured by the Dorr-Oliver slurry granulation process.

Major emissions from wet-process phosphoric acid manufacture are composed of gaseous fluorides in the form of silicon tetrafluoride (SiF_4) and hydrogen fluoride (HF). The source of fluorides is phosphate rock, which contains from 3.5% to 4.0% (w/w) fluorine. The fluorine is generally precipitated out with gypsum, leached out with phosphoric acid product, or vaporized in the reactor or evaporator. The reactor where phosphate rock is contacted with sulfuric acid is the primary source of emissions. Vacuum flash cooling of the reactor slurry will minimize these emissions as the system is closed. During acid concentration, 20%–40% of the fluorine in the rock may vaporize.

Scrubbers (venturi, wet cyclonic, and semicross flow) are used to control emissions of fluorine. Leachate fluorine may settle in settling ponds, and if the water becomes saturated, it will be emitted to the air as fluorine gas. Thermal or furnace processing of phosphoric acid results in phosphoric acid mist, which is contained in the gas stream exiting the hydrator. A large amount of phosphorus pentoxide product may be present as liquid phosphoric acid particles suspended in the gas stream, so most plants attempt to control this loss. Control equipment includes venturi scrubbers, cyclonic separators with wire-mesh mist eliminators, fiber mist eliminators, high energy wire-mesh contactors, and electrostatic precipitators.

Normal superphosphate manufacture produces emissions of gaseous fluorides in the form of silicon tetrafluoride (SiF_4) and hydrogen fluoride (HF). Particulates composed of fluoride and phosphate material are also emitted. Sources include rock unloading and feeding, mixing operations, storage, and fertilizer handling.

Sources of emissions for triple superphosphates manufacture include the reactor, granulator, dryer, screens, cooler, mills, and transfer conveyors. Particulates may be emitted during unloading, grinding, storage, and transfer of ground phosphate rock. Baghouses, scrubbers, or cyclonic separators are used to control emissions.

3.3.4 Ammonium Sulfate

Ammonium sulfate ($(NH_4)_2SO_4$) is an inorganic salt with a number of commercial uses. The most common use is as a soil fertilizer since the chemical contains 21% (w/w) nitrogen and 24% (w/w) sulfur. The chemical has also been used in flame retardant chemicals because, as a flame retardant, it increases the combustion temperature of the material, decreases maximum weight loss rates, and causes an increase in the production of residue or char. Ammonium sulfate has been used as a wood preservative, but due to its hygroscopic nature, this use has been largely discontinued because of associated problems with metal fastener corrosion, dimensional instability, and finish failures.

Ammonium sulfate is produced by three different processes: (1) synthetic manufacture from pure ammonia and concentrated sulfuric acid, (2) as a coke oven by-product, (3) from ammonia scrubbing of tail gas at sulfuric acid (H_2SO_4) plants, and (4) as a by-product of caprolactam ($(CH2)_5COHN$) production,

Typically, ammonium sulfate is produced by combining anhydrous ammonia and sulfuric acid:

$$2NH_3 + H_2SO_4 \rightarrow (NH_4)_2SO_4$$

In the process, a mixture of ammonia gas and water vapor is introduced into a reactor that contains a saturated solution of ammonium sulfate and about 2%–4% (v/v) of free sulfuric acid at 60°C (140°F); the heat of the reaction maintains the desired temperature. Concentrated sulfuric acid is added to keep the solution acidic and to retain its level of free acid. Dry, powdered ammonium sulfate may be formed by spraying sulfuric acid into a reaction chamber filled with ammonia gas. The heat of reaction evaporates all of the water present in the system with the resulting formation of as dry powdery salt.

Ammonium sulfate crystals are formed by circulating the solution through an evaporator where it thickens. A centrifuge separates the crystals from the mother liquor. The crystals contain 1%–2.5% moisture and are dried in a fluidized bed or rotary drum dryer. Dryer exhaust gases are sent to a particulate collection system (e.g., wet scrubber) to control emissions and recover residual product. Coarse and fine granules are separated by screening before they are stored or shipped.

Ammonium sulfate also is manufactured from gypsum ($CaSO_4 \cdot 2H_2O$). In this process, finely divided gypsum is added to a solution of ammonium carbonate causing calcium carbonate precipitates as a solid, leaving ammonium sulfate in the solution:

$$(NH_4)_2CO_3 + CaSO_4 \rightarrow (NH_4)_2SO_4 + CaCO_3$$

After formation of the ammonium sulfate solution, manufacturing operations of each process are similar. Ammonium sulfate crystals are formed by circulating the ammonium sulfate liquor through an evaporator. Evaporation of the water thickens the solution and ammonium sulfate crystals are separated from the liquor in a centrifuge.

Ammonium sulfate also occurs naturally as the rare mineral mascagnite in volcanic fumaroles and due to coal fires on some dumps. A fumarole (or *fumarole*; smoke hole) is an opening in the crust of the Earth and is often found in areas surrounding volcanoes, which emits steam (forms when superheated water vaporizes as its pressure drops when it emerges from the ground) and gases such as carbon dioxide, sulfur dioxide, and hydrogen sulfide.

Particulate ammonium sulfate is the air emission occurring in the largest amount from manufacture of this fertilizer. Dryer exhaust is the primary source of the particulates, and emission rates are dependent on gas velocity and particle size distribution. Particulate rates are higher for fluidized bed dryers than for the rotary drum type of dryer. Most plants use baghouses to control particulates of ammonium sulfate, although venturi and centrifugal wet scrubbers are better suited for this purpose.

Some volatile carbon emissions may be present in caprolactam plants where ammonium sulfate is produced as a by-product.

3.3.5 Brine Purification

By way of definition, brine is a solution of salt (usually sodium chloride, NaCl) in water with varying properties, depending on the concentration of the sodium chloride (Table 3.3). In different contexts, brine may refer to salt solutions ranging from about 3.5% (w/w) (a typical concentration of seawater or the lower end of solutions used for preserving food) up to

Table 3.3 Properties of Brine Solutions (Sodium Chloride in Water)

NaCl % (w/w)	Freezing Point (°C)	Density (g/cm^3)	Refractive Index at 589 nm	Viscosity (cP)
0	0	0.99984	1.333	1.002
0.5	−0.3	1.0018	1.3339	1.011
1	−0.59	1.0053	1.3347	1.02
2	−1.19	1.0125	1.3365	1.036
3	−1.79	1.0196	1.3383	1.052
4	−2.41	1.0268	1.34	1.068
5	−3.05	1.034	1.3418	1.085
6	−3.7	1.0413	1.3435	1.104
7	−4.38	1.0486	1.3453	1.124
8	−5.08	1.0559	1.347	1.145
9	−5.81	1.0633	1.3488	1.168
10	−6.56	1.0707	1.3505	1.193
12	−8.18	1.0857	1.3541	1.25
14	−9.94	1.1008	1.3576	1.317
16	−11.89	1.1162	1.3612	1.388
18	−14.04	1.1319	1.3648	1.463
20	−16.46	1.1478	1.3684	1.557
22	−19.18	1.164	1.3721	1.676

about 26% (w/w) (a typical saturated solution, depending on the temperature). Other levels of concentration (the degree of water salinity) are called by different names:

Fresh water: <0.05% (w/w) dissolved salts

Brackish water: 0.05%–3.0% (w/w) dissolved salts

Saline water: 3%–5% (w/w) dissolved salts

Brine: >5% (w/w) dissolved salts

Not only brine is regarded as tools for pressure support in the reservoir, but also the chemical properties of brine have been also taken into consideration, and the use of brine has also proved to be an asset as a starting material in the inorganic chemicals industry. In addition, knowledge of the properties of brine is important in the development and design of desalination systems. Literature contains many data for the properties of seawater, but only a few sources provide full coverage for these properties (Riley and Skirrow, 1975; Fabuss, 1980; El-Dessouky and Ettouney, 2002; Sharqawya et al., 2010).

Brine plays an important role in the hydrogen and chlorine industries; approximately 4% of the hydrogen gas produced worldwide is created by electrolysis, and most of this hydrogen produced through electrolysis is a side product (along with sodium hydroxide (caustic soda, NaOH)) in the production of chlorine:

$$2NaCl(aq) + 2H_2O(l) \rightarrow 2NaOH(aq) + H_2(g) + Cl_2(g)$$

Prior to the introduction of brine to the electrolytic cells, brine solutions are typically treated with several chemicals to remove impurities prior to input to the electrolytic cells. In the case of membrane cell systems, the brine is first acidified with HCl to remove dissolved chlorine after which sodium hydroxide (NaOH) and sodium carbonate (Na_2CO_3) are added to precipitate calcium and magnesium ions as calcium carbonate and magnesium hydroxide. Barium carbonate ($BaCO_3$) is then added to remove sulfates that precipitate out as barium sulfate ($BaSO_4$). The precipitants are removed from the brine solution by settling and filtration.

Impurities primarily consist of calcium, magnesium, barium, iron, aluminum, sulfate derivatives, and trace metals that can significantly reduce the efficiency of any enduing process. Thus, the necessary removal of impurities accounts for a significant portion of the overall costs of chlor-alkali production, especially in the membrane process. In addition to the dissolved natural impurities, chlorine must be removed from the recycled brine solutions used in mercury and membrane processes. Dissolved chlorine gas entering the anode chamber in the brine solution will react with hydroxide ions formed at the cathode to form chlorate, which reduces product yields. In addition, chlorine gas in the brine solution will cause corrosion of pipes, pumps, and containers during further processing of the brine.

In a typical chlorine plant, hydrochloric acid is added to the brine solution leaving the cells to liberate the chlorine gas. A vacuum is applied to the solution to collect the chlorine gas for further treatment. To further reduce the chlorine levels, sodium sulfite (Na_2SO_3) or another reducing agent is added to remove the final traces of chlorine after which the dechlorinated brine is then resaturated with solid salt before further treating to remove impurities.

Brine solution is typically heated before treatment to improve reaction times and precipitation of impurities. Calcium carbonate impurities are precipitated out through treatment with sodium carbonate; magnesium, iron, and aluminum are precipitated out through treatment with sodium hydroxide; and sulfates are precipitated out through the addition of calcium chloride or barium carbonate. Most trace metals are also precipitated out through these processes. Flocculants are sometimes added to the clarifying equipment to improve settling. The sludge generated in this process is washed to recover entrained sodium chloride. Following the clarification steps, the brine solution is typically passed through sand filters followed by polishing filters. The brine passing through these steps will contain less than 4 ppm calcium and 0.5 ppm magnesium, which is sufficient purification for the diaphragm and mercury cell processes.

Brine preparation and caustic evaporation processes release emissions through the combustion of fuels in process heaters and in boilers that produce process steam. When operating in an optimum condition and burning cleaner fuels (e.g., natural gas and refinery gas), these heating units create relatively low emissions of SO_x, NO_x, CO, particulates, and volatile hydrocarbon emissions.

Ion exchange wash water from membrane cell processes usually contains dilute hydrochloric acid with small amounts of dissolved calcium, magnesium, and aluminum chloride. This wastewater is usually treated along with other acidic wastewaters by neutralization. Wastewater streams are generated from mercury cells during the chlorine drying process, brine purge, and from other sources. Mercury is present in the brine purge and other sources (floor sumps and cell wash water) in small amounts. This mercury is generally present in concentrations ranging from 0 to 20 ppm, and is precipitated using sodium hydrosulfide (NaSH) to form mercuric sulfide (HgS). The mercuric sulfide is removed from the aqueous phase by passage through a filter before the water is discharged.

Brine purification results in brine mud, one of the largest waste streams from the chlor-alkali industry. About 30 kg of brine mud is generated for every 1000 kg of chlorine produced, but this varies with the purity of the salt used to produce the brine. Prepurified salts, for example, will generate only about 0.7–6.0 kg per 1000 kg of chlorine produced. The brine mud contains a variety of compounds, typically magnesium hydroxide and calcium carbonate formed during the addition of compounds to purify the brine. The sludge or brine mud containing these impurities must be disposed of in a landfill. If a mercury process is being used, the brine mud may contain trace levels of mercury. In this case, the sludge is treated with sodium sulfide to create mercury sulfide, an insoluble compound. The sludge is further treated by casting it into concrete blocks, which are treated for leachability and sent to a controlled landfill.

Mercury cell brine muds may also contain mercury in elemental form or as mercuric chloride. These muds are considered hazardous and must be disposed of in a RCRA Subtitle C landfill after treatment with sodium sulfide, which creates an insoluble sulfide compound. Other brine muds are segregated and stored in lagoons, which are periodically dredged or drained and covered over, and the waste is usually landfilled off-site (subject to local environmental regulations). The processing of hydrogen gas from mercury cells also creates hazardous waste. In this process, small amounts of mercury present in the hydrogen gas are extracted by cooling the gas. A large part of the condensed mercury is removed in this fashion and returned to the electrolytic cell. Some facilities use activated carbon treatment to further purify the hydrogen of mercury, and the spent carbon is shipped off-site for disposal as a hazardous waste.

3.3.6 Caustic Soda Processing

Caustic soda (NaOH, more formally known as sodium hydroxide and less formally known as lye) is a white solid and is available in pellets, flakes, granules, and as prepared solutions at different concentrations. Sodium hydroxide forms an approximately 50% (w/w) saturated solution with water. It is also soluble in ethanol (ethyl alcohol, C_2H_5OH) and methanol (methyl alcohol, CH_3OH). Caustic soda is deliquescent and readily absorbs moisture and carbon dioxide when stored unprotected in air.

Caustic soda solution generated from chlor-alkali processes is typically processed to remove impurities and, in the case of the diaphragm and membrane processes, is concentrated either to a 50% or 73% (v/v) water-based solution or to anhydrous caustic soda. The water vapor from the evaporators is condensed in barometric condensers and, in the case of the diaphragm process, will primarily contain about 15% (v/v) caustic soda solution and high concentrations of salt.

If sodium sulfate is not removed during the brine purification process, salt recovered from the evaporators is often recrystallized to avoid sulfate buildup in the brine. If the salt is recrystallized, the wastewater from sodium hydroxide processing will also contain sodium sulfate. Significant levels of copper may also be present in the wastewater due to corrosion of pipes and other equipment. Wastewater from the membrane process contains caustic soda solution and virtually no salt or sodium sulfate. Caustic soda processing wastewater is typically neutralized with hydrochloric acid, sent to a lagoon, and then discharged directly to a receiving water or land disposed. The caustic soda generated from the mercury process only requires filtration to remove mercury droplets that are typically recovered for reuse.

Caustic evaporation, where the sodium hydroxide solution is concentrated to a 50% or 70% solution, evaporates about 5 tons of water per ton of 50% caustic soda produced. The water vapor from the evaporators is condensed, and in the case of the diaphragm process, will contain about 15% caustic soda solution and a relatively high salt content of 15%–17%. If sodium sulfate is not removed during the brine purification process, salt recovered from evaporators may be recrystallized to avoid buildup of sulfate in the brine. If the salt is recrystallized, the wastewater may contain sodium sulfate.

3.3.7 Chlor-Alkali Chemicals

The term chlor-alkali chemicals refer to the two chemicals (chlorine and an alkali) that are simultaneously produced as a result of the electrolysis of a salt-water. The most common chlor-alkali chemicals are chlorine and sodium hydroxide (caustic soda) but can include potassium hydroxide and hydrochloric. Thus, chlor-alkali chemical industry is a basic raw material industry producing caustic soda, chlorine, and hydrogen from salt and electricity. Industries producing sodium carbonate (or soda ash) and its derivatives and compounds based on calcium oxide (or lime) are usually included under this category. As both sodium hydroxide and chlorine have a common raw material, sodium chloride, they are produced in quantities that reflect their equal molar ratio, irrespective of the market for either product. Since these chemicals are produced by electrolysis, they require a cheap source of brine and electricity:

$$2NaCl + 2H_2O \rightarrow 2NaOH + 2HCl$$

Most processes are based on the electrolysis of a sodium chloride solution, but some plants operate with the molten salt. Three different cell types are used in electrolysis in water: mercury cells, diaphragm cells, and membrane cells. Membrane cells are replacing the other two types in modern units, but it may not be economically feasible to convert older plants

The chlor-alkali industry produces mainly chlorine, caustic soda (sodium hydroxide), soda ash (sodium carbonate), sodium bicarbonate, potassium hydroxide, and potassium carbonate. Chlorine and caustic soda are coproducts produced in about equal amounts primarily through the electrolysis of salt (brine). Much of the domestic chlorine production is used in the manufacturing of organic chemicals including: vinyl chloride monomer ($CH_2{=}CHCl$), ethylene dichloride ($CH_2ClCH2Cl$), glycerine

($CH_2OHCHOHCH_2OH$), glycol (CH_2OHCH_2OH), chlorinated solvents, and chlorinated methane derivatives (CH_xCl_y, where $x+y=4$). Other major uses are disinfection treatment of water and the production of hypochlorite (—OCl) derivatives. The primary uses of caustic soda are in industrial processes, neutralization, and off-gas scrubbing; as a catalyst; and in the production of alumina, propylene oxide, polycarbonate resin, epoxies, synthetic fibers, soaps, detergents, rayon, and cellophane. The pulp and paper industry uses caustic soda for pulping wood chips, and other processes. Caustic soda is also used in the production of soaps and cleaning products, and in the petroleum and natural gas extraction industry as a drilling fluid.

Since both sodium hydroxide and chlorine have a common raw material (sodium chloride), they are produced in quantities that reflect their equal molar ratio, irrespective of the market for either product. Since they are produced by electrolysis, they require a cheap source of brine and electricity:

$$2NaCl + 2H_2O \rightarrow 2NaOH + Cl_2 + H_2$$

Most processes are based on the electrolysis of a sodium chloride solution, but some plants operate with the molten salt. Three different cell types are used in electrolysis in water: (1) a mercury cell, (2) a diaphragm cell, and (3) a membrane cell. There has been a tendency to use a membrane cells in modern process units in place of the mercury cell and the diaphragm cell, which may not be economically feasible in older plants.

Sodium hydroxide and sodium carbonate are alternative sources of alkali, and their use has followed the availability of raw materials and the efficiency of processes developed for their production. Both feedstocks require sodium chloride and energy, and if limestone ($CaCO_3$) deposits are also available, sodium carbonate may be produced by the Solvay process. Limestone can also be used to produce calcium oxide (quicklime, CaO) and calcium hydroxide (slaked lime, $Ca(OH)_2$). Calcium oxide may also be obtained by heating limestone at 1200–1500°C (2190–2732°F) limestone, while calcium hydroxide, which is more convenient to handle, is obtained by adding water to calcium oxide:

$$CaCO_3 \rightarrow CaO + CO_2$$
$$CaO + H_2O \rightarrow Ca(OH)_2$$

The principal use of calcium oxide is in steelmaking, and it is also used for the manufacture of chemicals, water treatment units, and pollution control. In the Solvay process, calcium carbonate and sodium chloride are used to produce calcium chloride and sodium carbonate with ammonia (which is recycled) as a medium for dissolving and carbonating the sodium chloride and calcium hydroxide for precipitating calcium chloride from the solution.

The Solvay process results in the production of soda ash (sodium carbonate, Na_2CO_3) from brine (as a source of sodium chloride, NaCl) and from limestone (as a source of calcium carbonate, $CaCO_3$):

$$2NaCl + CaCO_3 \rightarrow Na_2CO_3 + CaCl_2$$

In the first step in the process, carbon dioxide (CO_2) is passed through a concentrated aqueous solution of sodium chloride (NaCl) and ammonia (NH_3):

$$NaCl + CO_2 + NH_3 + H_2O \rightarrow NaHCO_3 + NH_4Cl$$

The reaction is carried out by passing concentrated brine (salt water) through two towers: (1) In the first tower, ammonia bubbles up through the brine and is absorbed by it, while (2), in the second tower, carbon dioxide bubbles up through the ammoniated brine and sodium bicarbonate (baking soda, NaHCO3) precipitates out of the solution; in a basic solution, sodium bicarbonate is less water-soluble than sodium chloride. The ammonia (NH_3) buffers the solution at a basic pH (>7.0), but without the ammonia, a hydrochloric acid by-product would render the solution acidic and interfere with (halt) the precipitation.

The ammonia for the reaction is reclaimed in a later step, and relatively, little ammonia is consumed. The carbon dioxide required for the reaction is produced by heating (calcination) of the limestone at 950–1100°C (1740–2010°F).

Briefly, calcination (also called *roasting*, *firing*, or *burning*) is the process of subjecting a substance to the action of heat, but without fusion, for the purpose of causing some change in its physical or chemical constitution. The object of the process is (i) to drive off water, present as absorbed moisture, as water of crystallization, or as water of constitution; (ii) to drive off carbon dioxide, sulfur dioxide, or other volatile constituents; and (iii) to oxidize a part or the whole of the substance. The process is carried on in furnaces, retorts, or kilns, and very often the material is raked over or stirred, during the process, to secure uniformity in the product.

At the high temperature of the calcination process, the calcium carbonate ($CaCO_3$) in the limestone is partially converted to quicklime (calcium oxide, CaO) and carbon dioxide:

$$CaCO_3 \rightarrow CO_2 + CaO$$

The sodium bicarbonate ($NaHCO_3$) that precipitates out in reaction is filtered out from the hot ammonium chloride (NH_4Cl) solution, and the solution is then reacted with the quicklime (calcium oxide, CaO) left over from heating the limestone:

$$2NH_4Cl + CaO \rightarrow 2NH_3 + CaCl_2 + H_2O$$

Calcium oxide produced a strongly basis aqueous solution and the ammonia this reaction is recycled back to the initial brine solution of reaction.

The sodium bicarbonate ($NaHCO_3$) precipitate is then converted to the final product, sodium carbonate (washing soda, Na_2CO_3), by heating at 160–230°C (320–610°F), which also produces water and carbon dioxide:

$$2NaHCO_3 \rightarrow Na_2CO_3 + H_2O + CO_2$$

The carbon dioxide from step is recovered for reuse.

Sodium hydroxide has many different uses in the chemical industry. Considerable amounts are used in the manufacture of paper and to make sodium hypochlorite for use in disinfectants and bleaches. Chlorine is also used to produce vinyl chloride, the starting material for the manufacture of polyvinyl chloride (PVC) and in water purification. Hydrochloric acid, which may be prepared by the direct reaction of chlorine and hydrogen gas or by the reaction of sodium chloride and sulfuric acid, is used as a chlorinating agent for metals and organic compounds.

In certain regions of the world, there are salt deposits or brines that have been enriched by bromine. Commercially, bromine may be extracted by treating the brines with chlorine and removing it by steam:

$$2Br^- + Cl_2 \rightarrow Br_2 + 2Cl^-$$

Bromine is used in water disinfection, in bleaching fibers and silk, and in the manufacture of medicinal bromine compounds and dyestuffs.

Air emissions from brine electrolysis include chlorine gas emissions (both fugitive and point source) and other vapors. Fugitive emissions arise from cells, scrubbers, and vents throughout the system. While individual leaks may be minor, the combination of fugitive emissions from various sources can be substantial. In 1995, nearly 3 million pounds of chlorine fugitive and point source emissions were reportedly released by the inorganic chemical industry. These emissions are controlled through leak-resistant equipment modifications, source reduction, and programs to monitor such leaks.

Diaphragm cells and membrane cells release chlorine as fugitive emissions from the cell itself and in process tail gases, which are scrubbed with aqueous soda ash (sodium carbonate, Na_2CO_3) or aqueous sodium hydroxide (NaOH) to remove chlorine.

$$Na_2CO_3 + Cl_2 \rightarrow NaCl + NaOCl + CO_2$$

$$2NaOH + Cl_2 \rightarrow NaCl + NaOCl + H_2O$$

The spent solution from this wash is neutralized and then discharged to water treatment facilities. Mercury cells release small amounts of mercury vapor and chlorine gas from the cell itself. Process tail gases from chlorine processing, caustic soda processing, and hydrogen processing also release small amounts. Mercury is removed from the hydrogen gas stream by cooling followed by absorption with activated carbon.

3.3.8 Chlorine Processing

Chlorine is a yellow-green gas at room temperature and is an extremely reactive element as well as strong oxidizing agent. It has the highest electron affinity and the third-highest electronegativity after oxygen and fluorine. The most common compound of chlorine, sodium chloride (common salt, NaCl) has been known since ancient times. The high oxidizing potential of elemental chlorine led to the development of commercial bleach and disinfectants and it is reagent for many processes in the chemical industry. Chlorine is used in the manufacture of a wide range of consumer products, about two-thirds of them organic chemicals such as PVC, and many intermediates for the production of plastics and other end products that do not contain the element.

Elemental chlorine is commercially produced from brine by electrolysis. The chlorine gas produced by electrolytic processes is saturated with water vapor. Chlorine gas from the diaphragm process also contains liquid droplets of sodium hydroxide and salt solution. In the process, chlorine gas recovered from electrolytic cells is cooled to remove water vapor. The condensed water is usually recycled as brine make-up although some facilities combine this waste stream with other waterborne waste streams prior to treatment.

The remaining water vapor is removed by scrubbing the chlorine gas with concentrated sulfuric acid. The chlorine gas is then compressed and cooled to form liquid chlorine. The majority of the spent sulfuric acid waste is shipped off-site for refortification to concentrated sulfuric acid or for use in other processes. The remainder is used to control effluent pH and/or is discharged to water or land disposed. The process of purifying and liquefying impure chlorine gas involves the absorption of the chlorine in a stream of carbon tetrachloride. The chlorine is subsequently removed in a stripping process in which the carbon tetrachloride either is recovered and reused or is vented to the atmosphere.

The first steps in processing the chlorine to a usable product consists of cooling the chlorine to less than 10°C and then passing it through demisters or electrostatic precipitators to remove water and solids. In the next step, chlorine is passed through packed towers with concentrated sulfuric acid flowing countercurrently. The water vapor is absorbed by the sulfuric acid and the dry chlorine gas is then passed through demisters to remove sulfuric acid mist. If the chlorine is to be liquefied, liquid chlorine is then added to the gas to further purify the chlorine and to prechill it prior to compression. Prechilling is primarily carried out to prevent the temperature from reaching the chlorine-steel ignition point during compression.

During chlorine gas processing, water vapor is removed by scrubbing with concentrated sulfuric acid. Most of this wastewater is shipped off-site for processing into concentrated sulfuric acid or for use in other processes. The remainder is used for pH control or discharged to water treatment facilities for disposal.

3.3.9 Hydrochloric Acid

Hydrochloric acid (HCl, also known as muriatic acid) is a colorless corrosive, strong mineral acid with many industrial uses among which, when it reacts with an organic base it forms a hydrochloride salt. Hydrochloric acid was historically produced from rock salt and green vitriol and later from the chemically similar common salt (NaCl) and sulfuric acid:

$$2NaCl + H_2SO_4 \rightarrow 2HCl + Na_2SO_4$$

Hydrochloric acid is a versatile chemical that hydrochloric acid is used in the chemical industry as a chemical reagent in the large-scale production of vinyl chloride ($CH_2{=}CHCl$) for PVC plastic, and polyurethane. It has numerous other industrial uses such as (i) hydrometallurgical processing, for example, production of alumina and/or titanium dioxide; (ii) chlorine dioxide synthesis; (iii) hydrogen production; (iv) activation of petroleum wells; (v) miscellaneous cleaning/etching operations including metal cleaning (e.g., steel pickling); and (vi) being used by masons to clean finished brick work. In the context of this book, the term *hydrogen chloride* is used to describe the gaseous form while the term *hydrogen acid* is used to describe a solution of hydrogen chloride in water.

Hydrochloric acid may be manufactured by several different processes; however, most of the hydrochloric acid (hydrogen chloride) produced in the United States is a by-product of the chlorination reaction. After leaving the chlorination process, the gas stream containing hydrogen chloride proceeds to the absorption column, where concentrated liquid hydrochloric acid is produced by absorption of hydrogen chloride vapors into a weak solution of hydrochloric acid. The hydrogen chloride-free chlorination gases are removed for further processing. The liquid acid is then either sold or used elsewhere in the plant. The final gas stream is sent to a scrubber to remove the remaining hydrogen chloride prior to venting.

In the chlor-alkali industry, brine solution (a mixture of sodium chloride and water) is electrolyzed to produce chlorine (Cl_2), sodium hydroxide, and hydrogen (H_2):

$$2NaCl + 2H_2O \rightarrow Cl_2 + 2NaOH + H_2$$

The pure chlorine gas can be combined with hydrogen to produce hydrogen chloride in the presence of UV light:

$$Cl_2(g) + H_2(g) \rightarrow 2HCl(g)$$

The resulting hydrogen chloride gas is absorbed in deionized water to produce chemically pure hydrochloric acid.

3.3.10 Hydrofluoric Acid

Hydrofluoric acid is a solution of hydrogen fluoride (HF) in water and is a precursor to almost all fluorine compounds. It is a colorless solution that is highly corrosive, capable of dissolving many materials, especially oxide and its ability to dissolve glass has been known since the 17th century. Because of the high reactivity toward glass and moderate reactivity toward many metals, hydrofluoric acid is usually stored in plastic containers (although PTFE is slightly permeable to it). Hydrogen fluoride gas is an acute poison that may immediately and permanently damage lungs and the corneas of the eyes.

Aqueous hydrofluoric acid is a contact-poison with the potential for deep, initially painless burns and ensuing tissue death.

Thus, hydrogen fluoride (HF) is produced in two forms, as anhydrous hydrogen fluoride (the gaseous form, HF(g)) and as aqueous hydrofluoric acid (the aqueous solution form, HF(aq)). The predominant form manufactured is anhydrous hydrogen fluoride, a colorless or gas that fumes on contact with air and is water-soluble. Traditionally, hydrofluoric acid (HF) has been used to etch and polish glass. Currently, the largest use for HF is in aluminum production. Other HF uses include uranium processing, petroleum alkylation, and stainless steel pickling. Hydrofluoric acid is also used to produce fluorocarbons used in aerosol sprays and in refrigerants. Although fluorocarbons are heavily regulated due to environmental concerns, other applications for fluorocarbons include manufacturing of resins, solvents, stain removers, surfactants, and pharmaceuticals.

In the process, hydrogen fluoride is manufactured by the reaction of acid-grade fluorspar (calcium fluoride, CaF_2) with sulfuric acid (H_2SO_4):

$$CaF_2 + H_2SO_4 \rightarrow CaSO_4 + 2HF$$

This endothermic reaction requires 30–60 min in horizontal rotary kilns externally heated to 200–250°C (390–480°F).

To accomplish this, dry fluorspar and a slight excess of sulfuric acid are fed continuously to the front end of a stationary prereactor (to ensure thorough contact by mixing the components prior to charging to the rotary kiln.) or directly to the kiln by a screw conveyor. Calcium sulfate ($CaSO_4$) is removed through an air lock at the opposite end of the kiln, and the gaseous reaction products—hydrogen fluoride and excess sulfuric acid from the primary reaction and silicon tetrafluoride (SiF_4), sulfur dioxide (SO_2), carbon dioxide (CO_2), and water produced in various secondary reactions—are removed from the front end of the kiln along with entrained particulate matter. The particulate matter is removed from the gas stream by a dust separator and returned to the kiln, while sulfuric acid and water are removed by a precondenser. Hydrogen fluoride vapors are then condensed in refrigerant condensers as a crude grade of the product, which is removed to intermediate storage tanks.

The remaining gas stream passes through a sulfuric acid absorption tower or acid scrubber where most of the remaining hydrogen fluoride and some residual sulfuric acid are removed and are also placed in intermediate storage. The gases exiting the scrubber then pass through water scrubbers, where the silicon tetrafluoride (SiF_4) and remaining hydrogen fluoride are recovered as fluorosilicic acid (H_2SiF_6). The tail gases from the water scrubber are passed through a caustic scrubber before being released to the atmosphere but only after further treatment has removed any potential environmental contaminants. The hydrogen fluoride and sulfuric acid are delivered from intermediate storage tanks to distillation columns where the hydrofluoric acid is extracted to a purity of 99.98% (v/v). Weaker concentrations of the hydrofluoric acid (typically 70%–80%, v/v) are produced by measured and careful dilution with water.

3.3.11 Hydrogen Production

Hydrogen is the lightest element in the periodic table (Chapter 2). The most common isotope of hydrogen, termed *protium* (a name that is rarely used), has one proton and no neutrons. Hydrogen gas (molecular hydrogen, H_2) is highly flammable (even explosively flammable) and will burn in air at a very wide range of concentrations between 4% and 75% (v/v). These reactions may be triggered by spark, heat, or sunlight. The hydrogen autoignition temperature (the temperature of spontaneous ignition in air) is 500°C (932°F).

The hydrogen produced in the various electrolytic processes contains small amounts of water vapor, sodium hydroxide, and salt that is removed through cooling. Condensed salt water and sodium hydroxide solution is either recycled as brine make-up or treated with other waterborne waste streams. Hydrogen produced during the mercury cell process (used to produce chlorine-free sodium hydroxide), however, also contains small amounts of mercury that must be removed prior to liquefaction. Most of the entrained mercury is extracted by cooling the gas after which the condensed mercury is then returned to the electrolytic cells.

The hydrogen gas may be purified further purify by using an activated carbon process in which the gas is passed through a bed of activated carbon to remove any further impurities. However, the spent activated carbon is typically shipped from the site to a recognized hazardous waste disposal facility.

3.3.12 Nitric Acid and Nitrogen Compounds

3.3.12.1 Nitric Acid

Nitric acid (HNO_3, also known as *aqua fortis* and spirit of niter) is a highly corrosive mineral acid. The pure compound is colorless, but older samples tend to acquire a yellow cast due to decomposition into nitrogen oxides (NO_x) and water. Most commercially available nitric acid has a concentration of 68% (v/v) in water. When the solution contains more than 86% (v/v) nitric acid, it is referred to as *fuming nitric acid*, which, depending on the amount of nitrogen dioxide (NO_2) present, is further characterized as (i) white fuming nitric acid or (2) red fuming nitric acid at concentrations above 95%. Nitric acid is the primary reagent used for nitration—the addition of a nitro group (—NO_2), typically to an organic molecule. While some of the resulting nitro compounds are shock-sensitive and thermally sensitive explosives (such as nitroglycerin and trinitrotoluene (TNT)), some are sufficiently stable to be used in munitions and demolition, while others are still more stable and used as pigments in inks and dyes. Nitric acid is also commonly used as a strong oxidizing agent.

Nitric acid is used as an intermediate in the manufacture of ammonium nitrate (NH_4NO_3), which is primarily used in to manufacture fertilizers. Another use for nitric acid is in the oxidation process for the manufacture of adipic acid ($HO_2CCH_2CH_2CH_2CH_2CO_2H$) that is a dicarboxylic acid used in the production of nylon.

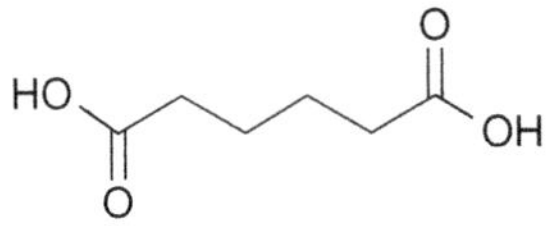

HO, O, OH, O

Adipic acid

Nitric acid is also used in organic oxidation to manufacture terephthalic acid ($C_6H_4(CO_2H)_2$) and other organic compounds.

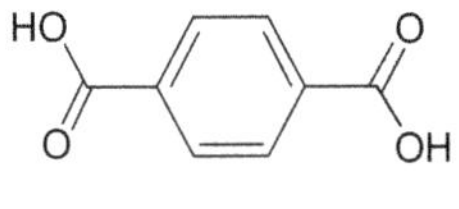

HO, O, O, OH

Nitric acid is also used in the manufacture of explosives, such as nitrobenzene derivatives, dinitrotoluene derivative, and TNT derivatives, and for producing other chemical intermediates.

Nitric acid is produced by two methods: (1) The first method utilizes oxidation, condensation, and absorption to produce a weak nitric acid that can have concentrations ranging from 30% to 70% (v/v) nitric acid, and (2) the second method combines dehydrating, bleaching, condensing, and absorption to produce high-strength nitric acid from weak nitric acid; high-strength nitric acid generally contains more than 90% (v/v) nitric acid.

The process typically consists of three steps: (1) ammonia oxidation, (2) nitric oxide oxidation, and (3) absorption. Each step corresponds to a distinct chemical reaction. For the ammonia oxidation step, a 1:9 ammonia/air mixture is oxidized at a temperature of 750–800°C (1380–1470°F) as it passes through a catalytic converter:

$$4NH_3 + 5O_2 \rightarrow 4NO + 6H_2O$$

The most commonly used catalyst is composed of platinum (90%, w/w) and rhodium (10%, w/w) gauze, and under these conditions. the oxidation of ammonia to nitric oxide proceeds in an exothermic reaction with a yield range on the order of 93%–98% (v/v). Higher catalyst temperatures increase reaction selectivity toward the production of nitric oxide (NO), while lower catalyst temperatures tend to be more selective toward the production of nitrogen (N_2) and nitrous oxide (N_2O). Nitric oxide is considered to be a criteria pollutant, and nitrous oxide is known to be a greenhouse gas. The nitrogen dioxide/dimer mixture then passes through a waste heat boiler and a platinum filter.

The nitric oxide formed during the ammonia oxidation is oxidized in another separate step. In this step, the process stream is passed through a cooler/condenser and cooled to 38°C (100°F) or less at pressures up to 116 psi, and the nitric oxide reacts noncatalytically with residual oxygen to form nitrogen dioxide and the liquid dimer, nitrogen tetroxide; this slow, homogeneous reaction is temperature and pressure dependent:

$$4NO + 2O_2 \rightarrow 2NO_2 + N_2O_4$$

At low temperatures and high pressure, the maximum production of nitrogen dioxide occurs within a minimum reaction time.

Absorption is the final step in the process, and the nitrogen dioxide/dimer mixture is introduced into an absorption process after being cooled. The mixture is pumped into the bottom of the absorption tower, while liquid dinitrogen tetroxide is added at a higher point. Deionized water enters the top of the column, and both liquids flow in a countercurrent direction, thereby allowing the exothermic oxidation reaction to occur in the free space between the trays, while absorption occurs on the trays. The absorption trays are usually sieve or bubble cap trays:

$$3NO_2 + H_2O \rightarrow 2HNO_3 + NO$$

A secondary air stream is introduced into the column to reoxidize the nitric oxide, which is formed in this reaction. This secondary air stream also removes any nitrogen dioxide from the product acid. An aqueous solution of (typically) 55%–65% (v/v) nitric acid is withdrawn from the bottom of the tower; the acid concentration can vary from 30% to 70% (v/v) nitric acid. However, the acid concentration depends upon the temperature, pressure, and number of absorption stages as well as on the concentration of nitrogen oxides entering the absorber.

While configurations may differ somewhat between plants, three essential steps are commonly employed. In the first step, ammonia is oxidized to nitric oxide (NO) in a catalytic convertor over a platinum catalyst (90% platinum and 10% rhodium gauze). The reaction is exothermic (heat releasing) and produces nitric oxide in yields on the order of 93%–98%. The reaction proceeds at high temperatures ranging from 750°C to 900°C (1380–1650°F). The resulting mixture from this reaction is then sent to a waste heat boiler where steam is produced. In the second step, nitric oxide is oxidized by passage through a cooler/condenser, where it is cooled to temperatures on the order of 38°C (100°F) or less, at pressures of up to 116 psia. During this stage, the nitric oxide reacts with residual oxygen to form nitrogen dioxide and nitrogen tetroxide. The final step introduces this mixture of nitrogen oxides into an absorption process where the mixture flows countercurrent to deionized water and additional liquid dinitrogen tetroxide. The tower is packed with sieve or bubble cap distillation type trays. Oxidation takes places in between the trays in the tower; absorption occurs on the trays. An exothermic reaction between NO_2 and water occurs in the tower to produce nitric acid and NO. Air is introduced into the tower to reoxidize the NO that is being formed and to remove NO_2 from the nitric acid. A weak acid solution (of 55%–65 %) is withdrawn from the bottom of the absorption tower.

High strength nitric acid (98%–99%, v/v) can be obtained by concentrating weak nitric acid (30%–70% concentration) using extractive distillation. The weak nitric acid cannot be concentrated by simple fractional distillation. The distillation must be carried out in the presence of a dehydrating agent. Concentrated sulfuric acid (typically 60%, v/v sulfuric acid) is most commonly used for this purpose. The nitric acid concentration process consists of feeding strong sulfuric acid and 55%–65% (v/v) nitric acid into the top of a packed dehydrating column at approximately atmospheric pressure. The acid mixture flows downward and concentrated nitric acid leaves the top of the column as 99% (v/v) vapor that contains a small amount of nitrogen dioxide and oxygen that result from dissociation of nitric acid. The concentrated acid vapor leaves the column and goes to a bleacher and a countercurrent condenser system to achieve the condensation of strong nitric acid and the separation of oxygen and nitrogen oxide by-products. These by-products then flow to an absorption column where the nitric oxide mixes with auxiliary air to form nitrogen dioxide, which is recovered as weak nitric acid. Inert and unreacted gases are vented to the atmosphere from the top of the absorption column.

Emissions from the manufacture of nitric acid include mostly nitrogen oxides (NO and NO_2), and trace amounts of ammonia and nitric acid mist. The tail gas from the acid absorption tower is the largest source of nitrogen oxide emissions. These emissions can increase when insufficient air is supplied to the oxidizer and absorber, under low absorber pressure conditions, and during high-temperature conditions in the cooler/condenser and absorber. Other factors may contribute, such as high throughputs, very high-strength products, or faulty compressors or pumps.

Control of emissions from nitric acid plants is usually accomplished through either extended absorption or catalytic reduction. Extended absorption works by increasing the efficiency of the absorption process. Catalytic reduction oxidizes nitrogen oxides in the tail gas and reduces them to nitrogen. While catalytic reduction is more energy-intensive, it achieves greater emission reductions than the extended absorption method. Less used control options include wet scrubbers or molecular sieves, both of which have higher capital and operating costs than the other options.

Solid wastes from nitric acid manufacture include spent catalysts that are either returned to the manufacturer or disposed of. Dust from the catalyst may settle out in the equipment, but if it contains precious metals, it is recovered and sent for reprocessing to an outside vendor. Precious metals (e.g., platinum) lost from the ammonia oxidation catalyst are captured by a recovery gauze (getter), which must be replaced periodically and is reprocessed by a gauze manufacturer. Filters used for ammonia/air filtration must also be replaced.

3.3.12.2 Nitrogen Compounds

In general, chemicals containing nitrogen are manufactured from ammonia produced by the Haber process. However, since molecular nitrogen is inert, its reaction with hydrogen requires very severe conditions and a catalyst. High pressure favors the formation of products, but an increase in temperature will shift the equilibrium in the opposite direction.

Nitrogen is obtained from the air and hydrogen and can be produced by the shift reaction as well as from or from hydrocarbon reforming (the best example of which is by natural gas reforming or by methane reforming):

$$CO + H_2O \rightarrow CO_2 + H_2$$
$$CH_4 + 2H_2O \rightarrow CO_2 + 4H_2$$

However, further process stages are required to assure complete conversion and production of hydrogen as well as to remove carbon dioxide or carbon monoxide from the gas mixture (Mokhatab et al., 2006; Speight, 2007, 2014, 2017a; Kidnay et al., 2011; Bahadori, 2014). A mixture of ammonia and synthesis gas ($CO + H_2$) results from the reaction with nitrogen so the two must be separated and the synthesis gas recycled.

Most of the ammonia that is produced is employed as fertilizer or used to manufacture other fertilizers, such as urea, ammonium sulfate, ammonium nitrate, or diammonium hydrogen phosphate. Ammonia is also used in the Solvay process, and it is a starting material for the manufacture of cyanides and nitriles (which are used to make polymers such as nylon and acrylics) as well as aromatic compounds containing nitrogen, such as pyridine and aniline:

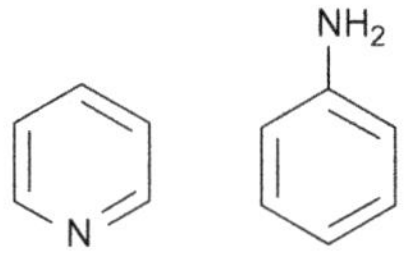

Pyridine Aniline

The other source of nitrogen compounds in the chemical industry is nitric acid, obtained from the oxidation of ammonia:

$$4NH_3 + 5O_2 \rightarrow 4NO + 6H_2O$$
$$6NO + 3O_2 \rightarrow 6NO_2$$
$$3NO_2 + H_2O \rightarrow 2HNO_3 + NO$$

The first reaction is run over platinum-rhodium catalysts at around 900°C (1652°F). In the second and third stages, a mixture of nitric oxide and air circulates through condensers, where it is partially oxidized. The nitrogen dioxide is absorbed in a tower, and nitric acid sinks to the bottom. Nitric acid is mainly used to make ammonium nitrate, most of it for fertilizer although it also goes into the production of explosives. Nitration is used to manufacture explosives such as nitroglycerin and TNT as well as many important chemical intermediates used in the pharmaceutical and dyestuff industries.

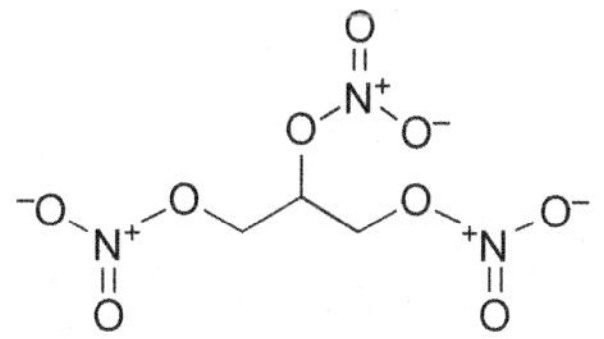

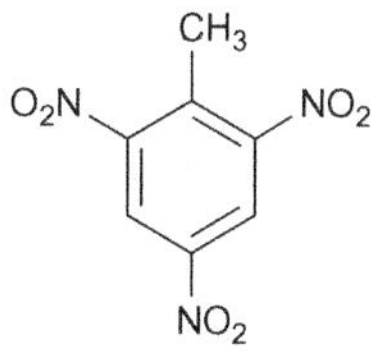

3.3.13 Phosphoric Acid and Phosphorus Compounds

3.3.13.1 Phosphoric Acid

Phosphoric acid (H_3PO_4, also known as orthophosphoric acid or phosphoric (V) acid) is a mineral inorganic acid. Orthophosphoric acid refers to *phosphoric acid* in which the prefix *ortho* is used to distinguish the acid from related phosphoric acids, called polyphosphoric acids. Orthophosphoric acid, when pure, is a solid at room temperature and pressure. The most common source of phosphoric acid is an 85% aqueous solution that is colorless and nonvolatile but is sufficiently acidic to be corrosive. Because of the high percentage of phosphoric acid in this reagent, at least some of the orthophosphoric acid is condensed into polyphosphoric acids. For the sake of labeling and simplicity, the 85% represents the acid as if it was all orthophosphoric acid. Dilute aqueous solutions of phosphoric acid exist in the orthoform.

Phosphoric acid (H_3PO_4) can be manufactured using either a thermal or a wet process. However, the majority of phosphoric acid is produced using the wet-process method. Wet-process phosphoric acid is used for fertilizer production. Thermal process phosphoric acid is commonly used in the manufacture of high-grade chemicals, which require a much higher purity. The production of wet-process phosphoric acid generates a considerable quantity of acidic cooling water with high concentrations of phosphorus and fluoride. This excess water is collected in cooling ponds that are used to temporarily store excess precipitation for subsequent evaporation and to allow recirculation of the process water to the plant for reuse.

In the *wet process*, phosphoric acid is produced by reacting sulfuric acid (H_2SO_4) with naturally occurring phosphate rock. The phosphate rock is dried, crushed, and then continuously fed into the reactor along with sulfuric acid. The reaction combines calcium from the phosphate rock with sulfate, forming calcium sulfate (gypsum, $CaSO_4$), which is separated from the reaction solution by filtration. Some facilities generally use a dihydrate process that produces gypsum in the form of calcium sulfate with two molecules of water (calcium sulfate dihydrate, $CaSO_4 \cdot 2H_2O$). Other facilities may use a hemihydrate process that produces calcium sulfate with the equivalent of a half molecule of water per molecular of calcium sulfate ($2CaSO_4 \cdot H_2O$). The one-step hemihydrate process has the advantage of producing wet-process phosphoric acid with a higher concentration of phosphorus pentoxide (P_2O_5) and less impurities than the dihydrate process. A simplified reaction for the dihydrate process is as follows:

$$Ca_3(PO_4)_2 + 3H_2SO_4 + 6H_2O \rightarrow 2H_3PO_4 + 3(CaSO_4 \cdot 2H_2O)$$

In order to make the strongest phosphoric acid possible and to decrease evaporation costs, 93% (v/v) sulfuric acid is normally used. During the reaction, gypsum crystals are precipitated and separated from the acid by filtration. The separated crystals must be washed thoroughly to yield at least a 99% (v/v) recovery of the filtered phosphoric acid. After washing, the slurried gypsum is pumped into a gypsum pond for storage. Water is siphoned off and recycled through a surge cooling pond to the phosphoric acid process. Wet-process phosphoric acid normally contains 26%–30% (w/w) phosphorus pentoxide, and in most cases, the acid must be further concentrated to meet phosphate feed material specifications for fertilizer production. Depending on the types of fertilizer to be produced, phosphoric acid is usually concentrated to 40%–55% (w/w) phosphorus pentoxide by using two or three vacuum evaporators.

In the *thermal process*, the raw materials for the production of phosphoric acid are elemental (yellow) phosphorus, air, and water. The process involves three major steps: (1) combustion, (2) hydration, and (3) demisting. In the combustion step, the liquid elemental phosphorus is burned (oxidized) in ambient air in a combustion chamber at temperatures of 1650–2760°C (3000–5000°F) to form phosphorus pentoxide:

$$4P + 5O_2 \rightarrow 2P_2O_5$$

The phosphorus pentoxide is then hydrated with dilute phosphoric acid (H_3PO_4) or water to produce strong phosphoric acid liquid.

$$P_2O_5 + 6H_2O \rightarrow 4H_3PO_4$$

The final step is a demisting step that is applied to removal of the phosphoric acid mist from the combustion gas stream before releasing to the atmosphere, which is usually accomplished by use of high-pressure-drop demisters. As always, release to the atmosphere can only be accomplished if the demisted product is a clean and nonpolluting stream.

The concentration of phosphoric acid (H_3PO_4) produced from the thermal process normally ranges from 75% to 85% (v/v). This concentration is required for high-grade chemical production and other nonfertilizer product manufacturing. Efficient plants recover approximately 99.9% (w/w) of the elemental phosphorus burned as the phosphoric acid product.

In phosphoric acid production, the fluorine released from reactors and evaporators is usually recovered as a by-product that can be sold. The remainder is passed to the condenser that produces a liquid effluent with mostly fluoride and small amounts of phosphoric acid. Closed systems recycle this effluent; in other cases, it is discharged to open waters.

The manufacture of phosphoric acid produces a gypsum slurry that is sent to settling ponds to allow the solids to settle out. About 5 lbs of phosphor-gypsum is generated per pound of phosphoric acid. This phosphor-gypsum contains trace elements from phosphate rock, such as cadmium and uranium. Pond systems are usually fitted with lining systems and collection ditches to maintain control of trace elements and avoid contamination of ground water.

The major source of phosphorus in the world is apatite, which is a group of phosphate minerals, usually referring to hydroxylapatite, fluorapatite and chlorapatite, with high concentrations of hydroxyl (OH^-) ions, fluoride (F^-) ions, and chloride (Cl^-) ions, respectively, in the crystal apatite ($Ca_5(PO_4)_3(F,Cl,OH)$). Commercially, the most important is fluoroapatite, a calcium phosphate that contains fluorine. This fluorine must be removed for the manufacture of phosphoric acid, but it also can be used to produce hydrofluoric acid and fluorinated compounds.

3.3.13.2 Phosphorus Compounds

Phosphoric acid is the starting material for most of the phosphates that are produced industrially. It is obtained from the reaction of the apatite mineral with sulfuric acid.

Silica is present in the mineral as an impurity, and it reacts with hydrofluoric acid to yield silicon tetrafluoride, which can be converted to fluorosilicic acid, an important source of fluorine. More than half of the phosphoric acid that is produced by the reaction of phosphates with sulfuric acid is converted directly to sodium or ammonium phosphates to be used as fertilizer; thus, purity is not a concern.

For products that require high purity, such as detergents and foodstuffs, phosphoric acid is produced from elemental phosphorus (at about four times the cost). An electric furnace operating at 1400–1500°C (2552–2732°F) is used to form a molten mass of apatite and silica that reacts with coke and reduces the phosphate mineral:

$$2Ca_3(PO_4)_2 + 6SiO_2 + 10C \rightarrow 4P + 6CaSiO_3 + 10CO$$

Concentrating phosphoric acid leads to polyphosphoric acid, a mixture of several polymeric species, a good catalyst and dehydrating agent. Polyphosphate salts are used as water softeners in detergents or as buffers in food. Small quantities of elemental phosphorus are used to make matches, and phosphorus halides to prepare specialty chemicals for the pharmaceutical and agrochemical industries.

3.3.14 Sodium Carbonate

Sodium carbonate (Na_2CO_3, also called soda ash) is one of the largest volume mineral products in the inorganic chemicals industry. Sodium carbonate commonly occurs as a crystalline decahydrate ($Na_2CO_3{\cdot}10H_2O$), which readily effloresces to form a white powder, the monohydrate ($Na_2CO_3{\cdot}H_2O$). Pure sodium carbonate is a white, odorless powder that is hygroscopic (absorbs moisture from the air) and that forms a moderately basic solution in water. The two processes presently used to produce natural soda ash differ only in the recovery and primary treatment of the raw material used. The raw material for Wyoming soda ash is mined Trona ore, while California soda ash is derived from sodium carbonate-rich brine extracted from Searles Lake.

There are four distinct methods used to mine the Wyoming Trona ore: (1) solution mining, (2) room-and-pillar mining, (3) longwall mining, and (4) shortwall mining. In the solution mining process, dilute sodium hydroxide (NaOH) is injected into the Trona to dissolve it. This solution is treated with carbon dioxide gas in carbonation towers to convert the sodium carbonate (Na_2CO_3) in solution to sodium bicarbonate ($NaHCO_3$), which precipitates and is retrieved by filtration. The crystals are again dissolved in water, the sodium bicarbonate is then precipitated with carbon dioxide, and filtered. The product is calcined to produce dense soda ash. Brine extracted from below the Searles Lake in California is treated similarly.

For the room-and-pillar mining, longwall mining, and shortwall mining methods, the conventional blasting agent is prilled (pelletized) ammonium nitrate and fuel oil (ANFO, ammonium nitrate-fuel oil) explosive. Beneficiation is accomplished with either of two methods: (1) the sesquicarbonate process and (2) the monohydrate processes. In the sesquicarbonate process, the Trona ore is first dissolved in water and then treated as brine. The liquid is filtered to remove insoluble impurities before the sodium sesquicarbonate ($Na_2CO_3{\cdot}NaHCO_3{\cdot}2H_2O$) is precipitated by using one or more vacuum crystallizers. The product result is centrifuged to remove remaining water and can be sold as a finished product or further calcined to yield soda ash having a light to intermediate density.

In the monohydrate process, the crushed trona is calcined in a rotary kiln, yielding dense soda ash and carbon dioxide and water as by-products. The calcined material is combined with water to allow settling out or filtering of impurities such as shale and is then concentrated by triple-effect evaporators and/or mechanical vapor recompression crystallizers to precipitate sodium carbonate monohydrate ($Na_2CO_3{\cdot}H_2O$). Impurities such as sodium chloride (NaCl) and sodium sulfate (Na_2SO_4) remain in solution. The crystals and liquor are centrifuged, and the recovered crystals are calcined again to remove any remaining water after which the product is cooled, screened, and prepared for transportation.

During the production of sodium carbonate, particulate emissions are created from ore calciners, soda ash coolers and dryers, ore crushing, screening and transporting, and product handling and shipping. Combustion products (SO_x, NO_x, CO, particulates, and volatile hydrocarbons) are also emitted from direct-fired process heating units (ore-calcining kilns and soda ash dryers).

Production of sodium carbonate from trona ore also creates emissions of carbon dioxide, a suspected greenhouse gas. Additional carbon dioxide may be emitted as sodium carbonate is processed in other manufacturing processes (glassmaking, water treatment, flue gas desulfurization, soap and detergent production, and pulp and papermaking). Data are listed on releases of carbon dioxide from these processes.

Emissions of particulates from calciners and dryers are most often controlled by venturi scrubbers, electrostatic precipitators, or cyclones. The high moisture content of exiting gases makes it difficult to use baghouse-type filters. Control of particulates from ore and product handling systems, however, is often accomplished by baghouse filters or venturi scrubbers. These are essential to the cost-effectiveness of the process as they permit capture and recovery of valuable product.

Sodium carbonate manufacture creates significant volumes of wastewater that must be treated prior to discharge or recycling to the process. These may contain both mineral (e.g., shale) and salt impurities. Limitations for toxic or hazardous compounds contained in these wastewaters are given by the US Environmental Protection Agency in 40 CFR, Chapter 1, Part 415, which was originally promulgated in 1974 and has been revised several times since then. The chemicals in the chlor-alkali industry and sodium carbonate are covered under Subparts F and O. Specific limitations for restricted compounds and total suspended solids are shown in Tables 6.8–6.10. "BPT Standards" refers to the use of the best practicable control technology currently available. "BAT" refers to the best available technology economically achievable. "NSPS" refers to new source performance standards (NSPS) that apply to new process water impoundment or treatment facilities.

The dissolving and clarification steps in sodium carbonate production create a waste sludge containing nonhazardous impurities, such as salts and minerals.

3.3.15 Sulfuric Acid

Sulfuric acid (H_2SO_4) (the historical name *oil of vitriol*) is an inorganic chemical that is a highly corrosive strong mineral acid that is a pungent-ethereal, colorless to slightly yellow viscous liquid that is soluble in water at all concentrations. Sometimes, the acid may be sold as a dark brown liquid (dye added during production) to alert purchases the hazards of handling this acid.

Sulfuric acid is manufactured in large quantities on a world scale with the production of the chemical often being linked to the stage of development of a country, owing to the large number of transformation processes in which it is used. Sulfuric acid (H_2SO_4) is a basic raw material used in a wide range of industrial processes and manufacturing operations. A high proportion of the manufactured sulfuric acid is used in the production of phosphate fertilizers and other uses include copper leaching, inorganic pigment production, petroleum refining, paper production, and industrial organic chemical production.

Sulfuric acid is manufactured from elemental sulfur in a three-stage process:

$$S + O_2 \rightarrow SO_2$$
$$2SO_2 + O_2 \rightarrow 2SO_3$$
$$SO_3 + H_2O \rightarrow H_2SO_4$$

Since the reaction of sulfur with dry air is exothermic, the sulfur dioxide must be cooled to remove excess heat and avoid reversal of the reaction.

The combustion of elemental sulfur is the predominant source of sulfur dioxide used to manufacture sulfuric acid. The combustion of hydrogen sulfide from waste gases, the thermal decomposition of spent sulfuric acid or other sulfur-containing materials, and the roasting of pyrites are also used as sources of sulfur dioxide. Sulfuric acid may be manufactured commercially by either the *lead chamber process* or the *contact process* with a modern leaning toward the contact process.

In the contact process, the process plants are generally characterized according to the raw materials charged to them: (1) combustion of elemental sulfur, (2) combustion of spent sulfuric acid and hydrogen sulfide, and (3) combustion of metal sulfide ores and smelter gas burning. More specifically, the contact process incorporates three basic operations, each of which corresponds to a distinct chemical reaction. First, the sulfur in the feedstock is oxidized (burned) to sulfur dioxide:

$$S + O_2 \rightarrow SO_2$$

The resulting sulfur dioxide is fed to a process unit (often referred to as the *converter*) where it is catalytically oxidized to sulfur trioxide:

$$2SO + 2O_2 \rightarrow 2SO_3$$

Finally, the sulfur trioxide is absorbed in a strong sulfuric acid (98%) solution:

$$SO_3 + H_2O \rightarrow H_2SO_4$$

In the Frasch process, elemental sulfur is melted, filtered to remove ash, and sprayed under pressure into a combustion chamber where the sulfur is burned in clean air that has been dried by scrubbing with 93%–99% (v/v) sulfuric acid. The gases from the combustion chamber are cool by passing through a waste heat boiler and then enter the catalyst (vanadium pentoxide, V_2O_5) converter. Typically, 95%–98% (v/v) of the sulfur dioxide from the combustion chamber is converted to sulfur trioxide, with an accompanying large evolution of heat. After being cooled, again by generating steam, the converter exit gas enters an absorption tower. The absorption tower is a packed column where acid is sprayed in the top and the sulfur trioxide enters from the bottom. The sulfur trioxide is absorbed in the 98%–99% (v/v) sulfuric acid where the sulfur trioxide combines with the water in the acid and forms more sulfuric acid. If oleum (a solution of uncombined sulfur trioxide dissolved in sulfuric acid) is produced, sulfur trioxide from the converter is first passed to an oleum tower that is fed with 98% (v/v) acid from the absorption system. The gases from the oleum tower are then pumped to the absorption column where the residual sulfur trioxide is removed. The single absorption process uses only one absorber as the name implies, but many plants have installed a dual absorption step.

In the dual absorption step, the sulfur trioxide gas formed in the primary converter stages is sent to an interpass absorber where most of the sulfur trioxide is removed to form sulfuric acid. The remaining unconverted sulfur dioxide is forwarded to the final stages in the converter to remove much of the remaining sulfur dioxide by oxidation to sulfur trioxide, from whence it is sent to the final absorber for removal of the remaining sulfur trioxide.

If oleum (fuming sulfuric acid, simply represented as $H_2SO_4{\cdot}SO_3$) is produced (a mixture of excess sulfur trioxide and sulfuric acid), sulfur trioxide from the converter is passed to an oleum tower that is fed with 98% (v/v) acid from the absorbers. The gases from this tower are then pumped to the absorption column where sulfur trioxide is removed. Various concentrations of oleum can be produced. Common ones include 20% oleum (20%, v/v sulfur trioxide in 80%, v/v sulfuric acid, with no water), 40% oleum, and 60% oleum.

Sulfur dioxide is the primary emission from sulfuric acid manufacture and is found primarily in the exit stack gases. Conversion of sulfur dioxide to sulfur trioxide is also incomplete during the process, which gives rise to emissions. Dual absorption is considered the best available control technology (BACT) for meeting NSPS for sulfur dioxide. In addition to stack gases, small amounts of sulfur dioxide are emitted from storage and tank-truck vents during loading, from sulfuric acid concentrators, and from leaking process equipment.

Acid mists may also be emitted from absorber stack gases during sulfuric acid manufacture. The very stable acid mist is formed when sulfur trioxide reacts with water vapor below the dew point of sulfur trioxide. Typical control devices include vertical tube, vertical panel, and horizontal dual pad mist eliminators.

During the production of sulfuric acid, a sludge is produced in the carbon dioxide removal unit used to absorb solvent gas. A hydrocarbon solvent is used in the unit, which breaks down into a hydrocarbon sludge during the process. This sludge is usually combusted in another part of the process. Sulfuric acid manufacture also produces a solid waste containing the heavy metal vanadium, when the convertor catalyst is regenerated or screened. This waste is sent to an off-site vendor for reprocessing. Additional solid wastes from sulfuric acid production may contain both vanadium and arsenic, depending on the raw materials used, and care must be taken to dispose of them properly in landfills.

3.3.16 Sulfur Recovery

Sulfur is a chemical element (symbol, S; referred to as *brimstone* in the Bible) that is an abundant, multivalent nonmetal that, under normal conditions, exists as cyclic octa-atomic molecules (S_8). Elemental sulfur is a bright yellow crystalline solid at room temperature. Although elemental sulfur occurs naturally as the element (referred to as *native sulfur*), it most commonly occurs in combined forms as sulfide and sulfate minerals. In the modern industry, most elemental sulfur is produced as a by-product of removing sulfur-containing contaminants from natural gas and crude oil (Mokhatab et al., 2006; Speight, 2007, 2014, 2017a; Bahadori, 2014). The greatest commercial use of the element is the production of sulfuric acid for sulfate and phosphate fertilizers and other chemical processes.

Hydrogen sulfide is a by-product of processing natural gas and refining high-sulfur crude oils (Mokhatab et al., 2006; Speight, 2007, 2014, 2017a; Bahadori, 2014). The term *sulfur recovery*, as used in this context, refers to the conversion of hydrogen sulfide (H_2S) to elemental sulfur and the most common conversion method used is the Claus process (Mokhatab et al., 2006; Speight, 2007, 2014, 2017a; Bahadori, 2014).

The Claus process is not so much a gas cleaning process but a process for the recovery of sulfur by the disposal of hydrogen sulfide, a toxic gas that originates in natural gas as well as during crude oil processing such as in the coking, catalytic cracking, hydrotreating, and hydrocracking processes. Burning hydrogen sulfide as a fuel gas component or as a flare gas component is precluded by safety and environmental considerations since one of the combustion products is the highly toxic sulfur dioxide (SO_2), which is also toxic. As described above, hydrogen sulfide is typically removed from the refinery light ends gas streams through an olamine process after which application of heat regenerates the olamine and forms an acid gas stream. Following from this, the acid gas stream is treated to convert the hydrogen sulfide elemental sulfur and water. The conversion process utilized in most modern refineries is the Claus process or a variant thereof. The Claus process typically recovers 95%–97% (v/v) of the hydrogen sulfide in the feed stream.

In the process, hydrogen sulfide (H_2S), a by-product of crude oil and natural gas processing, is recovered and converted to elemental sulfur. The process consists is a multistage catalytic oxidation of hydrogen sulfide to produce sulfur:

$$2H_2S + O_2 \rightarrow 2S + 2H_2O$$

Each catalytic stage consists of a gas reheater, a catalyst chamber, and a condenser. The Claus process involves burning one-third of the hydrogen sulfide with air in a reactor furnace to form sulfur dioxide (SO_2):

$$2H_2S + 3O_2 \rightarrow 2SO_2 + 2H_2O + \text{Heat}$$

The furnace normally operates at combustion chamber temperatures ranging from 980°C to 1540°C (1800–2800°F) with pressures on the order of (10 psia). Before entering a sulfur condenser, the hot gas from the combustion chamber is quenched in a waste heat boiler that generates high to medium pressure steam. Liquid sulfur from the condenser runs through a seal leg into a covered pit from which it is pumped to trucks or railcars for shipment to end users. In this stage, approximately 65%–70% of the sulfur is recovered.

The cooled gases exiting the condenser are then sent to the catalyst beds where the remaining noncombusted two-thirds of the hydrogen sulfide undergoes the Claus reaction to form elemental sulfur

$$2H_2S + SO_2 \longleftrightarrow 3S + 2H_2O + \text{Heat}$$

The catalytic reactors operate at lower temperatures, ranging from 200°C to 315°C (400–600°F). Alumina (Al2O3) or bauxite is sometimes used as a catalyst. Bauxite is an aluminum ore that consists mostly of the minerals gibbsite ($Al(OH)_3$), boehmite (γ-AlO(OH)), and diaspore (α-AlO (OH)), mixed with the two iron oxides goethite (FeO(OH)) and hematite (F_e2O_3), the clay mineral kaolinite and small amounts of anatase (titanium dioxide, TiO_2) and ilmenite ($FeTiO_3$ or $FeO{\cdot}TiO_2$).

This reaction represents an equilibrium chemical reaction and, thus, it is not possible for a Claus plant to convert all the incoming sulfur compounds to elemental sulfur. Therefore, two or more stages are used in series to recover the sulfur. Each catalytic stage can recover half to two-thirds of the incoming sulfur. From the condenser of the final catalytic stage, the process stream passes to some form of tail-gas-treatment process. The tail gas, containing hydrogen sulfide, sulfur dioxide, sulfur vapor, and traces of other sulfur compounds formed in the combustion section, escapes with the inert gases from the tail end of the plant. Thus, it is frequently necessary to follow the Claus unit with a tail-gas cleanup unit to achieve higher recovery (Mokhatab et al., 2006; Speight, 2007, 2014, 2017a; Bahadori, 2014). In addition to the oxidation of hydrogen sulfide to sulfur dioxide and the reaction of sulfur dioxide with hydrogen sulfide in the reaction furnace, many other side reactions can and do occur in the furnace such as the following:

$$CO + H_2S \rightarrow COS + H_2O$$

$$COS + H_2S \rightarrow CS_2 + H_2O$$

$$2H_2S \rightarrow S_2 + 2H_2$$

$$CH_4 + 4S \rightarrow CS_2 + 2H_2S$$

Emissions from the Claus process may be reduced by: (1) extending the Claus reaction into a lower-temperature liquid phase, (2) adding a scrubbing process to the Claus exhaust stream, or (3) incinerating the hydrogen sulfide gases to form sulfur dioxide.

Currently, there are five processes available that extend the Claus reaction into a lower-temperature liquid phase including the BSR/Selectox, Sulfreen, Cold Bed Absorption, Maxisulf, and IFP-1 processes. These processes take advantage of the enhanced Claus conversion at cooler temperatures in the catalytic stages. All of these processes give higher overall sulfur recoveries of 98%–99% when following downstream of a typical two- or three-stage Claus sulfur recovery unit and therefore reduce sulfur emissions.

Sulfur emissions can also be reduced by adding a scrubber at the tail end of the plant. Here are essentially two generic types of tail-gas-scrubbing processes: oxidation tail-gas scrubbers and reduction tail-gas scrubbers. The first scrubbing process is used to scrub SO_2 from incinerated tail gas and recycle the concentrated SO_2 stream back to the Claus process for conversion to elemental sulfur.

There are at least three oxidation scrubbing processes: the Wellman-Lord, Stauffer Aquaclaus, and IFP-2. Only the Wellman-Lord process has been applied successfully to US refineries. The Wellman-Lord process uses a wet generative process to reduce stack gas sulfur dioxide concentration to less than 250 ppm by volume and can achieve approximately 99.9% sulfur recovery.

Claus plant tail gas is incinerated, and all sulfur species are oxidized to form SO_2 in the Wellman-Lord process. Gases are then cooled and quenched to remove excess water and to reduce gas temperature to absorber conditions. The rich SO_2 gas is then reacted with a solution of sodium sulfite (Na_2SO_3) and sodium bisulfite ($NaHSO_3$) to form the bisulfite:

$$SO_2 + Na_2SO_3 + H_2O \rightarrow 2NaHSO_3$$

The off-gas is reheated and vented to the atmosphere. The resulting bisulfite solution is boiled in an evaporator-crystallizer, where it decomposes to SO_2 and water (H_2O) vapor and sodium sulfite is precipitated:

$$2NaHSO_3 \rightarrow Na_2SO_3 + H_2O + SO$$

Sulfite crystals are separated and redissolved for reuse as lean solution in the absorber. The wet SO_2 gas is directed to a partial condenser where most of the water is condensed and reused to dissolve sulfite crystals. The enriched SO_2 stream is then recycled back to the Claus plant for conversion to elemental sulfur.

In the second type of scrubbing process, sulfur in the tail gas is converted to H_2S by hydrogenation in a reduction step. After hydrogenation, the tail gas is cooled, and water is removed. The cooled tail gas is then sent to the scrubber for H_2S removal prior to venting. There are at least four reduction scrubbing processes developed for tail-gas sulfur removal: Beavon, Beavon MDEA, SCOT, and ARCO. In the Beavon process, hydrogen sulfide is converted to sulfur outside the Claus unit using a lean H_2S-to-sulfur process (the Stretford process). The other three processes utilize conventional amine scrubbing and regeneration to remove hydrogen sulfide and recycle back as Claus feed.

Emissions from the Claus process may also be reduced by incinerating sulfur-containing tail gases to form sulfur dioxide. In order to properly remove the sulfur, incinerators must operate at a temperature of 650°C (1200°F) or higher if all the H_2S is to be combusted. Proper air-to-fuel ratios are needed to eliminate pluming from the incinerator stack. The stack should be equipped with analyzers to monitor the SO_2 level.

3.3.17 Titanium Dioxide

Titanium dioxide (TiO_2, also known as titanium (IV) oxide or titania) is the naturally occurring oxide of titanium. When used as a pigment, it is called *titanium white*, *Pigment White 6* (PW6), or *CI 77891*. Generally, it is produced from ilmenite, rutile, and anatase. It has a wide range of applications, from paint to sunscreen lotion to food coloring. The production method depends on the feedstock.

Titanium dioxide (TiO_2) is by far the most important titanium compound. It can be purified by dissolving in sulfuric acid and precipitating the impurities. The solution is then hydrolysed, washed, and calcined. Alternatively, ground rutile is chlorinated in the presence of carbon, and the resulting titanium tetrachloride is burned in oxygen to produce the chloride.

The most common method for the production of titanium dioxide utilizes the mineral ilmenite. The mineral ore is mixed with sulfuric acid, which reacts to remove the iron oxide group in the ilmenite. The by-product iron sulfate (ferrous sulfate, iron (II) sulfate) is crystallized and removed by filtration to yield only the titanium salt in the digestion solution (synthetic rutile). This product is further processed in a similar way to rutile to give the titanium dioxide product.

3.4 INORGANIC POLYMERS

The previous section has dealt with the industrial production of inorganic compounds, but there is another class of industrial inorganic compounds that also are worthy of mention—inorganic polymers that are polymers with a skeletal structure that does not include carbon atoms in the main backbone of the polymer. Inorganic polymers offer some properties not found in organic materials including low-temperature flexibility, electric conductivity, and nonflammability.

Traditionally, the area of inorganic polymers focuses on materials where the backbone is composed exclusively of main group elements. For example, polymeric forms of the group IV elements are well known; the premier materials are the silicon-based (i) polysiloxane derivatives, sometimes called polysilicone derivatives; (ii) polysilane derivatives, which are analogous to the carbon-based polyethylene products and related organic polymers; and (iii) polyphosphazene derivatives.

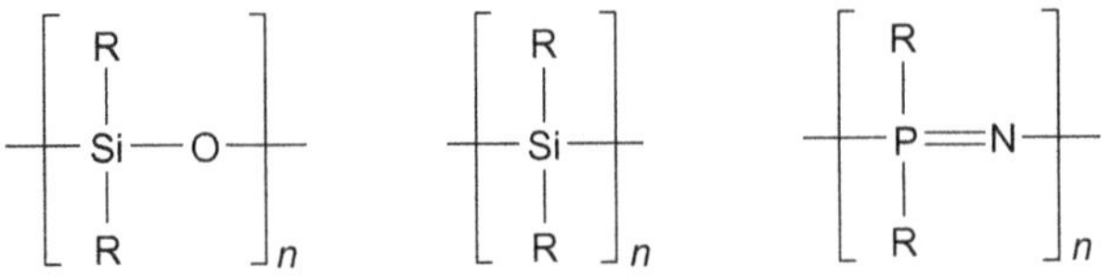

$$\left[-\underset{R}{\overset{R}{\mid}}\!\!Si-O- \right]_n \qquad \left[-\underset{R}{\overset{R}{\mid}}\!\!Si- \right]_n \qquad \left[-\underset{R}{\overset{R}{\mid}}\!\!P{=}N- \right]_n$$

Polysiloxanes Polysilanes Polyphosphazenes

The polysilane derivatives are more fragile than the organic analogs and, because of the longer silicon-silicon bonds, carry higher molecular weight substituents. Inorganic polymers typically focus on one-dimensional polymers, not heavily cross-linked materials such as silicate minerals. The predominant types of polymers are (i) homochain polymers and (ii) the more complex heterochain polymers.

Homochain polymers have only one kind of atom in the main chain or backbone. One member is polymeric sulfur, which forms reversibly upon melting any of the cyclic allotropes, such as S_8. Organic polysulfide derivatives and polysulfane derivatives feature short chains of sulfur atoms, capped, respectively, with an alkyl group and/or a hydrogen atom. On the other hand, *heterochain polymers* have more than one type of atom in the main chain. Typically, two types of atoms alternate along the main chain such as a the polysiloxane derivatives where the main chain features silicon (Si) and oxygen (O) centers, (i.e., —Si—O—Si—O—) in which each silicon atoms can carry two substituents, which are usually methyl substituents or phenyl substituents (Fig. 3.1). Any other bonds, such as silicon-carbon (Si—C) bonds, are not within the backbone.

Linear and cyclic silicones are produced by the reaction of water with organochlorosilane derivatives (R_2SiCl_2) followed by a polymerization reaction that occurs by the elimination of a molecule of water from two hydroxyl groups of adjacent $R_2Si(OH)_2$ molecules:

$$R_2SiCl_2 + 2H_2O \longrightarrow R_2Si(OH)_2 + 2HCl$$

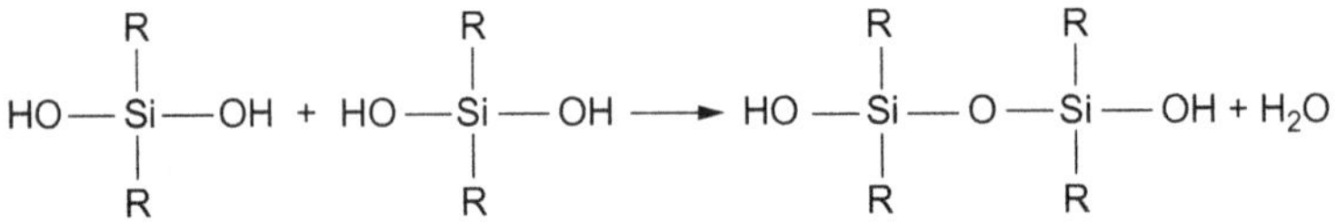

$$HO-\underset{R}{\overset{R}{\mid}}\!\!Si-OH + HO-\underset{R}{\overset{R}{\mid}}\!\!Si-OH \longrightarrow HO-\underset{R}{\overset{R}{\mid}}\!\!Si-O-\underset{R}{\overset{R}{\mid}}\!\!Si-OH + H_2O$$

These types of polymers generally have a high stability because of the presence of strong silicon-oxygen bonds. A general formula for silicones is $(R_2SiO)x$, where R can be any one of a variety of organic groups that are substituent groups or atoms that do not lie within the —Si—O—Si—O— backbone. Silicones may be linear, cyclic, or cross-linked polymers (Fig. 3.1).

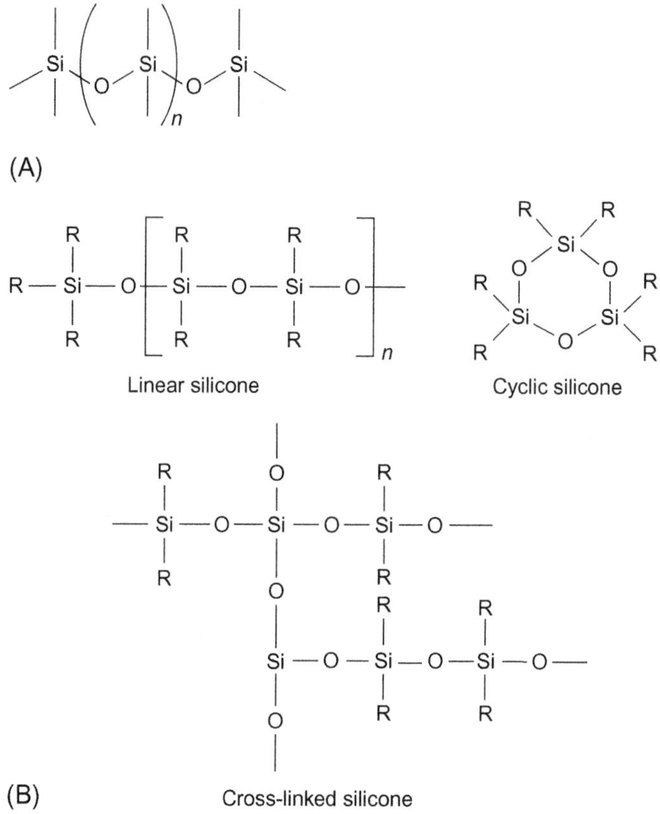

(A)

(B)

Fig. 3.1 General structures for silicone polymers. (A) Chemical backbone of silicone polymers. (B) Alternate structures.

Other examples of inorganic polymers include polydimethylsiloxane derivatives (PDMS, $(Me_2SiO)_n$), polymethylhydrosiloxane derivatives (PMHS, $(MeSi(H)O)_n$), and polydiphenylsiloxane derivatives ($(Ph_2SiO)_n$). Related to the siloxanes are the polysilazane derivatives that are built around the silicon-nitrogen-silicon backbone (e.g., Si—N—Si—N—). One example is perhydridopolysilazane (PHP) in which the valency of the silicon atom (valency = 4) and the valency of the nitrogen atom (valency = 3) are fulfilled by hydrogen atoms. In the polysilane polymers, the polymer backbone contains only silicon atoms. Most uses are in the electronics industry as photoresists and precursors to silicon-containing materials.

A related family of well-studied inorganic polymers is the polyphosphazene derivatives that feature the backbone phosphorus-nitrogen-phosphorus backbone (i.e., —P—N—P—N—). With two substituents on phosphorus, they are structurally similar related to the polysiloxane derivatives. These types of materials are generated by the ring-opening polymerization of hexachlorophosphazene followed by substitution of the P—Cl groups by alkoxide groups.

The polythiazyl derivatives have the backbone sulfur-nitrogen-sulfur backbone (i.e., (—S—N—S—N—)), but unlike most inorganic polymers, these materials lack substituents on the main chain atoms. Such materials exhibit high electric conductivity.

To the inorganic purist, many of these polymers would not be considered to be truly inorganic chemicals because of the presence of organic functional groups in some of the polymers. A simple qualifying definition is that the true organic polymers are those inorganic polymers with a skeletal structure that does not include carbon atoms in the backbone. This leaves open the question of inorganic backbone structures that carry other (organic) substituents. Such containing polymers composed of inorganic chemicals (the backbone) and organic chemicals (the substituents) components are often referred to as hybrid polymers. Thus, many of the so-called inorganic polymers are, in fact, hybrid polymers.

3.5 INORGANIC PIGMENTS

A pigment is a material that changes the color of reflected or transmitted light as the result of wavelength-selective absorption. This physical process differs from fluorescence, phosphorescence, and other forms of luminescence, in which a material emits light. Naturally occurring pigments such as ocher (a natural earth pigment containing hydrated iron oxide—iron(III) hydroxide $Fe(OH)_3$, which is also known as hydrated iron oxide or yellow iron oxide—which ranges in color from yellow to deep orange or brown), and iron oxides have been used as colorants since prehistoric times (Cornell and Schwertmann, 2003).

Inorganic pigments are generally metal oxides and/or synthetic chemicals, some of which are very simple in composition but find wide use in paints and coatings. Previously, the inorganic pigments were all naturally occurring colored chemicals. Examples of inorganic pigments include lead oxide, cobalt blue, cadmium yellow, and titanium yellow (Table 3.4). As new environmental laws are very strict about toxicity, a few of these heavy metal pigments are no longer in use. However, the surviving inorganic pigments fill a play double-duty role fillers that provide a greater benefit than simple coloration of a formulation; they also impact physical properties of the

Table 3.4 List of the Common Inorganic Pigments[a]

Purple Pigments

Aluminum pigments
Ultramarine violet; silicate of sodium and aluminum containing sulfur ($Na_{8-10}Al_6Si_6O_{24}S_{2-4}$)
Copper pigments
Han purple; $BaCuSi_2O_6$
Cobalt pigments
Cobalt violet; cobaltous orthophosphate, Co3(PO4)2
Manganese pigments
Manganese violet; manganese ammonium pyrophosphate, $NH_4MnP_2O_7$

Blue Pigments

Aluminum pigments
Ultramarine; a complex naturally occurring pigment of sodium-silicate ($Na_{8-10}Al_6Si_6O_{24}S_{2-4}$)
Cobalt pigments
Cobalt blue and cerulean blue; cobalt(II) stannate, $CoO_n{\cdot}SnO_2$ ($n=1$ or 2)
Copper pigments
Egyptian blue; a synthetic pigment of calcium copper silicate ($CaCuSi_4O_{10}$)
Han blue; a barium copper silicate, $BaCuSi_4O_{10}$
Azurite; cupric carbonate hydroxide ($Cu_3(CO_3)_2(OH)_2$)
Iron pigments
Prussian blue; a synthetic pigment of ferric hexacyanoferrate ($Fe_7(CN)_{18}$)
Manganese pigments
YInMn blue; a synthetic pigment ($YIn_{1-x}Mn_xO_3$).

Green Pigments

Cadmium pigments
Cadmium green; a light green pigment consisting of a mixture of cadmium yellow (CdS) and viridian (Cr_2O_3)
Chromium pigments
Chrome green; chromic oxide (Cr_2O_3)
Viridian; a dark green pigment of hydrated chromic oxide ($Cr_2O_3{\cdot}H_2O$)
Cobalt pigments
Cobalt green; also known as Rinmann's green or zinc green ($CoZnO_2$)
Copper pigments
Malachite; cupric carbonate hydroxide ($Cu_2CO_3(OH)_2$)
Paris green; cupric acetoarsenite ($Cu(C_2H_3O_2)_2{\cdot}3Cu(AsO_2)_2$)
Scheele's green (also called Schloss green); cupric arsenite ($CuHAsO_3$)
Verdigris; various poorly soluble copper salts, notably cupric acetate ($Cu(CH_3CO_2)_2$) and malachite ($Cu_2CO_3(OH)_2$)
Other pigments
Green earth; also known as *terre verte* and verona green $K((Al{\cdot}Fe^{III})(Fe^{II}{\cdot}Mg)(AlSi_3{\cdot}Si_4)O_{10}(OH)_2)$

Continued

Table 3.4 List of the Common Inorganic Pigments—cont'd

Yellow Pigments

Arsenic pigments
  Orpiment; natural monoclinic arsenic sulfide (As_2S_3)
Cadmium pigments
  Cadmium yellow; cadmium sulfide (CdS), which also occurs as the mineral greenockite
Chromium pigments
  Chrome yellow; lead chromate ($PbCrO_4$), which also occurs as the mineral crocoite
Cobalt pigments
  Aureolin (also called cobalt yellow); potassium cobaltinitrite ($K_3Co(NO_2)_6$)
Iron Pigments
  Yellow ochre; a naturally occurring clay of monohydrated ferric oxide ($Fe_2O_3\cdot H_2O$)
Lead pigments
  Naples yellow; lead antimonate, $Pb(SbO_3)_2\cdot Pb_3(SbO_4)_2$
  Lead-tin-yellow; $PbSnO_4$ or $Pb(Sn\cdot Si)O_3$
Titanium pigments
  Titanium yellow also called nickel antimony titanium yellow, nickel antimony titanium yellow rutile, $NiO\cdot S_b2O_3\cdot 20TiO_2$
Tin Pigments
  Mosaic Gold: stannic sulfide (SnS_2)
Orange Pigments
Cadmium pigments
  Cadmium orange; an intermediate between cadmium red and cadmium yellow; cadmium sulfoselenide ($CdS\cdot CdSe$)
Chromium pigments
  Chrome orange; a mixture of lead chromate and lead(II) oxide ($PbCrO_4\cdot PbO$)

Red Pigments

Arsenic pigments
  Realgar; an arsenic sulfide mineral (As_4S_4)
Cadmium pigments
  Cadmium red; cadmium selenide (CdSe)
Iron oxide pigments
  Sanguine, caput mortuum, indian red, venetian red, oxide red (Fe_2O_3)
  Red Ochre; anhydrous Fe_2O_3
  Burnt sienna; a pigment produced by heating raw sienna ($Fe_2O_3\cdot MnO_2\cdot nH_2O$)
Lead pigments
  Minium (pigment); also known as red lead, lead tetroxide (Pb_3O_4)
Mercury pigments
  Vermilion; synthetic and natural pigment; occurs naturally in mineral cinnabar (mercuric sulfide), (HgS)

Table 3.4 List of the Common Inorganic Pigments—cont'd

Brown Pigments

Clay earth pigments (naturally formed iron oxides)
Raw umber; a natural clay pigment consisting of iron oxide, manganese oxide and aluminum oxide; $Fe_2O_3 \cdot MnO_2 \cdot nH_2O \cdot Si \cdot AlO_3$. When calcined (heated), it is referred to as burnt umber and has more intense colors
Raw sienna ($Fe_2O_3 \cdot MnO_2 \cdot nH_2O$); a naturally occurring yellow-brown pigment from limonite clay

Black Pigments

Carbon pigments
Carbon black; produced by incomplete combustion of carbonaceous material
Ivory black; produced by charring ivory or bones
Vine black; produced by charring desiccated grape vines and stems
Lamp black (lampblack); produced by collecting soot from oil lamps
Iron pigments
Mars black (iron black); Fe_3O_4
Manganese pigments
Manganese dioxide; blackish or brown in color, used since prehistoric times (MnO_2)
Titanium pigments
Titanium black; Titanium(III) oxide (Ti_2O_3)

White Pigments

Antimony pigments
Antimony white; stibous oxide, antimony(III) oxide, Sb_2O_3
Barium pigments
Barium sulfate; $BaSO_4$
Lithopone; $BaSO_4 \cdot ZnS$
Lead pigments
Cremnitz white; basic lead carbonate ($(PbCO_3)_2 \cdot Pb(OH_2)$)
Titanium pigments
Titanium white; titanium dioxide (TiO_2)
Zinc pigments
Zinc white; zinc oxide (ZnO)

[a]Inorganic pigments of natural and synthetic origin

paint film or the coating film during application and throughout the product lifecycle.

A major property for pigments (or dispersing any solids) is that the solid has a higher surface tension than the liquid it is being introduced into. For solids that have a low surface tension, additives can be added that lower the surface tension of the liquid components, thereby allowing for improved wetting characteristics. Also, dry pigment powders can be modified via surface coatings (i) to provide specific traits to the product or (ii) to improve upon the ease of manufacturing coatings containing them. As an example, titanium dioxide can be coated with polysiloxanes (commonly, silicone oils) to modify the compressive strength and compressibility of the pigment.

The modern pigments for the coating manufacturer are available in specific particle size and tightness ranges. Reduction of particle size (i.e., severing chemical bonds) is not the main goal of pigment dispersion, but rather deagglomeration of pigment particles loosely bound by residual moisture remnant from the original particle size reduction. The manufacturer reduces the size of the pigment particles by milling for many hours or days on a three-roll mill or similar machinery. The pigment is then dried to remove the majority of the moisture; some solvent remains entrapped between the agglomerates, due to capillary forces and wetting phenomena, causing the pigment to form a cake. When the pigment is added to the coating formulation or to a solvent predispersion, the dry cake is quickly broken apart into coarse grains due to the high shear created by the disperser. This leave behind a collection of agglomerates—which can be further broken up with a reasonable energy input—and aggregates that would require a high (possibly 10-fold) energy input to further reduce to primary particle size.

REFERENCES

Bahadori, A., 2014. Natural Gas Processing: Technology and Engineering Design. Gulf Professional Publishing/Elsevier, Amsterdam.

Brock, W.H., 2000. The Chemical Tree: A History of Chemistry. Norton Publishers, New York, NY.

Buchel, K.H., Moretto, H.-H., Werner, D., 2008. Industrial Inorganic Chemistry, second ed. John Wiley & Sons Inc., Hoboken, NJ.

Budavari, S. (Ed.), 1996. The Merck Index: An Encyclopedia of Chemicals, Drugs, and Biologicals, 12th ed. Merck, Whitehouse Station, NJ.

Cornell, R.M., Schwertmann, U., 2003. The Iron Oxides: Structure, Properties, Reactions, Occurrences and Uses. Wiley-VCH/John Wiley & Sons Inc., Weinheim/Hoboken, NJ.

El-Dessouky, H., Ettouney, H., 2002. Fundamentals of Salt Water Desalination. Elsevier, New York, NY.

Fabuss, B.M., 1980. Principles of Desalination, second ed. Academic Press, New York, NY, pp. 746–799. Part B.

Heaton, A. (Ed.), 1994. The Chemical Industry, second ed. Blackie Academic & Professional, New York, NY.

Heaton, A. (Ed.), 1996. An Introduction to Industrial Chemistry, third ed. Blackie Academic & Professional, New York, NY.

Kidnay, A.J., Parrish, W.R., McCartney, D.G., 2011. Fundamentals of Natural Gas Processing, second ed. CRC Press/Taylor & Francis Group, Boca Raton, FL.

Mokhatab, S., Poe, W.A., Speight, J.G., 2006. Handbook of Natural Gas Transmission and Processing. Elsevier, Amsterdam.

Riley, J.P., Skirrow, G., 1975. Chemical Oceanography, second ed., vol. 1 Academic Press, New York, NY, pp. 557–589.
Schrödter, K., Bettermann, G., Staffel, T., Wahl, F., Klein, T., Hofmann, T., 2008. Phosphoric Acid and Phosphates. Ullmann's Encyclopedia of Industrial Chemistry. Wiley-VCH, Weinheim.
Sharqawya, M.H., Lienhard, J.H., Zubair, S.M., 2010. Thermophysical properties of seawater: a review of existing correlations and data. Desalin. Water Treat. 16, 354–380.
Speight, J.G., 2007. Natural Gas: A Basic Handbook. GPC Books, Gulf Publishing Company, Houston, TX.
Speight, J.G., 2014. The Chemistry and Technology of Petroleum, fifth ed. CRC Press/Taylor & Francis Group, Boca Raton, FL.
Speight, J.G., 2017a. Handbook of Petroleum Refining. CRC Press/Taylor & Francis Group, Boca Raton, FL.
Speight, J.G., 2017b. Environmental Organic Chemistry for Engineers. Butterworth-Heinemann/Elsevier, Cambridge, MA.

CHAPTER FOUR

Properties of Inorganic Compounds

4.1 INTRODUCTION

Scientists and engineers have wondered and continue to wonder about the properties of matter especially in the light the new phenomena have been discovered but perhaps not satisfactorily explained. For example, the concepts of superfluids and superconductivity have become new areas of advance in understanding. However, the laws of thermodynamics have not changed, and the properties of, say, copper metal are the same as they have been since the metal was first discovered and worked as the metal used to make bronze.

In the modern world, an inorganic chemical can be considered as a compound that does not contain a carbon-hydrogen bond (Chapter 2). Moreover, inorganic chemicals tend to occur as minerals or geologically based compounds that do not contain carbon-to-hydrogen bonds (Chapters 1 and 2). Thus, there are numerous compounds that fall into the realm (or subdiscipline) of inorganic. In fact, most compounds within the known universe are inorganic in nature, and for this reason, inorganic compounds have an overwhelming amount of applications and practical uses in the real world of planet Earth. The result is that the Earth is inundated by inorganic chemicals that can take on a host of forms and possess many different characteristics (Galasso et al., 1970).

Thus, due to ionic bonding typically found in inorganic compounds (Chapter 2), they are held together very rigidly and possess extremely high melting and boiling points. Another distinct characteristic of inorganic compounds is their color. Transition metal inorganic compounds, even sitting on a benchtop, are usually highly colored, and this is, again, due to the configuration of the "d-block" electrons. Because inorganic compounds display a unique color when burned, this can be used as a "marker" to identify the metal involved. Also, inorganic compounds are typically highly soluble in

Environmental Inorganic Chemistry for Engineers
http://dx.doi.org/10.1016/B978-0-12-849891-0.00004-7

water, and another revealing characteristic of inorganic compounds is the ability of many inorganic chemicals to form crystals.

Briefly, a crystal is a chemical in which the constituents, such as atoms, molecules, or ions, are arranged in a highly ordered microscopic structure. These constituents are held together by interatomic forces (chemical bonds) such as metallic bonds, ionic bonds, covalent bonds, and van der Waals bonds. The crystalline state of matter is the state with the highest order, that is, with very high internal correlations and at the greatest distance range. This is reflected in their properties: anisotropic and discontinuous. Crystals usually appear as unadulterated, homogeneous, and with well-defined geometric shapes when they are well formed.

In addition, since many inorganic compounds contain some type of metal (alkali, alkaline, transition, etc.), they tend to be able to conduct electricity. For example, while in the solid state, inorganic compounds may be poor conductors of electricity, but in the liquid phase, many (if not all) inorganic compounds are highly conductive because of the free movement of electrons in this phase, and this movement of electrons is manifested as electricity.

Furthermore, there are two basic types of properties that are associated with matter: (i) chemical properties, which are properties that change the chemical nature of matter, and (ii) physical properties, which are properties that do not change the chemical nature of matter. All inorganic chemicals have chemical and physical properties that can be used as a means of identification of the behavior of the chemical (Table 4.1). The more properties can be used for identification of an inorganic substance, the better the understanding of the nature of that chemical substance that can further assist in an understanding of the way in which the substance will behave under various (in the context of this book) environmental conditions. In addition, the critical properties of a chemical are part of the property assessment and behavior of the chemical (Section 4.4).

In fact, the assessment of the physicochemical properties is an early step in alternative assessment because physical hazards, environmental fate, and intrinsic human health hazards and ecotoxicity are directly related to a chemical's intrinsic physicochemical properties. However, unlike the hazardous organic constituents (Speight, 2017b), metals cannot be degraded or readily detoxified. The presence of metals among wastes can pose a long-term environmental hazard (Table 4.2). Metals may be found in the elemental form, but more often, they are found as salts mixed in the soil. Asbestos fibers require special care to prevent their escape during handling

Table 4.1 The Various Physical and Chemical Properties of Chemicals

Property	Rationale for Inclusion
Flammability	Associated with flammability hazard
Corrosivity	Associated with ability to gradually destroy materials by chemical reactions
Oxidizing ability	Associated with ability to give off oxidizing substances or oxidize combustible materials, increasing fire, or explosion hazards
Melting and boiling point	Impacts environmental fate and transport and potential bioavailability
Vapor pressure	Impacts environmental fate and transport and potential bioavailability
Acidity (pK_a)	Determines ionization state in the environment and in biological compartments
Aqueous solubility	Reflects ability to partition into aquatic environment
Octanol-water partition coefficient (log P)	Important determinant of human/mammalian oral and skin bioavailability and relevance to acute and chronic aquatic toxicity (narcosis) and directly related to bioconcentration
Henry's law constant (log $P_{w/g}$)	Relevance to environmental partitioning and transport and human/mammalian alveolar absorption
Molecular electronic dipole moments, μ, and dipole polarizabilities, α	Important in determining the energy, geometry, and intermolecular forces of molecules, often related to biological activity
Biodegradation	Indicator of persistence, and persistence is tied to ecotoxicity
Bioconcentration factor (BCF)	Bioconcentration enhances the hazard potential of lipophilic chemicals and provides a comparative basis for assessing the potential for a chemical to have effects that resonate through the food chain

and disposal; permanent containment must be provided. Treatment options for radioactive materials (Table 4.3) are limited to volume reduction/concentration and immobilization. Implementation of remediation technologies should consider the potential for radiological exposure; the degree of exposure is based on the radionuclide(s) present (Table 4.4) and the type

Table 4.2 Properties and Behavior of Selected Metals in the Environment[a]

Arsenic	Arsenic (As) exists in the soil environment as arsenate, As(V), or as arsenite, As(III). Both are toxic; however, arsenite is the more toxic form and arsenate is the most common form. The behavior of arsenate in soil seems analogous to that of phosphate because of their chemical similarity Like phosphate, arsenate is fixed to soil and thus is relatively immobile. Iron (Fe), aluminum (Al), and calcium (Ca) influence this fixation by forming insoluble complexes with arsenate. The presence of iron in soil is most effective in controlling arsenate's mobility Arsenite compounds are 4–10 times more soluble than arsenate compounds. Under anaerobic conditions, arsenate may be reduced to arsenite. Arsenite is more subject to leaching because of its higher solubility. The adsorption of arsenite is also strongly pH-dependent
Barium	Barium (Ba) metal does not occur in nature. The most common ores are the sulfate (barite) and the carbonate (witherite). Barium is released to water and soil in the discharge and disposal of drilling wastes, from the smelting of copper, and the manufacture of motor vehicle parts and accessories In water, the more toxic barium salts are likely to precipitate out as the less toxic insoluble sulfate or carbonate. Barium is not very mobile in most soil systems. Adsorption of barium was measured in a sandy soil and a sandy loam soil at levels closely corresponding to those expected for field conditions. In general, sludge solutions appeared to increase the mobility of elements in a soil
Cadmium	Cadmium (Cd) most often occurs in small quantities associated not only with zinc ores but also with copper and lead ores. Cadmium oxide and sulfide are relatively insoluble, while the chloride and sulfate salts are soluble The adsorption of cadmium onto soils and silicon or aluminum oxides is strongly pH-dependent, increasing as conditions become more alkaline. When the pH is below 6–7 (i.e., in acidic media), cadmium is desorbed from these materials. Cadmium has considerably less affinity for the absorbents tested than do copper, zinc, and lead and might be expected to be more mobile in the environment than these materials The mode by which cadmium is sorbed to the sediments is important in determining its disposition toward remobilization. Cadmium found in association with carbonate minerals, precipitated as stale solid compounds, or coprecipitated with hydrous iron oxides would be less likely to be mobilized by resuspension of sediments or biological activity. Cadmium absorbed to mineral surfaces (e.g., clay) or organic materials would be more easily bioaccumulated or released in the dissolved state when sediments are disturbed, such as during flooding

Table 4.2 Properties and Behavior of Selected Metals in the Environment—cont'd

Chromium	Chromium (Cr) can exist in soil in three forms: the trivalent Cr(III) form, Cr^{+3}, and the hexavalent Cr(VI) forms, $(Cr_2O_7)^{-2}$ and $(CrO_4)^{-2}$. Hexavalent chromium is the major chromium species used in industry; wood preservatives commonly contain chromic acid, a Cr(VI) oxide. The two forms of hexavalent chromium are pH-dependent; hexavalent chromium as a chromate ion $(CrO_4)^{-2}$ predominates above a pH of 6; dichromate ion $(Cr_2O_7)^{-2}$ predominates below a pH of 6. The dichromate ions present a greater health hazard than chromate ions, and both Cr(VI) ions are more toxic than Cr(III) ions
	Because of its anionic nature, Cr(VI) associates only with soil surfaces at positively charged exchange sites. This association decreases with increasing soil pH. Iron and aluminum oxide surfaces adsorb the chromate ion at an acidic or neutral pH
	Chromium (III) is the stable form of chromium in soil. Cr(III) hydroxy compounds precipitate at pH 4.5, and complete precipitation of the hydroxy species occurs at pH 5.5. In contrast to Cr(VI), Cr(III) is relatively immobile in soil. Chromium (III) does, however, form complexes with soluble organic ligands, which may increase its mobility
	Regardless of pH, Cr(VI) in soil is reduced to Cr(III). The reduction reaction in the presence of organic matter proceeds at a slow rate under normal environmental pH and temperatures, but the rate of reaction increases with decreasing soil pH
Copper	Copper (Cu) is retained in the soil through exchange and specific adsorption. Copper adsorbs to most soil constituents more strongly than any other toxic metal, except lead (Pb). Copper, however, has a high affinity to soluble organic ligands; the formation of these complexes may greatly increase its mobility in soil. Copper has high toxicity to aquatic organisms
Lead	Lead (Pb) is a heavy metal that exists in three oxidation states, O, +2(II), and +4(IV). Lead is generally the most widespread and concentrated contaminant present at a lead battery recycling site (i.e., battery breaker or secondary lead smelter)
	Lead tends to accumulate in the soil surface, and concentration tends decrease with depth. Insoluble lead sulfide (PbS) is typically immobile in soil as long as reducing conditions are maintained
	The capacity of soil to adsorb lead increases with pH, cation exchange capacity, organic carbon content, soil/water Eh (redox potential), and phosphate levels. Lead exhibits a high degree of adsorption on clay-rich soil. Only a small percent of the total lead is leachable; the major portion is usually solid or adsorbed onto soil particles. Surface runoff, which can transport soil particles containing adsorbed lead, facilitates migration and subsequent desorption from contaminated soils. On the other hand, groundwater (typically low in suspended soils and leachable lead salts) does not normally create a major pathway for lead migration

Continued

Table 4.2 Properties and Behavior of Selected Metals in the Environment—cont'd

	Lead compounds are soluble at low pH and at high pH, such as those induced by solidification/stabilization treatment. Several other metals are also amphoteric, which strongly affects leaching
Mercury	Mercury (Hg) is extremely toxic and very mobile in the environment. In soils and surface waters, volatile forms (e.g., metallic mercury and dimethylmercury) evaporate to the atmosphere, whereas solid forms partition to particulates. Mercury exists primarily in the mercuric (Hg^{2+}) and mercurous (Hg^{+}) forms as several complexes with varying water solubility
	In soils and sediments, sorption is one of the most important controlling pathways for removal of mercury from solution; sorption usually increases with increasing pH. Other removal mechanisms include flocculation, coprecipitation with sulfides, and organic complexation. Mercury is strongly sorbed to humic materials
	Inorganic mercury sorbed to soils is not readily desorbed; therefore, freshwater and marine sediments are important repositories for inorganic mercury
Selenium	Selenium (Se) occurs in nature usually in the sulfide ores of the heavy metals and constitutes about 0.09 ppm of the earth's crust. It is the most strongly enriched element in coal, being present as an organoselenium compound, a chelated species, or as an adsorbed element. Selenium is used extensively in the manufacture and production of glass, pigments, rubber, metal alloys, textiles, petroleum, medical therapeutic agents, and photographic emulsions. Selenium dioxide is the most widely used selenium compound in industry. It is used as an oxidizing agent in drug and other chemical manufacture, a catalyst in organic syntheses, and an antioxidant in lubricating oils (Note, selenium is not a true metal; however, it is included here as it is one of the eight RCRA metals.)
	The toxicity of selenium depends on whether it is in the biologically active oxidized form. In alkaline soils and oxidizing conditions, selenium may be oxidized sufficiently to maintain the availability of its biologically active form and cause plant uptake of the metal to be increased. In acidic or neutral soils, it tends to remain relatively insoluble, and the amount of biologically available selenium should steadily decrease. Selenium volatilizes from soils when converted to volatile selenium compounds (e.g., dimethyl selenide) by microorganisms
Silver	Silver (Ag) occurs naturally and in ores such as argentite (Ag_2S) and horn silver (AgCl). Lead, lead-zinc, copper, gold, and copper-nickel ores are the principal source
	Silver itself is not considered to be toxic; most of its salts are poisonous due to the anions present. Silver compounds can be absorbed in the circulatory system and reduced silver deposited in the various tissues of the body

Table 4.2 Properties and Behavior of Selected Metals in the Environment—cont'd

Zinc	Zinc (ZN) is readily adsorbed by clay minerals. The greatest percentage of total zinc in polluted soil and sediment is associated with iron (Fe) and manganese (Mn) oxides Rainfall removes zinc from soil because the zinc compounds are highly soluble. As with all cationic metals, zinc adsorption increases with pH. Zinc hydrolyzes at a pH >7.7. These hydrolyzed species strongly adsorb to soil surfaces. Zinc forms complexes with inorganic and organic ligands, which will affect its adsorption reactions with the soil surface

[a]Listed alphabetically rather than on the basis of effects.

Table 4.3 Properties and Treatment Methods for Radionuclides

- Radionuclides are not destroyed in nature or by natural forces, and *ex situ* techniques require eventual disposal of residual radioactive wastes, which must meet the acceptable disposal criteria
- There are different disposal requirements associated with different types of radioactive waste. The technologies may not be applicable to spent nuclear fuel and, for the most part, are not applicable for high-level radioactive waste
- If a remediation technology results in the concentration of radionuclides, there may (of necessity) be a change in the classification of the waste, which impacts the requirements for disposal. Thus, waste classification requirements for the disposal of residual waste (if applicable) should be given full consideration when remediation technologies are being evaluated
- Disposal capacity for radioactive and mixed waste is limited, and a suitable remediation technology may not be available. In addition, mixed waste can be treated to address the hazardous characteristics of the soil, thereby allowing the waste to be considered or classified solely as a radioactive waste

Table 4.4 List of Typical Radionuclide Contaminants

Americium-241	Iodine-129, 131	Ruthenium-103,106
Barium-140	Krypton-85	Silver-110
Carbon-14	Molybdemum-99	Strontium-89, 90
Cerium-144	Neptunium-237	Technetium-99
Cesium-134, -137	Plutonium-238, 239, 241	Tellurium-132
Cobalt-60	Polonium-210	Thorium-228, 230, 232
Curium-242, 244	Radium-224, 226	Tritium
Europium-152, 154, 155	Radon-222	Uranium-234, 235, 238

and energy of radiation emitted (i.e., alpha particles, beta particles, gamma radiation, and neutron radiation), and there should be efforts to maintain exposure to zero.

More specifically, the fate of the metal depends on its physical and chemical properties, associated waste matrix, and soil. Significant downward transportation of metals from the soil surface occurs when the metal retention capacity of the soil is overloaded or when metals are solubilized (e.g., by low pH, i.e., high acidity). As the concentration of metals exceeds the ability of the soil to retain them, the metals will travel downward with the leaching waters. Surface transport through dust and erosion of soils is a common transport mechanism. The extent of vertical contamination intimately relates to the soil solution and surface chemistry.

Physicochemical properties such as those indicative of physical hazards could be used to eliminate a chemical from consideration and prioritize chemicals for further screening for human and ecotoxicological effects. Several properties can be informative to alternative assessment, as described in detail in this chapter, and a high-priority data set is also defined. Property data can be obtained from experimental or in silico methods (in-glass or in-the-lab methods). In fact, state-of-the-art methodologies are making in-laboratory methods increasingly reliable, low-cost approaches. The suggested uses of the derived physicochemical property data are (i) to identify the potential for direct physical hazards posed by the chemical or material, (ii) to determine the environmental compartment(s) into which the chemicals will partition, (iii) to estimate the potential for bioconcentration and bioavailability, (iv) to estimate the likely route(s) of mammalian exposure and bioavailability and the likelihood for high aquatic toxicity, and (v) to estimate the potential for inducing floral and faunal human toxicity.

By way of explanation, the *critical temperature* of an inorganic chemical is (i) the temperature above which a gas cannot be liquefied, regardless of the pressure applied; (ii) the temperature at which a chemical becomes a superconductor; and (iii) the temperature at which the property of a chemical, such as its magnetism, changes. On the other hand, the critical pressure is (i) the minimal pressure required to liquefy a gas at the critical temperature or (ii) the pressure of a gas or the saturated vapor pressure of a chemical in its critical state. Furthermore, at the *critical point*, the particles in a closed container are thought to be vaporizing at such a rapid rate that the density of liquid and vapor is equal and thus forms a supercritical fluid.

Every gas has a critical temperature above which it cannot be liquefied by the application of pressure alone. The critical pressure is that required to liquefy a gas at its critical temperature. As a consequence, liquefied gases may be stored fully refrigerated, with the liquid at its bubble point at near atmospheric pressure; fully pressurized, that is, at ambient temperature; or semirefrigerated with the temperature below ambient but the vapor pressure above atmospheric pressure. Gases with critical temperatures below ambient must be maintained under refrigeration to keep them in the liquid phase. If the temperature remains constant, the pressure within any cylinder containing liquefied gas will remain constant as gas is drawn off (i.e., more liquid simply evaporates), so the quantity of gas remaining cannot be deduced from the pressure.

A supercritical fluid has the properties between those of a gas and a liquid. Although supercritical fluids are common in the subdiscipline of organic chemistry (Speight, 2017a), they are less common in the subdiscipline of inorganic chemistry—carbon dioxide, nitrous oxide (N_2O), and water being the most commonly used supercritical fluids (Table 4.5).

Because of the high rates of change, the surface tension of the liquid eventually disappears. The temperature and pressure corresponding to the critical point are the critical temperature and critical pressure. The condensation of a gas will never occur above the critical point. High pressures can be applied to a gas in a closed container, and it may become highly dense, but will not exhibit a meniscus. Molecules at critical temperatures possess high kinetic energy, and as a result, the intermolecular forces in the molecules are weakened. In addition, when a fluid is present in two phases and a critical point is near establishment, contact with the imminently forming third phase does not occur. This phenomenon can be accounted for by examining the other two existing phases. The third phase does not immediately form because one of the other two phases wets the third phase, causing it to be eliminated. This wetting phase will continually occur when a phase is not entirely stable.

Table 4.5 Critical Properties of Carbon Dioxide, Water, and Nitrous Oxide

Solvent	Molecular Weight, g/mol	Critical Temperature, K	Critical Pressure, MPa (atm)	Critical Density, g/cm^3
Carbon dioxide (CO_2)	44.01	304.1	7.38 (72.8)	0.469
Nitrous oxide (N_2O)	44.013	306.57	7.35 (72.5)	0.452
Water (H_2O)	18.015	647.096	22.064 (217.755)	0.322

The properties of inorganic compounds (Speight, 2017a), as with any chemical, are an important aspect of handling the chemicals and the properties as they influence disposal of the chemical into the environment. At this point, a general comparison of the inorganic compounds and the properties of organic compounds is worthwhile to set the stage for the relative behavior of inorganic compounds.

Inorganic compounds form ionic bonds have high melting points and are composed of either (i) single elements or (ii) compounds that do not include carbon and, in some cases, do not contain hydrogen. This latter qualification, the lack of carbon and hydrogen, is open to serious questioning because compounds such as sodium bicarbonate ($NaHCO_3$) contain carbon and hydrogen and compounds such as lithium aluminum hydride ($LiAlH_3$) are inorganic chemicals that, in the first case, contain carbon and hydrogen and, in the second case, contain hydrogen. Caution is always advised when applying general (umbrella) definitions to any chemical.

In solution, inorganic compounds break down into ions (charged entities) that conduct electricity. In addition, inorganic compounds are typically (but not always) nonvolatile, while organic chemicals, on the other hand, organic compounds have a carbon-based structure with covalent bonding (nonpolar bonding) and are often volatile in nature. Moreover, even in liquid state, organic compounds do not conduct electricity unless they are salts that are formed because of interactions between the organic compound with inorganic acids and bases.

Because of the covalent bonding in organic compounds (Chapter 2) (Speight, 2017b), most organic compounds are nonpolar, although some organic compounds may ionize in water or produce ions when subjected to electrolysis. The volatility of organic compounds is due to the weakness of the covalent bonding relative to the strength of ionic bonds. In fact, the volatility of organic compounds is increased in situations in which the main attractive force between the molecules is due to weak bonds, such as hydrogen bonds and Van der Waals forces. As a result of the volatility, many organic compounds, including the hydrocarbons, butane, hexane, propane, and octane, are used as fuels. Overall, the properties of organic chemicals are somewhat easier to estimate than the properties of inorganic chemicals.

In fact, the estimation of environmental properties of inorganic species has been difficult. Properties such as aqueous solubility, bioconcentration, and acute toxicity need to be estimated for inorganic compounds using existing relationships. Furthermore, for complex solution chemistry, the accuracy of the estimations improves when a more complete description of the solution species is available, and the toxicity of inorganic chemicals also depends on an estimation of the bioactive amount and the molecular configuration of the chemicals (Hickey, 1999).

Thus, the primary purpose of this chapter is to present an overview of the environmentally important physical and chemical properties of inorganic chemicals; a full description of the properties would require several volumes. Like the "organics" book and the various chapters (Speight, 2017b), this chapter describes an overview of the available information dealing with the properties of inorganic chemicals to support the primary objective of the book, which is to assist engineers (many of whom may not have a detailed knowledge or understanding of inorganic environmental chemistry) in overcoming the common problem of property data gaps and developing timely responses to environmental issues that arise on a regular basis. Moreover, this chapter book covers environmentally important properties and presents not only generic discussion of these properties but also a summary of environmentally important data for the most common elements and pollutants.

4.2 BOND LENGTHS AND BOND STRENGTHS

The *molecular geometry* of a molecule (sometimes referred to as the *molecular shape* of a molecule) is the specific three-dimensional arrangement of atoms and the positions of the atomic nuclei in a molecule. This shape is dependent on the preferred spatial orientation of the moieties that form the molecule. The three-dimensional shape or configuration of any molecule is an important characteristic of that molecule and has an influence on the properties of the molecule. Various instrumental techniques such as X-ray crystallography and other experimental techniques can be used to derive information about the locations of atoms in a molecule. Thus, molecular geometry is associated with the specific orientation of bonding atoms. However, the best way to study the three-dimensional shapes of molecules is by using molecular models, and many kinds of model kits are available for use by chemists and engineers.

Furthermore, *molecular symmetry* is a means of describing the symmetry present in a molecule and the classification of the molecule according to the symmetry. Molecular symmetry is a fundamental concept in chemistry, as it can predict or explain many of the chemical properties of the molecule, such as the dipole moment. Many techniques for the practical assessment of molecular symmetry exist, including X-ray crystallography and various forms of spectroscopy (Table 4.6).

Table 4.6 Common Spectroscopic Methods for Identification of Inorganic Chemicals

UV-Vis absorption spectroscopy
The number, energies, and intensities of a transition metal compound's absorption bands in the UV-Vis and near IR can be used to determine the general type of atom bound to a metal and the geometry about the metal. There are some complicating factors, which we will discuss in lecture, but there is enough published about the spectroscopy of transition metal compounds to help you sort things out
IR absorption spectroscopy
In simple compounds, the number, energy, and intensity of the IR transitions are directly related to the geometry of compound and to which atoms are bound to which other atoms. Unfortunately, some metal-ligand vibrations and many of the vibrations for the heavier elements occur outside the frequency window of most commercial instruments. For complex compounds involving large organic moieties, the IR becomes more difficult to interpret. IR is useful to determine the presence of complex counter ions such as PF^{6-}, ClO^{4-}, and BF^{4-}, because they have distinctive absorptions in the infrared
Raman spectroscopy
Raman spectroscopy is complementary to infrared absorption spectroscopy. Both probe vibrations within a compound, but they have different selection rules. By considering what peaks are present or absent, in the two spectra of a compound, one can determine geometry, at least in simple cases. Raman is also useful because it can, depending on instrument design, scan to very low frequencies ($\sim$100 cm^{-1}), and thus observe transitions too low for infrared absorption
Nuclear magnetic resonance spectroscopy
Nuclear magnetic resonance is usually performed on diamagnetic compounds. The nuclear magnetic resonance spectra of inorganic compounds are often more complicated than organics because other nuclei also have nuclear magnetic moments. So, in addition to the familiar 1H ($I=1/2$) and 13C ($I=1/2$), there are $\sim$90 other elements that have at least one NMR-active nucleus. Although less widespread than the standard solution NMR, solid-state NMR and even single-crystal NMR have been used on materials that simply do not dissolve in any solvent
Electron paramagnetic resonance spectroscopy
While NMR is usually only for diamagnetic compounds, electron paramagnetic resonance (EPR or ESR) spectroscopy is typically used to identify paramagnetic compounds with an odd number of unpaired electrons. In this technique, a sample is irradiated with microwave radiation, and the field is swept until resonance occurs. The field at which resonance occurs depends on the number of unpaired electrons, the geometry about the metal center, and the metal ligands. In many ways, EPR and NMR are the same, and there is even a technique that combines them such as electron-nuclear double resonance spectroscopy (ENDOR)

Table 4.6 Common Spectroscopic Methods for Identification of Inorganic Chemicals—cont'd

X-Ray crystallography
Single-crystal X-ray diffraction is the most powerful X-ray technique for inorganic chemists. From precise measurement of the intensity and angles at which an X-ray beam diffracts off a crystal, the arrangement of the atoms can be reconstructed. Obviously, as a direct probe of structure crystallography is an invaluable characterization method for all types of compounds. Some inorganic compounds (such as rocks and minerals) cannot be obtained as single crystals. In these cases, the X-ray method can be used to obtain the dimensions of the unit cell for use in identification (there is a large, indexed catalog of lattice constants for many minerals)

A major determinant of the molecular shape is the *bond length* or *bond order*. The distances between centers of bonded atoms (*bond lengths* and *bond distances*) vary depending on several factors, but in general, they are very consistent. In addition, the bond order affects the bond length, but bond lengths of the same order for the same pair of atoms in different molecules are very consistent. The *bond order* is the number of electron pairs shared between two atoms in the formation of the bond; for example, using examples from organic chemistry (Speight, 2017b), the bond order for the carbon-carbon double bond (C=C) and the carbon oxygen double bond (C=O) is 2. The amount of energy required to break a bond is called *bond dissociation energy* or simply *bond energy*. Since bond lengths are consistent, bond energies of similar bonds are also consistent.

Bonds between the same type of atom are *covalent bonds*, and bonds between atoms when the electronegativity of the atoms differs slightly are also predominant covalent in character. Theoretically, even ionic bonds have some covalent character, and the so-called line-in-the-sand boundary between ionic and covalent bonds is not always a clear line of demarcation. Also, when various metals used as electrodes are arranged in the order of the increasing values of standard reduction potential on the hydrogen scale, the metals form an electrochemical arrangement (the *electrochemical series*) (Table 4.7).

The significance and applications of the *electrochemical series* relate to situations when various metals as electrodes are arranged in the order of the increasing values of standard reduction potential on the hydrogen scale, and then, the arrangement is called electrochemical series. In addition, standard electrode potential of any cell can be calculated using this series, which also can be used to indicate the relative ease of oxidation and reduction. For

Table 4.7 Electrochemical Series of Metals

Electrode	**Electrode Reaction**	**Reaction Potential (E^0 Values) Volts**
Li^+/Li	$Li + e^- \rightarrow Li$	−3.01
Mg^{2+}/Mg	$Mg^{2+} + 2e^- \rightarrow Mg$	−2.37
Pb^{2+}/Pb	$Pb^{2+} + 2e^- \rightarrow Pb$	−1.12
Zn^{2+}/Zn	$Zn^{2+} + 2e^- \rightarrow Zn$	−0.76
Fe^{2+}/Fe	$Fe^{2+} + 2e^- \rightarrow Fe$	−0.44
Sn^{2+}/Sn	$Sn^{2+} + 2e^- \rightarrow Sn$	−0.13
H^+/H	$H^+ + e^- \rightarrow H$	0.00
Cu^{2+}/Cu	$Cu^{2+} + 2e^- \rightarrow Cu$	+0.34
Ag^+/Ag	$Ag^+ + 2e^- \rightarrow Ag$	+0.80
Au^+/Au	$Au^+ + e^- \rightarrow Au$	+1.50

example, the higher the value of standard reduction potential (+ve value), the greater is the tendency for reduction since metals on the top having more negative (−ve) values are more easily ionized (oxidized), and the spontaneity of redox reaction can be predicted from the standard electrode potential values of the complete cell reaction. Furthermore, metals that lie higher in the series can displace those elements that lie below them in the series. Furthermore, the equilibrium constant for the cell can be calculated from the standard electrode potential, and metals having a more negative potential in the series will displace hydrogen from acid solutions.

For covalent bonds, bond energies and bond lengths depend on many factors: electron affinities, sizes of atoms involved in the bond, differences in their electronegativity, and overall structure of the molecule. There is a general trend in that *the shorter the bond length, the higher the bond energy*, but there is no formula to show this relationship, because of the widespread variation in bond character.

The *atom radius* of an element is the shortest distance between like atoms. It is the distance of the centers of the atoms from one another in metallic crystals, and for these materials, the atom radius is often called the metal radius, except for the lanthanide series of elements and also for the actinide series of elements (Fig. 4.1). Furthermore, one of the major factors in determining the structures of the substances that can be thought of as made up of cations and anions packed together is ionic size, which can be assessed by use of the periodic table of the elements (Fig. 4.1). It is obvious from the nature of wave functions that no ion has a precisely defined radius. However, with the insight afforded by electron density maps and with a large base of data, new efforts to establish tables of ionic radii have been made.

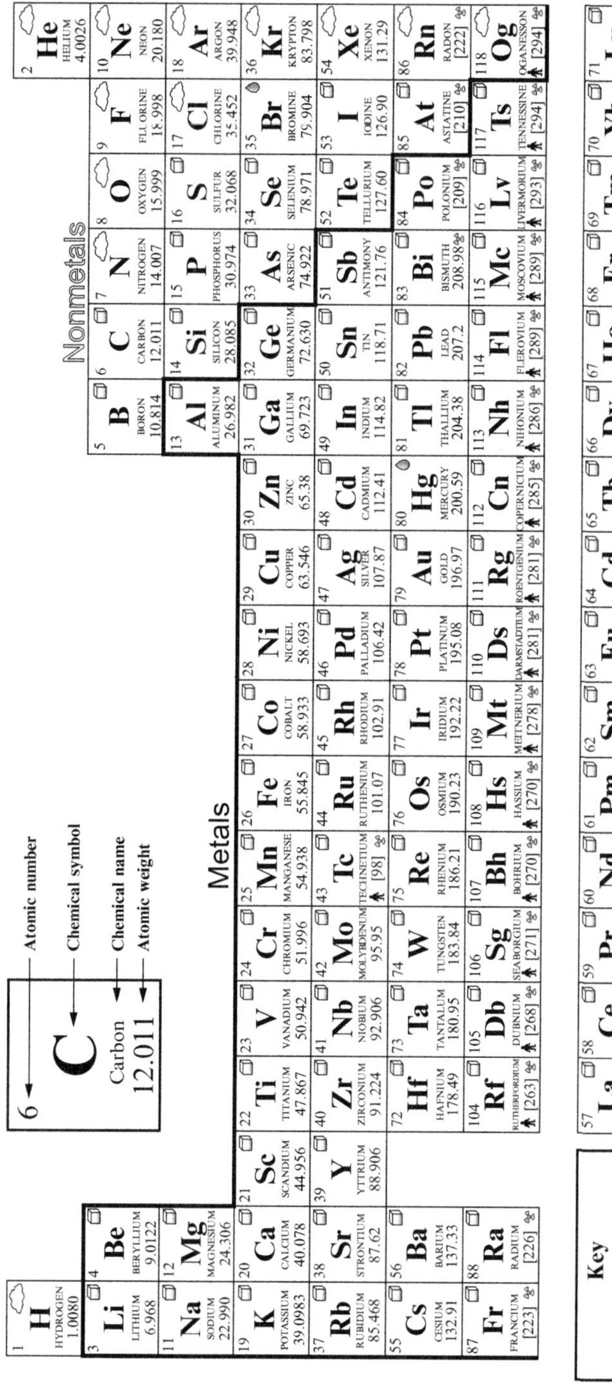

Fig. 4.1 The periodic table of the elements.

The *bond dissociation energy* (*enthalpy change*) for a bond A-B, which is broken through a reaction (e.g., $AB \rightarrow A + B$), is the standard-state enthalpy change for the reaction at a specified temperature, here at 298 K. That is,

$$\Delta Hf_{298} = \Delta Hf_{298}(A) + \Delta Hf_{298}(B) - \Delta Hf_{298}(AB)$$

The *dipole moment* is the mathematical product of the distance between the centers of charge of two atoms multiplied by the magnitude of that charge. The dipole moment arises because not all atoms attract electrons with the same force. The amount of attraction that an atom exerts on the surrounding electrons gives rise to the electronegativity of the atom. As a result of the electronegativity of the atom, an atom with a high electronegativity—such as fluorine, oxygen, and nitrogen—exerts a greater attractive force on electrons than an atom with a lower electronegativity. In an inorganic chemical bond, this leads to unequal sharing of electrons between the atoms (i.e., the bonding electrons are not at the so-called midway point between the two atoms of the bond) because the electrons are drawn closer to the atom with the higher electronegativity. Furthermore, because the electrons in an inorganic bond (and for that matter in an organic bond also but typically to a lesser extent) have a negative charge, the unequal sharing of electrons within a bond leads to the formation of an electric dipole, which is a separation of a positive and a negative electric charge. Because the amount of charge separated in such dipoles is usually smaller than a fundamental charge, they are often referred to as partial charges, which are denoted as $\delta+$ (delta plus) and $\delta-$ (delta minus). The bond dipole moment is calculated by multiplying the amount of charge separated and the distance between the charges. These dipoles within molecules can interact with dipoles in other molecules, creating dipole-dipole intermolecular forces.

Mathematically, the bond dipole is modeled as $+\delta$—$\delta-$ with a distance r between the partial charges $+\delta$ and $\delta-$. It is a vector, parallel to the bond axis, pointing from minus to plus, as is conventional for electric dipole moment vectors. Some chemists draw the vector pointing from plus to minus, but only in situations where the direction is not important. This vector can be physically interpreted as the movement undergone by electrons when the two atoms are placed a distance apart, r, and allowed to interact, the electrons will move from their free state positions to be localized more around the electronegative atom. Thus, the dipole moment (m) of a compound or molecule is

$$m = Q \times r$$

Q is the magnitude of the electric charge(s) that are separated by the distance r; the unit of measurement is the Debye (D). All bonds between equal atoms are given zero values. Because of their symmetry, methane and ethane molecules are nonpolar. The principle of bond moments thus requires that the CH_3 group moment equals one H—C moment. Hence, the substitution of any aliphatic hydrogen by a methyl group does not alter the dipole moment, and all saturated hydrocarbons have zero moments if the tetrahedral angles are maintained.

Furthermore, the polarity of a chemical bond is typically divided into three groups that are arbitrarily based on the difference in electronegativity between the two bonded atoms. Accordingly, (i) *nonpolar bonds* generally occur when the difference in electronegativity between the two atoms is less than 0.4, (ii) *polar bonds* generally occur when the difference in electronegativity between the two atoms is approximately between 0.4 and 1.7, and (iii) *ionic bonds* generally occur when the difference in electronegativity between the two atoms is greater than 1.7. This scale was based on the *partial ionic character* of a bond, which is an approximate function of the difference in electronegativity between the two bonded atoms (Pauling, 1960; Jensen, 2009).

In the present context of inorganic chemistry, the manner in which molecular inorganic molecules and ions pack in the solid state not only is important for the formation of crystals but also contributes significantly to the conduction, magnetic, and nonlinear optical properties of these molecular inorganic compounds in the solid state. Furthermore, inorganic compounds present coordination geometries different from those found for carbon. For example, although four-coordinate carbon is nearly always tetrahedral, both tetrahedral and square planar shapes occur for four-coordinate compounds of both metals and nonmetals. When metals are the central atoms, with anions or neutral molecules bonded to them (frequently through nitrogen, oxygen, or sulfur), these are usually designated as (called) coordination complexes; when carbon is the element directly bonded to metal atoms or ions, the chemicals are organometallic compounds.

4.3 PHYSICAL PROPERTIES

As already noted, inorganic compounds form ionic bonds have high melting points and are made from either single elements or compounds that do not include carbon and hydrogen. In solutions, they break down into ions that conduct electricity. Organic compounds have a carbon-based structure with covalent bonding and are often volatile in nature. Even in liquid state, they do not conduct electricity unless they are salts formed with inorganic acids and bases.

Due to the type of bonding, most inorganic compounds are polar compared with organic chemicals (Speight, 2017b) and have a strong tendency to ionize in water or when subjected to electrolysis. The low volatility of inorganic compounds is due to the strength of the intramolecular bonding; the property of low volatility is increased in situations in which other attractive forces within (and between) the molecules add to the strength of the various bonding arrangements due to weak bonds, such as hydrogen bonds and van der Waals forces.

Examples of properties that might be considered to be relevant to identification and characterization of inorganic compounds include melting point and/or boiling point, crystal shape, and color. Inorganic compounds tend to have high melting points, and while some inorganic compounds are solids with accessible melting points and some are liquids with reasonable boiling points; they are not the exhaustive tabulations of melting point data and boiling point data for inorganic compounds that exist for organic chemicals (Speight, 2017b). In general, melting and boiling points are not useful in identifying an inorganic compound, but they can be used to assess its purity, if they are accessible.

Another aspect of the identification of inorganic chemicals is that there is information about the arrangement of the particles in the solid from the shape of a well-formed crystal or by observing the visual changes in the crystal when it is rotated under a polarizing microscope. In fact, crystal shape is used as a means of mineral identification, but mineralogists have the advantage over chemists in that their crystals are the result of very long, slow crystallization processes that often result in large, well-formed crystals. On the other hand, while inorganic compounds exhibit a variety of colors—in contrast to many organic compounds—color alone is not a reliable means of characterizing inorganic chemicals but may be a useful property when taken into consideration with other properties.

Thus, knowledge of the physical properties of inorganic chemicals can lead to an assessment of the original chemical and the behavior of the chemical after disposal into an ecosystem. Thus, the data lead to an assessment (i) of the inherent hazard of a chemical, such as its capacity to interfere with normal biological processes, and its physical hazards and environmental fate (degradation and persistence) are determined by its intrinsic physicochemical properties and the system with which it is interacting; for inorganic chemicals, these intrinsic properties are determined by molecular structure, while for materials, they are determined by composition, size, structure, and morphology, and (ii) the physical properties can be used to eliminate from consideration chemicals that are likely to exhibit particular physical or toxicological hazards (Table 4.1). As important as these data are, obtaining them is relatively fast and inexpensive and can be readily done at the initial stages of the alternatives assessment.

4.3.1 Colligative Properties

In addition, *colligative properties* are properties of chemicals that are not dependent on the chemical species but instead on, for example, the ratio of the number of solute particles to the number of solvent molecules in a solution. Examples of colligative properties include lowering of vapor pressure, elevation of boiling point, and depression of freezing point.

Chemicals exist as gases, liquids, or solids. On the bulk scale, the shape of gases and liquids are perceived as having the shape of the container in which they are contained. However, on the molecular scale, solids have definite shapes and volume and are held together by strong intermolecular and interatomic forces and have specific physical properties (Table 4.8). For many substances, these forces are strong enough to maintain the atoms in definite ordered arrays (*crystals*), or some solids (specifically *amorphous solids*) have little or no crystal structure.

Thus, because of the weaker molecular forces that are manifested as weaker attractive forces between individual molecules, gases diffuse rapidly and assume the shape of the container in which the gas is contained. Furthermore, the molecules in a gas can be separated by (relatively) large distances unless the gas is subjected to high pressure. The volume of a gas is easily affected by temperature and pressure, and the behavior of any gas is dependent on several laws that are based upon the properties of volume, pressure, and temperature.

4.3.2 Ideal Gas Law

The *ideal gas law* is the equation of state of a hypothetical ideal gas and offers an approximation of the behavior of many gases under many conditions and is written as a combination of the empirical Boyle's law, Charles' law, and Avogadro's law:

$$PV = nRT$$

In this equation, P is the pressure of the gas; V is the volume of the gas; n is the amount of gas (in moles); R is the ideal or universal gas constant, equal

Table 4.8 Examples of Important Common Physical Properties of Gases, Liquids, and Solids

Gases
Density
Critical temperature, critical pressure
Solubility in water
Solubility in organic solvents
Odor threshold
Color
Diffusion coefficient
Liquids
Vapor pressure-temperature relationship
Density, specific gravity
Viscosity
Miscibility with water
Miscibility with organic solvents
Odor
Color
Coefficient of thermal expansion
Interfacial tension
Solids
Melting point
Density
Odor
Solubility in water
Solubility in organic solvents
Coefficient of thermal expansion
Hardness/flexibility
Particle size distribution
Physical form (powder, granules, pellets, lumps)
Porosity

to the product of Boltzmann's constant and Avogadro's constant; and T is the absolute temperature of the gas in degrees Kelvin. For convenience, the Kelvin temperature conversion formulas are

	From Degrees Kelvin	To Degrees Kelvin
Celsius	(°C) = (°K) − 273.15	(°K) = (°C) + 273.15
Fahrenheit	(°F) = (°K) × 9/5 − 459.67	(°K) = (°F) + 459.67 × 5/9
Rankine	(°R) = (°K) × 9/5	(°K) = (°R) × 5/9

For temperature intervals, rather than specific temperatures, 1°K = 1°C = 9/5°F = 9/5°R.

In liquids, the molecules of liquids are separated by relatively small distances, and the attractive forces between the molecules tend to hold firm within a definite liquid volume at a fixed temperature. Molecular forces also result in the phenomenon of interfacial tension. The repulsive forces between molecules exert a sufficiently powerful influence that volume changes caused by pressure changes can be neglected, that is, liquids are incompressible. A useful property of liquids is their ability to dissolve gases, other liquids, and solids. The solutions produced may be end products, or the process itself may serve a useful function, such as the removal of pollutant gas from air by absorption or the leaching of an environmental contaminant from a bulk solid, such as soil. However, the properties of a solution can differ significantly from the individual constituents of the solution, that is, the liquid solvent and the solute. In the current context of inorganic chemical solutes, solvents are compounds in which molecules are much closer together than in a gas, and the intermolecular forces are therefore relatively strong. When the molecules of a solute are physically and chemically similar to the molecules of a liquid solvent, the intermolecular forces of each are the usually equivalent, and the solute and solvent will usually mix readily with each other. The quantity of solute in solvent is often expressed as a concentration, for example, in grams/liter.

4.3.3 Boiling Point

The normal *boiling point* (*boiling temperature*) of a chemical is the temperature at which the vapor pressure of the substance is equal to atmospheric pressure. At the boiling point, a substance changes its state from liquid to gas. A stricter definition of boiling point is the temperature at which the liquid and vapor (gas) phases of a substance can exist in equilibrium. When heat is applied to a liquid, the temperature of the liquid rises until the *vapor pressure* of the liquid equals the pressure of the surrounding atmosphere (gases). At this point, there is no further rise in temperature, and the additional heat energy supplied is absorbed as *latent heat* of vaporization to transform the liquid into gas. This transformation occurs not only at the surface of the liquid (as in the case of *evaporation*) but also throughout the volume of the liquid, where bubbles of gas are formed. The boiling point of a liquid is lowered if the pressure of the surrounding atmosphere (gases) is decreased.

On the other hand, if the pressure of the surrounding atmosphere (gases) is increased, the boiling point is raised. For this reason, it is customary when the boiling point of a substance is given to include the pressure at which it is observed, if that pressure is other than standard, that is, 760 mm of mercury or 1 atm (standard temperature and pressure, STP). The boiling point of a solution is usually higher than that of the pure solvent; this boiling-point elevation is one of the colligative properties common to all solutions. *Boiling point* is given at atmospheric pressure (1 atm, 760 mm of mercury, and 101.325 Pa) unless otherwise indicated. Thus, 8215 mm indicates that the boiling point is 82°C (179.6°F) when the pressure is 15 mm of mercury. Also, *subl. 550* indicates that the compound sublimes at 550°C (1020°F). Occasionally, decomposition products are mentioned.

In addition, if the heat of vaporization and the vapor pressure of a liquid at a certain temperature are known, the boiling point can be calculated by using the Clausius-Clapeyron equation:

$$T_B = \left[1/T_0 - (R \ln P/P_0)/\Delta H_{vap}\right]^{-1}$$

T_B is the boiling point at the pressure of interest, R is the ideal gas constant, P is the vapor pressure of the liquid at the pressure of interest, P_0 is the pressure where the corresponding temperature, T_0 is known from data available usually at 1 atm or 100 kPa, ΔH is the heat of vaporization of the liquid, and ln is the natural logarithm.

If a liquid near its boiling point at one pressure is "let down" to a reduced pressure, vapor flashing will occur. This will cease when the liquid temperature is reduced, due to removal of the latent heat of vaporization, to a temperature below the saturation temperature at the new pressure. Thus, (i) flashing of vapor containing entrained mist may occur on venting equipment or vessels containing volatile liquids. This may create a toxic or flammable hazard depending on the chemical; with steam, the risk is of scalding, (ii) escape, or spillages of liquefied petroleum gas or chlorine or ammonia that rapidly generate a vapor cloud; (iii) the loss of containment, for example, due to a crack or open valve, from beneath the liquid level in a liquefied gas vessel is potentially more serious than if it occurs from the gas space because the mass flow rate is greater, and (iv) absorption of heat (autorefrigeration) and consequent temperature reduction on flashing may have a serious effect on associated heat transfer media, upon the strength of materials of construction, and result in frosting at the point of leakage.

4.3.4 Formula Weight

In chemistry, the *formula weight* is a quantity computed by multiplying the atomic weight (in atomic mass units) of each element in a formula by the number of atoms of that element present in the formula and then adding all of these products together. The simplest example is the formula weight of water (H_2O) that is two times the atomic weight of hydrogen plus one times the atomic weight of oxygen. The formula weights are, thus, based on the international atomic weights and are computed to the nearest hundredth when justified. The actual significant figures are given in the atomic weights of the individual elements. Each element has either a stable isotope or a characteristic natural isotopic composition that typically represented by one of the commonly known radioisotopes of that element, which is identified by mass number and relative atomic mass.

More specific to inorganic chemistry, ionic compounds are made up of a large collection of anions and cations. The basic rule is that the ratio of cation to anions is such that the positive charges balance the negative charges. This ratio of cations to anions is known as the formula unit. For example, magnesium chloride has the formula unit $MgCl_2$, which illustrates that for every magnesium cation, there are two chloride anions in the crystal. Also, the *molar mass*, *M*, is a physical property that is the mass of inorganic chemicals (a chemical element or chemical compound) divided by the amount of the chemical. The base SI unit for molar mass is kilograms per mole (kg/mol), but for historical reasons, molar masses are almost always expressed in g/mol. Again, using water (H_2O) as the example, the molar mass is 18 g/mol. The same logic presented in these last two paragraphs also applies for organic chemicals.

4.3.5 Dernsity

Density is the mass of a substance contained in a unit volume. In the SI system of units, the ratio of the density of a substance to the density of water at 15°C (59°F) is known as the *specific gravity* (*relative density*). Various units of density, such as kg/m^3, lb-mass/ft^3, and g/cm^3 are commonly used. In addition, molar densities or the density divided by the molecular weight is often specified. *Density* values are given at room temperature unless otherwise indicated by the superscript figure; for example, 2.48715 indicates a density of 2.487 g/cm^3 for the substance at 15°C (59°F). A superscript 20 over a subscript 4 indicates a density at 20°C (68°F) relative to that of water at 4°C (39°F), which is also known as the specific gravity or relative density, suing water as the base. For gases, the values are given as grams per liter (g/L). On the other hand, the *vapor density* of a gas is the ratio between the mass of certain volume of the gas and the mass of the same volume of hydrogen under the same conditions of temperature and pressure. The relative vapor density may also be given using air as the reference in which air is assigned the arbitrary value of 1.000.

Thus, with this definition, the vapor density would indicate whether a gas is denser (greater than one) or less dense (less than one) than air. In terms of discharge of a chemical into ecosystem, the density value has implications for container storage and personnel safety; if a container can release a dense gas, its vapor could sink and, if the chemical is flammable, collect until it is at a concentration sufficient for ignition. Even if not flammable, the chemical could collect in the lower floor or level of a confined space and displace air, possibly presenting an asphyxiation hazard to individuals entering the lower part of that space.

As an approximation, at constant pressure, the density of a gas can be related to the relative molecular mass:

$$\text{density of a gas/vapor} = (\text{relative molecular mass})/(\text{absolute temperature})$$

Since few inorganic chemicals (such as hydrogen, methane, and ammonia) have a molecular weight less than that of air, under ambient conditions, most gases or vapors are heavier than air (Table 4.9) (Carson and Mumford, 1988, 1995, 2002). Thus, gases and vapors heavier than air tend to spread and will accumulate in pits or depressions in ground. This may promote a fire/explosion hazard, or a toxic hazard, or cause an oxygen-deficient atmosphere to form, depending on the chemical. In addition, a heavy vapor can remain in a so-called empty vessel after draining out liquid and venting via the top with similar associated hazards. In addition, gases and vapors that are less dense than air at ambient temperature may tend to spread at low level when cold (e.g., vapor from liquid ammonia or liquefied natural gas spillages). On the other hand, gases and vapors that are less dense than air

Table 4.9 Density of Inorganic Gases and Vapors Relative to Air

Gas/Vapor	Relative Density[a]	Relative Molecular Mass
Bromine		160
Phosgene	3.53	99
Chlorine	2.46	71
Sulfur dioxide	2.22	64
Hydrogen cyanide	0.94	27
Hydrogen fluoride	0.69	20
Ammonia	0.59	17

[a]Air = 1.0.

may rise upward through equipment or buildings and, if unvented, will tend to accumulate at high level. Hot gases rise by thermal lift. Hence, in the open air, they will disperse. Within buildings, this is a serious cause of toxic and asphyxiation hazards if smoke and hot gases are able to spread without restriction (or venting) to upper levels.

4.3.6 Melting Point

The *melting point* of an inorganic chemical (sometimes referred to as the *liquefaction point* in organic chemistry) is the temperature at which the chemical changes the physical state from solid to liquid at atmospheric pressure and may be dependent upon the symmetry of the molecule; high molecular symmetry is associated with high melting point (Brown and Brown, 2000). This general rule can apply to different categories of crystal for different reasons, which can be explained by thermodynamic analysis. If the crystal is ordered, high melting point is usually due to high enthalpy change of fusion. If the crystal is disordered, high melting point is due to low entropy change of fusion.

At the melting point, the solid and liquid phases exist in equilibrium. The melting point of a substance depends on pressure and is usually specified at standard pressure. When considered as the temperature of the reverse change from liquid to solid, it is referred to as the freezing point or crystallization point. Because of the ability of some substances to supercool, the freezing point is not always considered as one of the characteristic physical property of a chemical. When the characteristic freezing point of a substance is determined, in fact, the actual methodology almost always involves the principle of observing the disappearance rather than the formation of solid, which is, in fact, the melting point. Confusion can arise when using such definitions!

Thus, to recap, the *melting point* of a solid is the temperature at which the vapor pressure of the solid and the liquid is the same, and the pressure totals one atmosphere, and the solid and liquid phases are in equilibrium.

4.3.7 Freezing Point

For a pure substance, the *melting point* is equal to the *freezing point*, which is the temperature at which a liquid becomes a solid at normal atmospheric pressure. The *melting point* may be recorded in a certain case as *250d* and in some other cases as *d250*, the distinction being made in this manner to indicate that the former is a melting point with decomposition at 250°C (482°F), while in the latter, decomposition only occurs at 250°C (482 in organic chemistry F) and higher temperatures. Where a value such as *$-6H_2O$, 150* is given, it indicates a loss of 6 moles of water per formula weight of the compound at a temperature of 150°C (482°F). For mineral hydrates, the temperature stated represents the compound melting in its water of hydration.

Thermodynamically, at the melting point, the change in Gibbs free energy (ΔG) of the chemical is zero, but the enthalpy (H) and the entropy (S) of the chemical are increasing (ΔH, $\Delta S > 0$). The phenomenon of melting happens when the Gibbs free energy of the liquid becomes lower than the solid for that material. At various pressures, this happens at a specific temperature. Thus,

$$\Delta S = \Delta H / T$$

In this equation, ΔS is the change of entropy of melting, ΔH is the change of enthalpy of melting, and T is the temperature at the melting point.

The melting point is sensitive to extremely large changes in pressure, but generally, the sensitivity is orders of magnitude less than that for the boiling point because the solid-liquid transition represents only a small change in volume. If a chemical is denser in the solid than in the liquid state, the melting point will increase with an increase in pressure. Otherwise, the reverse behavior occurs.

4.3.8 Triple Point

In relation to the melting point, the *triple point* of a material occurs when the vapor, liquid, and solid phases are all in equilibrium (Table 4.10). This is the point on a *phase diagram* where the solid-vapor, solid-liquid, and liquid-vapor equilibrium lines all meet. A *phase diagram* is a diagram that shows the state of a substance at different temperatures and pressures.

The triple point of water is used to define the *Kelvin*, which is the base unit of thermodynamic temperature in the International System of Units (SI). The value of the triple point of water is fixed by definition rather than by measurement. The triple points of several substances are used to define points in the International Temperature Scale of 1990 (ITS-90) from the triple point of hydrogen (13.8033°K) to the triple point of water (273.16 K, 0.01°C, or 32.018°F).

Table 4.10 Triple Points of Selected Inorganic Chemicals

Chemical	*T*, K (°C)	*p*, kPa[a] (atm)
Ammonia	195.40 K (−77.75°C)	6.076 kPa (0.05997 atm)
Argon	83.81 K (−189.34°C)	68.9 kPa (0.680 atm)
Arsenic	1090 K (820°C)	3628 kPa (35.81 atm)
Carbon (graphite)	4765 K (4492°C)	10,132 kPa (100.00 atm)
Carbon dioxide	216.55 K (−56.60°C)	517 kPa (5.10 atm)
Carbon monoxide	68.10 K (−205.05°C)	15.37 kPa (0.1517 atm)
Hydrogen	13.84 K (−259.31°C)	7.04 kPa (0.0695 atm)
Hydrogen chloride	158.96 K (−114.19°C)	13.9 kPa (0.137 atm)
Iodine	386.65 K (113.50°C)	12.07 kPa (0.1191 atm)
Krypton	115.76 K (−157.39°C)	74.12 kPa (0.7315 atm)
Mercury	234.2 K (−39.0°C)	1.65×10^{-7} kPa
Neon	24.57 K (−248.58°C)	43.2 kPa (0.426 atm)
Nitric oxide	109.50 K (−163.65°C)	21.92 kPa (0.2163 atm)
Nitrogen	63.18 K (−209.97°C)	12.6 kPa (0.124 atm)
Nitrous oxide	182.34 K (−90.81°C)	87.85 kPa (0.8670 atm)
Oxygen	54.36 K (−218.79°C)	0.152 kPa (0.00150 atm)
Palladium	1825 K (1552°C)	3.5×10^{-3} kPa
Platinum	2045 K (1772°C)	2.0×10^{-4} kPa
Radon	202 K (−71°C)	70 kPa (0.69 atm)
Sulfur dioxide	197.69 K (−75.46°C)	1.67 kPa (0.0165 atm)
Titanium	1941 K (1668°C)	5.3×10^{-3} kPa
Uranium hexafluoride	337.17 K (64.02°C)	151.7 kPa (1.497 atm)
Water	273.16 K (0.01°C)	0.611657 kPa (0.00603659 atm)
Xenon	161.3 K (−111.8°C)	81.5 kPa (0.804 atm)
Zinc	692.65 K (419.50°C)	0.065 kPa (0.00064 atm)

[a]Typical atmospheric pressure = 101.325 kPa (1 atm).

4.3.9 Solubility

Solubility refers to the ability of an inorganic chemical (the solute) to dissolve in a solvent (i.e., water or an aqueous solution). However, the primary measurement of interest in the assessment of the behavior of inorganic chemicals is the solubility of the chemical in water.

An important aspect of solubility is the solubility of gases in liquids (gas-liquid solubility). For a dilute solution, the partial pressure exerted by a dissolved liquid (a solute) "a" in a liquid solvent is given by

$$pa = Hxa$$

H is Henry's law constant for the system, and xa is the mole fraction of solute. A different value of H is applicable to each gas-liquid system.

In summary, the solubility of a gas generally decreases with any increase in temperature. If a solution in a closed receptacle is heated above the filling temperature during transport or storage, loss of gas can result on opening or liquid discharge. If the gas is sparingly soluble in the liquid, a much-higher partial pressure of that gas is in equilibrium with a solution of a given concentration than is the case with a highly soluble gas. Furthermore, exposure of a solution to any atmosphere will lead to the take-up or release of gas until equilibrium is eventually attained and, particularly import, if rapid absorption of a gas in a liquid in an inadequately-vented vessel can result in implosion, that is, collapse inward due to a partial vacuum.

4.3.10 Partition Coefficient

Related to solubility, the terms solvation properties and solution properties are often used interchangeably. Both can be related to the phenomenon of *phase partitioning* that occurs when two immiscible phases compete for a chemical. A partition coefficient or distribution coefficient is defined mathematically as the ratio of concentrations of a given compound across two mixed, immiscible phases at equilibrium. In the context of a chemical alternative assessment, important partition coefficients are often measured in the liquid phase.

Though partitioning can be measured across a range of solvents and phases, the phase partition coefficient most often encountered when assessing physicochemical properties is from a system where one solvent is water or an aqueous phase and the second is organic and hydrophobic, such as 1-octanol. In a mixture of oil and water, some chemicals dissolve in the oil more readily than water and vice versa. Similarly, some chemicals preferentially dissolve in the solvent, octanol, more readily than in water (i.e., the octanol/water partition coefficient, K_{ow} represented by P):

$$\text{Partition coefficient}(P) = \frac{\text{concentration in organic medium}}{\text{concentration in aqueous medium}}$$

In addition to the partition coefficient, P, there are other media-specific partition coefficients that can provide valuable information about environmental fate, such as a chemical's phase partition coefficient in soil and water (K_d) and in water and air ($K_{w/g}$, Henry's law). These coefficients provide insight into environmental partitioning of the molecule and the potential for bioaccumulation. As with other physicochemical properties, some of these values must be directly measured, and some may be estimated.

4.3.11 Vapor Pressure

The *vapor pressure* is the pressure exerted by a pure component at equilibrium, at any temperature, when both liquid and vapor phases exist and thus extends from a minimum at the triple point temperature to a maximum at the critical temperature (the critical pressure), and is the most important of the basic thermodynamic properties affecting liquids and vapors (Table 4.11). Except at very high total pressures, there is little-to-no effect of total pressure on vapor pressure. If such an effect is present, a correction can be applied. The pressure exerted above a solid-vapor mixture may also be called vapor pressure but is normally only available as experimental data for common compounds that sublime. In addition, the *equilibrium vapor pressure* is an indication of the evaporation rate of a liquid and relates to the tendency of particles to escape from the liquid (or a solid). As the temperature of a liquid increases, the kinetic energy of its molecules also increases. As the kinetic energy of the molecules increases, the number of molecules transitioning into a vapor also increases, thereby increasing the vapor pressure.

The Antoine equation is a mathematical expression of the relation between the vapor pressure and the temperature of pure liquid or solid substances. Thus,

$$\log P = A - (B/C + T)$$

P is the absolute vapor pressure of the chemical; T is the temperature of the chemical; A, B, and C are chemical-specific coefficients; and log is typically either $\log_{10}$ or $\log_e$.

The vapor pressure of any substance increases nonlinearly with temperature according to the Clausius-Clapeyron equation, and the boiling point of a liquid at atmospheric pressure (normal boiling point) is the temperature at which the vapor pressure equals the ambient atmospheric pressure. With any incremental increase in that temperature, the vapor pressure becomes sufficient to overcome atmospheric pressure and lifts the liquid to form vapor bubbles inside the bulk of the substance. Bubble formation in a liquid chemical requires a higher pressure and therefore higher temperature, because the fluid pressure increases above the atmospheric pressure as the depth increases.

Table 4.11 Vapor Pressures of Common Inorganic Compounds

	Temperature (Degrees Centigrade) for the Indicated Pressure									
	1 Pa 0.00001 0.00015	10 Pa 0.0001 0.00145	100 Pa 1 0.0154	1 kPa 0.0099 0.145	2 kPa 0.0197 0.290	5 kPa 0.049 0.725	10 kPa 0.099 1.450	20 kPa 0.197 2.901	50 kPa 0.493 7.252	101.325 kPa 1.000 atm 14.696 psi
Aluminum	1209	1359	1544	1781	1867	1987	2091	2157	2367	2516
Aluminum bromide				112	127	149	168	190	224	256
Aluminum chloride	58	77	97	121	129	140	148	158	171	181
Aluminum fluoride	744	819	906	1008	1062	1102	1130	1167	1217	1257
Aluminum iodide				218			285			385
Aluminum oxide			2122	2351	2437	2547	2629	2727	2867	2987
Ammonia	−139	−127	−112	−95	−89	−80	−71	−61	−47	−33
Ammonium bromide	121	154	195	246	264	289	310	333	367	396
Ammonium chloride	91	121	159	205	221	244	263	284	314	340
Ammonium cyanide				−31	−24	−13	−4	−5	20	32
Ammonium iodide	125	159	201	253			318	342	377	406
Antimony				947	1022	1132	1218	1322	1467	1617
Antimony(III) bromide				137	152	175	197	216	248	275
Antimony(III) chloride				81	94	116	135	157	189	219
Antimony(III) iodide				215	235	266	292	321	365	401
Antimony(III) oxide	426	478	539	610	703	807	907	1027	1231	1427
Argon		−226	−220	−212	−210	−205	−202	−198	−192	−196
Arsenic	280	323	373	433	450	481	508	535	576	610
Arsenic(III) bromide				49	95	117	137	158	191	220
Arsenic(III) chloride			−8	21	31	49	63	81	108	131
Arsenic(III) fluoride						−3	8	21	40	56
Arsenic(III) iodide				187			261			367

Table 4.11 Vapor Pressures of Common Inorganic Compounds—cont'd

Arsenic (III) oxide	134	163	197	236	270	297	319	354	409	457
Bromine	−88	−72	−53	−29	−21	−9	2.5	17	39	59
Cadmium	770	866	983	475	508	555	597	642	710	772
Cadmium bromide	373	435	509							
Cadmium chloride	412	471	541	640	678	733	768	827	902	967
Cadmium fluoride				1257	1307	1392	1461	1537	1652	1757
Cadmium iodide	296	344	406	498	532	581	622	668	737	797
Cadmium oxide				1128	1182	1257	1314	1382	1477	1562
Calcium	591	683	798	954	1007	1092	1170	1252	1377	1492
Carbon dioxide				−122	−117	−109	−103	−96	−87	−78
Carbon monoxide				−216	−214	−210	−207	−203	−197	−191
Chlorine	−145	−134	−120	−104	−97	−86	−76	−65	−49	−34
Chromium	1383	1534	1718	1950	1987	2097	2257	2307	2457	2597
Cobalt	1517	1687	1892	2150			2482			2925
Cobalt(II) chloride					757	812	818	912	987	1057
Copper	1236	1388	1577	1816	1927	2047	2131	2267	2437	2587
Copper(I) bromide				697	752	837	917	1012	1177	1357
Copper(I) chloride		459	543	675	737	832	914	1027	1227	1492
Copper(I) iodide				636	692	782	864	967	1147	1337
Fluorine	−235	−230	−223	−215	−212	−208	−204	−200	−194	−188
Gold	1373	1541	1748	2008	2117	2247	2347	2497	2687	2847
Helium		−271							−270	−269
Hydrogen				−262	−261	−260	−259	−257	−255	−253
Hydrogen bromide		−153	−140	−124	−118	−109	−102	−93	−80	−67.0
Hydrogen chloride				−138	−133	−125	−118	−110	−97	−85
Hydrogen cyanide				−51	−43	−31	−22	−12	3	16
Hydrogen fluoride				−71	−61	−46	−34	−20	1	20

Continued

Table 4.11 Vapor Pressures of Common Inorganic Compounds—cont'd

Hydrogen iodide	−146	−135	−121	−102	−94	−84	−76	−66	−51	−36
Hydrogen peroxide			13	45	57	74	89	105	129	150
Iodine	−13	9	36	69	80	96	108	128	157	184
Iron	1455	1617	1818	2073	2107	2227	2406	2447	2617	2757
Iron(II) chloride				685	722	775	821	875	956	1027
Iron(III) chloride	118	153	190	229	241	256	268	280	297	311
Krypton	−214	−208	−199	−189	−185	−179	−175	−169	−161	153
Lead	705	815	956	1139	1211	1311	1387	1487	1632	1757
Lead(II) bromide	374	431	502	597	631	682	726	777	852	917
Lead(II) chloride			541	637	670	721	765	817	887	957
Lead(lid) fluoride				865	932	997	1054	1117	1212	1297
Lead(II) iodide			470	558	591	640	682	727	802	872
Lead(II) oxide	724	816	928	1065	1117	1187	1241	1307	1397	1477
Lead(II) sulfide	656	741	838	953	992	1047	1088	1137	1207	1267
Lithium	524	612	722	871	927	1002	1064	1142	1257	1357
Lithium bromide		630	733	868	917	987	1049	1117	1222	1312
Lithium chloride		649	761	905	962	1037	1101	1172	1282	1387
Lithium fluoride	801	896	1024	1188	1247	1327	1395	1472	1587	1682
Lithium iodide	545	619	710	824	867	927	972	1027	1107	1172
Magnesium	428	500	588	698	752	812	859	932	1022	1102
Magnesium chloride			762	908	962	1042	1111	1192	1312	1422
Manganese	955	1074	1220	1418	1507	1617	1682	1807	1967	2117
Manganese(Il) chloride				760	872	937	933	1057	1147	1232
Mercury	42	77	120	176	196	225	250	278	320	357
Mercury(II) bromide	71	98	132	174	190	212	227	251	288	320
Mercury(II) chloride	64	95	131	175	188	209	229	246	273	297
Mercury(II) iodide	85	116	152	198	212	236	255	278	318	254

Table 4.11 Vapor Pressures of Common Inorganic Compounds—cont'd

Molybdenum	2469	2721	3039	3434	3587	3787	3939	4137	4407	4627
Molybdenum(VI) fluoride	−98	−82	−64	−41	−36	−23	−13	−1	16	36
Molybdenum(VI) oxide				801	822	882	935	992	1077	1157
Neon	−261	−260	−172 e	−258	−254	−253	−255	−250	−248	−246
Nickel	1510	1677	1881	2137	2227	2357	2468	2587	2757	2907
Nickel(II) chloride	534	592	662	747	787	822	852	887	932	972
Nitric acid			−37	−9			28			82
Nitric oxide	−201	−195	−188	−179	−176	−173	−168	−164	−158	−152
Nitrogen	−236	−232	−227	−220	−218	−214	−211	−207	−201	−196
Nitrogen dioxide				−41	−34	−25	−17	−7	12	29
Nitrogen pentoxide	−71	56	−40	−20	−13	−4	4	12	24	34
Nitrous oxide	−167	−157	−145	−131			−113			−88
Oxygen				−212	−209	−204	−201	−196	−189	−183
Ozone	−189	−182	−172	−158	−153	−146	−140	−132	−121	−111
Potassium	200	257	328	424	467	521	559	621	702	777
Potassium bromide	597	674	773	947	997	1067	1117	1182	1277	1387
Potassium chloride	625	704	804	945	997	1077	1137	1207	1317	1407
Potassium fluoride			869	1017	1072	1152	1216	1292	1402	1501
Potassium hydroxide	520	601	704	842	917	972	1035	1112	1227	1327
Potassium iodide			731	866	917	992	1052	1122	1232	1327
Silicon	1635	1829	2066	2363	2417	2567	2748	2827	3037	3217
Silicon dioxide (silica)				1712	1777	1867	1937	2017	2137	2227
Silver	1010	1140	1302	1509	1592	1697	1782	1887	2037	2167
Silver(l) bromide	569	656	765	905			1093			1359

Continued

Table 4.11 Vapor Pressures of Common Inorganic Compounds—cont'd

Silver(I) chloride	670	769	873	1052	1112	1197	1264	1347	1462	1567
Silver(l) iodide	594	686	803	959	1017	1107	1177	1262	1392	1507
Sodium	281	344	424	529	573	631	673	737	822	902
Sodium bromide			791	931	982	1057	1120	1192	1297	1392
Sodium chloride	653	733	835	987	1052	1127	1182	1267	1372	1467
Sodium cyanide				962	1022	1107	1182	1267	1392	1497
Sodium fluoride				1222	1277	1357	1426	1502	1612	1707
Sodium hydroxide	513	605	722	874	932	1012	1080	1162	1277	1382
Sodium iodide			753	883	932	1002	1058	1122	1222	1307
Sulfur	102	135	176	235	255	288	318	349	299	445
Sulfur dioxide			−98	−80	−74	−62	−52	−41	−25	−10
Sulfuric acid	72	103	140	187	204	228	248	270	303	331
Sulfur trioxide				−20	−13	−2	7	16	31	45
Tin	1224	1384	1582	1834	1937	2077	2165	2327	2527	2717
Tin(IV) bromide				67	82	104	122	143	175	205
Tin(II) chloride				387	415	457	492	532	593	649
Tin(IV) chloride				5	17	34	49	65	90	113
Tin(IV) iodide				167	193	224	250	279	322	361
Uranium	2052	2291	2586	2961			3454			4129
Uranium hexafluoride				−17	−9		14	25	41	56
Vanadium	1828	2016	2250	2541			2914			3406
Water	−61	−42	−20	7	17	33	46	60	81	100
Xenon	−190	−181	−170	−156	−151	−143	−137	−129	−118	−108
Zinc	337	397	477	579	618	672	717	767	847	912
Zinc chloride	305	356	419	497	524	563	596	632	685	733

Table 4.11 Vapor Pressures of Common Inorganic Compounds—cont'd

Zinc fluoride	731	813	911	1048	1102	1132	1237	1312	1417	1512
Zinc iodide	301	351	409	488			598			750
Zirconium	2366	2618	2924	3302			3780			4405
Zirconium(IV) bromide	136	167	203	245	259	279	295	311	336	357
Zirconium(IV) chloride	117	146	181	2225	238	257	272	289	312	331
Zirconium(IV) iodide	187	220	259	305	321	343	361	380	408	431

4.3.12 Refractive Index

The *refractive index* n is the ratio of the velocity of light in a chemical substance to the velocity of light in vacuum (Table 4.12). Values reported refer to the ratio of the velocity in air to that in the substance saturated with air. Usually, the yellow sodium doublet lines are used; they have a weighted mean of 589.26 nm and are symbolized by D. When only a single refractive index is available, approximate values over a small temperature range may be calculated using a mean value of 0.00045 per degree for dn/dt and remembering that n_D decreases with an increase in temperature, but if a transition point lies within the temperature range, extrapolation is not reliable for determination of the refractive index.

The *specific refraction* r_D is expressed by the Lorentz and Lorenz equation:

$$r_D = \frac{n_D^2 - 1}{n_D^2 + 2} \cdot \frac{1}{\rho}$$

In this equation, r is the density at the same temperature as the refractive index and is independent of temperature and pressure. The molar refraction is equal to the specific refraction multiplied by the molecular weight, and typically, the molar refraction is an additive property of the groups or elements comprising the compound.

The refractive index of moist air can be calculated from the expression

$$(n-1) \times 10^6 = \frac{103.49}{T} p_1 + \frac{177.4}{T} p_2 + \frac{86.26}{T} \left(1 + \frac{5748}{T}\right) p_3$$

Table 4.12 Example of Refractive Indices

Chemical	n
Vacuum	1 (by definition)
Air at STP	1.000277
Air at 0°C and 1 atm	1.000293
Bromine	1.661
Carbon dioxide	1.001
Halite (rock salt)	1.516
Helium (gas)	1.000036
Helium (liquid)	1.025
Hydrogen	1.000132
Sodium chloride	1.544
Water (ice)	1.31
Water (liquid)	1.33

In this equation, p_1 is the partial pressure of dry air (in mm and Hg), p_2 is the partial pressure of carbon dioxide (in mm and Hg), $p3$ is the partial pressure of water vapor (in mm and Hg), and T is the temperature (in degrees Kelvin).

4.3.13 Thermal Conductivity

The *thermal conductivity* is a measure of the effectiveness of a material as a thermal insulator. The energy transfer rate through a body is proportional to the temperature gradient across the body and the cross-sectional area of the body. In the limit of infinitesimal thickness and temperature difference, the fundamental law of heat conduction is

$$Q = \lambda A dT/dx$$

Q is the heat flow, A is the cross-sectional area, dT/dx is the temperature/thickness gradient, and λ is the thermal conductivity. A substance with a large thermal conductivity value is a good conductor of heat; one with a small thermal conductivity value is a poor heat conductor, that is, a good insulator.

4.3.14 Viscosity

The *viscosity* of a fluid is a measure of the resistance to flow of the fluid leading to gradual deformation by shear stress or by tensile stress. Viscosity is a property in which an induced stress (such as a pressure difference between the two ends of a tube through which the fluid is flowing) is needed to overcome the internal friction to keep the fluid moving. For a given velocity pattern, the stress required is proportional to the viscosity of the fluid. A fluid that has no resistance to shear stress is known as an *ideal fluid*. However, zero viscosity is observed only at very low temperatures; otherwise, all fluids have a positive viscosity. A liquid is said to be *viscous* if the viscosity of the fluid is substantially greater than that of water. The fluid may also be described as *mobile* if the viscosity is noticeably less than water, whereas a fluid with a relatively high viscosity may appear to be a solid.

Thus, *viscosity* is the shear stress per unit area at any point in a confined fluid divided by the velocity gradient in the direction perpendicular to the direction of flow. If this ratio is constant with time at a given temperature and pressure for any species, the fluid is called a Newtonian fluid. The *absolute viscosity* (m) is the sheer stress at a point divided by the velocity gradient at that point; the most common unit is the poise (1 kg/m s). As many common fluids have viscosities in the hundredths of a poise, the centipoise (cP) is often used. One centipoise is then equal to 1 mPa/s.

The *kinematic viscosity* (*n*) is the ratio of the absolute viscosity to density at the same temperature and pressure. The most common unit corresponding to the poise is the stoke (1 cm^2/s), and the SI unit is m^2/s. The molecules in a gas-liquid interface are in tension and tend to contract to a minimum surface area. This tension may be quantified by the surface tension (*s*), which is the force in the plane of the surface per unit length.

4.3.15 Electrical Properties

As of the electric properties of inorganic materials, there are several properties that are of interest, such as the chemicals that are semiconductors. By way of explanation, semiconductors are crystalline or amorphous solids that have distinct electric characteristics. For example, semiconductors are of high resistance—higher than typical resistance materials but of much lower resistance than insulators. The resistance of a semiconductor decreases with an increase in temperature; this behavior is directly opposite to the property of a metal insofar as the resistance of increases with an increase in temperature. Finally, the conducting properties of a semiconductor may be altered in useful ways by deliberation but controlled introduction of impurities (doping) into the crystal structure. This lowers the resistance of the semiconductor but also permits the creation of semiconductor junctions between differently doped regions of the semiconductor crystal. The behavior of charge carriers that include electrons and ions at these junctions is the basic concept behind the creation of diodes, transistors, and many (if not all) modern electronic systems.

Typically, a semiconductor is an electric conductor with electric resistance in the range of about 10^4–10^8 Ω. A typical semiconductor is a super-high grade silicon that is manufactured on a large scale and is widely used for information processing devices such as computers and energy conversion devices such as solar cells. Very large-scale integrated circuits (VLSIs) are printed on wafers made from silicon single crystals that are free or almost frees of defects silicon single crystals. Memory chips with a very high degree of integration and highly efficient computer chips have recently been realized.

Although silicon semiconductors currently represent 90 or more of all semiconductors, isoelectronic 1:1 compounds of groups II–VI or groups III–V form compound semiconductors and are also used for optical or ultra-high-speed electronic devices. For example, chemicals such as zinc sulfide (ZnS) and cadmium sulfide (CdS) are typical chemicals that are semiconductors. Light-emitting diodes (LED) or semiconductor lasers are important applications of compound semiconductors.

The historical term *dielectric constant*, now referred to as the *relative permittivity*, *K*, is the ratio of the permittivity of the material to the permittivity of free space and is the property of a material that determines the relative speed with which an electric signal will travel in that material. Thus,

$$K = -\epsilon_T / -\epsilon_0$$

The permittivity may be quoted either as a static property or as a frequency-dependent variant. The relative permittivity of air changes with temperature, humidity, and barometric pressure. Sensors can be constructed to detect changes in capacitance caused by changes in the relative permittivity. Most of this change is due to effects of temperature and humidity as the barometric pressure is usually stable. Using the capacitance change, along with the measured temperature, the relative humidity can be obtained using engineering formulas.

Permittivity is typically associated with dielectric materials, and metals are described as having an effective permittivity, with a real relative permittivity equals to one. In the low-frequency region, which extends from radio frequencies to the far infrared and terahertz region, the plasma frequency of the electron gas is much greater than the electromagnetic propagation frequency, so the complex index n of a metal is practically a purely imaginary number; expressed in terms of effective relative permittivity, it has a low imaginary value (loss) and a negative real value (high conductivity).

The term *dielectric loss* quantifies the inherent dissipation of electromagnetic energy (such as heat) in a dielectric material. It can be parameterized in terms of either the loss angle, Δ, or the corresponding loss tangent, tan Δ. This *dielectric loss factor* is the tangent of the loss angle, and the *loss tangent* (tan Δ) is defined by the relationship below:

$$\tan \Delta = 2s/\epsilon v$$

s is the electric conductivity, ϵ is the dielectric constant, and ν is the frequency.

The *electromotive force* (emf, ϵ; measured in volts) is the voltage developed by any source of electric energy such as a battery or dynamo and is the electric potential for a source in a circuit. The electromotive force converts chemical, mechanical, and other forms of energy into electric energy, and the device that supplies electric energy is the *seat* of the electromotive force. In electromagnetic induction, the electromotive force around a closed loop is the electromagnetic work that would be done on a charge if it travels once around that loop. For a time-varying magnetic flux linking a loop, the electric potential scalar field is not defined due to circulating electric vector field, but nevertheless, an electromotive force does work that can be measured as a virtual electric potential around that loop. A source of the electromotive force acts to move positive charge from a point of low potential through its interior to a point of high potential. By chemical, mechanical, or other means, the source of an electromotive force performs work (dW) on that charge to move it to the high-potential terminal. The electromotive force (ϵ) of the source is the work (dW) done per charge dq:

$$\epsilon = dW/dq$$

Conductance is due to the presence of electrolytes. An *electrolyte* is a substance that produces an electrically conducting solution when dissolved in a polar solvent, such as water. The dissolved electrolyte separates into cations (atoms or molecules with a net negative charge) and anions (atoms or molecules with a net positive charge), which disperse uniformly through the solvent. Electrically, such a solution is neutral, but if an electric potential (voltage) is applied to such a solution, the cations of the solution will be drawn to the electrode that has an abundance of electrons, whereas the anions will be drawn to the electrode that has a deficit of electrons. The movement of anions and cations in opposite directions within the solution amounts to a *current*. This includes most soluble salts, acids, and bases (Table 4.13). Some gases, such as hydrogen chloride (HCl), under conditions of high temperature or low pressure can also function as electrolytes. Electrolyte solutions can also result from the dissolution of some biological polymers (such as deoxyribonucleic acid, e.g., DNA, and polypeptides) and synthetic polymers (such as polystyrene sulfonate), termed *polyelectrolytes*, which contain charged functional groups. A substance that dissociates into ions in solution acquires the capacity to conduct electricity. Sodium, potassium, calcium, chloride, and phosphate are examples of electrolytes. In medicine, electrolyte replacement is needed when a patient has prolonged vomiting or loss of bodily fluids and as a response to strenuous athletic activity.

Electrolyte solutions are normally formed when a salt is placed into a solvent such as water, and the individual components dissociate due to the thermodynamic interactions between solvent and solute molecules (solvation). For

Table 4.13 Examples and Properties of Common Bases

Base	Properties
Sodium hydroxide (NaOH) (caustic soda)	White deliquescent solid. Sticks, flakes, pellets Dissolution in water is highly exothermic Strongly basic Severe hazard to skin tissue
Potassium hydroxide (KOH) (caustic potash)	White deliquescent solid. Sticks, flakes, pellets Dissolution in water is highly exothermic Strongly basic Severe hazard to skin tissue
Calcium hydroxide [$Ca(OH)_2$]	White powder soluble in water yielding lime water (slaked lime) Alkaline
Ammonium hydroxide (NH_4OH) (aqueous ammonia solution)	Weakly alkaline Emits ammonia gas Severe eye irritant

example, when table salt (sodium chloride, NaCl) is placed in water, the salt (a solid) dissolves into its component ions (Na^+ and Cl^-):

$$NaCl(s) \rightarrow Na^+(aq) + Cl^-(aq)$$

It is also possible for substances to react with water, producing ions. Molten salts can also be electrolytes, because, for example, when sodium chloride is molten, the liquid conducts electricity. In particular, ionic liquids, which are molten salts with melting points below 100°C (212°F), are a type of highly conductive nonaqueous electrolytes and thus have found more and more applications in fuel cells and batteries.

The standard unit of conductance is electrolytic conductivity (formerly called specific conductance) k, which is defined as the reciprocal of the resistance (Ω^{-1}) of a 1 m cube of liquid at a specified temperature (Ω^{-1} m^{-1}). At low concentrations, it is necessary to subtract the conductivity of the pure solvent from that of the solution to obtain the conductivity due to the electrolyte.

Also, *resistivity (specific resistance)* is

$$\rho = 1/k$$

And the *conductance* of an electrolyte solution is given by

$$1/R = kS/d$$

S is the surface area of the electrode or the mean cross-sectional area of the solution (m^2), and d is the mean distance between the electrodes (in meters).

4.3.16 Magnetic Properties

Magnetic materials are divided into hard (permanent magnets) and soft magnetic materials. Permanent magnets are indispensable to machines using motors and MRI, which requires a high magnetic field. Alnico magnets with iron (Fe), nickel (Ni), and aluminum (Al) as the main constituents; ferrite magnets composed of solid solutions of cobalt and iron oxides (CoO and Fe_2O_3) and magnetite (Fe_3O_4); cobalt rare-earth magnets such as samarium-cobalt (SmCo); and niobium-iron-boron (Nb-Fe-B) magnets are especially useful. Moreover, since soft magnetic materials are strongly magnetized in weak magnetic fields, they are most suitable for use as core materials in transformers. Hard magnetic properties are necessary for the stable maintenance of information, whereas soft magnetic properties are required for recording and overwriting information in magnetic recording materials such as magnetic tapes, floppy disks, and hard disks. Although a form of ferric oxide (γ-Fe_2O_3) is a typical magnetic powder used for these purposes, cobalt oxide or crystalline chromium (IV) oxide (chromium dioxide, CrO_2) is added to improve the magnetic properties of the ferric oxide.

4.3.17 Optical Properties

Many inorganic chemicals are used as materials for optical applications. In particular, the optical fiber has been used for optical communications on a large scale and has had a major social influence in information communication. A necessary property of good optical glass materials is the transmission of information to distant places with little optical loss. Silica fibers (SiO_2 fibers) are manufactured by lengthening silica glass rods produced from silica grains. The silica is made from ultrapure silicon tetrachloride ($SiCl_4$), which is oxidized in the vapor phase by an oxyhydrogen flame. As the optical loss along fibers obtained by this method has already reached its theoretical limit, fluoride glasses are being used in the search for materials with lower levels of loss.

Finally, despite the myriad of paper studies, reliable property estimation is rarely successful for inorganic chemicals. The more difficult species to estimate accurately include the least soluble compounds, aluminates, borates, and some of the organometals. Improved solubility estimation for the difficult classes is likely as solution species are better defined. It is also likely that there is another variable needed to explain the activity of (and so estimate more accurately) the highly insoluble salts (Hickey, 1999).

4.3.18 Specific Heat

The *specific heat* is the amount of heat per unit mass required to raise the temperature by 1°C. The relationship between heat and temperature change is usually expressed in the form shown below where c is the specific heat. The relationship does not apply if a phase change is encountered, because the heat added or removed during a phase change does not change the temperature;

$$Q = \mathrm{cm}\Delta T$$

That is, the heat added is equal to the specific heat multiplied by the mass (weight) multiplied by the temperature difference ($\Delta T = t_{\mathrm{final}} - t_{\mathrm{initial}}$).

4.4 CRITICAL PROPERTIES

All inorganic chemicals have chemical and physical properties that can be used as a means of identification of between chemicals. Briefly, there are two basic types of properties that are associated with matter: (i) chemical properties, which are properties that change the chemical nature of matter, and (ii) physical properties, which are properties that do not change the chemical nature of matter. Thus, the more properties can be used for identification of an inorganic chemical, the better the understanding of the nature of the chemical, which can also assist in understanding how the chemical will behave under various conditions (i.e., process conditions and the conditions present in an ecosystem). A part of this understanding, knowledge of the critical properties is part of the assessment of the behavior of any inorganic chemical. Furthermore, after an overview of the basic properties and a reminder of the importance of measurement, the scientist or engineer can assess, in teen, the behavior of gases, liquids, and solids, as well as any associated phase changes (Fig. 4.2).

The associated term *phase transition* is most commonly used to describe transitions between phase changes between gaseous, liquid, and solid states of matter (see also Fig. 4.2). Thus,

	To:		
From:	**Solid**	**Liquid**	**Gas**
Solid	Solid-solid transformation	Melting	Sublimation
Liquid	Freezing	–	Boiling/evaporation
Gas	Deposition	Condensation	–

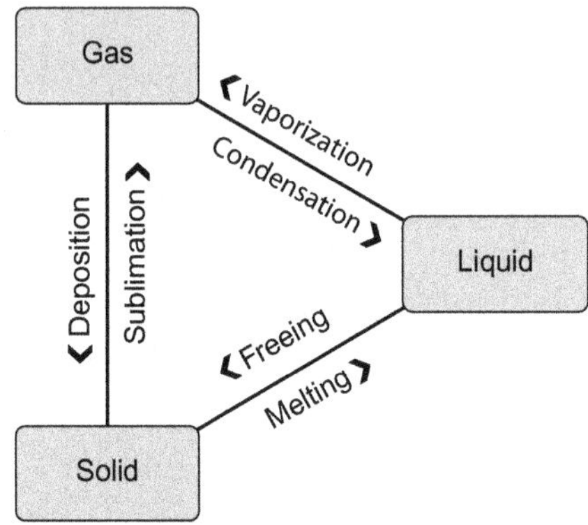

Fig. 4.2 Representation of phase changes.

A phase of a thermodynamic system and the states of matter have uniform physical properties. However, during a phase transition of a chemical, certain properties of the medium changes, often discontinuously, because of the change of some external condition, such as temperature and pressure. For example, a liquid may become gas upon heating to the boiling point, thereby resulting in a significant change in volume of the chemical. The measurement of the external conditions at which the transformation occurs is termed the phase transition.

In terms of critical properties, critical temperature (T_c), critical pressure (P_c), and critical volume (V_c) represent three widely used pure component constants. These critical constants are very important properties in chemical engineering field because almost all other thermochemical properties are predictable from boiling point and critical constants with using corresponding state theory. Therefore, precise prediction of critical constants is very necessary.

The *critical temperature* of a compound is the temperature above which a liquid phase cannot be formed, no matter what the pressure on the system. The critical temperature is important in determining the phase boundaries of any compound and is a required input parameter for most phase equilibrium thermal property or volumetric property calculations using analytic equations of state or the theorem of corresponding states. Critical temperatures are predicted by various empirical methods according to the type of compound or mixture being considered.

A simpler method for estimating the critical temperature of pure compounds requires the normal boiling point, the relative density, and the compound family:

$$\log T_c = A + B\log_{10}(\text{relative density}) + C\log T_b$$

T_c and T_b are the critical and normal boiling temperatures, respectively, expressed in degrees Kelvin. The relative density of the liquid at 15°C is 0.1 MPa. The regression constants *A*, *B*, and C are available by family. For pure inorganic compounds, the method only requires the normal boiling point as input. Thus,

$$T_c = 1.64\,T_b$$

The *critical pressure* of a compound is the vapor pressure of that compound at the critical temperature. Below the critical temperature, any compound above its vapor pressure will be a liquid. The *critical volume* of a compound is the volume occupied by a specified mass of a compound at its critical temperature and critical pressure.

The *critical compressibility factor* of a compound is calculated from the experimental or predicted values of the critical properties:

$$Z_c = (P_c V_c)/(RT_c)$$

Critical compressibility factors are used as characterization parameters in corresponding state methods to predict volumetric and thermal properties. The factor varies from approximately 0.23 for water, 0.26–0.28 for most hydrocarbons, and the factor is more than 0.30 for low-molecular-weight gases.

A *proton transfer reaction* is a reaction in which the main feature is the intermolecular or intramolecular transfer of a proton from one binding site to another. In the detailed description of proton transfer reactions, especially of rapid proton transfers between electronegative atoms, it should always be specified whether the term is used to refer to the overall process, including the more-or-less *encounter-controlled* formation of a hydrogen-bonded complex and the separation of the products or, alternatively, the proton transfer event (including solvent rearrangement) by itself. For the general proton transfer reaction $HB = H^+ + B$, the acidic dissociation constant is

Table 4.14 Examples and Uses of Common Acids

Acid	Main Uses
Hydrochloric acid (HCl)	Chemical manufacture, chlorine, food and rubber production, metal cleaning, petroleum well activation
Nitric acid (HNO_3)	Ammonium nitrate production for fertilizers and explosives, miscellaneous chemical production
Phosphoric acid (H_3PO_4)	Detergent builders and water treatment, foods, metal industries
Sulfuric acid (H_2SO_4)	Alkylation reactions, caprolactam, copper leaching, detergents, explosives, fertilizers, inorganic pigments, textiles
Hydrofluoric acid (HF)	Etching glass

$$K_a = [(H^+)(B)]/[HB]$$

The most common charge types for the acid HB and its conjugate base B are

$$HSO_4^- = H^+ + SO_4^{2-} \text{(hydrogen sulfate ion, sulfate ion)}$$
$$NH_4^+ = H^+ + NH_3 \text{(ammonium ion, ammonia)}$$

Acids (Table 4.14) that have more than one acidic hydrogen ionize in steps, as shown for phosphoric acid:

$$H_3PO_4 = H^+ + H_2PO_4\text{-} \quad pK_1 = 2.148 \quad K_1 = 7.11 \times 10^{-3}$$
$$H_2PO_4\text{-} = H^+ + HPO_4{}^{2-} \quad pK_2 = 7.198 \quad K_2 = 6.34 \times 10^{-8}$$
$$HPO_4{}^{2-} = H^+ + PO_4{}^{3-} \quad pK_3 = 11.90 \quad K_3 = 1.26 \times 10^{-12}$$

If the basic dissociation constant K_b for the equilibrium such as $NH_3 + H_2O = NH_4 + OH$ is required, pK_b may be calculated from the relationship

$$pK_b = pK_w - pK_a$$

In general, for an organic acid, a useful estimate of its pK_a value can sometimes be obtained by making a comparison with recognizably similar compounds for which pK_a values are known.

4.5 REACTIVE CHEMICALS

The term *reactive chemicals* refer (in the context of this book) to a class of inorganic chemicals that displays a broad range of reactions. This category includes explosives, oxidizers, reducers, acid-sensitive and air-sensitive chemicals, unstable chemicals, and water-sensitive chemicals. These chemicals substances can produce toxic gases, explosive mixtures, being explosive, and reacting violently with water violently. Reactive chemicals exhibit moderate to extremely rapid reaction rates (without little or no much inhalation) and include materials capable of rapid release of energy by themselves (self-reaction or polymerization) and/or rates of reaction that may be increased by heat or pressure or by contact with incompatible substances.

Finally, for a chemical reaction, the *equilibrium constant* is the value of the reaction quotient for a system at equilibrium. The reaction quotient is the ratio of molar concentrations of the reactants to those of the products, each concentration being raised to the power equal to the coefficient in the equation. For the hypothetical chemical reaction,

$$A + B \leftrightarrow C + D,$$

The equilibrium constant, K, is

$$K = [C][D]/[A][B]$$

The notation [A] signifies the molar concentration of species A. An alternative expression for the equilibrium constant can involve the use of partial pressures. The equilibrium constant can be determined by allowing a reaction to reach equilibrium, measuring the concentrations of the various solution-phase or gas-phase reactants and products and substituting these values into the relevant equation. The *dissociation constant* of an acid K_a may conveniently be expressed in terms of the pK_a value where p$K_a = -\log_{10}$ (K_a, mol/dm^3).

4.5.1 General Aspects

Reactive chemicals may be broadly classified into two groups: (i) those chemicals that may explode and (ii) those chemicals that do not explode. However, the reactivity of individual chemicals in specific chemical classes (such as the class known as *alkali metals*) varies considerably. Moreover, the inclination of a chemical to dangerous reactivity and the rate of the reaction may also vary because of aging of the chemical or contamination of the chemical. Reactive chemicals may be further subdivided and placed into eight classes based upon their chemical behavior.

Class I chemicals are those chemicals that are typically unstable and which readily undergo violent change without detonating. This class of chemicals includes (i) pyrophoric chemicals, which undergo spontaneous ignition when in contact with air—examples include metal alkyls, phosphorus, finely divided metal powders such as magnesium, aluminum, and zinc; ignition can be prevented by storage and use in an inert environment so that, for example, contact with air or water is prevented; (ii) oxidizing chemicals, which undergo violent reaction when in contact with organic materials or strong reducing agents—examples, (i) perchloric acid, $HClO_4$; (ii) chromic acid, H_2CrO_4; and fuming nitric acid, HNO_3-NO_2 and reaction can be prevented by use of the chemical in minimum amounts for and prescribed procedure and not keeping excessive amounts of any material in the vicinity of process and proper storage so that the chemical is not in contact (or close proximity to) from organic materials, flammable materials, and reducing chemicals.

Class II chemicals are those chemicals that react violently with water and which (i) cause a large evolution of heat when in contact with water, (ii) decompose in moist air, and (iii) violently decompose with liquid water—examples are (i) sulfuric acid, H_2SO_4; (ii) oleum also known as fuming sulfuric acid, H_2SO_4-SO_3; (iii) chlorosulfonic acid, in which a hydroxyl function of sulfuric acid is replaced by a chlorine atom, HSO_3Cl; (iv) phosphorous trioxide, P_2O_3 that is sometimes shown as P_4O_6; (v) phosphorus pentoxide P_2O_5 that may also be shown as P_4O_{10}; (vi) phosphorus halides, PX_3 and PX_5, where X is a halogen atom; and (vii) titanium tetrachloride, $TiCl_4$. These chemicals are corrosive and should be kept away from moisture; fuming in moist air can result in exposure to corrosive and/or toxic gases.

Class III chemicals are chemicals that form potentially explosive mixtures with water. These chemicals decompose violently when contact with water; there is a considerable evolution of heat that is accompanied by flammable gases, which may ignite (sometime explosively) if exposed to an ignition source. In fact, the evolution of heat when contacted with water may be sufficient to cause autoignition (and explosion)—examples of these chemicals are (i) alkaline metals, such as lithium, Li; sodium, Na; and potassium, K; (ii) alkaline earth metals, such as magnesium, Mg and calcium, Ca; (iii) alkali metal hydrides, such as lithium hydride, LiH; sodium hydride, NaH; and potassium hydride, KH; and (iv) the alkaline metal nitrides, such as lithium nitride, Li_3N; sodium nitride, Na_3N; and potassium nitride, K_3N. Handling these chemicals requires that they should be provided with adequately vented storage to disperse flammable gases, and contact with water should be strenuously avoided.

Class IV chemicals are chemicals that, when mixed with water, generate toxic gases, vapors, or fumes in quantity sufficient to present a danger to the floral and faunal species (including human fauna) in the environment. These chemicals react rapidly with water with the production of toxic gases—examples, (i) the alkali metal phosphides; (ii) phosphorus halides, such as PX_3 and PX_5, where X is a halogen atom; and (iii) aluminum phosphide, AlP. These chemicals should be contained in a sealed container and provided with adequate ventilation during storage, and contact with water should be avoided.

Class V chemicals consist of the cyanide- or sulfide-bearing chemicals. These acid-sensitive chemicals that produce extremely toxic hydrogen cyanide (HCN) has hydrogen sulfide (H_2S) gas on contact with acids or materials, which form acids in the presence of moisture or liquid water—examples include (i) metal cyanide salts and (ii) metal sulfide salts. These chemicals should not be stored in cabinets with acids and oxidizers and must be isolated from other reactive chemicals. In addition, the sulfide salts should be protected from moisture, and adequate ventilation must be provided due to the severe inhalation hazard of hydrogen cyanide and hydrogen sulfide and the acute toxic effects from skin contact with hydrogen cyanide.

The *Class VI chemicals* are those chemicals that are capable of detonating or exploding if subjected to a strong initiating source or if heated under confinement. For example, these chemicals can detonate or explode when they are heated above ambient temperature or if exposed to an initiating source such as shock, mechanical shock, spark or flame, or a catalyst that accelerates decomposition—examples of these chemicals include (i) lead amide, $Pb(NH_2)_2$; (ii) sodium amide, $NaNH_2$; (iii) thallous nitride, Ti_3N; (iv) metal azide derivatives such as sodium azide, Na_3N; (v) ammonium chromate, $(NH_4)_2CrO_4$ and ammonium dichromate, $(NH_4)_2Cr_2O_7$; (vi) metal periodate derivatives such as sodium metaperiodate, which has the formula $NaIO_4$, sodium orthoperiodate or sodium hydrogen periodate ($Na_2H_3IO_6$), and fully reacted sodium salt, Na_5IO_6; (vii) ammonium nitrate, NH_4NO_3; and (viii) ammonium chlorate, NH_4ClO_3. The containers of these chemicals must be protected from physical damage, heat, and incompatible chemicals. The chemicals in this class exhibit a wide range of other properties, such as flammability, sensitivity to acids, sensitivity to water, and sensitivity to light sensitivity.

The *Class VII Chemicals* are chemicals that are readily capable of detonation, explosive decomposition, or reaction at standard temperature and pressure under ambient temperature and pressure without any external initiating source—examples, (i) ammonium chlorate, NH_4ClO_3 and metal azide derivatives such as sodium azide, Na_3N. When in storage, these chemicals should be evaluated periodically to determine whether deterioration has occurred, which, if so, the chemical should be sent for immediate disposal by the recommended procedures for the particular chemical.

Finally, the Class VIII Chemicals consist of chemicals within the subclass known as *forbidden explosives*: Class A explosives and Class B explosives as defined in the US Code of Federal Regulations (49 CFR 173). Forbidden explosives are capable of detonation or explosive decomposition under ambient conditions, considered too dangerous for transportation; the well-known inorganic chemical is mercury fulminate, $Hg(CNO)_2$, which can rearrange to the also dangerous but structurally different mercy cyanate and mercury isocyanate that have the identical empirical formula but a different atomic arrangement; the cyanate and the fulminate anions are isomers:

$$Hg(CNO)_2 \rightarrow Hg(OCN)_2(\text{cyanate and/or isocyanate})$$

However, for the most part, the Class VIII chemicals are organic chemicals such as unstabilized nitroglycerin, stabilized nitroglycerin, nitrocellulose, stabilized nitrocellulose, mercury fulminate, diazodinitrophenol, and lead 2,4-dinitroresorcinate. These chemicals should only be handled by experienced and properly equipped personnel.

4.5.2 Water-Sensitive Chemicals

Within the context of this book—the environmental inorganic chemistry—the most likely reactivity aspect of inorganic chemicals is the reactivity with or sensitivity to water, which can or will occur whenever an inorganic chemical is released into the environment, be it into the atmosphere, the aquasphere, or the geosphere. It is, therefore, worth adding to the description of water sensitivity (reactivity with water) in more detail.

Water-sensitive chemicals (also called *water-reactive chemicals*) are chemicals that are reactive and dangerous when wet because of a chemical reaction with water (Table 4.15). This reaction may release a gas that either is flammable or presents a toxic health hazard or both. In addition, the heat generated when water contacts such materials is often enough for the item to spontaneously combust and even explode. Examples of water-reactive chemicals are the

Table 4.15 Examples of Inorganic Chemicals That React Dangerously With Water

Chemical	Reaction With Water
Aluminum bromide	Violent hydrolysis
Aluminum chloride	Violent decomposition forming hydrogen chloride gas
Boron tribromide	Violent or explosive reaction when water added
Butyl lithium	Ignites on contact with water
Calcium carbide	Gives off explosive acetylene gas
Calcium hydride	Hydrogen gas liberated
Chlorosulfonic acid	Highly exothermic violent reaction
Lithium aluminum hydride	Releases and ignites hydrogen gas
Lithium hydride	Violent decomposition
Lithium metal	Powder reacts explosively with water
Phosphorus pentachloride	Violent reaction with water
Phosphorus pentachloride	Violent reaction
Phosphorus pentoxide	Violent exothermic reaction
Phosphorus tribromide	Reacts violently with limited amounts of warm water
Phosphorus trichloride	Violent reaction releasing flammable diphosphane
Phosphoryl chloride	Slow reaction that may become violent
Potassium hydride	Releases hydrogen gas
Potassium metal	Forms potassium hydroxide and hydrogen gas
Potassium hydroxide	Highly exothermic reaction
Silicon tetrachloride	Violent reaction producing silicic acid
Sodium amide	Generates sodium hydroxide and ammonia (flammable)
Sodium azide	Violent reaction with strongly heated azide
Sodium hydride	Reacts explosively with water
Sodium hydrosulfite	Heating and spontaneous ignition
Sodium hydroxide	Highly exothermic reaction
Sodium metal	Generates flammable hydrogen gas
Sodium peroxide	Reacts violently or explosively
Strontium metal	Violent reaction
Sulfuric acid	May boil and spatter
Tetrachlorosilane	Violent reaction
Thionyl chloride	Violent reaction which forms hydrochloric acid and sulfur dioxide gas
Titanium tetrachloride	Violent reaction that produces hydrogen chloride gas
Trichlorosilane	Releases toxic and corrosive fumes
Triethyl aluminum	Explodes violently in water
Triisobutyl aluminum	Violent reaction with water
Zirconium tetrachloride	Violent reaction with water

alkali metals (lithium, sodium, potassium, rubidium, and cesium, sometimes spelled caesium) and the alkaline earth metals, anhydrides, carbides, hydrides, sodium hydrosulfite, and similar chemicals. An example of the chemical reaction of sodium metal with water is

$$2Na(s) + 2H_2O \rightarrow 2Na^+(aq) + 2HO^-(aq) + H_2(g)$$

The heat generated by this reaction is generally sufficient to ignite the hydrogen gas (H_2) that evolved in the reaction and can result in an explosion, depending on the amount and surface area of the alkali metal; that is, these types of chemical are classed as *pyrophoric chemicals*. Elemental potassium and cesium are particularly dangerous in this regard.

A pyrophoric chemical is a chemical with the ability to spontaneously ignite, without the influence of heat or fire, in air at temperatures of 54° C (130°F) or below can be in the gas phase, the liquid phase, or the solid phase. Pyrophoric gases, pyrophoric liquids, and pyrophoric solids all share the property of spontaneous ignition. Some pyrophoric chemicals, such as the gaseous silane (SiH_4) derivatives—which are a nonmetallic hydride derivative—can ignite immediately upon exposure to air. Typically, pyrophoric gases are stored in compressed gas cylinders, while pyrophoric liquids such as tertiary-butyl lithium are often metal alkyl derivatives, metal aryl derivatives, metal vinyl derivatives, metal carbonyl derivatives, or metal hydride derivatives. Pyrophoric liquids are often sold and stored in flammable hydrocarbon solvents, such as ethyl ether, tetrahydrofuran (THF), pentane, or heptane. Pyrophoric solids such as lithium are often alkali metals and stored under kerosene or oil.

Another example of a dangerous when wet substance is aluminum phosphide—a highly toxic inorganic chemical that is used as a wide bandgap semiconductor and also as a fumigant—which reacts with water to release the highly toxic phosphine gas, PH_3. This chemical reaction is commercially exploited to kill moles and related pests:

$$2AlP(s) + 3H_2O \rightarrow Al_2O_3(s) + 2PH_3(g)$$

Thus, it is critical that water-reactive substances be stored in dry areas and kept off the floor by the use of pallets or rack storage. Dangerous when wet materials should never be stored directly beneath active water sprinklers and should be isolated by a waterproof or water-resistant barrier (e.g., plastic sheeting or a watertight secondary container) to protect the materials from water in the event the sprinkler system is activated elsewhere in the facility.

Water-reactive chemical pose serious safety and health risks. The material safety data sheet (MSDS) will provide information about these risks and the precautions that should be taken when handling the material. Laboratory personnel should ensure that water-reactive materials are well-marked so that fire fighters and other personnel are aware of the danger in an emergency situation and be sure to have the correct type of fire extinguisher on hand.

4.6 CATALYSTS

Catalysts reduce the activation energy of reactions and enhance the rate of specific reactions (Table 4.16) (Busca, 2014). Therefore, they are crucially important in chemical industry, exhaust gas treatment, and other chemical reactions. While the chemical essence of catalysis is obscure, practical catalysts have been developed based on the accumulation of empirical knowledge. However, while gradually we have come to understand the mechanisms of homogeneous catalysis through the development of inorganic chemistry, our understanding of surface reactions in solid catalysts is also deepening.

In the presence of a catalyst, a lower free energy of the reaction is required to reach the transition state, but the total free energy from reactants to products does not change. A catalyst may participate in multiple chemical transformations, but the effect of a catalyst may vary due to the presence of *inhibitors* or *poisons* (which reduce the catalytic activity) or *promoters* (which increase the activity).

Thus, catalyzed reactions have lower activation energy (the rate-limiting free energy of activation) than the corresponding uncatalyzed reaction, which results in a higher reaction rate at the same temperature

Table 4.16 Characteristics of Inorganic Catalysts

Function	Increase the rate of a chemical reaction
Molecular weight	Typically, a low-molecular-weight compound
Nature	Simple inorganic molecule
Alternate terms	Inorganic catalyst
Reaction rates	Variable, depending upon reaction
Specificity	Not always specific; can produce byproducts
Conditions	Variable temperature and variable pressure
C—C and C—H bonds	Absent

and for the same concentration of reaction(s). However, the detailed mechanics of catalysis is complex, but on a kinetic basis, catalytic reactions are typical chemical reactions insofar as the reaction rate depends on the frequency of contact of the reactants in the rate-determining step. Usually, the catalyst participates in this slowest step, and rates are limited by amount of catalyst and its activity. Although catalysts are not consumed by the reaction itself, they may be inhibited, deactivated, or destroyed by secondary processes.

A catalyst can be positive (increasing reaction rate) or negative (decreasing reaction rate) in character. They react with reactants in a chemical reaction to give rise to intermediates that eventually release the product and regenerate the catalyst. For example, a typical catalytic chemical reaction can be represented in the simplest form as

$$A + C \rightarrow AC$$

$$B + AC \rightarrow BC$$

$$ABC \rightarrow PC$$

$$PC \rightarrow P + C$$

In these reactions, C is a catalyst, A and B are reactants, and P is the product. The catalyst is regenerated in the last step even though in the intermediate steps it had integrated with reactants.

In the current context, catalysis impacts the environment by increasing the efficiency of industrial processes, but catalysis also plays a direct role in the environment. For example, there is the catalytic role of chlorine free radicals in the breakdown of ozone. These radicals are formed by the action of ultraviolet radiation on chlorofluorocarbons (CFCs):

$$Cl\cdot + O_3 \rightarrow ClO\cdot + O_2$$

$$ClO\cdot + O\cdot \rightarrow Cl\cdot + O_2$$

Moreover, catalysts play key roles in (i) the production of clean fuels, (ii) the conversion of waste, (iii) the conversion of biomass into energy, (iv) the clean combustion engines including control of nitrogen oxides, NO_x, (v) the control of soot production, (vi) the reduction of greenhouse gases, and (vii) the production of clean water. Catalysts are also of prime importance in the developing hydrogen and synthesis gas (syngas) production technology (Chadeesingh, 2011; Speight, 2013, 2014a, 2017c; Luque and Speight, 2015), aimed at producing clean fuels for the coming decades. And, furthermore, catalysts can be recycled.

4.6.1 Homogeneous Catalysis

Homogeneous catalysis is catalysis that typically occurs in a solution by a soluble catalyst or a reaction where the catalyst is in the same phase as the reactants, so homogeneous catalysis applies to reactions in the gas phase and even in a solid. Heterogeneous catalysis is the alternative to homogeneous catalysis, where the catalysis occurs at the interface of two phases, typically a gas-solid reaction. The term is used almost exclusively to describe solutions, and it is often used to imply that the reaction is catalyzed by organometallic compounds.

Thus, in homogeneous catalysis, the catalysts are present in the same phase as the substances are going into the reaction phase. The most widely used industrial unit operation is either homogeneous or heterogeneous catalysts as these are better suited for the industrial operations that are undertaken. Whether we are looking at heterogeneous or homogeneous forms of catalysts the catalyst does not undergo chemical changes, but the physical states of these substances undergo substantial changes in the size of the particles or the changes in colors. It is also evident that, unlike other forms of catalysts, the homogeneous form of catalysts supported reactions, the rate of catalytic reactions is proportional to the concentration of catalysts like what we see in cane sugar inversion process. The chemistry of catalysts that are soluble in solvents has developed remarkably over the past several decades since World War II. In the past, the mechanism of catalytic reactions was generally not very clear. However, catalytic reactions are now established as a cycle of a combination of a few elementary steps that occur on the metals of catalyst complexes.

4.6.2 Heterogenous Catalysis

Heterogeneous catalysis (also called *solid-state catalysis*) involves the use of, for example, a catalyst that promotes the reaction in a gaseous phase or a liquid phase, while the catalyst remains in a difference phase, that is, the solid phase (Galwey, 2015). Since adsorption of reactants on the catalyst surface is the initial step, a large surface area is required for good efficiency of catalysis. Polyphase systems, which carry active catalysts on materials such as zeolites with small pores of molecular sizes and gamma alumina and silica gel with large surface area, are often used.

In heterogeneous catalysis, the diffusion of reagents to the surface and diffusion of products from the surface can be rate determining. Thus, heterogeneous catalysis offers the advantage that products are readily separated

Table 4.17 Illustration of the Differences Between Homogeneous Catalysis and Heterogeneous Catalysis

Homogeneous Catalysis	Heterogeneous Catalysis
The catalyst is usually in the same phase as the reactants	These are found in liquid phase, gas phase, and solid phase
Operative temperature is generally low except when under high pressure	Operative temperature for heterogeneous catalysis is higher when compared with homogeneous process
Diffusivity is high	Diffusivity is low
Heat transfer is high; all of the reactants and catalysts are in same phase	Heat transfer is relatively low; the reactants and catalysts are in different phase
The active site is very well defined	The active site is not very well defined
Modification of the catalyst is relatively easy and depends upon the electronic and steric properties on metal	Modification of the catalyst is relatively difficult and depends on the particle site and the active size at the molecular level
Reaction mechanism is relatively easy to define	Reaction mechanism is relatively difficult to define
Selectivity of catalyst is high	Selectivity of catalyst is low

from the catalyst, and heterogeneous catalysts are often more stable and degrade much slower than homogeneous catalysts. However, heterogeneous catalysts are difficult to study, so their reaction mechanisms are often unknown (Table 4.17). Previously, heterogeneous catalysis was explained as arising from a mysterious activation of reactants due to adsorption, but it has become increasingly clear that catalysis is ascribable to surface chemical reactions. Namely, the action of solid-state catalysts depends on activation of reactants by surface acids or bases and by coordination to the metal surface. However, in heterogeneous catalysis (using the refining industry as the example), typical secondary processes include coking where the catalyst becomes covered by higher-molecular-weight carbonaceous byproducts (Speight, 2014a, 2017c). Additionally, heterogeneous catalysts can dissolve into the solution in a solid-liquid system or sublimate in a solid-gas system.

Since mechanisms of homogeneous catalysis have been clarified considerably, solid surface reactions can also be analyzed by introducing concepts such as "surface complexes" or "surface organometallic compounds." However, unlike homogeneous catalysis, in which only one or a few metal centers participate, many active sites are involved in solid-state catalysis.

Since surface homogeneity and reproducibility are difficult to maintain, major parts of reaction mechanisms are obscure even for such simple reactions as ammonia synthesis.

4.7 CORROSIVITY

Corrosion is the deterioration of a material as a result of its interaction of the material with the surroundings and can occur at any point or at any time during petroleum and natural gas processing. Although this definition is applicable to any type of material, it is typically reserved for metallic alloys. Furthermore, corrosion processes not only influence the chemical properties of a metal or metals alloy but also generate changes in the physical properties and the mechanical behavior (Speight, 2014b).

Corrosion is a natural phenomenon and is the deterioration of a material as a result of its interaction of the material with the surroundings (Fontana, 1986; Garverick, 1994; Shreir et al., 1994; Jones, 1996; Shalaby et al., 1996; Peabody, 2001; Bushman, 2002). Although this definition is applicable to any type of material, it is typically reserved for metallic alloys; of the known chemical elements, approximately 80 are metals (Fig. 4.1), and approximately 50% of these metals can be alloyed with other metals although the physical, chemical, and mechanical properties and propensity for corrosion are all dependent upon the composition of the alloys. Furthermore, corrosion processes not only influence the chemical properties of a metal or metals alloy but also generate changes in the physical properties and the mechanical behavior.

Pure metals and metal alloys interact gradually with the elements of a corrosive medium to form stable compounds, and the resulting metal surface is then *corroded* (Carson and Mumford, 1988, 1995, 2002). Generally, when dissimilar metals are used in contact with each other and are exposed to an electrically conducting solution, combinations of metals should be chosen that are as close as possible to each other in the series. Coupling two metals widely separated in the series will generally produce accelerated attack on the more active metal. Often, however, protective passive oxide films and other effects will tend to reduce corrosion. Insulating the metals from each other can prevent corrosion. The dual action of stress and corrodent may result in stress corrosion cracking or corrosion fatigue. Corrosion may be uniform or be intensely localized and characterized by pitting. The mechanisms can be direct oxidation, such as when a metal is heated in an oxidizing environment, or electrochemical. Many inorganic salts are corrosive to common materials of construction (Table 4.18)

Table 4.18 Common Salts That are Corrosive to Carbon Steel

Aluminum sulfate, $Al_2(SO_4)_3$	Lithium chloride, LiCl
Ammonium bisulfite, NH_4HSO_3	Mercuric chloride, $HgCl_2$
Ammonium bromide, NH_4Br	Nickel chloride, $NiCl_2$
Antimony trichloride, $SbCl_3$	Nickel sulfate, $NiSO_4$
Beryllium chloride, $BeCl_2$	Potassium bisulfate, $KHSO_4$
Cadmium chloride, $CdCl_2$	Potassium bisulfite, $KHSO_3$
Calcium hypochlorite, $Ca(OCl)_2$	Potassium sulfite, K_2SO_3
Copper nitrate, $Cu(NO_3)_2$	Silver nitrate, $AgNO_3$
Copper sulfate, $CuSO_4$	Sodium bisulfate, $NaHSO_4$
Cupric chloride, $CuCl_2$	Sodium hypochlorite, NaOCl
Cuprous chloride, CuCl	Sodium perchlorate, $NaCLO_4$
Ferric chloride, $FeCl_3$	Sodium thiocyanate, NaSCN
Ferric nitrate, $Fe(NO_3)_3$	Stannic chloride, $SnCl_4$
Ferrous chloride, $FeCl_2$	Stannous chloride, $SnCl_2$
Ferrous sulfate, $FeSO_4$	Zinc chloride, $ZnCl_2$
Lead nitrate, $PbNO_3$	

The consumption of oxygen due to atmospheric corrosion of sealed metal tanks may cause a hazard, due to oxygen deficiency affecting persons on entry. Stresses may develop resulting from the increased volume of corrosion products, such as rust formation that involves a sevenfold increase in volume. Corrosion may be promoted or accelerated by traces of contaminants. Whereas corrosion of metals is due to chemical or substantial electrochemical attack, the deterioration of plastics and other nonmetals that are susceptible to swelling, cracking, crazing, softening, etc. is essentially physicochemical rather than electrochemical.

Typically, corrosion prevention is achieved by correct choice of material of construction, by physical means (such as paints or metallic, porcelain, plastic or enamel linings, or coatings) or by chemical means (such as alloying or coating). Some metals (e.g., aluminum) are rendered passive by the formation of an inert protective film. Alternatively, a metal to be protected may be linked electrically to a more easily corroded metal (e.g., magnesium) to serve as a sacrificial anode.

REFERENCES

Brown, R.J.C., Brown, R.F.C., 2000. Melting point and molecular symmetry. J. Chem. Educ. 77 (6), 724.

Busca, G., 2014. Heterogeneous Catalytic Materials: Solid State Chemistry, Surface Chemistry and Catalytic Behavior. Elsevier BV, Amsterdam.

Bushman, J.B., 2002. Corrosion and Cathodic Protection Theory. Bushman & Associates Inc., Medina, OH.

Carson, P., Mumford, C., 1988. The Safe Handling of Chemicals in Industry, vols. 1 and 2 John Wiley & Sons Inc, New York, NY.
Carson, P., Mumford, C., 1995. The Safe Handling of Chemicals in Industry, vol. 3 John Wiley & Sons Inc., New York, NY.
Carson, P., Mumford, R., 2002. Hazardous Chemicals Handbook, second ed. Butterworth-Heinemann, Oxford.
Chadeesingh, R., 2011. The Fischer-Tropsch process. In: Speight, J.G. (Ed.), The Biofuels Handbook. The Royal Society of Chemistry, London, pp. 476–517. Part 3 (Chapter 5).
Fontana, M.G., 1986. Corrosion Engineering, third ed. McGraw-Hill, New York, NY.
Galasso, F.S., Kurti, N., Smoluchowski, R., 1970. Structure and Properties of Inorganic Solids. Elsevier BV, Amsterdam.
Galwey, A.K., 2015. Solid state reaction kinetics, mechanisms and catalysis: a retrospective rational review. React. Kinet. Mech. Catal. 114 (1), 1–29.
Garverick, L. (Ed.), 1994. Corrosion in the Petrochemical Industry. ASM International, Materials Park, OH.
Hickey, J.P., 1999. Estimating the Environmental Behavior of Inorganic and Organometal Contaminants: Solubilities, Bioaccumulation, and Acute Aquatic Toxicities. Contribution No. 1066, USGS Great Lakes Science Center, Ann Arbor, MI http://toxics.usgs.gov/pubs/wri99-4018/Volume2/sectionD/2512_Hickey/pdf/2512_Hickey.pdf.
Jensen, W.B., 2009. The origin of the "Delta" symbol for fractional charges. J. Chem. Educ. 86, 545.
Jones, D.A., 1996. Principles and Prevention of Corrosion, Second ed. Prentice Hall, Upper Saddle River, NJ.
Luque, R., Speight, J.G. (Eds.), 2015. Gasification for Synthetic Fuel Production: Fundamentals, Processes, and Applications. Woodhead Publishing/Elsevier, Cambridge.
Pauling, L., 1960. The Nature of the Chemical Bond and the Structure of Molecules and Crystals: An Introduction to Modern Structural Chemistry, third ed. Oxford University Press, Oxford, pp. 98–100.
Peabody, A.W., 2001. Control of Pipeline Corrosion, second ed. NACE International, Houston, TX.
Shalaby, H.M., Al-Hashem, A., Lowther, M., Al-Besharah, J. (Eds.), 1996. Industrial Corrosion and Corrosion Control Technology. Kuwait Institute for Scientific Research, Safat.
Shreir, L.L., Jarman, R.A., Burstein, G.T. (Eds.), 1994. In: Corrosion, vols. 1 and 2. Butterworth-Heinemann, Oxford.
Speight, J.G., 2013. The Chemistry and Technology of Coal, third ed. CRC Press/Taylor & Francis Group, Boca Raton, FL.
Speight, J.G., 2014a. The Chemistry and Technology of Petroleum, fifth ed. CRC Press/Taylor & Francis Group, Boca Raton, FL.
Speight, J.G., 2014b. Oil and Gas Corrosion Prevention. Gulf Professional Publishing/Elsevier, Oxford.
Speight, J.G. (Ed.), 2017a. Lange's Handbook of Chemistry, 17th ed. McGraw-Hill Education, New York, NY.
Speight, J.G., 2017b. Environmental Organic Chemistry for Engineers. Butterworth-Heinemann/Elsevier, Cambridge, MA.
Speight, J.G., 2017c. Handbook of Petroleum Refining. CRC Press/Taylor & Francis Group, Boca Raton, FL.

FURTHER READING

NRC, 2014. Physicochemical Properties and Environmental Fate: A Framework to Guide Selection of Chemical Alternatives. National Research Council, Washington, DC.

CHAPTER FIVE

Sources and Types of Inorganic Pollutants

5.1 INTRODUCTION

It is doubtful if anyone (even though there may be claims to the contrary) can state with any degree of accuracy when the Earth was last pristine and unpolluted—although there is always someone who can make a statement with a high degree of uncertainty! Yet, to attempt to return the environment to such a mythical time might have a severe effect on the current indigenous life, perhaps a form of pollution in reverse! There is always the possibility that changing an ecosystem to a pristine unpolluted system could have a deleterious effect on the flora and fauna of that ecosystem. Alternatively, the introduction of foreign floral and/or faunal species into a pristine or undisturbed ecosystem could also have a deleterious effect on the flora and fauna of that ecosystem through immediate effects or through prolonged and persistent effects. In order to develop and implement an effective international policy for pollutants' management, it is important, among other factors, to understand their decomposition mechanisms. To accomplish this understanding, it is necessary to understand the sources of the pollutants and the chemistry involved in the generation of the pollutants as well as the potential for the pollutants to react and change (transform) once they are released into the environment (Chapter 7).

By way of definition, environmental pollution is the contamination of the physical and biological components of the Earth system (atmosphere, aquasphere, and geosphere) to such an extent that the normal environmental processes are adversely affected. Thus, inorganic pollution is the introduction of inorganic chemicals into the environment that cause harm or discomfort to the floral and faunal species or that damage the environment, which can come in the form of inorganic chemicals or by-products arising from these chemicals such as radiant energy and noise, heat, or light. The pollutants can take the form of overloading the environment (or an

Environmental Inorganic Chemistry for Engineers
http://dx.doi.org/10.1016/B978-0-12-849891-0.00005-9

ecosystem) by returning (or disposing of) naturally occurring chemicals to the environment in amounts that exceed the natural abundance of the chemicals and that are considered contaminants when in excess of natural levels. Inorganic pollution is also the addition of any substance or form of energy (such as radioactivity) to the environment at a rate faster than the environment can accommodate it by dispersion, breakdown, recycling, or storage in some benign form.

Furthermore, pollution by inorganic chemicals is another case of habitat destruction (Speight, 2017a) and involves (predominantly) chemical destruction rather than the more obvious physical destruction. The overriding theme of the definition is the ability (or inability) of the environment to absorb and adapt to changes brought about by human activities. Thus, environmental pollution occurs when the environment is unable to accept, process, and neutralize any harmful by-product of human activities such as the gases that contribute to acid rain such as the various gases from the combustion of fossil fuels (Table 5.1), which cannot be disposed of without any damage to the environmental system and the transformation of the gases in the atmosphere to acid rain (Chapter 7):

$$2[C]_{\text{fossil fuel}} + O_2 \rightarrow 2CO$$
$$[C]_{\text{fossil fuel}} + O_2 \rightarrow CO_2$$

Table 5.1 Inorganic Air Contaminants and Major Sources

Carbon dioxide (CO_2)	Combustion emissions
Carbon monoxide (CO)	Combustion emissions
Fly ash	Combustion emissions
Hydrogen sulfide (H_2S)	Combustion, volcanic emissions
Hydroxyl (OH)	Photochemistry
Hydrogen peroxide (H_2O_2)	Photochemistry
Mercury (Hg)	Combustion emissions
Mineral and soil dust	Dust emissions
Nitric acid (HNO_3)	Gas phase chemistry
Nitric oxide (NO)	Combustion
Nitrogen dioxide (NO_2)	Photochemistry
Nitrous acid (HNO_2)	Photochemistry
Radionuclides	Nuclear accidents
Ozone (O_3)	Photochemistry
Sulfur dioxide (SO_2)	Photochemistry, combustion, volcanic emissions
Transition metals	Combustion
Volcanic ash	Volcanic emissions

$$2[N]_{\text{fossil fuel}} + O_2 \rightarrow 2NO$$
$$[N]_{\text{fossil fuel}} + O_2 \rightarrow NO_2$$
$$[S]_{\text{fossil fuel}} + O_2 \rightarrow SO_2$$
$$2SO_2 + O_2 \rightarrow 2SO_3$$
$$SO_2 + H_2O \rightarrow H_2SO_3 \text{ (sulfurous acid)}$$
$$SO_3 + H_2O \rightarrow H_2SO_4 \text{ (sulfuric acid)}$$
$$NO + H_2O \rightarrow HNO_2 \text{ (nitrous acid)}$$
$$3NO_2 + 2H_2O \rightarrow HNO_3 \text{ (nitric acid)}$$

Thus, inorganic pollution occurs when (i) the natural environment is incapable of decomposing the unnaturally generated chemicals (*anthropogenic pollutants*) and (ii) when there is a lack of knowledge on the means by which these chemicals can treated for disposal. This leaves the environment (or any ecosystem) subjected to any negative impacts on crucial environmental chemistry.

Although pollution had been known to exist for a very long time (at least since people started using fire thousands of years ago), it had seen the growth of truly global proportions only since the onset and expansion of the industrial revolution during the 19th century. In fact, the industrial revolution brought with it technological progress such as discovery of crude oil (and uses for crude oil products). The technological progress facilitated the manufacture of chemicals—with the accompanying chemical waste—that became one of the main causes of serious deterioration of natural resources and pollution of the environment. At the same time, of course, development of natural sciences and various applied sciences led to the better understanding of negative effects produced by pollution on the environment.

Pollution of the environment by inorganic chemicals is caused by the release of inorganic chemical waste that causes detrimental effects on the environment. Environmental pollution is often divided into pollution of (i) the air, (ii) the water, and (iii) the land or soil. Water supplies the atmosphere and the soil. Because the field of organic chemicals is broad and often refers to hundreds or thousands of several different types of pollutants, including toxic and high concentrations of normally innocuous compounds (Tables 5.2 and 5.3). While the blame for much of the pollution is caused by the inorganic chemical industry, domestic sources of pollution include waste and automobile exhaust. Individuals and chemical and petroleum companies contribute to the pollution of the atmosphere by releasing inorganic and

Table 5.2 Examples of Inorganic Pollutants

Pollutant	Formula	Sources	Polluted Medium
Sulfur dioxide	SO_2	Volcanoes	Air
Industry		Air	
Transports		Air	
Nitrogen dioxide	NO_2	Volcanoes	Air
Industry		Air	
Transports		Air	
Carbon monoxide	CO	Transports	Air
Ammonia	NH_3	Industry	Air, soil
		Agriculture	Air, soil
Hydrogen sulfide	H_2S	Industry	Soil, water
		Anaerobic fermentation	Soil, water
Hydrogen chloride	HCl	Industry	Air
		Transports	Air
Hydrogen fluoride	HF	Industry	Air
Ammonium salts	NH_4^+	Farms, factories	Soil, water
Nitrate salts	NO_3^-	Farms, factories	Soil, water
Nitrite salts	NO_2^-	Farms, factories	Soil, water
Lead salts[a]	Pb	Heavy industry	Air
		Transports	Air
Mercury salts[a]	Hg	Industry	Soil, water
Zinc salts[a]	Zn	Industry	Air, water

[a]May also appear as the metal.

organic gases and particulates into the air. These chemicals can react with tissues in the body and change the structure and function of the organ, cause abnormal growth and development of the individual, or bind with the genetic material of cells and cause cancer. One of the central tenets of the study of such effects (toxicology) is that the dose of a chemical determines its overall effects and that most chemicals can be dangerous at high exposures (Goyer and Clarkson, 1996; Gallo, 2001).

By way of clarification, an inorganic pollutant is an inorganic chemical that is released into an ecosystem and which causes pollution (however temporary or permanent) insofar as the chemical is harmful to or is destructive to the flora and/or the fauna of the ecosystem (Tables 5.2 and 5.3). Typically and by the name, a pollutant is a chemical that is not indigenous to the ecosystem. However, if the discharged chemical is indigenous to the ecosystem (i.e., the inorganic chemical is a naturally occurring compound), it can be (should be) classed as a pollutant when it is released into the system in amounts that are more than the natural concentration of the inorganic

Table 5.3 General Classification of Inorganic Chemical Pollutants[a]

Pollutants	Source
Metals and their salts	Usually from mining and smelting activities and disposal of mining wastes
Inorganic fertilizers (e.g., nitrates and phosphates)	Used largely in agriculture and gardening If present in large amounts in water, they can be harmful to human health and usually trigger algae blooming events
Sulfides (e.g., pyrite and FeS_2)	Usually mined minerals Once disposed in the environment, may generate sulfuric acid in the presence of precipitation water and microorganisms
Ammonia	A poisonous gas If released in higher amounts, may cause blindness followed by death
Oxides of nitrogen	Common air pollutants resulting from vehicle emissions, industrial processes, and other human activities
Oxides of sulfur	Common air pollutants resulting from vehicle emissions, industrial processes, and other human activities
Acids and bases	Used in a variety of industrial applications and in chemical laboratories Less problematic chemicals because their effect can be neutralized in the environment If spread in large amounts, will pose a threat to environment and human health
Perchlorates	Includes the perchloric acids and its various salts Used in a variety of applications including rocket fuel, explosives, military operations, fireworks, road flares, inflation bags Problematic because of environmental persistence May damage thyroid function in humans

[a]Chemicals of mineral origin and not produced by living organisms.

chemical in the ecosystem, and by this increased concentration, the chemical can cause harm to (or is destructive to) the flora and/or the fauna of the ecosystem.

More specifically, inorganic pollutants are elements or compounds found in water supplies and may be natural in the geology or caused by activities of man through mining, industry, or agriculture. It is common to have trace amounts of many inorganic contaminants in water supplies, but amounts of these pollutants that are above the maximum contaminant levels may cause a variety of damaging effects to an ecosystem depending upon the inorganic contaminant and level of exposure.

Persistent inorganic pollutants, like any inorganic chemical pollutant, can enter an ecosystem through the gas phase, the liquid phase, or the solid phase and that can resist degradation and are mobile over considerable distances (especially in the gas phase or through transportation in river systems) before being redeposited in a location that is remote to the location of their introduction into the ecosystem. Furthermore, persistent inorganic pollutants can be present as vapors in the atmosphere or bound to (adsorbed on) the surface of soil or mineral particles and have variable solubility in water.

Many persistent inorganic pollutants are currently (or were in the past) arose from the extensive use of agrochemicals (agricultural chemicals). Although some persistent inorganic pollutants arise naturally, for example, from various natural pathways, most are products of human industry and tend to have higher concentrations and are eliminated more slowly. If not removed and because of their properties, persistent inorganic pollutants will bioaccumulate and have significant impacts on the flora and fauna of the environment. The most frequently used measure of the potential for bioaccumulations and persistence of an inorganic compound in the environment is the result of the physicochemical properties (such partition coefficients and reaction rate constants).

However, there is a possibility that, through the judicious use of resources and the application of the principles of environmental science, environmental engineering, and environmental analysis (disciplines involved in the study of the environment and determining the *purity* of the environment), a state can be reached where pollution is minimal and does not pose a threat to the existing floral and faunal species. Such a program must, of necessity, involve not only well-appointed suites of analytic tests but also subsequent studies that cover the effects of changes to the environmental conditions in the ecosystem on the flora and fauna of that system. These suites of studies should include the relevant aspects of chemistry, chemical engineering, biochemistry, microbiology, hydrology, and climatology as they can be applied to solve environmental problems.

The potential for pollution of inorganic chemicals starts during the production stage when not only the desired products are produced but also there is the potential (often the reality in many processes) to produce various by-products. These products must be converted to useful products or discarded, to the detriment of the process, and (in the past) to the detriment of the environment. The typical inorganic chemical synthesis process involves combining one or more feedstocks in one or more unit process operations. Typically, commodity chemicals tend to be synthesized in a continuous reactor (one unit process), while specialty chemicals are usually produced in one or more batch reactors (perhaps involving two or more unit processes). The yield of the inorganic chemical and the efficiency of the chemical process will determine the type and quantity of the by-products.

Many of the process reactions (i) take place at high temperatures, (ii) include one or two additional reaction components, and (iii) involve the use of catalysts that may be based on inorganic chemicals (Chapter 4). Thus, many specialty inorganic chemicals may require a series of two or three reaction steps, each involving a different reactor system and each capable of producing one or more by-products. Once the reaction is complete, the desired product must be separated from the by-products, often by a separate unit operation in which any one (or more) of several separation techniques such as settling, distillation, or refrigeration may be used. In addition, to produce the saleable item, the final product may be further processed using process techniques such as spray drying or pelletizing. Frequently, by-products are also sold, and their value can influence the production efficiency and economics of the process.

Despite numerous safety protocols that have been adopted by industry and that are in place and the care taken to avoid environmental incidents that are harmful to the environment, every industry suffers accidents that lead to contamination of the environment by chemicals. It is therefore often helpful to be knowledgeable of the nature (the chemical and physical properties) of the chemical contaminants and the products of chemical transformations (when the ecosystem parameters interact with the chemicals) in order to understand not only the nature of the immediate and continuing chemical contamination but also the chemical changes to the contaminants from which cleanup methods can be chosen (Chapter 4).

In the past, the existence and source of such information was generally unknown and, if known, was not always used to determine any potential environmental issues (contamination and cleanup). When the existence and sources of the relevant information are known, decisions must be made for environmental scientists and engineers to make an informed, and often quick, decision on the next steps, even if it is decided later not to use the information for a site, on the basis that not two sites are exactly alike.

However, on the basis that *it is better to know than to not know*, knowledge of the relevant data gives investigators and analysts the ability to assess whether a chemical discharge into the environment should be addressed and made to halt the discharge of the chemical or whether the environment can take care of itself through biodegradation of the chemical. This is especially true for scientists and engineers involved in (i) site cleanup operations, (ii) assessment of ecological risk, (iii) assessment of ecological damage, and (iv) any steps necessary to protect the environment. Modern databases relating to the properties of chemicals, especially inorganic chemicals (which are the reason for the current text), and there can be no reasons (or excuses) for not knowing or understanding the fundamental aspects of the behavior of inorganic chemicals that are discharged to the environment.

Furthermore, the capacity of the environment to absorb the effluents and other impacts of process technologies is not unlimited, as some would have us believe. The environment should be considered as an extremely limited resource, and discharge of chemicals into it should be subjected to severe constraints. Indeed, the declining quality of raw materials dictates that more material must be processed to provide the needed chemical products. Moreover, the growing magnitude of the products and effluents from industrial processes has moved above the line where the environment has the capability to absorb such process effluents without disruption.

As a commencement to this process of data examination and data use, this chapter introduces the terminology of environmental technology as it pertains to the sources and types of inorganic pollutants. Briefly, a *contaminant*, which is not usually classified as a pollutant unless it has some detrimental effect, can cause deviation from the normal composition of an environment. A *receptor* is an object (animal, vegetable, or mineral) or a locale that is affected by the pollutant. A *chemical waste* is any solid, liquid, or gaseous waste material that, if improperly managed or disposed of, may pose substantial hazards to human health and the environment. At any stage of the management process, a chemical waste may be designated by law as a *hazardous waste*. Improper disposal of these waste streams, such as inorganic solvents, in the past has created hazards to human health and the need for very expensive cleanup operations. Correct handling of these chemicals and dispensing with many of the myths related to chemical processing can mitigate some of the problems that will occur, especially problems related to the flammability of inorganic liquids, when incorrect handling is practiced. Chemical waste is also defined and classified into various subgroups.

5.2 SOURCES

Environmental pollutants arise from a variety of sources that are constituent parts of the pollution process. The sources of the actual causative agents of environmental pollution can occur in *gaseous*, *liquid*, or *solid* form. Most important, inorganic pollutants (or any pollutants for that matter) are transboundary insofar as they (the pollutants) do not recognize boundaries. In addition, many inorganic pollutants cannot be degraded by living organisms and therefore stay in the ecosphere for many years, even decades. Inorganic pollutants also destroy biota and the habitat, and because of this, it is emphasized that inorganic pollutants can present a serious long-term global problem that affects more or less every country and, therefore, can only be solved by a coordinated set of actions and the necessary commitment of nations of the Earth (with no exceptions) to international environmental agreements.

The contamination of the environment by inorganic chemicals (Table 5.4) can occur from several process sources (Chapter 3), and the effects of the chemical on the environment are determined by the chemical and physical properties of the chemical (Chapter 4): for example, most toxic pollutants originate from human-made sources, such as mobile sources (such as cars, trucks, and buses) and stationary sources (such as factories, refineries, and power plants), as well as indoor sources (such as building materials and activities such as cleaning).

Table 5.4 Summary of the Most Common Air, Water, and Soil Pollutants

Air pollutants
Sulfur dioxide
Nitrogen oxides
Ammonia
Carbon monoxide
Ozone
Persistent inorganic pollutants (POPs)
Airborne particles
Toxic metals
Radioactive pollutants
Water pollutants
Heavy metals
Acid rain
Chemical waste
Fertilizers (inorganic) from agricultural use
Silt from construction sites, logging, slash, and burn operation
Soil pollutants
Pesticides (inorganic)
Lead and other heavy metals

Fossil fuels that are sources of pollutants include coal, natural gas and crude oil, and more familiar fuels refined from crude oil refining, which include gasoline, diesel fuel, and fuel oil (Speight, 2014, 2017b). The burning, or combustion, of fossil fuels is a major source of pollutants, which contribute to acid rain. The pollutants that come from the combustion of fossil fuels include sulfur dioxide (SO_2), nitrogen oxides (NO_x), particulate matter (PM), carbon monoxide (CO), and carbon dioxide (CO_2), and various heavy metals. There are also several organic pollutants that are not the subject of this book and have been described elsewhere (Speight, 2017a). The inorganic pollutants are a result of the use of air, which is composed of nitrogen (78%) and oxygen (21%), in the combustion process, and the presence of impurities such as sulfur within the fuel.

Climate change on a global scale has been attributed to increased emissions of carbon dioxide (CO_2), a greenhouse gas, although the actual contribution of anthropogenic effects is not fully understood or known (Speight and Islam, 2016). A global average temperature rise of only 1°C (1.8°F) has been estimated to produce serious climatological implications. Possible consequences include melting of polar ice caps, an increase in sea level, and increases in precipitation and severe weather events like hurricanes, tornadoes, heat waves, floods, and droughts. Other atmospheric effects of air pollution include urban smog and reduced visibility, associated with ozone-forming nitrogen oxides and volatile organic compound emissions. Sulfur dioxide (SO_2) and nitrogen oxides (NO_x) combine with water in the atmosphere to cause acid rain, which is detrimental to forests and other vegetation, soil, lakes, and aquatic life. Acid rain also causes monuments and buildings to deteriorate.

Renewable energy is derived from resources (such as biomass) that are replenished naturally on a human timescale. Such resources include biomass, geothermal heat, sunlight, water, and wind. However, of these sources, some are more suited to certain locations than others, for instance. For example, some of the renewable sources only produce electricity intermittently (when the sun is shining in the case of solar), though they can be accompanied paired with energy storage solutions to provide reliable electricity 24 h a day throughout the year. Others sources, such as biomass, hydropower, and geothermal, can be used as baseload generation, producing a constant, predictable supply of electricity. But, none of these sources can effectively meet all of the electricity needs of a modern society, but once production is increased to the desired level, these alternate energy sources not only can supplement energy from fossil fuels but also may, one day, replace fossil fuels as the prime energy sources.

On the downside, biomass as proposed for use as a fuel (or fuel source) can also be a source of environmental pollution (Table 5.5). Knowledge of the inorganic components of biomass feedstock is important for process control and for handling coproducts and wastes resulting from energy and fuel utilization of biomass. Analytic survey of forestry thinnings (wood chips), agricultural residues (rice straw, wheat straw, and corn stover), and dedicated perennial grass crops (switchgrass, wheatgrass, and miscanthus) shows that, potentially, the whole periodic table may be present in biomass. The main effect of ashing is bonding of oxygen in the ash mainly as silicate, oxides, hydroxides, phosphates,

Table 5.5 Sources and Effects of Common Air pollutants

Pollutant	Anthropogenic Sources	Environmental Effects
Ozone (O_3)	Secondary pollutant formed by chemical reaction of NO_x in the presence of sunlight	Damages crops, forests, and other vegetation; damages rubber, fabric, and other materials; smog reduces visibility
Nitrogen oxides (NO_x)	Burning of gasoline, natural gas, coal, oil	Ozone (smog) effects, precursor of acid rain; aerosols can reduce visibility
Carbon monoxide (CO)	Burning of gasoline, natural gas, coal, oil	Converts to carbon dioxide and then to acid rain
Particulate matter	Emitted as particles, burning of wood and other fuels, industrial processes, agriculture (plowing, field burning)	Source of haze that reduces visibility, Ashes, smoke, soot, and dust can dirty and discolor structures and property, including clothes and furniture
Sulfur dioxide (SO_2)	Burning of coal and oil and industrial processes (paper manufacturing, metal smelting)	Precursor of acid rain, which can damage trees, lakes, and soil; aerosols can reduce visibility
Mercury	Fossil-fuel combustion, waste disposal, industrial processes (incineration, smelting, chlor-alkali plants), mining	Accumulates in food chain

and carbonate residual minerals. Carbon is partially retained as carbonates and graphite (char). Nitrogen is dominantly released to the flue gas, while sulfur is mostly retained in the ash as sulfates. Small losses (approximately 19%, w/w) for both sulfur and chlorine were detected during ashing at 575°C (1070°F). The majority of the alkali metals (lithium, Li; sodium, Na; potassium, K; and rubidium, Rb) (Fig. 5.1) will substantially modify soil if applied as a fertilizer. Only magnesium (Mg), calcium (Ca), and strontium (Sr) of the alkali earth metals; manganese (Mn), copper (Cu), and zinc (Zn) of the period four transition metals; and molybdenum (Mo) and cadmium (Cd) of the period five transition metals may exceed regulatory limits if used as a fertilizer. The heavy elements occur in concentrations too low to cause concern except for selenium, Se. The high alkali content of some biomass ash thus makes them good candidates for use as fertilizers if they are applied in low proportions (<50%) to soil. Ash of wood material is a carrier of many of the alkali elements (Li, K, Rb, Mg, Ca, Sr, and Ba) and some transition elements (Mn, Cu, Zn, Mo, Ag, and Cd). In contrast, ash of herbaceous plant material is in addition to K only variably enriched: Li, Na, Se, and Mo in wheatgrass; Mg and Ca in switch grass; Mg, Ca, and Cd in corn stover; Mn in rice straw; Mo in wheat straw; and Mn and Cd in miscanthus. Water leaching results in significant losses for anionic chlorine and sulfur and for most of the alkali metals, thus making resulting ash from such treated feedstock less attractive as potassium-containing fertilizers although fuel properties are enhanced for thermal conversion (Thy et al., 2013).

In general, there are several types of environmental pollution, but the most important ones are (i) air pollution, (ii) water pollution, and (iii) soil pollution. Most notable among the air pollutants are sulfur dioxide (SO_2), nitrogen dioxide (NO_2), carbon monoxide (CO), ozone (O_3), volatile organic compounds (VOCs), airborne particles (particulate matter), and radioactive pollutants—probably the most destructive form of pollution, especially when produced by a nuclear explosion.

5.2.1 Air pollution

Fossil fuels are a contributor to health-harming air pollution in the form of the inorganic gases that are produced during fossil-fuel combustion and use. The emission of carbon monoxide gases from cars and trucks is the most commonly cited example of an inorganic pollutant, but it is not the only one. The combustion process also creates nitrogen oxides that lead to the creation of smog. Power-plant- and transportation-related activities are about equally responsible for nitrogen oxide emissions.

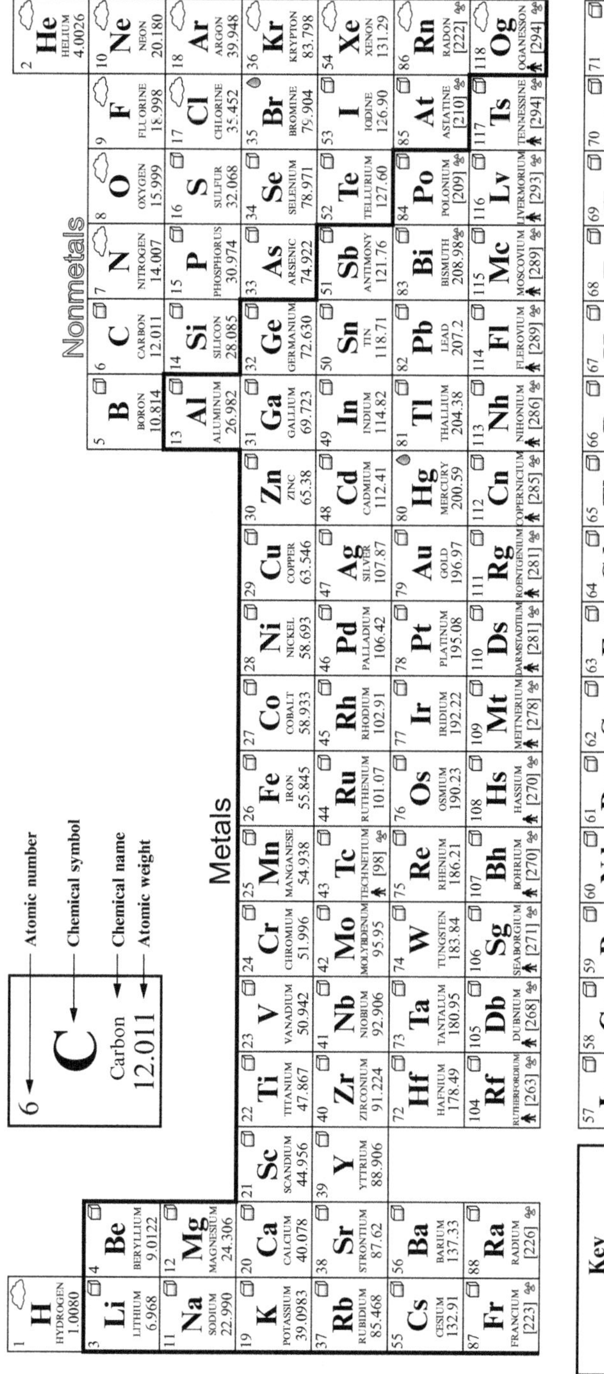

Fig. 5.1 The periodic table of the elements.

However, air pollution is caused by many types of sources of every size and include biogenic (i.e., natural) sources, such as volcanoes or forest fires, and anthropogenic (human) sources. The latter includes mobile sources (related to transportation) and stationary sources (nonmoving sources). Within the category of stationary sources, there are point sources such as a factory smokestack and area sources that include clusters of smaller sources (e.g., dry cleaners, print shops, spray painters, and woodstoves) that may not emit very much individually, but together contribute significantly to air pollution levels. While industrial sources can be either point or area sources, the term is often associated with larger operations. These can include any type of industry, such as agricultural operations, factories that manufacture goods, power plants, chemical plants, and waste disposal incinerators.

Air pollution is caused by harmful gases (Table 5.6) and particulate matter in the air that originate from various sources. In fact, the sources of air pollution can impact many areas even though the vast majority of air pollution is created elsewhere, even outside of national boundaries. There are four main types of air pollution sources: (i) mobile sources, such as cars, buses, planes, trucks, and trains; (ii) stationary sources, such as power plants, oil refineries, industrial facilities, and factories; (iii) area sources, such as agricultural areas, cities, and wood burning fireplaces; and (iv) natural sources, such as windblown dust, wildfires, and volcanoes. Mobile sources account for more than half of all the air pollution in the United States, and the primary mobile source of air pollution is the automobile, according to US Environmental Protection Agency. Stationary sources, like power plants, emit large amounts of pollution from a single location; these are also known as point sources of pollution.

A point source is a single, stationary source of pollution, such as an industrial facility, that typically operates under government authorization (e.g., a permit, approval, or regulation). A nonpoint source includes stationary and mobile sources that are individually small compared with point sources but collectively large, such as wood stoves, motor vehicles, and lawnmowers. It

Table 5.6 Pollutant Gases

Element	**Atmospheric Forms**
Oxygen	O_3
Carbon	CO, CO_2
Chlorine	Cl_2, $Cl\cdot$, Cl^-
Nitrogen	NH_3, N_2O, NO, NO_2, N_2O_5
Sulfur	H_2S, SO_2, SO_3

also includes sources where the emissions are spread out over a broad area, such as prescribed burning. Area sources are made up of several, (even, many) lots of smaller pollution sources that are not a big deal by themselves but when considered as a group can be. Natural sources of pollutants include forest fires, windblown dust, and volcanoes. Human sources include emissions from fossil-fuel burning (oil, gas, and coal), wood burning, and stirred-up dust from vehicles or construction. Natural sources can sometimes be significant but do not usually create ongoing air pollution problems such as created by the other sources.

Air pollution occurs in many forms but can generally be thought of as gaseous and particulate contaminants that are present in the atmosphere of the Earth. Air pollution can further be classified into two sections: (i) visible air pollution and (ii) invisible air pollution. Another way of looking at air pollution could be any substance that holds the potential to hinder the atmosphere or the well-being of the living beings surviving in it. The sustainment of all things living is due to a combination of gases that collectively form the atmosphere; the imbalance caused by the increase or decrease of the percentage of these gases can be harmful for survival. Furthermore, a physical, biological, or chemical alteration to the air in the atmosphere can be termed as pollution that occurs when any harmful gases, dust, and smoke enter into the atmosphere and make it difficult for the floral and faunal species to survive.

From the perspective of inorganic chemicals, virtually all metals are present in the atmosphere at low levels. Particulate matter emitted from combustion of fossil fuels contains trace metals that were present in the original fuel sample. The greatest health hazard is from aerosols that are smaller than 2.5 microns (μm) in diameter and contains lead, beryllium, mercury, cadmium, and chromium. These particles are removed from the atmosphere by settling out over time or by precipitation events. Mercury is the only metal that exists as a gas and is therefore the only metal that exists in a steady-state concentration in air. All other metals are emitted to the atmosphere from natural or anthropogenic sources and then removed by settling or precipitation events.

However, gaseous inorganic substances remain in the atmosphere for long residence times and would continue to build up much higher levels than observed today if it were not for the gas phase reactions that convert these substances to water-soluble species. Gaseous pollutants include, but are not limited to, carbon monoxide (CO), nitrogen oxides (NO_x), ozone (O_3), and sulfur dioxide (SO_2).

Carbon monoxide is a product of the incomplete combustion of fossil fuels and is often listed in urban air quality measures. As much as 20% of the carbon monoxide released to the atmosphere, each year comes from natural sources, but the greatest health problem is in metropolitan areas near high densities of vehicular traffic. Urban air may contain carbon monoxide at concentrations more than of 100 ppm during rush-hour traffic. Catalytic converters and air pumps have been installed on all motor vehicles built since the early 1970s to reduce carbon monoxide emissions. Carbon monoxide is transformed to carbon dioxide when exhaust gases are combined with air and passed over the surface of a catalyst. Carbon monoxide has a 4-month lifetime in the atmosphere, where it reacts with the hydroxyl radical to form carbon dioxide. The sequence of reactions that transform carbon monoxide to carbon dioxide and regenerate the hydroxyl radical are the following:

$$CO + HO\cdot \rightarrow CO_2 + H\cdot$$
$$H\cdot + O_2 + M \rightarrow HOO\cdot + M$$
$$HOO\cdot + HOO\cdot \rightarrow H_2O_2 + O_2$$
$$H_2O_2 + \text{energy} \rightarrow 2HO\cdot$$

Carbon monoxide, generated by the incomplete combustion of hydrocarbons, displaces and prevents oxygen from binding to hemoglobin (also spelled hemoglobin) in the blood and causes asphyxiation.

Complete combustion of heptane (a fuel constituent):

$$C_7H_{16} + 11O_2 \rightarrow 7CO_2 + 8H_2O$$

Incomplete combustion of heptane to produce carbon monoxide, water, and carbon:

$$C_7H_{16} + 6O_2 \rightarrow 4CO + 8H_2O + 3C$$

Also, carbon monoxide binds with metallic pollutants and causes them to be more mobile in air and water. These pollutants are emitted from large stationary sources such as fossil-fuel-fired power plants, smelters, industrial boilers, petroleum refineries, and manufacturing facilities and from area and mobile sources. They are corrosive to various materials, which causes damage to cultural resources, can cause injury to ecosystems and organisms, aggravate respiratory diseases, and reduce visibility. Atmospheric carbon dioxide levels are determined by a long-term equilibrium between carbon dioxide in the air and carbon dioxide dissolved in the oceans and surface water, releases of carbon dioxide from natural and anthropogenic sources, and losses by plant growth.

Although *ozone* (O_3) is a desirable substance in the stratosphere, it is a major environmental hazard at ground level. Ozone is a by-product of photochemical smog and reacts with hydrocarbons to form peroxy-nitrate derivatives (salts of the unsayable peroxy nitric acid, HNO_4) that damage sensitive biological tissues in the eyes, nasal passages, throat, and lungs. Excessive ozone levels in the troposphere have been blamed for killing plants through reactions with chlorophyll. Ozone is formed naturally when oxygen molecules are photochemically dissociated into oxygen atoms that can then react with a second oxygen molecule to make ozone. The presence of nitrogen oxides (NO, NO_2) leads to higher than normal background levels of ozone through several well-understood photochemical reactions.

Nitrogen oxides and sulfur dioxide can combine with water to form acids, which not only irritate the lungs but also contribute to the long-term destruction of the environment due to the generation of acid rain. Thus, *acid rain* is formed when sulfur dioxide and nitrogen oxides react with water vapor and other chemicals in the presence of sunlight to form various acidic compounds in the air. The principle sources of acid rain-causing pollutants, sulfur dioxide, and nitrogen oxides are from fossil-fuel combustion and from the combustion of fossil-fuel-derived fuels (Chapter 7):

$$2[C]_{\text{fossil fuel}} + O_2 \rightarrow 2CO$$
$$[C]_{\text{fossil fuel}} + O_2 \rightarrow CO_2$$
$$2[N]_{\text{fossil fuel}} + O_2 \rightarrow 2NO$$
$$[N]_{\text{fossil fuel}} + O_2 \rightarrow NO_2$$
$$[S]_{\text{fossil fuel}} + O_2 \rightarrow SO_2$$
$$2SO_2 + O_2 \rightarrow 2SO_3$$

In the modern industrialized world, fossil fuels (coal, crude oil, and natural gas) have been ascribed the blame (with justification) for virtually all of the imaginable detrimental effects of air pollution. Not only does modern society use fossil fuels for our obvious everyday needs and in the power-generating industry, but also fossil fuels (especially crude oil) are used as starting materials for products such as plastics, solvents, detergents, lubricating oils, and asphalt as well as a wide range of chemicals for industrial use. The combustion of fossil fuels produces extremely high levels of *air pollution* and is widely recognized as one of the most important areas for reduction and control of environmental pollution. Fossil fuels also contribute to *soil contamination* and *water pollution*. For example, when crude oil is transported from the point of its production to further destinations by pipelines, an oil leak from the pipeline may occur and pollute soil and subsequently groundwater. When crude oil is transported by tankers by ocean, an oil spill may occur and pollute ocean water.

Among other pollution sources, *agriculture* (livestock farming) cannot be ignored since this industry is the major generator of ammonia emissions. Chemicals such as pesticides and fertilizers are also widely used in agriculture, which may lead *water pollution* and *soil* contamination. Also, *trading activities* that involve packaging of products sold in supermarkets and other retail outlets generate large quantities of solid waste that ends up either in landfills or municipal incinerators leading to *soil contamination* and *air pollution*. Also, the *residential sector* is another significant source of pollution generating solid municipal waste that may end up in landfills or incinerators leading to *air pollution* and *soil contamination*.

In the current context, *toxic air pollutants* are a class of inorganic chemicals that may potentially cause health problems in a significant way. The sources of toxic air pollutants include power plants, various industries including the inorganic chemicals and organic chemicals industries, pesticide application, and contaminated windblown dust. Persistent toxic pollutants, such as mercury, are of particular concern because of their global mobility and ability to accumulate in the food chain.

Primary air pollutants are those inorganic pollutants that are emitted directly into the air from pollution sources. On the other hand, *secondary air pollutants* are formed when primary pollutants undergo chemical changes in the atmosphere. Ozone is an example of a secondary pollutant and is formed when nitrogen oxides (NO_x) and volatile organic compounds (VOCs) are mixed and warmed by sunlight.

Each toxic air pollutant comes from a slightly different source, but many are created in chemical plants or are emitted when fossil fuels are burned. Many toxic air pollutants can also enter the food and water supplies. Gases that stay in the air for prolonged periods warm up the planet by trapping sunlight (*greenhouse effect*), and some of the important greenhouse gases are carbon dioxide, methane, and nitrous oxide. Carbon dioxide is the most important greenhouse gas and arises from the burning of fossil fuels in cars, power plants, houses, and industry, although the process contribution of this gas to the so-called global climate change is not known because of analytic uncertainties (Speight and Islam, 2016).

The major inorganic air pollutants are (i) carbon monoxide, (ii) nitrogen dioxide, (iii) ozone, (iv) sulfur dioxide, and (v) particulate matter (Table 5.7).

Table 5.7 Types of Air Pollutants[a] and Their Effects

Pollutant	Source
Ammonia	A poisonous gas if released in higher amounts and may cause blindness followed by death
Carbon monoxide	Formed during the incomplete combustion of fuel from vehicles and engines—when fuels have too little oxygen to burn completely. It spews out in car exhausts, and it can also build up to dangerous levels inside the home if there is a poorly maintained gas boiler, stove, or fuel-burning appliance Reduces the amount of oxygen reaching body organs and tissues Aggravates heart disease, resulting in chest pain and other symptoms
Carbon dioxide	This gas is central to everyday life and is not normally considered a pollutant: it is produced as a product of breathing and plants such as crops and trees need this gas for survival. However, carbon dioxide is also a greenhouse gas released by engines and power plants. Since the beginning of the industrial revolution, it is been building up in atmosphere of the Earth and cited as a contributor to the problem of global warming and climate change
Lead and heavy metals	Lead and other toxic heavy metals can be spread into the air either as toxic compounds or as aerosols (when solids or liquids are dispersed through gases and carried through the air by them) in such things as exhaust fumes and the fly ash (contaminated waste dust) from incinerator smokestacks
Nitrogen dioxide	Fuel combustion (electric utilities, big industrial boilers, vehicles) and wood burning Worsens lung diseases leading to respiratory symptoms Increased susceptibility to respiratory infection
Oxides of nitrogen	Common air pollutants resulting from vehicle emissions and industrial processes Nitrogen dioxide (NO_2) and nitrogen oxide (NO) are pollutants produced as an indirect result of combustion, when nitrogen and oxygen from the air react together. Nitrogen oxide pollution comes from vehicle engines and power plants and plays an important role in the formation of acid rain, ozone, and smog. Nitrogen oxides are also *indirect greenhouse gases* insofar as they contribute to global warming by producing ozone, which is a greenhouse gas

Continued

Table 5.7 Types of Air Pollutants and Their Effects—cont'd

Pollutant	Source
Oxides of sulfur	Common air pollutants resulting from vehicle emissions and industrial processes
Ozone	Also called trioxygen; this is a type of oxygen gas whose molecules are made from three oxygen atoms joined together (so it has the chemical formula O_3), instead of just the two atoms in conventional oxygen (O_2) Secondary pollutant at ground level Formed from volatile organic compounds (VOCs) and NO_x in the presence of sunlight Decreases lung function and causes respiratory symptoms In the stratosphere (upper atmosphere), a band of ozone (the ozone layer) protects us by screening out harmful ultraviolet radiation (high-energy blue light) beaming down from the Sun. At ground level, it is a toxic pollutant that can damage health. It forms when sunlight strikes a cocktail of other pollution and is a key ingredient of smog
Particulate matter	May be carbonaceous material or mineral material Formed through chemical reactions, fuel combustion (including wood burning) Can cause heart or lung disease and sometimes premature deaths Particulates of different sizes are often referred to by the letters PM followed by a number, so PM_{10} means soot particles of less than 10 microns (10 millionths of a meter or 10 μm in diameter). In cities, most particulates come from traffic fumes
Smog	A combination of the words *smoke* and *fog* forms when sunlight acts on a cocktail of pollutant gases such as nitrogen and sulfur oxides, unburned hydrocarbons, and carbon monoxide; that is why it is sometimes called *photochemical smog* (the energy in light causes the chemical reaction that makes smog). One of the most harmful constituents of smog is a toxic form of oxygen called ozone, which can cause serious breathing difficulties and even, sometimes, death. When smog is rich in ozone, it tends to be a bluish color; otherwise, it's more likely to be brown; although smog can happen in any busy city where the local climate (influenced by the ocean and neighboring mountains) regularly causes what is known as a temperature inversion; normally, air gets colder at higher levels of the atmosphere, but in a temperature inversion the opposite happens: a layer of warm air traps a layer of cold air nearer the ground. This acts like a lid over a cloud of smog and stops it from rising and drifting away. Largely because of their traffic levels, smog afflicts many of the busiest cities of the world

Table 5.7 Types of Air Pollutants and Their Effects—cont'd

Pollutant	Source
Sulfur dioxide	Coal, petroleum, and other fuels are often impure and contain sulfur and organic (carbon-based) compounds. When sulfur burns with oxygen from the air, sulfur dioxide (SO_2) is produced Also, originates from volcanoes Aggravates asthma and makes breathing difficult Contributes to particle formation with associated health effects

[a]Any substance in air that could, in high enough concentration, harm animals, humans, vegetation, and/or materials. Such pollutants may be present as solid particles, liquid droplets, or gases. Air pollutants fall into two main groups: (1) those emitted from identifiable sources and (2) those formed in the air by interaction between other pollutants.
The Clean Air Act requires EPA to set national ambient air quality standards (NAAQS) established by EPA for maximum allowable concentrations of six "criteria" pollutants in outdoor air. The six pollutants are carbon monoxide, lead, ground level ozone, nitrogen dioxide, particulate matter, and sulfur dioxide. The standards are set at a level that protects public health with an adequate margin of safety for six common air pollutants (also known as *criteria air pollutants*).

Carbon monoxide (CO) is a colorless and odorless gas that is generated during the combustion (mostly in automobiles) of the typical hydrocarbon fuels derived from crude oil. Carbon monoxide is released when engines burn fossil fuels, and emissions of this gas are higher when engines are not tuned correctly and when fuel there is incomplete combustion in the engine; complete combustion of hydrocarbon fuels results in the production of carbon dioxide (CO_2) and water (H_2O). Industrial and domestic furnaces and heaters can also emit high concentrations of carbon monoxide if they are not properly maintained.

Nitrogen dioxide (NO_2) is a reddish-brown pungent gas that comes from the burning of fossil fuels that is produced when a nitrogen-containing fuel is combusted or when nitrogen in the air reacts with oxygen at very high temperatures, such as in certain types of industrial furnaces. Nitrogen dioxide can also react in the atmosphere to form ozone, acid rain, and particles.

Ozone (O_3) is a desirable substance in the stratosphere, but it is a major environmental hazard at ground level (the troposphere) where it contributes to the formation of smog. This harmful ozone in the lower atmosphere should not be confused with the protective layer of ozone in the upper atmosphere (stratosphere), which screens out harmful ultraviolet rays. Ozone is formed naturally when oxygen molecules are photochemically dissociated into oxygen atoms that can then react with a second oxygen molecule to make ozone. The presence of nitrogen oxides (nitric oxide, NO and nitrogen dioxide, NO_2) leads to higher than normal background levels of ozone through several well-understood photochemical reactions. Nitrogen oxides (including nitrogen dioxide) arise from the burning gasoline of various fossil-fuel-derived flues. As an example, of the harmful reactivity of ozone, it reacts with hydrocarbons to form peroxy-nitrate derivatives (Chapter 1).

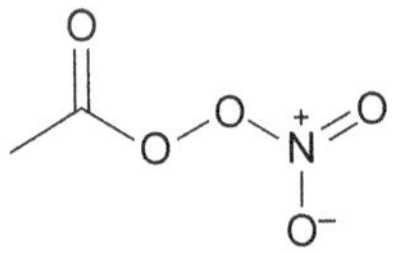

Peroxyacetyl nitrate

These chemicals damage sensitive biological tissues in the eyes, nasal passages, throat, and lungs. Moreover, excessive ozone levels in the troposphere have been blamed for killing plants through reactions with chlorophyll.

Sulfur dioxide (SO_2) is a pungent corrosive gas that arises predominantly from the burning of coal or crude oil in power plants and from factories that produce chemicals, paper, or fuel. Similar to nitrogen dioxide, sulfur dioxide reacts in the atmosphere to form acid rain.

Particulate matter—the term is sometimes used instead of particulates—refers to the mixture of solid particles and liquid droplets found in the air and is a solid matter or liquid matter (aerosol-type matter), that is suspended in the air—to remain in the air; particles usually must be less than 0.1 mm wide and can be as small as 0.00005 mm.

Particulate matter can be either carbonaceous matter (such as soot) or noncarbon matter (such as mineral matter). More generally, particulate matter can be divided into two types: (i) coarse particles and (ii) fine particles. Coarse particles are formed from sources such as road dust, sea spray, and construction dust. Fine particles are formed when fuel is burned in automobiles and power plants and includes sulfate (SO_4) derivatives and nitrate (NO_3) derivatives.

Chlorine is another atmospheric pollutant, and because of its structure, the chlorine atom can form a reactive species (a free radical):

$$Cl_2 \rightarrow 2Cl\cdot$$

However, the most common atmospheric form of chlorine, however, is chloride (Cl^-) that is water-soluble and washes out of the atmosphere quickly. Inorganic chloride derivatives, therefore, have very short lifetimes and do not always constitute a threat to the environment.

Chlorofluorocarbons (CFCs) and some volatile organic compounds (VOCs) contain covalently bonded chlorine atoms and, being relatively inert substances, will remain in the atmosphere for many years. By remaining in the atmosphere for such long times, these molecules migrate across the troposphere/stratosphere boundary where they become exposed to much higher levels of ultraviolet radiation than experienced at ground level. Ultraviolet radiation can cause covalent bonds to separate into individual atoms containing unpaired electrons (free radicals, such as the chlorine free radical, Cl·) that can react further to create environmental pollutants.

Controlling and curtailing air pollution from industrial sources is essential to improving air quality. Achieving meaningful reductions in industrial emissions has proved difficult due to divided constitutional jurisdiction and preference of government and business to deal with the problem through voluntary measures. There are also significant scientific challenges: due to difficulties associated with determining safe emission levels and predicting the interactive or synergistic effects of multiple pollutants, such as the formation of acid rain.

5.2.2 Water Pollution

Water pollution is the contamination of the various bodies of water (e.g., ponds, lakes, rivers, oceans, and aquifers and groundwater), which occurs when (in the context of this book) inorganic chemicals are directly or indirectly discharged into the water bodies without adequate pretreatment to remove harmful or hazardous inorganic chemicals. Furthermore, pollution of water bodies affects the biosphere—the floral and faunal species that live in or rely upon these sources of water for sustenance. In almost all cases, the effect is damaging not only to individual floral and faunal species but also to the natural biological communities.

Water pollution by inorganic chemicals (Tables 5.8 and 5.9) can be defined in many ways, but typically, the term water pollution means that one or more inorganic chemicals have built up in water to such an extent that they cause problems for the water-dependent flora and fauna (including humans). Thus, inorganic water pollutants include (i) acidity, which is caused by industrial discharges, especially sulfur dioxide from coal-burning or crude-oil-burning power plants; (ii) ammonia from food processing waste; (iii) chemical waste, which results from by-products of industrial process, fertilizers, which contain nutrients such as nitrate derivatives and phosphate derivatives and are contained in storm water runoff from agriculture and

Table 5.8 Types of Inorganic Water Pollutants[a]

Pollutant Type	Comments
Acidity	Industrial discharges, especially sulfur dioxide from power plants
Ammonia	Food processing waste
Chemical waste	Chemical industrial by-products
Metals and their salts	Usually from mining and smelting activities and disposal of mining wastes
Inorganic fertilizers	Such as nitrate derivatives and phosphate derivatives Unused largely in agriculture and gardening If present in large amounts, can be harmful to human health Can trigger algal blooming
Heavy metals	Motor vehicles, acid mine drainage
Silt	Runoff from construction sites, logging, land clearing sites
Sulfides	Such as pyrite, FeS_2 Usually mined minerals and once disposed in the environment May generate sulfuric acid in the presence of precipitation water and microorganisms
Acids and bases	Used in a variety of industrial applications and in chemical laboratories May be less problematic chemicals because their effect can be easily neutralized If spread in large amounts, may pose a threat to environment and human health
Perchlorate	Includes the perchloric acids and its various salts Used in a variety of applications including explosives Problematic because it is persistent and may damage thyroid function in humans
Shipwrecks	Derelict ships
Trash	Paper, plastic, or food waste

[a]Those chemicals of mineral origin not produced by living organisms.

commercial and residential use; (iv) heavy metals from motor vehicles via urban storm water runoff and from acid mine drainage; and (v) silt or sediment—fine grained powdery materials—in runoff from construction sites, logging, slash-and-burn practices, or land clearing sites. Another form of pollution, not usually considered as pollution by inorganic chemicals, is the disturbance of the oxygen balance (i.e., the inorganic chemical balance) within a water system. For example, *thermal pollution* from factories and power plants also causes problems in rivers and lakes. By raising the temperature of the water, the amount of oxygen dissolved in the water is reduced, thereby affecting (reducing) the level of aquatic life that the water system can support.

Table 5.9 Primary Inorganic Chemical Contaminants in Water

Chemical	Sources
Antimony	Occurs naturally in the ground and is often used in the flame retardant industry
Arsenic	A semimetal element in the periodic table, odorless, and tasteless
Asbestos	A naturally occurring mineral
Barium	Occurs naturally in some aquifers that serve as sources of groundwater
Beryllium	Occurs naturally in the ground
Cadmium	Food and the smoking of tobacco are common sources of general exposure; a contaminant in the metals used to galvanize pipe
Chromium	Occurs naturally in the ground and is often used in the electroplating of metals
Fluoride	Occurs naturally in some water supplies
Mercury	Used in electric equipment and some water pumps
Nickel	Occurs naturally in the ground and is often used in electroplating, stainless steel, and alloy products
Nitrate	Used in fertilizer and is found in sewage and wastes from human and/or farm animals and generally gets into drinking water from those activities
Nitrite	Used in fertilizers and is found in sewage and wastes from humans and/or farm animals
Selenium	Found naturally in food and soils and is used in electronics, photocopy operations, manufacture of glass, chemicals, drugs, and as a fungicide and a feed additive
Thallium	Occurs naturally in soils and is used in electronics, pharmaceuticals, manufacture of glass and alloys

There is also pollution that is caused by pollution involves the disruption of sediments (fine-grained powders) that flow from a river into the sea. Dams built for hydroelectric power or water reservoirs can reduce the sediment flow, thereby causing (i) a reduction in the formation of beaches; (ii) an increase in coastal erosion, i.e., the natural destruction of cliffs by the sea; and (iii) a reduction in the flow of nutrients from rivers into seas, which can reduce coastal fish stocks. An increased in the amount of sediment deposition can also present a problem since, during construction work, the sedimentary material (carried in the waterway as a fine powder) can enter nearby rivers in large quantities, causing the water to become turbid (muddy or silted). The extra sediment affects the well-being of the fauna (such as blocking the gills of fish) and seriously affects the floral species by preventing access to oxygen.

Oceans, lakes, rivers, and other water bodies can (in theory) naturally clean up a certain amount of pollution by dispersing it harmlessly and by lowering the concentration of the chemical when measured against the total water body. At such low levels of concentration, the chemicals would not present any real problem, but this is a nonvalid assumption when applied to all bodies of water. The effect of the chemical or chemical on the flora or fauna may take the form a domino effect, thereby spreading (without any form or degree of mitigation) throughout the water body. This would undoubtedly affect the existence of all of the flora and fauna that were dependent on the unpolluted water body.

Thus, water pollution is the contamination of any water body such as lakes, rivers, and oceans as well as aquifers and groundwater. This form of pollution occurs when chemical or other pollutants are directly or indirectly discharged into the water body without adequate treatment to remove the harmful chemical(s) compounds. Moreover, water pollution affects the entire floral and faunal communities living in these bodies of water. In almost all cases, the effect is damaging not only to individual floral and/or faunal species and population, but also to any natural biological communities. Typically, water is referred to as polluted when it is impaired by anthropogenic contaminants after which the either does not support a human use (e.g., as drinking water) or undergoes a marked shift in its ability to support its constituent floral and faunal biotic communities. It must also be recognized that natural phenomena such as volcanic activity, algae bloom activity, storms, and earthquakes also cause major changes in water quality and the ability of the water to support any ecological communities therein.

The specific contaminants leading to pollution in groundwater include, among other effects, a wide range of inorganic chemicals (Table 5.10) and physical changes such as elevated temperature and discoloration due to dissolution of the chemicals. In fact, inorganic chemicals have been reported to constitute by far the greatest proportion of chemical contaminants in drinking water (Fawell, 1993). These chemicals are present in greatest quantity as a consequence of natural processes, but several important contaminants are present because of human activities. While many of the chemicals (e.g., calcium, sodium, iron, and manganese) are regulated and may be indigenous to the water body or to the region, from which these chemicals could be transported into the water from mineral deposit, it is the concentration of these chemicals in the water that is often the key in determining what is a natural component of water and what is a contaminant. Indeed, a high concentration (higher than the indigenous concentration) of even a naturally occurring chemical can have a negative impact on aquatic flora and fauna. In addition, some inorganic chemicals undergo chemical change (Chapter 7)

Table 5.10 Examples of Contaminants Found in Groundwater and the Sources of the Chemicals

Contaminant	Sources to Groundwater
Aluminum	Occurs naturally in some rocks and drainage from mines
Antimony	Enters the environment from natural weathering, industrial production, municipal waste disposal, and manufacturing of flame retardants, ceramics, glass, batteries, fireworks, and explosives
Arsenic	Enters the environment from natural processes, industrial activities, pesticides, and industrial waste, smelting of copper, lead, and zinc ore
Barium	Occurs naturally in limestone, sandstone, and soil
Beryllium	Occurs naturally in soils, groundwater, and surface water; enters the environment from mining operations, processing plants, and improper waste disposal
Cadmium	Found in low concentrations in rocks, coal, and crude oil; enters the groundwater and surface water when dissolved by acidic waters; also enters from industrial discharge, mining waste, metal plating, water pipes, batteries, paints and pigments, plastic stabilizers, and landfill leachate
Chloride	May be associated with the presence of sodium in drinking water when present in high concentrations; often enters the environment from saltwater intrusion, mineral dissolution, industrial, and domestic waste
Chromium	Enters the environment from old mining operations runoff and leaching into groundwater, fossil-fuel combustion, cement-plant emissions, mineral leaching, and waste incineration
Copper	Enters the environment from metal plating, industrial and domestic waste, mining, and mineral leaching
Cyanide	Enters the environment from improper waste disposal
Fluoride	Occurs naturally or as an additive to municipal water supplies
Iron	Occurs naturally as a mineral from sediment and rocks or from mining, industrial waste, and corroding metal
Lead	Enters environment from industry, mining, plumbing, gasoline, coal, and as a water additive
Manganese	Occurs naturally as a mineral from sediment and rocks or from mining and industrial waste
Mercury	Occurs as an inorganic salt and as organic mercury compounds; enters the environment from industrial waste, mining, pesticides, coal, electric equipment (batteries, lamps, and switches), smelting, and fossil-fuel combustion
Nickel	Occurs naturally in soils, groundwater, and surface water; used in electroplating, stainless steel and alloy products, mining, and crude oil refining

Continued

Table 5.10 Examples of Contaminants Found in Groundwater and the Sources of the Chemicals—cont'd

Contaminant	Sources to Groundwater
Nitrate	Occurs naturally in mineral deposits, soils, seawater, freshwater systems, the atmosphere, and biota; enters the environment from fertilizer, feedlots, and sewage
Nitrite	Enters the environment from fertilizer, sewage, and human or farm-animal waste
Selenium	Enters the environment from naturally occurring geologic sources, sulfur, and coal
Silver	Enters the environment from ore mining and processing, product fabrication, and disposal
Sodium	Derived from leaching of surface and underground deposits of salt and decomposition of various minerals
Sulfate	Elevated concentrations may result from saltwater intrusion, mineral dissolution, domestic, or industrial waste
Thallium	Enters environment from soils; used in electronics, pharmaceuticals manufacturing, glass, and alloys
Zinc	Found naturally in water, most frequently in areas where it is mined. Enters environment from industrial waste, metal plating, and plumbing, and is a major component of sludge

over long periods of time in groundwater reservoirs or in aquifers where the chemicals contact clay minerals that can serve as adsorbents of the chemicals.

Generally, groundwater pollution is much more difficult to abate than surface pollution because groundwater can move great distances through unseen aquifers. On the beneficial side, nonporous aquifers such as clay minerals can partially purify water of bacteria or various contaminants by simple filtration (adsorption and absorption), dilution and, in some cases, chemical reactions through and biological activity. However, on the adverse side, the pollutants may be transformed from water contaminants to soil contaminants. Furthermore, groundwater that moves through (natural) open fractures is not filtered and can be transported as easily as surface water. As part of the transportation process, there is a variety of secondary effects that are not due to the presence of the original pollutant but from a chemical derivative that has occurred as part of chemical transformation process (Chapter 7), a derivative condition.

If there is any doubt about the positive effect of discharged chemical on an ecosystem, it is environmentally safer to assume that the chemical substance is toxic; the chemical is guilty until is it proved innocent! A chemical can alter the physical chemistry of the water, and any one or more of the following effects can occur: change in acidity (change in pH), electric conductivity, temperature, and eutrophication are possible. Eutrophication is an increase in the concentration of chemical nutrients in an ecosystem to an extent that increases in the primary productivity of the ecosystem. Depending on the degree of eutrophication, subsequent negative environmental effects such as anoxia (oxygen depletion) and severe reductions in water quality may occur, affecting fish and other animal populations.

Heavy metals are naturally occurring elements that have a high atomic weight and a density at least five times greater than that of water. Their multiple industrial, domestic, agricultural, medical, and technological applications have led to their wide distribution in the environment, raising concerns over their potential effects on human health and the environment. Their toxicity depends on several factors including the dose, route of exposure, and chemical species, as well as the age, gender, genetics, and nutritional status of exposed individuals. Because of their high degree of toxicity, arsenic, cadmium, chromium, lead, and mercury rank among the priority metals that are of public health significance. These metallic elements are considered systemic toxicants that are known to induce multiple organ damage, even at lower levels of exposure.

Thus, heavy metals are defined as metallic elements that have a relatively high density compared to water (Fergusson, 1990). With the assumption that density of the metal and toxicity are interrelated, heavy metals also include metalloids, such as arsenic, that are able to induce toxicity at low level of exposure. In recent years, there has been an increasing ecological and global public health concern associated with environmental contamination by these metals. Also, human exposure has risen dramatically as a result of an exponential increase of their use in several industrial, agricultural, domestic, and technological applications (Bradl, 2002). Reported sources of heavy metals in the environment include geogenic, industrial, agricultural, pharmaceutical, domestic effluents, and atmospheric sources (He et al., 2005). Environmental pollution is very prominent in point source areas such as mining, foundries and smelters, and other metal-based industrial operations (Fergusson, 1990; Bradl, 2002; He et al., 2005).

Although heavy metals are naturally occurring elements that are found throughout the earth's crust, most environmental contamination and human exposure result from anthropogenic activities such as mining and smelting operations, industrial production and use, and domestic and agricultural use of metals and metal-containing compounds (Shallari et al., 1998; Herawati et al., 2000; He et al., 2005). Environmental contamination can also occur through metal corrosion, atmospheric deposition, soil erosion of metal ions and leaching of heavy metals, sediment resuspension, and metal evaporation from water resources to soil and groundwater. Natural phenomena such as weathering and volcanic eruptions have also been reported to significantly contribute to heavy-metal pollution (Nriagu, 1989; Fergusson, 1990; Shallari et al., 1998; Bradl, 2002; He et al., 2005). Industrial sources include metal processing in refineries, coal-burning in power plants, petroleum combustion, nuclear power stations and high tension lines, plastics, textiles, microelectronics, wood preservation, and paper processing plants (Pacyna, 1996; Arruti et al., 2010).

Heavy metals are also considered as trace elements because of their presence in trace concentrations (ppb range to less than 10 ppm) in various environmental matrices (Kabata-Pendias and Pendias, 2001). Their bioavailability is influenced by physical factors such as temperature, phase association, adsorption, and sequestration. It is also affected by chemical factors that influence speciation at thermodynamic equilibrium, complexation kinetics, lipid solubility, and octanol/water partition coefficients. Biological factors such as species characteristics, trophic interactions, and biochemical/physiological adaptation also play an important role (Hamelink et al., 1994). Heavy-metal-induced toxicity and carcinogenicity involves many mechanistic aspects, some of which are not clearly elucidated or understood. However, each metal is known to have unique features and physical-chemical properties that confer to its specific toxicological mechanisms of action. Thus, in the case of pollution by heavy metals, metals, such as lead, cadmium, and mercury, are particularly toxic. Lead was once commonly used in gasoline (as antiknock additive) although the use of lead is now restricted in many countries. Mercury and cadmium are still used in batteries, although some brands now use other metals instead.

Pollution by silt or sediments involves the disruption of sediments that flow from rivers into the sea. This reduces the formation of beaches, increases coastal erosion (the natural destruction of cliffs by the sea), and reduces the flow of nutrients from rivers into seas (potentially reducing coastal fish stocks). Increased sediments can also present an environmental problem insofar as during construction work, soil, rock, and other fine powders sometimes enter nearby rivers in large quantities, causing it to become turbid (muddy or silted). The extra sediment can block the gills of fish, effectively causing suffocation.

In addition, many chemicals undergo partial or full chemical change after entry into the environment (leading to new hazardous chemicals; Chapter 7), especially over long periods of time in groundwater reservoirs.

5.2.3 Land Pollution

Soils are formed by the decomposition of rock and organic matter over many years, and the properties of soil vary from place to place with differences in bedrock composition, climate, and other factors. Certain chemical elements occur naturally in soils as components of minerals yet may be toxic at some concentrations, while other potentially harmful substances can end up in soils through human activities. Moreover, the properties of soil are affected by past land use, current activities on the site, and nearness to pollution sources. Human activities have intentionally added substances such as pesticides, fertilizers, and other amendments to soils. Accidental spills and leaks of chemicals used for commercial or industrial purposes have also been the sources of contamination, while other contaminants are moved through the air and deposited as dust or by precipitation.

The distribution of contaminants released to soils by human activities is related to how and where they are added. For example, the amounts of contaminants in the soil of an industrially contaminated site can be expected to vary depending on the activities conducted on the site. The movement of air and water will also affect how soil contaminants move throughout a site. Chemicals may be carried by winds and deposited on the surface of soils; tilling can then mix these surface deposits into the soil. The movement of groundwater or surface water may also affect how contaminants spread from the source.

Generally, soil pollution (soil contamination) as part of land degradation is caused by the presence of anthropogenic chemicals (human-made or xenobiotic chemicals) or other alteration in the natural environment (Table 5.11). This form of pollution is typically caused by industrial activity, by agricultural chemicals, or by improper disposal of waste chemical products, and pollution can typically be correlated with the degree of industrialization and intensity of chemical usage. One example is surface mining and processing of coal, which disturbs larger areas of land than is caused by underground mining. Coal-fired power plants also release pollutants that contaminate nearby soils; in addition to the gases formed during coal combustion, other examples include fly ash, bottom ash, and slag.

Table 5.11 Example of the Sources of Soil Pollution

Industrial activity
Industrial activity has been the biggest contributor to the problem in the last century, especially since the amount of mining and manufacturing has increased. Whether it is iron ore or coal, the by-products are contaminated, and they are not disposed in a manner that can be considered safe. As a result, the industrial waste can linger in the soil surface for a years (if not, decades)
Agricultural activity
Chemical utilization has gone up tremendously since chemical technology provided pesticides and fertilizers. These chemicals that are not produced in nature and cannot be broken down by natural force and reduce the fertility of the soil. Some chemicals damage the composition of the soil and make it easier to erode by water and air
Acid rain
Acid rain is caused when pollutants present in the air mixes up with the rain and fall back on the ground. The polluted water could dissolve away some of the important nutrients found in soil and change the structure of the soil

Moreover, the discharge of chemicals on land also makes it dangerous for the ecosystem. These chemicals are consumed by the flora and fauna within an ecosystem, and the process of consumption by the flora and fauna can result in biomagnification (also known as bioamplification or biological magnification; is the increasing concentration of a substance, such as a toxic chemical, in the tissues of organisms at successively higher levels in a food chain) and is a serious threat to the ecology.

In addition, pollution of the soil can also lead to water pollution if toxic chemicals leach into groundwater or if contaminated runoff reaches streams, lakes, or oceans. Soil also naturally contributes to air pollution by releasing volatile compounds into the atmosphere. Nitrogen escapes through ammonia volatilization and denitrification. The decomposition of organic materials in soil can release sulfur dioxide and other sulfur compounds, causing the formation and release of acidic gases into the atmosphere than can, through facile chemical transformations (Chapter 7), lead to acid rain.

Heavy metals and other potentially toxic elements are the most serious soil pollutants in sewage. Sewage sludge contains heavy metals, and if applied repeatedly or in large amounts, the treated soil may accumulate heavy metals and consequently become unable to even support plant life. Heavy metals include transition metals such as cadmium, mercury, and lead (Fig. 5.1), all of which can contribute to brain damage. Inorganic pollutants such as hydrochloric acid, sodium chloride, and sodium carbonate change the acidity, salinity, or alkalinity of the water, making it undrinkable or unsuitable for the support of animal and plant life. These effects can result in serious consequences for the flora and fauna (especially for the higher mammals, including humans).

Heavy metals occur naturally in the soil environment from the pedogenetic processes of weathering of parent materials at levels that are regarded as *trace* (<1000 mg kg^{-1}) and rarely toxic (Pierzynski et al., 2000; Khan et al., 2008). Due to the disturbance and acceleration of nature's slowly occurring geochemical cycle of metals by man, most soils of rural and urban environments may accumulate one or more of the heavy metals above defined background values high enough to cause risks to human health, plants, animals, ecosystems, or other media. The heavy metals essentially become contaminants in the soil environments because (i) their rates of generation via man-made cycles are more rapid relative to natural ones, (ii) they become transferred from mines to random environmental locations where higher potentials of direct exposure occur, (iii) the concentrations of the metals in discarded products are relatively high compared with those in the receiving environment, and (iv) the chemical form (species) in which a metal is found in the receiving environmental system may render it more bioavailable (D'Amore et al., 2005). A simple mass balance of the heavy metals in the soil can be expressed as follows (Alloway, 1995; Lombi and Gerzabek, 1998):

$$M_{total} = \left(M_p + M_a + M_f + M_{ag} + M_{ow} + M_{ip}\right) - \left(M_{cr} + M_l\right)$$

In this equation, M is the heavy metal, p is the parent material, a is the atmospheric deposition, f is the fertilizer sources, ag are the agrochemical sources, ow are the organic waste sources, ip are other inorganic pollutants, cr is crop removal, and l is the losses by leaching and volatilization. It is projected that the anthropogenic emission into the atmosphere, for several heavy metals, is one-to-three orders of magnitude higher than natural fluxes (Sposito and Page, 1984). Heavy metals in the soil from anthropogenic sources tend to be more mobile, hence bioavailable than pedogenic, or lithogenic ones (Kuo et al., 1983; Kaasalainen and Yli-Halla, 2003). Metal-bearing solids at contaminated sites can originate from a wide variety of anthropogenic sources in the form of metal mine tailings, disposal of high metal wastes in improperly protected landfills, leaded gasoline and lead-based paints, land application of fertilizer, animal manure, biosolids (i.e., sewage sludge), compost, pesticides, coal combustion residues, petrochemicals, and atmospheric deposition (Basta et al., 2005; Khan et al., 2008).

Thus, soil pollution comprises the pollution of soils with materials, mostly chemicals that are out of place or are present at concentrations higher than normal, which may have adverse effects on humans or other organisms. It is difficult to define soil pollution exactly because different opinions exist on how to characterize a pollutant, while some observers consider the use of certain chemicals to be acceptable if the effect of these chemicals does not exceed the intended result, while other observers do not consider any use of any chemicals to be acceptable. The most important inorganic contaminants include nitrate derivatives, phosphate derivatives, and heavy metals such as cadmium, chromium and lead, and inorganic acids. However, soil pollution is also caused by means other than the direct addition of xenobiotic (man-made) chemicals such as agricultural runoff water, industrial waste material, acidic material, and (not forgetting or ignoring) radionuclides (radioactive substances and any radioactive fallout).

Inorganic residues in industrial waste cause serious problems in regard to their disposal. They contain metals that have high potential for toxicity. Industrial activity also emits large amounts of arsenic fluorides and sulfur dioxide (SO_2). Fluorides are found in the atmosphere from superphosphate, phosphoric acid, aluminum, and steel and ceramic industries. Sulfur dioxide emitted by factories and thermal plants may make soils very acidic. These metals cause leaf injury and destroy vegetation. Copper, mercury, cadmium, lead, nickel, and arsenic are the elements that can accumulate in the soil, if they get entry either through sewage, industrial waste, or mine washings. Some of the fungicides containing copper and mercury also add to soil pollution. Smokes from automobiles contain lead, which gets adsorbed by soil particles and is toxic to plants. The toxicity can be minimized by building up soil organic matter, adding lime to soils, and keeping the soil alkaline.

Heavy metals are elements having a density greater than five in the elemental form, and these metals mostly find specific absorption sites in the soil where they are retained very strongly on either the inorganic or organic colloids. They are widely distributed in the environment, soils, plants, animals, and in their tissues. These are essential for plants and animals in trace amounts. Mainly urban and industrial aerosols, combustion of fuels, liquid and solid from animals and human beings, mining wastes, industrial and agricultural chemicals, etc., are contributing heavy-metal pollution. Heavy metals are present in all uncontaminated soils as the result of weathering from various parent materials (Table 5.12).

Table 5.12 Heavy-Metal Concentration in the Land (Lithosphere[a]), Soil, and Plants (μg/gm Dry Matter)

Heavy Metal	Lithosphere	Soil Range	Plants
Cadmium (Cd)	0.2	0.01–0.7	0.2–0.8.2
Cobalt (Co)	40	1–40	0.05–0.5
Chromium (Cr)	200	5–3000	0.2–1.0
Copper (Cu)	70	2–100	4–15
Iron (Fe)	50,000	7000–100,000	140
Mercury (Hg)	0.5	0.01–0.3	0.015
Manganese (Mn)	1000	100–4000	15–100
Molybdenum (Mo)	2.3	0.2–5	1–10
Nickel (Ni)	100	10–1000	1
Lead (Pb)	16	2–200	0.1–10
Tin (Sn)	40	2–100	0.3
Zinc (Zn)	80	10–300	8–100

[a]The solid portion of the Earth.

In agriculture soil, the concentration of one or more of these elements may be significantly increased in several ways, such as through applications of chemicals, sewage sludge, and farm slurries. Increased doses of fertilizers, pesticides, or agricultural chemicals, over a period, add heavy metals to soils, which may contaminate them. Certain phosphate-containing fertilizers frequently contain trace amounts of cadmium, which may accumulate in these soils.

Thus, soils may also become polluted by the accumulation of heavy metals and metalloids through emissions from the rapidly expanding industrial areas, mine tailings, disposal of high metal wastes, leaded gasoline and paints, land application of fertilizers, animal manures, sewage sludge, pesticides, wastewater irrigation, coal combustion residues, spillage of petrochemicals, and atmospheric deposition (Pierzynski et al., 2000; Khan et al., 2008; Zhang et al., 2010; Wuana and Okieimen, 2011).

Mercury, cadmium, and arsenic are common constituents of pesticides, and all these heavy metals are toxic. At present, DDT and many organochlorine compounds used as pesticides have been declared harmful and banned in many countries due to the persistence of the DDT residues in soils for considerable time without losing their toxicity (Speight, 2017a). This has led to higher concentration of these pesticides in vegetation, in animal flesh, and in milk. Eventually, man has been affected. In view of their demerits, organochlorines have been replaced by organophosphate pesticides that are more toxic, but do not leave any residue. They do not pollute the soil. The rodenticides too add to soil pollution. A major method of checking this pesticidal pollution is to increase the organic matter content of the soil and choose such pesticides that are nonpersistent and leave no harmful residue.

Heavy metals constitute an ill-defined group of inorganic chemical hazards, and those most commonly found at contaminated sites are lead (Pb), chromium (Cr), arsenic (As), zinc (Zn), cadmium (Cd), copper (Cu), mercury (Hg), and nickel (Ni). Soils are the major sink for heavy metals released into the environment by the aforementioned anthropogenic activities, and unlike organic contaminants that are oxidized to carbon (IV) oxide by microbial action, most metals do not undergo microbial or chemical degradation, and their total concentration in soils persists for a long time after their introduction. Changes in their chemical forms (speciation) and bioavailability are, however, possible. The presence of toxic metals in soil can severely inhibit the biodegradation of organic contaminants. Heavy-metal contamination of soil may pose risks and hazards to humans and the ecosystem through direct ingestion or contact with contaminated soil, food chain (soil-plant-human or soil-plant-animal-human), drinking of contaminated groundwater, reduction in food quality (safety and marketability) via phytotoxicity, reduction in land usability for agricultural production causing food insecurity, and land tenure problems (Wuana and Okieimen, 2011).

However, it must always be remembered that heavy metals occur naturally in the soil environment from the ever-present processes of weathering of parent materials at levels that are regarded as trace amounts (<1000 mg kg^{-1}) and are rarely toxic (Kabata-Pendias and Pendias, 2001; Wuana and Okieimen, 2011). Due to the disturbance and acceleration of any natural cycles (such as the geochemical cycle of the metals) by human interference, most soils of rural and urban environments may accumulate one or more of heavy-metal contaminants above the naturally occurring values that are sufficiently high enough to cause risks to the floral and faunal inhabitants of various ecosystems. In short, human activities can cause the concentration of a heavy-metal to exceed the natural abundance, thereby endangering an ecosystem. The heavy metals essentially become contaminants in the soil of an ecosystem because (i) the rate of generation of the heavy-metal by means of anthropogenic cycles is more rapid relative to natural ones, (ii) the heavy-metal is transferred from mineral to random environmental locations where there is a higher potential for direct exposure to occur, (iii) the concentration of the heavy-metal in discarded products is relatively high compared with the natural concentration of the indigenous metal in the receiving environment, and (iv) the chemical form in which a metal exists in the receiving environmental system may render it more bioavailable (Basta et al., 2005).

The fate of heavy metals in soil will be controlled by physical and biological processes acting within the soil. Metal ions enter the soil solution from these various forms of combination in different rates; they may either remain in solution or pass into the drainage water or be taken up by plants growing on the soil or be retained by the soil in sparingly soluble or insoluble forms. The organic matter of this soil has a high affinity for heavy-metal cations that form stable complexes, thereby leading to reduced nutrient content.

The adequate protection and restoration of soil ecosystems contaminated by inorganic chemicals require the characterization and remediation of the chemical. Soil characterization can provide an insight into contaminant types and contaminant sources, but any attempt at remediation of contaminated soil will also entail knowledge of the source of contamination, basic chemistry, and environmental and associated health effects (risks) of the contaminants. Risk assessment is an effective scientific tool that enables scientists and engineers to manage sites so contaminated in a cost-effective manner while preserving public and ecosystem health (Speight and Singh, 2014).

5.3 FLY ASH AND BOTTOM ASH

Coal ash is the general term that refers to whatever waste is leftover after coal is combusted, usually in a coal-fired power plant, and typically contains arsenic, mercury, lead, and many other heavy metals (Table 5.13). Coal ash is commonly divided into two subcategories based on particle size: (i) fly ash and (ii) bottom ash. Fly ash, bottom ash, and boiler slag are presented here in a separate section because of the composition of these types of waste products, none of which can be classed as a single inorganic chemical.

Table 5.13 Chemical Composition of Coal Ash

Component, % (w/w)	Bituminous Coal	Subbituminous Coal	Lignite
SiO_2	20–60	40–60	15–45
Al_2O_3	5–35	20–30	20–25
Fe_2O_3	10–40	4–10	4–15
CaO	1–12	5–30	15–40

Fly ash (also known as *pulverized fuel ash*) in the United Kingdom is one of the coal combustion products and is composed of the fine particles that are driven out of the boiler with the flue gases. Ash that falls in the bottom of the boiler is called *bottom ash*. Fly ash, together with bottom ash removed from the bottom of the boiler, is more commonly known as coal ash. *Bottom ash* and *boiler slag* are the coarse, granular, incombustible by-products (Tables 5.14 and 5.15) that are collected from the bottom of furnaces that burn coal for the generation of steam, the production of electric power, or both. *Boiler slag* is the melted form of coal ash that can be found both in the filters of exhaust stacks and the boiler at the bottom of the stack. Most of these coal by-products are produced at coal-fired electric utility generating stations, although considerable bottom ash and boiler slag are also produced from many smaller industrial or institutional coal-fired boilers and from coal-burning-independent power production facilities. The type of by-product (i.e., bottom ash or boiler slag) produced depends on the type of furnace used to burn the coal.

5.3.1 Fly Ash

The most voluminous and well-known constituent is fly ash, which makes up more than half of the coal leftovers. Fly ash particles are the lightest kind

Table 5.14 Composition of Selected Bottom Ash and Boiler Slag Samples (%, w/w)

Ash Type	**Bottom Ash**					**Boiler Slag**		
Coal Type	**Bituminous**			**Subbituminous**	**Lignite**	**Bituminous**		**Lignite**
SiO_2	53.6	45.9	47.1	45.4	70.0	48.9	53.6	40.5
Al_2O_3	28.3	25.1	28.3	19.3	15.9	21.9	22.7	13.8
Fe_2O_3	5.8	14.3	10.7	9.7	2.0	14.3	10.3	14.2
CaO	0.4	1.4	0.4	15.3	6.0	1.4	1.4	22.4
MgO	4.2	5.2	5.2	3.1	1.9	5.2	5.2	5.6
Na_2O	1.0	0.7	0.8	1.0	0.6	0.7	1.2	1.7
K_2O	0.3	0.2	0.2	–	0.1	0.1	0.1	1.1

Table 5.15 Typical Properties of Bottom Ash and Boiler Slag

Property	**Bottom Ash**	**Boiler Slag**
Specific gravity	2.1–2.7	2.3–2.9
Dry unit weight	720–1600 kg/m^3 (45–100 lb/ft^3)	960–1440 kg/m^3 (60–90 lb/ft^3)
Plasticity	None	None
Absorption	0.8%–2.0%	0.3%–1.1%

of coal ash and pass upward from the combustor into the exhaust stacks of the power plant. Filters within the stacks capture about 99% w/w of the ash, attracting it with opposing electric charges, and the captured fly ash is recyclable. The fine particles bind together and solidify, especially when mixed with water, making them an ideal ingredient in concrete and wallboard. The recycling process also renders the toxic materials within fly ash safe for use.

In modern coal-fired power plants, fly ash is generally captured by electrostatic precipitators or other particle filtration equipment before the flue gases reach the chimneys. Depending upon the source and makeup of the coal being burned, the components of fly ash vary considerably, but all fly ash includes substantial amounts of silica (silicon dioxide, SiO_2) (both amorphous and crystalline), aluminum oxide (alumina, Al_2O_3), and calcium oxide (CaO), the main mineral compounds in coal-bearing rock strata. The constituents depend upon the specific coal-bed makeup but may include one or more of the following elements or substances found in trace concentrations (up to hundreds parts per millions): arsenic, beryllium, boron, cadmium, chromium, hexavalent chromium, cobalt, lead, manganese, mercury, molybdenum, selenium, strontium, thallium, and vanadium (Speight, 2013).

In the past, fly ash was generally released into the atmosphere, but air pollution control standards now require that it be captured prior to release by fitting pollution control equipment. In the United States, fly ash is generally stored at coal power plants or placed in landfills. About 43% (w/w) of the fly ash is recycled, often used as a pozzolana to produce hydraulic cement or hydraulic plaster and a replacement or partial replacement for Portland cement in concrete production. Pozzolanas ensure the setting of concrete and plaster and provide concrete with more protection from wet conditions and chemical attack. In the case that fly or bottom ash is not produced from coal, for example, when solid waste is used to produce electricity in an incinerator, this kind of ash may contain higher levels of contaminants than coal ash. In that case, the ash produced is often classified as hazardous waste.

Fly ash material solidifies while suspended in the exhaust gases and is collected by electrostatic precipitators or filter bags. Since the particles solidify rapidly while suspended in the exhaust gases, fly ash particles are generally spherical in shape and range in size from 0.5 to 300 μm. The major consequence of the rapid cooling is that few minerals have time to crystallize and that mainly amorphous, quenched glass remains. Nevertheless, some refractory phases in the pulverized coal do not melt (entirely) and remain crystalline. In consequence, fly ash is a heterogeneous material. SiO_2, Al_2O_3, Fe_2O_3, and occasionally CaO are the main chemical components present in fly ashes. The mineralogy of fly ashes is very diverse.

The main phases encountered are glass phase, together with quartz, mullite, and iron oxides: hematite, magnetite, and/or maghemite. Other phases often identified are cristobalite, anhydrite, free lime, periclase, calcite, sylvite, halite, portlandite, rutile, and anatase. The Ca-bearing minerals anorthite, gehlenite, akermanite, and various calcium silicates and calcium aluminates identical to those found in Portland cement can be identified in Ca-rich fly ashes. The mercury content can reach 1 ppm but is generally included in the range 0.01–1 ppm for bituminous coal. The concentrations of other trace elements vary as well according to the kind of coal combusted to form it. In fact, in the case of bituminous coal, with the notable exception of boron, trace element concentrations are generally similar to trace element concentrations in unpolluted soils.

Two classes of fly ash are defined: Class F fly ash and Class C fly ash (ASTM C618). The chief difference between these classes is the amount of calcium, silica, alumina, and iron content in the ash. The chemical properties of the fly ash are largely influenced by the chemical content of the coal burned (i.e., anthracite, bituminous coal, and lignite). Not all fly ashes meet ASTM C618 requirements; although depending on the application, this may not be necessary. Ash used as a cement replacement must meet strict construction standards, but no standard environmental regulations have been established in the United States. Seventy-five percent of the ash must have a fineness of 45 μm or less and have a carbon content, measured by the loss on ignition (LOI), of less than 4%. In the United States, LOI must be under 6%. The particle size distribution of raw fly ash tends to fluctuate constantly, due to the changing performance of the coal mills and the boiler performance. This makes it necessary that, if fly ash is used in an optimal way to replace cement in concrete production, it must be processed using beneficiation methods like mechanical air classification. But if fly ash is used also as a filler to replace sand in concrete production, nonbeneficiated fly ash with higher LOI can be also used. Especially important is the ongoing quality verification.

Class F fly ash is produced during the burning of harder, older anthracite, and bituminous coal typically produces Class F fly ash. This fly ash is pozzolanic in nature and contains less than 7% lime (CaO). Possessing pozzolanic properties, the glassy silica and alumina of Class F fly ash requires a cementing agent, such as Portland cement, quicklime, or hydrated lime, mixed with water to react and produce cementitious compounds. Alternatively, adding a chemical activator such as sodium silicate (water glass) to a Class F ash can form a geopolymer.

On the other hand, *Class C fly ash* is produced from the burning of younger lignite or subbituminous coal; in addition to having pozzolanic properties, it also has some self-cementing properties. In the presence of water, Class C fly ash hardens and gets stronger over time. Class C fly ash generally contains more than 20% lime (CaO). Unlike Class F, self-cementing Class C fly ash does not require an activator. Alkali and sulfate (SO_4) contents are generally higher in Class C fly ashes.

Fly ash contains trace concentrations of heavy metals and other substances that are known to be detrimental to health in sufficient quantities. Potentially, toxic trace elements in coal include arsenic, beryllium, cadmium, barium, chromium, copper, lead, mercury, molybdenum, nickel, radium, selenium, thorium, uranium, vanadium, and zinc. Approximately 10% of the mass of coals burned in the United States consists of unburnable mineral material that becomes ash, so the concentration of most trace elements in coal ash is approximately 10 times the concentration in the original coal.

Crystalline silica and lime along with toxic chemicals represent exposure risks to human health and the environment. Exposure to fly ash through skin contact, inhalation of fine particulate dust, and ingestion through drinking water may well present health risks. Fly ash contains crystalline silica, which is known to cause lung disease, in particular silicosis. Also, lime (CaO) reacts with water (H_2O) to form calcium hydroxide [$Ca(OH)_2$], giving fly ash a pH somewhere between 10 and 12, a medium to strong base. This can also cause lung damage if present in sufficient quantities.

5.3.2 Bottom Ash

The most common type of coal-burning furnace in the electric utility industry is the dry, bottom pulverized coal boiler. When pulverized coal is burned in a dry, bottom boiler, approximately 80% (w/w) of the unburned material or ash is entrained in the flue gas and is captured and recovered as fly ash. The remaining 20% (w/w) of the ash is dry bottom ash, dark gray, granular, porous, predominantly sand size minus 12.7 mm (½ in.) material that is collected in a water-filled hopper at the bottom of the furnace. When a sufficient amount of bottom ash drops into the hopper, it is removed by means of high-pressure water jets and conveyed by sluiceways either to a disposal pond or to a decant basin for dewatering, crushing, and stockpiling for disposal or use.

Bottom ash is the coarser component of coal ash, comprising about 10% of the waste. Rather than floating into the exhaust stacks, it settles to the bottom of the power plant's boiler. Bottom ash is not quite as useful as fly ash, although power plant owners have tried to develop options for beneficial use options, such as structural fill and road-base material. However, the bottom ash remains toxic when recycled and can leak heavy metals into the groundwater.

Bottom ash is part of the noncombustible residue of combustion in a furnace or incinerator. In an industrial context, it usually refers to coal combustion and comprises traces of combustibles embedded in forming clinkers and sticking to hot sidewalls of a coal-burning furnace during its operation. The portion of the ash that escapes up the chimney or stack is, however, referred to as *fly ash*. The clinkers fall by themselves into the bottom hopper of a coal-burning furnace and are cooled. The above portion of the ash is also referred to as *bottom ash*.

In a conventional water impounded hopper (WIH) system, the clinker lumps get crushed to small sizes by clinker grinders mounted under water and fall down into a trough from where a water ejector takes them out to a sump. From there, it is pumped out by suitable rotary pumps to dumping yard far away. In another arrangement, a continuous link chain scrapes out the clinkers from under water and feeds them to clinker grinders outside the bottom ash hopper. More modern systems adopt a continuous removal philosophy. Essentially, a heavy-duty chain conveyor (SSC) submerged in a water bath below the furnace, which quenches hot ash as it falls from the combustion chamber and removes the wet ash continuously up to a dewatering slope before onward discharge into mechanical conveyors or directly to storage silos.

Bottom ash can be extracted, cooled, and conveyed using dry ash technology from various companies. Dry ash handling has many benefits. When left dry, the ash can be used to make concrete and other useful materials. There are also several environmental benefits. For example, bottom ash may be used as raw alternative material, replacing earth or sand or aggregates, for example, in road construction and in cement kilns (clinker production). A noticeable other use is as growing medium in horticulture (usually after sieving). In the United Kingdom, it is known as furnace bottom ash (FBA), to distinguish it from incinerator bottom ash (IBA), the noncombustible elements remaining after incineration. An example of the use of bottom ash was in the production of concrete blocks for use in building construction.

Due to the salt content and, in some cases, the low pH of bottom ash (and boiler slag), the material can exhibit corrosive properties. Corrosivity indicator tests normally used to evaluate bottom ash or boiler slag are pH, electric resistivity, soluble chloride content, and soluble sulfate content. Materials are judged to be noncorrosive if the pH exceeds 5.5, the electric resistivity is greater than 1500 ohm-cm, the soluble chloride content is less than 200 ppm, or the soluble sulfate content is less than 1000 ppm.

5.3.3 Boiler Slag

Boiler slag is a by-product produced from a wet-bottom boiler, which a special type of boiler designed to keep bottom ash in a molten state before it is removed. These types of boilers (slag-tap and cyclone boilers) are much more compact than pulverized coal boilers used by most large utility generating stations and can burn a wide range of fuels and generate a higher proportion of bottom ash than fly ash (50–80%, w/w, bottom ash compared with 15–20%, w/w, bottom ash for pulverized coal boilers). With wet-bottom boilers, the molten ash is withdrawn from the boiler and allowed to flow into quenching water. The rapid cooling of the slag causes it to immediately crystallize into a black, dense, fine-grained glassy mass that fractures into angular particles, which can be crushed and screened to the appropriate sizes for several uses.

There are two types of wet-bottom boilers: (i) the slag-tap boiler and (ii) the cyclone boiler. The slag-tap boiler burns pulverized coal, and the cyclone boiler burns crushed coal. In each type, the bottom ash is kept in a molten state and tapped off as a liquid. Both boiler types have a solid base with an orifice that can be opened to permit the molten ash that has collected at the base to flow into the ash hopper below. The ash hopper in wet-bottom furnaces contains quenching water. When the molten slag comes in contact with the quenching water, it fractures instantly, crystallizes, and forms pellets. The resulting boiler slag, often referred to as *black beauty*, is a coarse, hard, black, angular, and glassy material.

When pulverized coal is burned in a slag-tap furnace, as much as 50% of the ash is retained in the furnace as boiler slag. In a cyclone furnace, which burns crushed coal, some 70–80% (w/w) of the ash is retained as boiler slag, with only 20–30% (w/w) leaving the furnace in the form of fly ash. Wet-bottom boiler slag is a term that describes the molten condition of the ash as it is drawn from the bottom of the slag-tap or cyclone furnaces. At intervals, high-pressure water jets wash the boiler slag from the hopper pit into a sluiceway, which then conveys it to a collection basin for dewatering, possible crushing or screening, and either disposal or reuse.

Since boiler slag is angular, dense, and hard, it is often used as a wear-resistant component in surface coatings of asphalt in road paving. Finer-sized boiler slag can be used as blasting grit and is commonly used for coating roofing shingles. Other uses include raw material for the manufacture of cement, and in colder climates, it is spread onto icy roads for traction control. Because there are so many uses and such a limited supply, most of the boiler slag produced in the United States is used and even imported some from other countries.

5.4 CHARACTERIZATION OF INORGANIC COMPOUNDS

In the analytic characterization of chemicals, there are two general expressions that are used: (i) qualitative analysis and (ii) quantitative analysis.

The term *quantitative analysis* is often used in comparison (or contrast) with *qualitative analysis*, which seeks information about the identity or form of substance present. Thus, the general expression *qualitative analysis* refers to analyses in which the chemicals are identified or classified based on the chemical or physical properties, such as chemical reactivity, solubility, molecular weight, melting point, radiative properties (emission and absorption), mass spectra, and nuclear half-life. The term *quantitative analysis* refers to analyses in which the amount or concentration of a chemical may be determined (estimated) and expressed as a numerical value in appropriate units. Both forms of analysis (qualitative analysis and quantitative analysis) may be used, but quantitative analysis does require the identification (qualification) of the chemical for which numerical estimates are given.

An example, if a sample is unknown, the investigator should use *qualitative* techniques—such as infrared (IR) spectroscopy of nuclear magnetic resonance (NMR) spectroscopy to identify the chemical (or chemicals, if the original sample is a mixture) compounds present and then quantitative techniques to determine the amount of each chemical in the sample. Careful procedures for recognizing the presence of different metal ions have been developed, although they have largely been replaced by modern instruments; these are collectively known as qualitative inorganic analysis, and similar test methods are available for identifying organic compounds by testing for different the presence of any of the known functional groups (Speight, 2017a). In addition to the two spectroscopic methods mentioned above, many techniques can be used for either qualitative or quantitative measurements.

The nature of the compounds precludes the use of some methods, but it also opens the investigation to the use of other methods. The following is a partial list of different physical methods, the information that each method produces, and the applicability to the characterization of inorganic chemicals. There are many techniques for characterizing inorganic compounds. These are not that important in every day work, but it is necessary to be aware of the various methods and the strengths and weaknesses of each method.

5.4.1 Physical Characteristics

The physical characteristics of inorganic chemicals are related the physical properties of inorganic chemicals that have a propensity to form ionic bonds, have high melting points, and are made from either single elements or compounds that do not include carbon and hydrogen. In solutions, they break down into ions that conduct electricity. The physical properties of a chemical can be observed or measured without changing the composition of matter and include appearance, texture, color, odor, melting point, boiling point, density, solubility, and polarity. The three states of matter are (i) solid, (ii) liquid, and (iii) gas; the melting point and boiling point are related to changes of the state of matter.

5.4.1.1 Melting Points and Boiling Points

The melting point (sometimes referred to as the *liquefaction point*) of an inorganic solid is the temperature at which it changes state from solid to liquid at atmospheric pressure. The melting point of a substance depends on pressure and is usually specified at standard pressure. When considered as the temperature of the reverse change from liquid to solid, it is referred to as the freezing point or the crystallization point. Because of the ability of some chemicals to supercool, the freezing point is not considered as a characteristic property of a substance.

The boiling point of an inorganic chemical is the temperature at which the vapor pressure of the liquid is equal to the pressure surrounding the liquid, and the liquid changes into a vapor. The boiling point of a liquid varies depending upon the surrounding environmental pressure. For example, water boils at 100°C (212°F) at sea level, but at 93.4°C (200.1°F) at 6600 ft altitude.

Inorganic compounds are often ionic and so have very high melting points. While some inorganic compounds are solids with accessible melting points and some are liquids with reasonable boiling points, there are not the exhaustive tabulations of melting/boiling point data for inorganic compounds that exist for organics. In general, melting and boiling points are not useful in identifying an inorganic compound, but they can be used to assess its purity, if they are accessible.

5.4.1.2 Color and Crystal Shape

Inorganic compounds, in contrast to many organic compounds, are very colorful. Unfortunately, color alone is not reliable indicator of a compound's identity, but it is useful when one is following a separation on a column.

Information can be obtained about the arrangement of the particles in the solid from the shape of a well-formed crystal or by observing the visual changes in the crystal when it is rotated under a polarizing microscope. Crystal shape is used as a means of mineral identification, but mineralogists have the advantage over chemists in that their crystals are the result of very long, slow crystallization processes that often result in large, well-formed crystals.

5.4.1.3 Elemental Analysis

Elemental analysis is a process where a sample of the inorganic chemical (such as soil, chemical waste, and minerals) is analyzed for the elemental composition. Elemental analysis can be qualitative (determining what elements are present), and it can be quantitative (determining how much of each are present).

Elemental analysis is one of the most useful methods available to characterize a compound. Elemental analysis using standard procedures (gravimetric, colorimetric, and AA) is also an option.

5.4.1.4 Chromatography

Chromatography is the collective term for a variety of laboratory techniques that are used for the separation of mixture. In the method, the mixture is dissolved in a fluid (*mobile phase*), which carries it through a structure holding another material (*stationary phase*). The various constituents of the mixture travel at different speeds, causing them to separate. The separation is based on differential partitioning between the mobile and stationary phases. Subtle differences in the rate of passage through the stationary phase result in differential retention on the stationary phase, thus changing the separation.

Chromatography is most often used to separate a product from a complex reaction mixture. If a known sample of the compound is available, it can be identified in a reaction mixture by spiking the analyte with the known. Most chromatography in inorganic chemistry is performed using solutions—such as column chromatography and high-performance liquid chromatography (HPLC, both normal and reverse-phase), because the high boiling point of many inorganic compounds precludes analysis by gas chromatography.

5.4.2 Spectroscopic and Structural Methods

Spectroscopy and spectrography are terms used to refer to the measurement of radiation intensity as a function of wavelength. Spectral measurement devices are referred to as spectrometers, spectrophotometers, spectrographs, or spectral analyzers. The focus of this section is on UV-Vis absorption spectroscopy, infrared absorption spectroscopy, Raman spectroscopy, nuclear magnetic resonance spectroscopy, electron paramagnetic resonance spectroscopy, and X-ray diffraction spectroscopy. Because these techniques use optical materials to disperse and focus the radiation, they often are identified as optical spectroscopies.

5.4.2.1 UV-Vis Absorption Spectroscopy

Ultraviolet-visible spectroscopy or ultraviolet-visible spectrophotometry (UV-vis spectroscopy or UV/vis spectroscopy) refers to absorption spectroscopy or reflectance spectroscopy in the ultraviolet-visible region of the spectrum. This technique uses light in the visible and adjacent (near-UV and near-infrared, NIR) ranges. The absorption or reflectance in the visible range directly affects the perceived color of the chemicals involved. In this region of the electromagnetic spectrum, atoms and molecules undergo electronic transitions. Absorption spectroscopy is complementary to fluorescence spectroscopy insofar as fluorescence deals with transitions from the excited state to the ground state while absorption spectroscopy transitions from the ground state to the excited state.

The number, energies, and intensities of a transition metal compound's absorption bands in the UV-vis and near-infrared spectroscopy can be used to determine the general type of atom bound to a metal and the geometry about the metal.

5.4.2.2 Infrared Absorption Spectroscopy

Infrared spectroscopy (*IR spectroscopy* or *vibrational spectroscopy*) involves the interaction of infrared radiation with inorganic chemicals and covers a range of techniques, mostly based on absorption spectroscopy. As with all spectroscopic techniques, it can be used to identify and study inorganic chemicals. For a given sample, which may be solid, liquid, or gaseous, the method or technique of infrared spectroscopy uses an infrared spectrometer to produce an infrared spectrum.

A basic infrared spectrum is a graph of infrared light absorbance (or transmittance) on the vertical axis vs frequency or wavelength on the horizontal axis. Typical units of frequency used in an infrared spectrum are reciprocal centimeters (sometimes referred to as wave numbers), with the symbol cm^{-1}. The units of the infrared wavelength are commonly presented in micrometers (μm, also referred to as *microns*), symbol, which are related to wave numbers as a reciprocal. A common laboratory instrument that uses this technique is a Fourier transform infrared (FTIR) spectrometer.

Infrared absorption spectroscopy can be used just like in organic chemistry to fingerprint a compound. In simple compounds, the number, energy, and intensity of the transitions in the infrared portion of the spectrum are directly related to the geometry of compound and to which atoms are bound to which other atoms. Unfortunately, some metal-ligand vibrations and many of the vibrations for the heavier elements occur outside the frequency window of most commercial instruments. For complex compounds involving large organic moieties, the IR becomes more difficult to interpret. IR is useful to determine the presence of complex counter ions like PF_6—, ClO_4—, and BF_4— because they have distinctive absorptions in the infrared spectrum.

5.4.2.3 Raman Spectroscopy

Raman spectroscopy is a spectroscopic technique used to observe vibrational, rotational, and other low-frequency modes in a system and is commonly used in chemistry to provide a fingerprint by which molecules can be identified. The technique relies on inelastic scattering (Raman scattering) of monochromatic light, usually from a laser in the visible, near-infrared, or near-ultraviolet range. The laser light interacts with molecular vibrations in the system, resulting in the energy of the laser photons being shifted up or down, which gives information about the vibrational modes in the system.

Raman spectroscopy is complementary to infrared absorption spectroscopy. Both probe vibrations within a compound, but they have different selection rules. By considering the various peaks that are present or absent in the two spectra of a compound, it is possible to determine molecular geometry, at least in simple cases. Raman spectroscopy is also useful because it can, depending on instrument design, scan to very low frequencies (approximately 100 cm^{-1}) and thus observe transitions too low for IR absorption. A variant technique, resonance Raman spectroscopy, can be used to assign vibrations and identify various types of ligands.

5.4.2.4 Nuclear Magnetic Resonance Spectroscopy

Nuclear magnetic resonance spectroscopy (NMR spectroscopy) is a technique that exploits the magnetic properties of certain atomic nuclei and can be used to determine the physical and chemical properties of atoms or the molecules in which they are contained. The technique can provide detailed information about the structure, dynamics, reaction state, and chemical environment of molecules. The intramolecular magnetic field around an atom in a molecule changes the resonance frequency, thus giving access to details of the electronic structure of a molecule and its individual functional groups.

When application of paramagnetic nuclear magnetic resonance spectroscopy is not impossible, nuclear magnetic resonance spectroscopy is usually performed on diamagnetic compounds. The nuclear magnetic resonance spectra of inorganic compounds are often more complicated than organics because other nuclei also have nuclear magnetic moments. Although less widespread than the standard solution nuclear magnetic resonance spectroscopy, solid-state nuclear magnetic resonance spectroscopy and even single-crystal nuclear magnetic resonance spectroscopy have been used on materials that simply do not dissolve in any solvent.

5.4.2.5 Electron Paramagnetic Resonance Spectroscopy

Electron paramagnetic resonance spectroscopy (EPR spectroscopy) or electron spin resonance spectroscopy (ESR spectroscopy) is a method for studying materials with unpaired electrons. The basic concepts of the technique are analogous to those of nuclear magnetic resonance spectroscopy, but it is the electron spins that are excited instead of the spins of atomic nuclei and is particularly useful for studying metal complexes or organic radicals.

While nuclear magnetic resonance spectroscopy is usually only for diamagnetic compounds, electron paramagnetic resonance spectroscopy (EPR) or electron spin resonance spectroscopy (ESR) is for paramagnetic compounds with an odd number of unpaired electrons. Electron paramagnetic resonance spectroscopy can be used when there are an even number of unpaired electrons. In this method, a sample is irradiated with microwave radiation, and the field is swept until resonance occurs. The field at which resonance occurs depends on the number of unpaired electrons, the geometry about the metal center, and the metal ligands. In many ways, electron paramagnetic resonance spectroscopy and nuclear magnetic resonance spectroscopy are similar, and there is even a technique that combines both techniques (ENDOR, electron-nuclear double resonance spectroscopy).

5.4.2.6 Mass Spectrometry

Mass spectrometry is an analytic technique that ionizes chemical species and separates the ions based on the mass-to-charge ratio. Thus, a mass spectrum measures the masses within a sample. The mass spectrum is a plot of the ion signal as a function of the mass-to-charge ratio and is used (i) to determine the elemental or isotopic signature of a sample, (ii) the masses of particles and of molecules, and (iii) to elucidate the chemical structures of molecules.

In inorganic chemistry, mass spectrometry is most often used to determine the molar mass of compounds. When mass spectrometry data are combined with an elemental analysis, the chemical formula of the substance can be determined. Analysis of a compound's fragmentation pattern can be used to gain structural information. This is not usually done because of the complex fragmentation patterns of inorganic compounds and because other methods are available for structure determination.

5.4.2.7 X-Ray Diffraction Crystallography

X-ray crystallography is a technique used for determining the atomic and molecular structure of a crystal in which the crystalline atoms cause a beam of incident X-rays to diffract into many specific directions. By measuring the angles and intensities of these diffracted beams, a three-dimensional picture of the density of electrons within a crystal can be reproduced. From this electron density, the mean positions of the atoms in the crystal can be determined, as well as (i) the chemical bonds, (ii) the disorder, and (iii) other molecular information.

Single-crystal X-ray diffraction is the most powerful X-ray technique for inorganic chemists. From precise measurement of the intensity and angles at which an X-ray beam diffracts off a crystal, the arrangement of the atoms can be reconstructed. Obviously, as a direct probe of structure crystallography is an invaluable characterization method for all types of compounds. Some inorganic compounds (such as rocks and minerals) cannot be obtained as single crystals, but in such cases, X-ray powder diffraction spectroscopy can be used to obtain the dimensions of the unit cell for use in identification of inorganic chemicals. In fact, there is a large, indexed catalog of lattice constants for many minerals that is available for identification purposes.

REFERENCES

Alloway, B.J., 1995. Heavy Metals in Soils. Blackie Academic and Professional Publishers, London.

Arruti, A., Fernández-Olmo, I., Irabien, A., 2010. Evaluation of the contribution of local sources to trace metals levels in urban $PM_{2.5}$ and PM_{10} in the Cantabria region (Northern Spain). J. Environ. Monit. 12 (7), 1451–1458.

Basta, N.T., Ryan, J.A., Chaney, R.L., 2005. Trace element chemistry in residual-treated soil: key concepts and metal bioavailability. J. Environ. Qual. 34 (1), 49–63.

Bradl, H. (Ed.), 2002. Heavy Metals in the Environment: Origin, Interaction and Remediation, vol. 6. Academic Press, New York.

D'Amore, J.J., Al-Abed, S.R., Scheckel, K.G., Ryan, J.A., 2005. Methods for speciation of metals in soils: a review. J. Environ. Qual. 34 (5), 1707–1745.

Fawell, J.K., 1993. The impact of inorganic chemicals on water quality and health. Ann. Ist Super Sanita. 29 (2), 293–303. https://www.ncbi.nlm.nih.gov/pubmed/8279720.

Fergusson, J.E. (Ed.), 1990. The Heavy Elements: Chemistry, Environmental Impact and Health Effects. Pergamon Press, Oxford.

Gallo, M., 2001. History and scope of toxicology. In: Klaasen, C.D. (Ed.), Casarett and Doull's Toxicology: The Basic Science of Poisons. sixth ed. McGraw-Hill, New York.

Goyer, R.A., Clarkson, T.W., 1996. Toxic effects of metals. In: Klaasen, C.D. (Ed.), Casarett and Doull's Toxicology: The Basic Science of Poisons. fifth ed. McGraw-Hill, New York.

Hamelink, J.L., Landrum, P.F., Harold, B.L., William, B.H. (Eds.), 1994. Bioavailability: Physical, Chemical, and Biological Interactions. CRC Press, Taylor & Francis Group, Boca Raton, FL.

He, Z.L., Yang, X.E., Stoffella, P.J., 2005. Trace elements in agroecosystems and impacts on the environment. J. Trace Elem. Med. Biol. 19 (2–3), 125–140.

Herawati, N., Suzuki, S., Hayashi, K., Rivai, I.F., Koyoma, H., 2000. Cadmium, copper and zinc levels in rice and soil of Japan, Indonesia and China by soil type. Bull. Environ. Contam. Toxicol. 64, 33–39.

Kaasalainen, M., Yli-Halla, M., 2003. Use of sequential extraction to assess metal partitioning in soils. Environ. Pollut. 126 (2), 225–233.

Kabata-Pendias, A., Pendias, H., 2001. Trace Metals in Soils and Plants, second ed. CRC Press, Taylor & Francis Group, Boca Raton, FL.

Khan, S., Cao, Q., Zheng, Y.M., Huang, Y.Z., Zhu, Y.G., 2008. Health risks of heavy metals in contaminated soils and food crops irrigated with wastewater in Beijing, China. Environ. Pollut. 152 (3), 686–692.

Kuo, S., Heilman, P.E., Baker, A.S., 1983. Distribution and forms of copper, zinc, cadmium, iron, and manganese in soils near a copper smelter. Soil Sci. 135 (2), 101–109.

Lombi, E., Gerzabek, M.H., 1998. Determination of mobile heavy metal fraction in soil: results of a pot experiment with sewage sludge. Commun. Soil Sci. Plant Anal. 29 (17, 18), 2545–2556.

Nriagu, J.O., 1989. A global assessment of natural sources of atmospheric trace metals. Nature 338, 47–49.

Pacyna, J.M., 1996. Monitoring and assessment of metal contaminants in the air. In: Chang, L.W., Magos, L., Suzuli, T. (Eds.), Toxicology of Metals. CRC Press, Taylor & Francis Group, Boca Raton, FL, pp. 9–28.

Pierzynski, G.M., Sims, J.T., Vance, G.F., 2000. Soils and Environmental Quality, second ed. CRC Press, Taylor & Francis Group, Boca Raton, FL.

Shallari, S., Schwartz, C., Hasko, A., Morel, J.L., 1998. Heavy metals in soils and plants of serpentine and industrial sites of Albania. Sci. Total Environ. 192 (09), 133–142.

Speight, J.G., 2013. The Chemistry and Technology of Coal, third ed. CRC Press, Taylor & Francis Group, Boca Raton, FL.

Speight, J.G., 2014. The Chemistry and Technology of Petroleum, fifth ed. CRC Press, Taylor & Francis Group, Boca Raton, FL.
Speight, J.G., 2017a. Environmental Organic Chemistry for Engineers. Butterworth-Heinemann, Elsevier, Cambridge, MA.
Speight, J.G., 2017b. Handbook of Petroleum Refining. CRC Press, Taylor & Francis Group, Boca Raton, FL.
Speight, J.G., Islam, M.R., 2016. Peak Energy—Myth or Reality. Scrivener, Salem, MA.
Speight, J.G., Singh, K., 2014. Environmental Management of Energy from Biofuels and Biofeedstocks. Scrivener Publishing, Beverly, MA.
Sposito, G., Page, A.L., 1984. Cycling of metal ions in the soil environment. In: Sigel, H. (Ed.), Metal Ions in Biological Systems. In: Circulation of Metals in the Environment, vol. 18. Marcel Dekker, New York, pp. 287–332.
Thy, P., Yu, C., Jenkins, B.M., Lesher, C.E., 2013. Inorganic composition and environmental impact of biomass feedstock. Energy Fuels 27 (7), 3969–3987.
Wuana, R., Okieimen, F.E., 2011. Heavy metals in contaminated soils: a review of sources, chemistry, risks and best available strategies for remediation. ISRN Ecol. 2011. Article Id. 402647. http://dx.doi.org/10.5402/2011/402647. https://www.hindawi.com/journals/isrn/2011/402647/.
Zhang, M.K., Liu, Z.Y., Wang, H., 2010. Use of single extraction methods to predict bioavailability of heavy metals in polluted soils to rice. Commun. Soil Sci. Plant Anal. 41 (7), 820–831.

FURTHER READING

Goyer, R.A., 1996. Toxic effects of metals. In: Klaasen, C.D. (Ed.), Cassarett and Doull's Toxicology: The Basic Science of Poisons. McGraw-Hill, New York, pp. 811–867.

CHAPTER SIX

Introduction Into the Environment

6.1 INTRODUCTION

Environmental pollution is a major issue for the modern world. Air, water, and soil are being polluted alike. Natural and human-induced chemicals can be found in all areas of the environment. For example, as groundwater flows through the ground, metals such as iron and manganese are dissolved and may later be found in high concentrations in the water. Industrial discharges, urban activities, agriculture, groundwater pumpage, and disposal of waste all can affect groundwater quality. Contaminants can be human-induced, as from toxic chemical spills and leakage from waste-disposal sites also can introduce inorganic chemicals into the water. Or, a well might have been placed in land that was once used for a garbage dump or a chemical dump site.

Thus, there is urgency in controlling all the forms of pollution in order to preserve the environment. Pollution may be defined as an undesirable change in the physical, chemical, and biological characteristics of air, water, and soil, which affect the flora and fauna of the various ecosystems. A pollutant is, in the current context, an inorganic chemical that adversely interferes with an ecosystem. Generally, most inorganic pollutants are introduced in the environment by waste and/or accidental discharge, or else, they are by-products or residues from the production process.

The phrase *chemical contamination*, in the context of this book, is used to indicate those situations where inorganic chemical contaminants (often referred to as IOCs) either are present where they should not be or are at higher concentrations than they would have occurred if the chemicals are indigenous to an ecosystem (Miller, 1984). IOCs can be found in mass-produced products. For many of these substances, accumulation into the various subdivisions of the environment (Fig. 6.1) can cause environmental problems, although some inorganic chemicals do not damage the environment, and for many inorganic chemicals, the consequences to the environment and the fate of the chemicals are often unknown. Inorganic chemicals

Environmental Inorganic Chemistry for Engineers
http://dx.doi.org/10.1016/B978-0-12-849891-0.00006-0

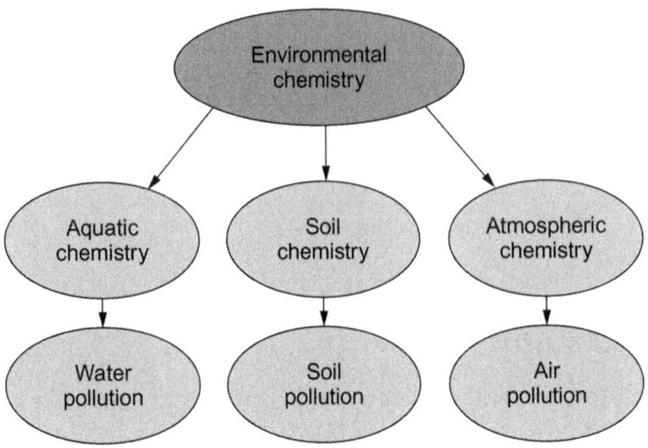

Fig. 6.1 The subdivisions of environmetnal chemistry.

(Table 6.1), especially the inorganic water-soluble contaminants, are often transported by water across the land, roads, and other impermeable surfaces. With little prior treatment, many of these contaminants may eventually discharge into rivers, lakes, oceans, and other waterways.

Thus, inorganic contaminants are elements or compounds found in water supplies and may be natural in the geology or caused by activities of man through mining, industry, or agriculture. It is common to have trace amounts of many inorganic contaminants in water supplies and that such remnants of fertilizers include derivatives of nitrogen (N), phosphorus (P), and potassium (K). Increases in these simple chemicals in waterways are nearly always a result of land-use activities such as fertilizer runoff or direct discharge from industrial operations. Inorganic contaminants also include metals and metal particles. These can be found in stormwater runoff from urban development and will accumulate in drainage systems or low lying areas of land, such as marshy areas and wetlands. Many of these contaminants eventually are the result of discharge into waterways with little prior treatment to remove chemicals. Some contaminants, such as mercury, may bioaccumulate in animal tissues and occur in fish and eventually end up as part of the human diet.

Chemicals that accumulate in living organisms so that their concentrations in body tissues continue to increase are called bioaccumulative. In fish and other aquatic organisms, bioaccumulation is sometimes called bioconcentration. The bioconcentration factor (BCF) is an expression of the extent to which the concentration of a chemical in a fish is higher than

Table 6.1 Examples of Common Inorganic Pollutants

Pollutant	Sources of Origin	Directly Polluted Medium	Biological Actions
SO_2	Volcanoes, industry, transports	Air	Expectoration, spasms, respiratory difficulties, bronchitis
NO_2	Volcanoes, industry, transports	Air	Methemoglobin that restrains the transport of oxygen to the tissues
CO	Transports	Air	Carboxyhemoglobin that generates dizziness, asphyxia due to oxygen deficiency
NH_3	Industry, agriculture	Air, soil	Irritations of the nasal mucous
H_2S	Industry, anaerobic fermentations with sulfur bacteria	Soil, water	Troubles of the nervous system functions and of the sanguine circulation
HCl	Industry, transports	Air	Respiratory diseases, cancerous effects
HF	Industry	Air	Bleedings, loss of the visual acuity, vomit, cerebral diseases
NH_4^+, NO_3^-, NO_2^-	Farms, factories that produce nitrogen	Soil, water	Respiratory diseases, cancer
Pb	Heavy industry, transports	Air	Intoxications, hemoglobin alteration, disturbances of the liver and kidney functions
Hg	Industry	Soil, water	Caught, insomnia, hallucinations
Zn	Industry	Air, water	Corrosive action on the tissues, muscular, cardiovascular, and nervous systems diseases

the concentration in the surrounding water. Very low concentrations of a bioaccumulative substance in water can result in markedly higher concentrations in the tissue of fish at higher levels of the aquatic food chain and in people or wildlife eating those fish. Concentrations of airborne bioaccumulative chemicals will also be magnified in air-breathing organisms.

The potential for a chemical to bioaccumulate can be predicted by examining whether the chemical preferentially dissolves in an organic solvent as opposed to water. If the concentration of a chemical in the solvent is more than 1000 times higher than its concentration in water when added to a mixture of solvent and water, the chemical is likely to bioaccumulate in organisms. If that concentration gradient is >5000, the chemical is highly likely to bioaccumulate. Many bioaccumulative chemicals are fat soluble so that they tend to reside primarily in fat deposits or in the fatty substances in blood. This explains why fat-soluble bioaccumulative chemicals are often found at elevated levels in fat-rich breast milk. But bioaccumulative substances may also be deposited elsewhere, including bone, muscle, or the brain.

Both the concentration of these chemicals and how they enter a waterway vary greatly. As with some contaminants, natural global cycling has always been a primary contributor to the presence of chemical elements in air, water, and soil (or sediments). This process can involve transfer of a contaminant between the atmosphere, the hydrosphere, and the lithosphere, where it may eventually be transported and deposited onto surface water and soil. Major anthropogenic sources of contaminants in the environment have been (i) mining operations; (ii) industrial processes; (iii) combustion of fossil fuels, especially charcoal; (iv) production of cement; and (v) incineration of municipal, chemical, and medical wastes. Alternatively, industries such as forest processing, meat processing, and dairy processing, wastewater treatment may discharge wastewater that can potentially contain IOCs: examples are bleach (hypochlorite derivative), curing agents (which include nitrate derivatives, $—NO_3$, and nitrite derivatives, $—NO_2$), and certain metals (like mercury, copper, chrome, zinc, iron, arsenic, and lead). Prior treatment of these inorganic chemicals before discharge is now strictly regulated and controlled via the resource consenting process and will vary depending on the type, quantity, and potential environmental reactivity of the inorganic chemical to be discharged.

If the case of the discharge of any inorganic chemical into the environment is proposed, it is necessary to set standards for acceptable concentrations in air, water, soil, and flora, as well as fauna—if there are any such acceptable concentrations. Monitoring of these concentrations of the chemicals in an ecosystem and any resultant biological effects must be undertaken to ensure that the standards as set in any regulation are realistic and provide protection of the environment from any adverse effects. Furthermore, considerable attention continues to be focused on regulation of the use of all inorganic chemicals, and a primary aspect involves the prediction of the behavior and effects of a chemical from the properties of that chemical (Chapter 4). Also, this concept that the molecular characteristics of the molecule (such as the three-dimensional structure and the presence of functional groups) govern the physical and chemical properties of the compound in turn influences the effects of the chemical on the ecosystem and the transformation and distribution of the chemical in the environment (Chapter 7). This suggests that the transformation and distribution in the environment and any effects on the floral and faunal species can be predicted from the physical properties and/or the chemical properties of the inorganic contaminant. However, the prediction of biological effects may also involve a complex set of chemical-floral and/or chemical-faunal interactions.

Industrial inorganic chemical manufacturers use and generate large numbers of chemicals and large quantities of chemicals. In the past, especially during the 19th century and the first half of the 20th century, when environmental legislation was not in place or, if in place, was not enacted, the inorganic chemical industry disposed of inorganic chemicals to all types of environmental ecosystems including air through both fugitive emissions (emissions of gas or vapor from pressurized equipment due to leaks and other unintended or irregular releases of gases) and direct emissions (emissions from sources that are owned or controlled by the reporting entity), water (direct discharge and runoff) and land (Table 6.2). However, the types of pollutants a single facility will release depend on (i) the type of process; (ii) the feedstocks to the process; (iii) the equipment used in the process, such as the reactor; and (iv) the equipment and process maintenance practices, which can vary over short periods of time (such as from hour to hour) and can also vary with the part of the process that is underway. For example, for batch reactions in a closed vessel, the chemicals are more likely to be emitted at the beginning and end of a reaction step (which are associated with reactor or treatment vessel loading and product transfer operations) than during the reaction. Fluidized bed reactors that are used in continuous process operations may emit chemicals at any part of the process.

The inorganic chemical pollutants that are most likely to present ecological risks are those that are (i) highly reactive and likely to react with any part of the ecosystem and undergo transformation; (ii) highly bioaccumulative, building up to high levels in floral and faunal tissues even when concentrations in the ecosystem remain relatively low; and (iii) highly toxic, which can cause adverse effects to the ecosystem itself and to the floral and faunal members of the ecosystem at comparatively low doses. In addition,

Table 6.2 Types of Releases From Industrial Processes

Release is an on-site discharge of a toxic chemical to the environment and includes (i) emissions to the air, (ii) discharges to bodies of water, (iii) releases at the facility to land, and (iv) the contained disposal into underground injection wells

Releases to air (point and fugitive air emissions)—these releases include all air emissions from industry activity; point emissions occur through confined air streams as found in stacks, ducts, or pipes, while fugitive emissions include losses from equipment leaks or evaporative losses from impoundments, spills, or leaks

Releases to water (surface water discharges)—these releases include any releases going directly to streams, rivers, lakes, oceans, or other bodies of water; any estimates for stormwater runoff and nonpoint losses must also be included

Releases to land—these releases include disposal of toxic chemicals in waste to on-site landfills, land treated or incorporation into soil, surface impoundments, spills, leaks, or waste piles; these activities must occur within the facility's boundaries for the inclusion in this category

Underground injection—this type of release is a contained release of a fluid into a subsurface well for the purpose of waste disposal

Transfer is a transfer of toxic chemicals in wastes to a facility that is geographically or physically separate from the facility reporting under the toxic release inventory. The quantities reported represent a movement of the chemical away from the reporting facility, and except for off-site transfers for disposal, these quantities of chemicals do not necessarily represent entry of the chemicals into the environment

Transfers to publicly owned treatment Works—these transfers include waste waters transferred through pipes or sewers to a publicly owned treatment works (POTW); treatment and chemical removal depend on the nature of the chemical and the treatment methods employed; chemicals that are not treated or destroyed by the publicly owned treatment works are generally released to surface waters or landfilled within the sludge

Transfers to recycling—these transfers include chemicals that are sent off-site for the purposes of regenerating or recovering still valuable materials; once these chemicals have been recycled, they may be returned to the originating facility or sold commercially

Transfers to energy recovery—these transfers include wastes combusted off-site in industrial furnaces for energy recovery; treatment of a chemical by incineration is not considered to be energy recovery

Transfers to treatment—these transfers are wastes moved off-site for either neutralization, incineration, biological destruction, or physical separation; in some cases, the chemicals are not destroyed but are prepared for further waste management

Transfers to disposal—these transfers are wastes taken to another facility for disposal generally as a release to land or as an injection underground

atmosphere-water interactions that control the input and outgassing of persistent inorganic pollutants in aquatic systems are critically important in determining the life cycle and residence time of an inorganic chemical in an ecosystem and the extent of any contamination and the resulting adverse effects.

Briefly, in the environment, many chemicals are degraded by sunlight, destroyed through reactions with other environmental substances, or metabolized by naturally occurring bacteria. Some chemicals, however, have features than enable them to resist environmental degradation. They are classified as *persistent* and can accumulate in soil and aquatic environments. Those that can evaporate into air (volatilize) or dissolve in water can migrate considerable distances from where they are released. Floral and faunal species are more likely to be exposed to a chemical if it does not easily degrade or is dispersed widely in the environment.

The structural characteristics that enable a chemical to persist in the environment can also help it to resist metabolic breakdown in people or wildlife. For example, synthetic chemicals that contain halogen atoms (particularly fluorine, chlorine, or bromine) are often resistant to degradation in the environment or within organisms.

Metals, such as lead, mercury, and arsenic, are always persistent, since they are basic elements and cannot be further broken down and destroyed in the environment. Although this discussion will focus on the synthetic organic chemicals, the potential health effects of exposure to metals should not be overlooked. For example, lead contamination of air, soil, or drinking water can ultimately result in significant exposures in fetuses, infants, and children, resulting in impaired brain development.

Although the effects of various types of inorganic chemical pollutants are usually evaluated independently, many ecosystems are subject to multiple pollutants, and their fate and impacts are intertwined. For example, the effects of nutrient deposition in an ecosystem can alter the methods by which the inorganic contaminants are assimilated and bioaccumulated and the means by which the floral and faunal organisms in the ecosystem are affected.

Of all the inorganic chemical pollutants released into the environment by anthropogenic activity, persistent inorganic pollutants (PIPs) are among the most dangerous inorganic chemicals and often require extreme measures for removal (Chapter 1). Persistent inorganic pollutants (PIPs) are chemical substances that persist in the environment, bioaccumulate through the food chains of the floral and faunal species, and pose a risk of eventually causing adverse effects to human health and the environment. Moreover, persistent inorganic pollutants can be transported across international boundaries far from their sources, even to regions where they have never been used or produced. Consequently, persistent inorganic pollutants can pose a serious threat not only to the environment but also to human health on a global scale.

Releases into the environment of inorganic chemicals that persist in an ecosystem (rather than undergo some form of biodegradation) lead to an exposure level that is subject not only to the length of time the chemical remains in circulation (in the environment) but also on the number of times that the inorganic chemical is recirculated before it is ultimately removed from the ecosystem. In addition, an inorganic chemical that is not a persistent pollutant may transform into a persistent inorganic pollutant in a manner that is not only dependent upon the chemical and physical properties of the pollutant but also dependent upon the influence of the ecosystem.

Typically, persistent inorganic pollutants are highly toxic and long lasting (hence the name *persistent*) and cause a wide range of adverse effects to environmental flora and fauna, including disease and birth defects in humans and animals; some of the severe human health impacts from persistent inorganic pollutants include (i) the onset of cancer, (ii) damage to the central nervous system, (iii) damage to the peripheral nervous system, (iv) damage to the reproductive system, and (v) disruption of the immune system. Moreover, persistent inorganic pollutants do not respect international borders, and the serious environmental and human health hazards created by these chemicals not only affect developing countries, where systems and technology for monitoring, tracking, and disposing of them can be weak or nonexistent, but also affect developed countries. As long as the chemical can be transported by air, water, and land, no region or country is immune from the effects of these chemicals.

Generally, persistent inorganic pollutants are pesticides; for example, borate derivatives, silicate derivatives, and sulfur derivative are minerals that are mined from the earth and ground into a fine powder. Some of these chemicals function as poisons, and others function by physically interfering with the pest; older inorganic pesticides included such highly toxic compounds as arsenic, copper, lead, and tin salts. On the other hand, the more modern inorganic pesticides are relatively low in toxicity and have a low environmental impact. Other contaminants include industrial chemicals or the unwanted by-products of industrial processes that have been used and subject to unmanaged disposal for decades prior to the inception of the various regulations and often without due regard for the environment but that have, more recently, been found to share several significant characteristics that need consideration before disposal is planned. These characteristics include (i) persistence in the environment insofar as these inorganic chemicals resist degradation in air, water, and sediments; (ii) bioaccumulation insofar as these inorganic chemicals accumulate in floral and faunal tissues at concentrations higher than those in the surrounding environment; and (iii) long-range transport insofar as these inorganic chemicals can travel great distances from the source of release through air, water, and the internal organs of migratory animals. Any of these characteristics can result in the contamination not only of local ecosystems but also of ecosystems that are significant distances (even up to thousands of miles) away from the source of the chemicals.

Briefly and by way of explanation, bioaccumulation is a process by which persistent environmental pollution leads to the uptake and accumulation of one or more contaminants, by organisms in an ecosystem. The amount of a pollutant available for exposure depends on its persistence and the potential for its bioaccumulation. Any chemical (including both organic chemicals and inorganic chemicals) can be considered to be capable of bioaccumulation if the chemical has a degradation half-life in excess of 30 days or if the chemical has a bioconcentration factor (BCF) >1000 or if the log K_{ow} (the octanol-water partition coefficient of the chemical, Chapter 4) is >4.2:

$$\mathrm{BCF} = (\text{concentration in biota})/(\text{concentration in ecosystem})$$

The octanol/water partition coefficient (K_{ow}) is the ratio of the concentration of an inorganic chemical in the octanol phase relative to the concentration of the chemical in the aqueous phase of a two-phase octanol/water system:

$$K_{ow} = (\text{concentration in octanol phase})/(\text{concentration in aqueous phase})$$

The bioconcentration factor (BCF) indicates the degree to which a chemical may accumulate in biota (flora and fauna). However, measurement of bioconcentration is typically made on faunal species and is distinct from food chain transport, bioaccumulation, or biomagnification. The bioconcentration factor is a constant of proportionality between the chemical concentration in flora and fauna in an ecosystem. It is possible, for many inorganic chemicals, to estimate the bioconcentration factor from the octanol-water partition coefficients (K_{OW}):

$$\mathrm{Log}(\text{bioconcentration factor}) = m \log K_{OW} + b$$

In terms of actual numbers, for many lipophilic inorganic chemicals, the bioconcentration factor can be calculated using the regression equation:

$$\log \text{BCF} = -2.3 + 0.76 \times (\log K_{ow})$$

Furthermore, empirical relationships between the octanol-water partition coefficients and the bioconcentration factor can be developed on a chemical-by-chemical basis.

On this note, it is worth defining the source of chemical contaminant insofar as chemical contaminants can originate from (i) a point source or (ii) a nonpoint source. The point source of pollution is a single identifiable source of pollution that may have a negligible extent, distinguishing it from other pollution source geometries. On the other hand, nonpoint source pollution generally results from land runoff, precipitation, atmospheric deposition, drainage, seepage, or hydrologic modification. Thus, nonpoint source (NPS) pollution, unlike pollution from industrial and sewage treatment plants, originates from many diffuse sources and is often caused by rainfall or snowmelt moving over and through the ground. As the runoff water moves, it picks up and transports human-made pollutants and natural pollutants away from the site, and the pollutants are ultimately deposited into the water systems such as groundwater, wetlands, lakes, rivers, wetlands, and coastal waters.

Indeed, an analysis of the amount of waste formed in processes for the manufacture of a range of fine chemicals and chemical intermediates has revealed that the amount of waste generated in some processes was in excess of the amount of the desired product and such overproduction of waste was not exceptional in the inorganic chemical industry. As a means of measuring the amount of waste vis-à-vis the amount of products, the E-factor (environmental factor) (kilograms of waste per kilogram of product) was introduced as an indication of the environmental footprint of the manufacturing process in various segments of the chemical industry (Chapter 7). This factor is derived from the chemical and fine chemical industries as a measure of the efficiency of the manufacturing process. Thus, simply

$$E = \text{kilograms of waste/kilogram of product}$$

The factor can be conveniently calculated from a knowledge of the number of tons of raw materials purchased and the number of tons of product sold, the calculation being for a particular product or a production site or even a whole company. A higher E-factor means more waste and, consequently, a larger environmental footprint; thus, since mass cannot be created, resulting in a negative **E-factor, the ideal E-factor for any process is zero.

However, in the context of environmental protection and to be all inclusive, the E-factor is the total mass of raw materials plus ancillary process requirements minus the total mass of product, all divided by the total mass of the product. Thus, the E-factor should represent the *actual amount* of waste produced in the process, defined as everything but the desired product and takes the chemical yield into account and includes reagents, solvent losses, process aids, and (in principle) even the fuel necessary for the process. Water has been generally excluded from the calculation of the E-factor since the inclusion of all process water could lead to exceptionally high E-factors in many cases. Thus, the exclusion of water from the calculation of the E-factor allowed meaningful comparisons of the technical factors (E-factors excluding water use) to be made for the various processes. Recent thinking, in this modern environmentally conscious era, has led to the conclusion that (water, being a reactant or reaction product) there is no reason for the water requirement or the water product to be omitted since the disposal of process water is an environmental issue. Moreover, the use of the E-factor has been widely adopted by many of the chemical industries and the pharmaceutical industries in particular. Thus, a major aspect of the process development recognized by process chemists and by process engineers is the need for determining an E-factor—whether or not it is called by that name (i.e., the E-factor)—but *chemicals in comparison with chemical out* have become a major yardstick in many of the chemical industries.

It is clear that the E-factor increases substantially when comparing bulk chemicals with fine chemicals. This is partly a reflection of the increasing complexity of the products, necessitating processes that not only use multi-step syntheses but also is a result of the widespread use of stoichiometric amounts of the reagents, i.e., the required amounts of the reagents (some observers would advocate the stoichiometric amounts of the reagents plus 10%) to accomplish conversion of the starting inorganic chemical to the product(s) (Chapters 2 and 3). A reduction in the number of steps of a process for the synthesis of inorganic chemicals will, in most cases (but not always), lead to a reduction in the amounts of reagents and solvents used and hence a reduction in the amount of waste generated. This has led to the introduction of the concepts of step economy and function-oriented synthesis (FOS) of some inorganic chemicals. The main issues behind the concept of function-oriented synthesis are that the structure of an active compound can be reduced to simpler structures designed for ease of synthesis while not exerting an adverse influence on the flora and fauna of an ecosystem (Chapter 7). This approach can provide practical access to new (designed) structures with novel activities while, hopefully, at the same time allowing for a relatively straightforward synthesis.

As noted above, a knowledge of the stoichiometric equation allows the process chemist or process engineer to predict the theoretical minimum amount of waste that can be expected (Chapters 2 and 3). This led to the concept of *atom economy* or *atom utilization* to quickly assess the environmental acceptability of alternatives to a particular product before any experiment is performed. It is a theoretical number; that is, it assumes a chemical yield of 100% and exactly stoichiometric amounts and disregards substances that do not appear in the stoichiometric equation. In short, the key to minimizing waste is precision or *selectivity* in inorganic synthesis, which is a measure of how efficiently a synthesis is performed. The standard definition of selectivity is the yield of product divided by the amount of substrate converted, expressed as a percentage.

Inorganic chemists distinguish between different categories of selectivity in two ways: (i) chemoselectivity, which relates to competition between different functional groups, and (ii) regioselectivity, which is the selective formation of one regioisomer. However, one category of selectivity was, traditionally, largely ignored by many inorganic chemists: the *atom selectivity* or *atom utilization* or *atom economy* and the virtual complete disregard of this important parameter by chemists and by engineers has been a major cause of the generation of large amounts of waste during the manufacture or inorganic chemicals. Quantification of the waste generated in the manufacture of inorganic chemicals, for example, by way of E-factors, served to illustrate the omissions related to the production of chemical waste and focus the attention of the companies that manufacture inorganic chemicals on the need for a paradigm shift from a concept of process efficiency, which was exclusively based on chemical yield, to a need that more focused on (or a more conscious process-related attitude to) the elimination of waste chemicals and maximization of the utilization of the raw materials used as process feedstocks.

6.2 MINERALS

Minerals (of which more than 3000 individual minerals are known) are substances formed naturally in the Earth. Minerals have a definite chemical composition and structure. Some are rare and precious such as gold (which occurs as the element) and diamond (a crystalline form of carbon), while others are more ordinary and ubiquitous, such as quartz (SiO_2).

A mineral is a naturally occurring inorganic solid that possesses an orderly internal structure and a definite chemical composition. The term *rock* is less specific, referring to any solid mass of mineral or mineral-like material. Common rocks are often made up of crystals of several kinds of minerals. There are some substances (*mineraloids*) that have the appearance of a mineral but lack any definite internal structure. The essential characteristics of a mineral are (i) it must occur naturally; (ii) it must be inorganic; (iii) it must be a solid; (iv) it must possess an orderly internal structure, that is, its atoms must be arranged in a definite pattern; and (v) it must have a definite chemical composition that may vary within specified limits. This can be confusing when coal, petroleum, and natural gas are often listed as part of the mineral resources of many states and countries.

Minerals are chemical compounds, sometimes specified by crystalline structure and by composition, which are found in rocks (or pulverized rocks or in sand). On the other hand, rocks consist of one or more minerals and fall into three main types depending on their origin and previous processing history: (i) igneous rocks, which are rocks that have solidified directly from a molten state, such as volcanic lava; (ii) sedimentary rocks, which are rocks that have been remanufactured from previously existing rocks, usually from the products of chemical weathering or mechanical erosion, without melting; and (iii) metamorphic rocks, which are rocks that have resulted from processing, by heat and pressure (but not melting), of previously existing sedimentary or igneous rocks.

On the other hand, sand is a naturally occurring granular inorganic material composed of finely divided rock and mineral particles that is defined by size, being finer than gravel and coarser than silt. The term *sand* can also refer to a textural class of soil or soil type, i.e., a soil containing more than 85% sand-sized particles by mass. The composition of sand varies, depending on the local rock sources and conditions, but the most common constituent of sand in inland continental settings and nontropical coastal settings is silica (silicon dioxide or SiO_2), usually in the form of quartz. The second most common type of sand is calcium carbonate ($CaCO_3$), for example, the mineral aragonite that has mostly been created, over the past half billion years, by various forms of life, such as coral and shellfish. For example, it is the primary form of sand apparent in areas where reefs have dominated the ecosystem for millions of years.

The most common minerals are the silica-based (SiO_2-based) minerals because of the abundance of silica in the crust of the Earth. There is a great variety of different minerals most of which contain silica but many of which contain other elements (Tables 6.3–6.5). In the crust of the Earth, ~20 of the minerals are common and fewer than 10 minerals account for over 90% of the crust by mass. Minerals are classified in many ways including properties such as (i) hardness, (ii) optical properties, and (iii) crystal structure. Non-silicate minerals constitute <10% of the Earth's crust, and the most common nonsilicate minerals are the carbonate minerals (—CO_3), the oxide minerals (—O), and the sulfide (—S) minerals. There are also naturally occurring phosphate minerals (—PO_4) and salts. There are some elements that occur in pure form, including gold, silver, copper, bismuth, arsenic, lead, and tellurium. Carbon is found in both graphite and diamond form.

Table 6.3 Elements in the Crust of the Earth

Element Name	Symbol	Percentage by Weight of the Earth's Crust
Oxygen	O	47
Silicon	Si	28
Aluminum	Al	8
Iron	Fe	5
Calcium	Ca	3.5
Sodium	Na	3
Potassium	K	2.5
Magnesium	Mg	2
All other elements		1

Table 6.4 Mineral Names and Chemical Composition

Mineral Name	Chemical Formula	Useful Element
Galena	PbS	Lead
Pyrite	FeS_2	Sulfur (pyrite is not used as an ore of iron)
Chalcopyrite	$CuFeS_2$	
Chalcocite	Cu_2S	
Bauxite	Al_2O_3	
Magnetite	Fe_3O_4	
Hematite	Fe_2O_3	
Rutile	TiO_2	

Table 6.5 Common Elements and Ores

Aluminum	The most abundant metal element in Earth's crust. Aluminum originates as an oxide called alumina. Bauxite ore is the main source of aluminum and must be imported from Jamaica, Guinea, Brazil, Guyana, etc. Used in transportation (automobiles), packaging, building/construction, electric, machinery, and other uses. The United States was 100% import reliant for its aluminum in 2012
Antimony	A native element, antimony metal is extracted from stibnite ore and other minerals. Used as a hardening alloy for lead, especially storage batteries and cable sheaths and also used in bearing metal, type metal, solder, collapsible tubes and foil, sheet and pipes, and semiconductor technology. Antimony is used as a flame retardant, in fireworks, and in antimony salts are used in the rubber, chemical, and textile industries, as well as medicine and glassmaking. The United States was 87% import reliant in 2012
Barium	A heavy metal contained in barite. Used as a heavy additive in oil well drilling; in the paper and rubber industries; as a filler or extender in cloth, ink, and plastics products; in radiography (barium milk shake); as a deoxidizer for copper; a sparkplug in alloys; and in making expensive white pigments
Bauxite	Rock composed of hydrated aluminum oxides. In the United States, it is primarily converted to alumina. See "aluminum." The United States was 100% import reliant in 2012
Beryllium	Used in the nuclear industry and to make light, very strong alloys used in the aircraft industry. Beryllium salts are used in fluorescent lamps, in X-ray tubes, and as a deoxidizer in bronze metallurgy. Beryl is the gemstones emerald and aquamarine. It is used in computers, telecommunication products, aerospace and defense applications, appliances and automotive, and consumer electronics. Also used in medical equipment. The United States was 10% import reliant in 2012
Chromite	The United States consumes about 6% of world chromite ore production in various forms of imported materials, such as chromite ore, chromite chemicals, chromium ferroalloys, chromium metal, and stainless steel. Used as an alloy and in stainless and heat-resisting steel products. Used in chemical and metallurgical industries (chrome fixtures, etc.) Superalloys require chromium. It is produced in South Africa, Kazakhstan, and India. The United States was 70% import reliant for chromium in 2012

Continued

Table 6.5 Common Elements and Ores—cont'd

Clay	Used in floor and wall tile as an absorbent, in sanitation, mud drilling, foundry sand bond, iron pelletizing, brick, lightweight aggregate, and cement. It is produced in 40 states. Ball clay is used in floor and wall tile. Bentonite is used for drilling mud, pet waste absorbent, iron ore pelletizing, and foundry sand bond. Kaolin is used for paper coating and filling, refractory products, fiberglass, paint, rubber, and catalyst manufacture. Common clay is used in brick, light aggregate, and cement. The United States was not import reliant in 2012
Cobalt	Used primarily in superalloys for aircraft gas turbine engines, in cemented carbides for cutting tools and wear-resistant applications, chemicals (paint dryers, catalysts, and magnetic coatings), and permanent magnets. The United States has cobalt resources in Minnesota, Alaska, California, Idaho, Missouri, Montana, and Oregon. Cobalt production comes principally from Congo, China, Canada, Russia, Australia, and Zambia. The United States was 78% import reliant in 2012
Copper	Used in building construction, electric, and electronic products (cables and wires, switches, plumbing, and heating); transportation equipment; roofing; chemical and pharmaceutical machinery; and alloys (brass, bronze, and beryllium alloyed with copper are particularly vibration resistant); alloy castings; electroplated protective coatings; and undercoats for nickel, chromium, zinc, etc. More recently, copper is being used in medical equipment due to its antimicrobial properties. The United States has mines in Arizona, Utah, New Mexico, Nevada, and Montana. Leading producers are Chile, Peru, China, the United States, and Australia. The United States was 35% import reliant in 2012
Feldspar	A rock-forming mineral; industrially important in glass and ceramic industries, patter and enamelware, soaps, bond for abrasive wheels, cements, insulating compositions, fertilizer, tarred roofing materials, and as a sizing or filler, in textiles and paper. In pottery and glass, feldspar functions as a flux. End uses for feldspar in the United States include glass (70%) and pottery and other uses (30%). The United States was 78% import reliant in 2012. The United States was not import reliant in 2012
Fluorite (fluorspar)	Used in the production of hydrofluoric acid, which is used in the pottery, ceramics, optical, electroplating, and plastic industries; in the metallurgical treatment of bauxite; as a flux in open-hearth steel furnaces and in metal smelting; in carbon electrodes; emery wheels; electric arc welders; toothpaste; and paint pigment. It is a key ingredient in the processing of aluminum and uranium. The United States was 100% import reliant in 2012

Gallium	Gallium is used in integrated circuits, light-emitting diodes (LEDs), photodetectors, and solar cells. It has a new use in chemotherapy for some types of cancer. Integrated circuits are used in defense applications, high-performance computers, and telecommunications. Optoelectronic devices were used in areas such as aerospace, consumer goods, industrial equipment, medical equipment, and telecommunications. Leading sources are Germany, the United Kingdom, China, and Canada. The United States was 99% import reliant in 2012
Gold	Used in jewelry and arts, dentistry and medicine, in medallions and coins, in ingots as a store of value, for scientific and electronic instruments, and as an electrolyte in the electroplating industry. Mined in Alaska and several western states. Leading producers are China, Australia, the United States, Russia, and Canada. The United States was not import reliant in 2012
Gypsum	Processed and used as prefabricated wallboard or an industrial or building plaster; used in cement manufacturing, agriculture, and other uses. The United States was 12% import reliant in 2012
Halite (sodium chloride—salt)	Used in human and animal diet, food seasoning, and food preservation; used to prepare sodium hydroxide, soda ash, caustic soda, hydrochloric acid, chlorine, and metallic sodium; used in ceramic glazes, metallurgy, curing of hides, mineral waters, soap manufacturing, home water softeners, highway deicing, photography, and in scientific equipment for optical parts. Single crystals used for spectroscopy, ultraviolet, and infrared transmission. The United States was 19% import reliant for salt in 2012
Indium	Indium tin oxide is used for electric conductivity purposes in flat-panel devices—most commonly in liquid crystal displays (LCDs). It is also used in solders, alloys, compounds, electric components, semiconductors, and research. Indium ore is not recovered from ores in the United States. China is the leading producer. It is also produced in Canada, Japan, and Belgium. The United States was 100% import reliant in 2012
Iron Ore	Used to manufacture steels of various types. Powdered iron used in metallurgy products, magnets, high-frequency cores, auto parts, and catalyst. Radioactive iron (iron 59) in medicine, tracer element in biochemical, and metallurgical research. Iron blue in paints, printing inks, plastics, cosmetics, and paper dyeing. Black iron oxide as pigment, in polishing compounds, metallurgy, medicine, and magnetic inks. Most US production is from Michigan and Minnesota. China, Australia, Brazil, and Russia are the major producers. The United States was not import reliant in 2012

Continued

Table 6.5 Common Elements and Ores—cont'd

Lead	Used in lead-acid batteries, gasoline additives (now being eliminated), and tanks and solders, seals, or bearing; electric and electronic applications; TV tubes and glass, construction, communications, and protective coatings; in ballast or weights; ceramics or crystal glass; X-ray and gamma radiation shielding; soundproofing material in construction industry; and ammunition. Industrial-type batteries are used as a source of uninterruptible power equipment for computer and telecommunications networks and mobile power. US mines lead mainly not only in Missouri but also in Alaska and Idaho. The United States was not import reliant in 2012
Lithium	Compounds are used in ceramics and glass, batteries, lubricating greases, air treatment, in primary aluminum production, in the manufacture of lubricants and greases, rocket propellants, vitamin A synthesis, silver solder, batteries, and medicine. Lithium ion batteries have become a substitute for nickel-cadmium batteries in handheld/portable electronic devices. There is one brine operation in Nevada. Australia, Chile, and China are major producers. The United States was more than 70% reliant for lithium in 2012
Manganese	Ore is essential to iron and steel production. Also used in the making of manganese ferroalloys. Construction, machinery, and transportation end uses account for most US consumption of manganese. Manganese ore has not been produced in the United States since 1970. Major producers are South Africa, Australia, China, Gabon, and Brazil. The United States was 100% import reliant in 2012
Mica	Micas commonly occur as flakes, scales, or shreds. Ground mica is used in paints, as joint cement and as a dusting agent; in oil-well-drilling muds; and in plastics, roofing, rubber, and welding rods. Sheet mica is fabricated into parts for electronic and electronic equipment. China and Russia are the leading producers. The United States was 100% import reliant in 2012
Molybdenum	Used in alloy steels to make automotive parts, construction equipment, gas transmission pipes, stainless steels, tool steels, cast irons, superalloys, and chemicals and lubricants. As a pure metal, molybdenum is used because of its high melting temperatures (4730 °F) as filament supports in light bulbs, metalworking dies, and furnace parts. Major producers are China, the United States, Chile, and Peru. The United States was not import reliant in 2012

Nickel	Vital as an alloy to stainless steel; plays key role in the chemical and aerospace industries. End uses were transportation, fabricated metal products, electric equipment, petroleum, chemical industries, household appliances, and industrial machinery. Major producers are the Philippines, Indonesia, Russia, Australia, and Canada. The United States was 49% import reliant in 2012
Perlite	Expanded perlite is used in building construction products like roof insulation boards, as fillers, for horticulture aggregate and filter aids. It is produced in New Mexico and other western states and is processed in over 20 states. Leading producers are the United States, Greece, and Turkey. The United States was 24% import reliant in 2012
Platinum group metals (PGM)	Includes platinum, palladium, rhodium, iridium, osmium, and ruthenium. Commonly occur together in nature and are among the scarcest of the metallic elements. Platinum is used principally in catalysts for the control of automobile and industrial plant emissions, in jewelry, and in catalysts to produce acids, organic chemicals, and pharmaceuticals. PGMs used in bushings for making glass fibers are used in fiber-reinforced plastic and other advanced materials, in electric contacts, in capacitors, in conductive and resistive films used in electronic circuits, and in dental alloys used for making crowns and bridge. South Africa, Russia, the United States, and Canada are major producers. The United States was over 50% import reliant for most PGMs in 2012
Phosphate rock	Used to produce phosphoric acid for ammoniated phosphate fertilizers, feed additives for livestock, elemental phosphorus, and a variety of phosphate chemicals for industrial and home consumers. US production occurs in Florida, North Carolina, Idaho, and Utah. The United States is a major producer. It was not import reliant in 2012
Potash	A carbonate of potassium, used as a fertilizer, in medicine, in the chemical industry, and to produce decorative color effects on brass, bronze, and nickel. The leading producers are Canada, Russia, and Belarus. The United States was 81% import reliant in 2012
Pyrite	Used in the manufacture of sulfur, sulfuric acid, and sulfur dioxide; pellets of pressed pyrite dust are used to recover iron, gold, copper, cobalt, nickel; used to make inexpensive jewelry

Continued

Table 6.5 Common Elements and Ores—cont'd

Quartz (silica)	As a crystal, quartz is used as a semiprecious gemstone. Crystalline varieties include amethyst, citrine, rose quartz, and smoky quartz. Cryptocrystalline forms include agate, jasper, and onyx. Because of its piezoelectric properties, quartz is used for pressure gauges, oscillators, resonators, and wave stabilizes; because of its ability to rotate the plane of polarization of light and its transparency in ultraviolet rays, it is used in heat-ray lamps, prism, and spectrographic lenses. Also used in manufacturing glass, paints, abrasives, refractory materials, and precision instruments
Rare-earth elements (lanthanum, cerium, praseodymium, neodymium, promethium, samarium, europium, gadolinium, terbium, dysprosium, holmium, erbium, thulium ytterbium, and lutetium)	Used mainly in petroleum fluid cracking catalysts, metallurgical additives and alloys, glass polishing and ceramics, permanent magnets and phosphors. It is estimated that 40 lb of rare earths are used in a hybrid car for rechargeable battery, permanent magnet motor, and the regenerative braking system. The United States now has one rare-earth (bastnaesite) mine in California. More than 85% of global production is in China. The United States was import reliant for most of its rare-earth metals in 2012
Silica	Aluminum and aluminum alloy producers and the chemical industry are major users of silicon metal. Silica is also used in manufacture of computer chips, glass, and refractory materials; ceramics; abrasives; water filtration; component of hydraulic cements; filler in cosmetics, pharmaceutical, paper, and insecticides; anticaking agent in foods; flatting agent in paints; thermal insulator; and photovoltaic cells. China is the leading producer. The United States was 36% reliant on metallurgical grade silicon metal in 2012
Silver	Used in coins and medals, electric and electronic devices, industrial applications, jewelry, silverware, and photography. The physical properties of silver include ductility, electronics conductivity, malleability, and reflectivity. Used in lining vats and other equipments for chemical reaction vessels, water distillation, etc.; a catalyst in manufacture of ethylene; mirrors; silver plating; table cutlery; dental, medical, and scientific equipment; bearing metal; magnet windings; brazing alloys, solder. Also used in catalytic converters, cell phone covers, electronics, circuit boards, and bandages for wound care and batteries. Silver is produced in the United States at over 30 bases and precious metal mines primarily in Alaska and Nevada. The leading global producers include Mexico, China, Peru, Chile, Australia, Bolivia, and the United States. The United States was 57% reliant in 2012

Sodium carbonate (soda ash or trona)	Used in glass container manufacture, in fiberglass, and specialty glass; also used in production of flat glass, in liquid detergents, in medicine, as a food additive, photography, cleaning and boiler compounds, and pH control of water. Most US production comes from Wyoming. The United States is a major producer
Sulfur	Used in the manufacture of sulfuric acid, fertilizers, petroleum refining, and metal mining. Elemental sulfur and by-product sulfuric acid were produced in over 100 operations in 26 states and the Virgin Islands. The United States, Canada, China, and Germany are major producers
Tantalum	A refractory metal with unique electric, chemical, and physical properties used to produce electronic components and tantalum capacitors (in auto electronics, pagers, personal computers, and portable telephones) and for high-purity tantalum metals in products ranging from weapon systems to superconductors, high-speed tools, catalyst, sutures and body implants, electronic circuitry, and thin-film components. Used in optical glass and electroplating devices. Leading producers are Mozambique, Brazil, and Congo. The United States was 100% reliant in 2012
Titanium	Titanium mineral concentrates are used primarily by titanium dioxide pigment producers. A small amount is used in welding rod coatings and for manufacturing carbides, chemicals, and metals. It is produced in Florida and Virginia. Leading producing countries are South Africa, Australia, Canada, and China. The United States was 77% reliant in 2012. Titanium and titanium dioxide are used in aerospace applications (in jet engines, airframes and space, and missile applications). It is also used in armor, chemical processing, marine, medical, power generation, sporting goods, and other nonaerospace applications. Titanium sponge metal was produced in three operations in Nevada and Utah. The leading global producers are China, Japan, Russia, and Kazakhstan
Tungsten	More than half of the tungsten consumed in the United States was used in cemented carbide parts for cutting and wear-resistant materials, primarily in the construction, metalworking, mining, and oil- and gas-drilling industries. The remaining tungsten was consumed to make tungsten heavy alloys for applications requiring high density; electrodes, filaments, wires, and other components for electric, electronic, heating, lighting, and welding applications; steels, superalloys, and wear-resistant alloys; and chemicals for various applications. China is by far the leading producer. Russia, Canada, Austria, and Bolivia also produce tungsten. The United States produces very little. It was 42% import reliant in 2012

Continued

Table 6.5 Common Elements and Ores—cont'd

Uranium	Nearly 20% of America's electricity is produced using uranium in nuclear generation. It is also used for nuclear medicine, atomic dating, powering nuclear submarines, and other uses in the US defense system. The United States received 83% of its uranium from other countries in 2012
Vanadium	Metallurgical use, primarily as an alloying agent for iron and steel, accounted for about 93% of the domestic vanadium consumption. Of the other uses for vanadium, the major nonmetallurgical use was in catalysts for the production of maleic anhydride and sulfuric acid. China, South Africa, and Russia are the largest producers. The United States was 96% reliant in 2012
Zeolites	Used in animal feed, cat litter, cement, and aquaculture (fish hatcheries for removing ammonia from the water); water softener and purification; in catalysts; odor control; and for removing radioactive ions from nuclear plant effluent. The United States was not import reliant in 2012
Zinc	Of the total zinc consumed in the United States, about 55% was used in galvanizing, 21% in zinc-based alloys, 16% in brass and bronze, and 8% in other uses. Zinc compounds and dust were used principally by the agriculture, chemical, paint, and rubber industries. Major coproducts of zinc mining and smelting, in order of decreasing tonnage, were lead, sulfuric acid, cadmium, silver, gold, and germanium. Zinc is used as protective coating on steel, as die casting, as an alloying metal with copper to make brass, and as chemical compounds in rubber and paints and used as sheet zinc and for galvanizing iron, electroplating, metal spraying, automotive parts, electric fuses, anodes, dry cell batteries, nutrition, chemicals, roof gutter, engravers' plates, cable wrappings, organ pipes, and pennies. Zinc oxide used in medicine, paints, vulcanizing rubber, and sun block. Zinc dust used for primers, paints, and precipitation of noble metals and removal of impurities from solution in zinc electrowinning. US production is in three states and 13 mines. Leading producers are China, Australia, Peru, and the United States. The United States was 72% import reliant in 2012

Source: The US Geological Survey, Facts About Minerals (National Mining Association); Mineral Information Institute; Energy Information Administration.

6.3 RELEASE INTO THE ENVIRONMENT

For the purposes of this text, it is assumed that any inorganic chemicals released into the environment are hazardous chemicals. It is not only safe but also necessary to assume that any inorganic chemicals (except chemicals that are indigenous to the ecosystem into which they exist but in quantities that do not exceed the indigenous amounts) have the potential to be hazardous to the environment and to human health (Table 6.6).

Contamination by chemicals is a global issue, and there is no single company that should shoulder all of the blame. Past laws and regulations (or the lack thereof) allowed unmanaged disposal of chemicals and discharge of chemicals into the environment. These companies were not breaking the law; it is a matter of there being insufficient laws (the fault of various level of government) enacted to protect the environment. Moreover, the inappropriate management of such chemical waste has resulted on negative impacts on the environment. Thus, inorganic chemicals (with varying levels of toxicity) are found practically in all ecosystems on earth along with the adverse effects of these chemicals (i) on biodiversity, (ii) on agricultural production, and/or (iii) on water resources. At the end of the life cycles of the various chemical, the chemicals are recycled or sent for disposal as part of waste.

Briefly, environmental pollution by inorganic chemical environmental pollution can be classified using the following criteria according to the origin

Table 6.6 Chemical Safety Guidelines to Be Followed When Working With Chemicals

1. Assume that any unfamiliar chemical is hazardous and treat it as such
2. Know all the hazards of the chemicals with which you work
3. Never underestimate the potential hazard of any chemical or combination of chemicals. Consider any mixture or reaction product to be at least as hazardous as its most hazardous component
4. Never use any substance that is not properly labeled
5. Date all chemicals when they are received and again when they are opened
6. Follow all chemical safety instructions, such as those listed in material safety data sheets or on chemical container labels, precisely
7. Minimize human exposure to any chemical regardless of the hazard rating of the chemical and avoid repeated exposure
8. Use personal protective equipment (PPE), as appropriate for that chemical
9. Always have a colleague present and do not work alone when working with hazardous chemicals

of the factors, (i) natural pollution resulting from volcanoes, hurricanes, earthquakes, sand storms, and (ii) artificial pollution resulting from human activities such as industry, agriculture, and domestic activities (Table 6.7). A second method of general classifications can be made according to the type of the pollutants, (i) physical pollution, such as radiation and (ii) chemical pollution, such as by combustion products (carbon monoxide, carbon dioxide and nitrogen oxides, sulfur compounds, nitrates, phosphates, and heavy metals). The third and final method of pollution is a subdivision according to the polluted medium, (i) air pollution resulting from gases, powders from factories, vehicle emissions, and odors from agricultural activities; (ii) water pollution, resulting from the discharges of industrial residues such as metals and salts; and (iii) soil pollution, resulting from nonecological tourism, waste grounds, car cemeteries, foams, and insecticides.

The potential for emissions from the manufacture and use of inorganic chemicals is high, but because of economic necessity, the potential for emissions is reduced by the recovery of the chemicals. In some cases, the

Table 6.7 Examples of Inorganic Pollutants

Pollutant	Formula	Sources	Polluted medium
Sulfur dioxide	SO_2	Volcanoes	Air
		Industry	Air
		Transports	Air
Nitrogen dioxide	NO_2	Volcanoes	Air
		Industry	Air
		Transports	Air
Carbon monoxide	CO	Transports	Air
Ammonia	NH_3	Industry	Air, soil
		Agriculture	Air, soil
Hydrogen sulfide	H_2S	Industry	Soil, water
		anaerobic fermentation	Soil, water
Hydrogen chloride	HCl	Industry	Air
		Transports	Air
Hydrogen fluoride	HF	Industry	Air
Ammonium salts	NH_4^+	Farms, factories	Soil, water
Nitrate salts	NO_3^-	Farms, factories	Soil, water
Nitrite salts	NO_2^-	Farms, factories	Soil, water
Lead salts[a]	Pb	Heavy industry	Air
		Transports	Air
Mercury salts[a]	Hg	Industry	Soil, water
Zinc salts[a]	Zn	Industry	Air, water

[a]May also appear as the metal.

manufacturing process is operated as a closed system that allows little or no emissions to escape to the environment (air, water, or land). Emission sources from chemical processes include heaters and boilers; valves, flanges, pumps, and compressors; storage and transfer of products and intermediates; wastewater handling; and emergency vents. Regular maintenance of the process equipment reduces the potential for these emissions. However, the emissions that do reach the atmosphere from the inorganic chemical industry are generally gaseous emissions that are controlled by a sequence of gas cleaning operations, including an adsorption process or an absorption process. In addition, emission of particulate matter, which could also lead to an environmental issue, since the particulate materials emitted is usually extremely small (typically a collection of particulates in the micron range), also requires (and is subject to) efficient treatment for removal (Mokhatab et al., 2006; Speight, 2007, 2014, 2017; Kidnay et al., 2011; Bahadori, 2014).

As already stated (Chapter 1), of all the pollutants released into the environment by human activity, persistent inorganic pollutants (PIPs) among the most dangerous to environmental flora and fauna are (i) inorganic pesticides, (ii) various industrial inorganic chemicals, or (iii) the unwanted by-products of industrial inorganic processes that have been used for decades but have more recently been found to share several disturbing characteristics. These characteristics include (i) persistence, which means that the chemicals resist degradation in air, water, and sediments; (ii) bioaccumulation, which means that the chemicals accumulate in living tissues at concentrations higher than those in the surrounding environment; and (iii) long-range transport, which means that the chemicals can travel a considerable distance from the source of release through air, water, and migratory animals, often contaminating areas miles away from any known source. On the environmental side, persistent inorganic pollutants are highly toxic and long lasting and cause an array of adverse effects on flora and fauna.

Thus, many inorganic chemicals that have toxic, carcinogenic, mutagenic, or teratogenic (causing developmental malformations) effects on environmental flora and fauna are designated either as (i) *acutely hazardous waste* or as (ii) *toxic waste* by the US Environmental Protection Agency (https://www.epa.gov/hw). Substances found to be fatal to humans in low doses or in the absence of data on human toxicity have been shown to have an oral LD_{50} toxicity (lethal dose at 50% concentration) of <2 mg/L or a dermal LD_{50} of <200 mg/kg or are otherwise capable of causing or significantly contributing to an increase in serious irreversible or incapacitating reversible illness and are designated as *acute hazardous waste* (https://www.epa.gov/hw). Materials containing any of the toxic constituents listed are to be considered hazardous waste unless after considering the following factors it can reasonably be concluded (by the US Department of Environmental Health and Safety) that (i) the waste is not capable of posing a substantial present or potential hazard to public health or (ii) the waste is not capable of posing a substantial present or potential hazard to the environment when improperly treated, stored, transported or disposed of, or otherwise managed.

However, despite the nature of the environmental regulations and the precautions taken by the inorganic chemicals industry, the accidental release of nonhazardous inorganic chemicals and hazardous inorganic chemicals into the environment has occurred and, without being unduly pessimistic, will continue to occur (by all industries—not wishing to select any single industry as the only industry that suffers accidental release of inorganic chemicals into the environment). It is a situation that, to paraphrase, *chaos theory: no matter how well the preparation, the unexpected is always inevitable.* It is at this point that the environmental scientist and engineer must identity (through careful analysis) the nature of the chemicals and their potential effects on the ecosystem(s). Thus, the predominance of one inorganic chemical or any particular class of inorganic chemicals may offer the environmental scientist or engineer an opportunity for predictability of behavior of the chemical(s) after consideration of the chemical and physical properties of the chemical (Chapter 4) and the effect of the chemical of the floral and faunal species in an ecosystem.

Thus, when a spill of inorganic chemicals occurs, the primary processes determining the fate of inorganic chemicals are (i) dispersion, (ii) dissolution, (iii) emulsification, (iv) evaporation, (v) leaching, (vi) sedimentation, (vii) spreading, and (viii) by wind. These processes are influenced by the physical and properties of the inorganic chemicals (especially if the inorganic chemicals are constituents of a mixture), spill characteristics, environmental conditions, and chemical and physical properties of the spilled material after it is undergone any form of chemical transformation.

6.3.1 Dispersion

For the purposes of this text, the term *dispersion* encompasses all phenomena that give rise to the proliferation of inorganic chemicals through the man-made and natural environment. Thus, the disposal of a chemical into the water column and subsequent transportation of the is referred to as dispersion.

This not only is often a result of water surface turbulence but also may result from the application of chemical agents (dispersants). The dispersed chemicals may remain in the water column or coalesce with other chemicals (the same chemical or different chemicals) and gain enough energy to be adsorbed by minerals, soil, or sediment. In the context of the geosphere, dispersion is a process that occurs in soil that is particularly vulnerable to erosion by water. In a soil layer where clay minerals are saturated with sodium ions (*sodic soil*), the soil can break down very easily into fine particles and wash away. This can lead to a variety of soil and water quality problems, including (i) large losses of soil losses by gully erosion and tunnel erosion; (ii) structural degradation of the soil, clogging, and sealing where dispersed particles settle; and (iii) turbidity in water due to suspended soil particles, which also cause transportation of nutrients from the land. Dispersive soil is more common in older landscapes where leaching and illuviation processes have had more time to show an effect.

Briefly, gulley erosion involves creation of a gully created by running water, eroding sharply into soil and typically occurs on a hillside. On the other hand, tunnel erosion occurs in some soils where channels and tunnels develop beneath the surface. It is insidious as it is not readily noticeable (apart from small sediment fans at the tunnel discharge point) until the surface itself is undermined. Illuvium is a material displaced across a soil profile, from one layer to another, by the action of rainwater.

The dispersion of chemicals released into the environment has been the focus of much attention because of the realization that the dispersion behavior of inorganic chemicals can be markedly different when different chemicals are considered. Accidents that involve inorganic chemicals give rise to a new class of problems in dispersion prediction for the following reasons: (i) The material is, in almost all cases, stored as a solid but may also emit a gas after a spill and exposure to the air; (ii) the modes of release can vary widely, and geometry of the source can take many forms, and the initial momentum of the spill may be significant; and (iii) in some cases, a chemical transformation also takes place as a result of reaction with water vapor in the ambient atmosphere.

Emissions of many chemicals of concern occur into the air initially, from where they are dispersed into other media. Many chemicals emitted into the air, for instance, from combustion processes, tend to become associated with particulate matter. Removal from the air occurs through a range of complex processes involving photodegradation and particle sedimentation and/or precipitation (known, respectively, as *dry deposition* and *wet deposition*).

Inorganic chemicals may undergo several cycles of physical and/or chemical transformation (Chapter 7), which can also make chemicals more accessible to photochemical or biodegradation.

In addition, the physical properties of the inorganic chemical may result in one or more interactions with the surrounding ecosystem, especially if the chemical is reactive and has the potential to react quickly with the air, water, or soil. This reactivity will influence the dispersibility of the chemical, and moreover, if the release occurs over a short timescale, compared with the steady-state release characteristic of many chemical release problems, this can give rise to the complication of predicting dispersion for time-varying releases. There is also the uncertainty of individual predictions resulting from variability about the behavior of a mixture. Also, the dispersing chemical, which is typically denser than air, may form a low-level cloud that is sensitive to the effects of either natural or man-made obstructions in the surrounding topography.

Wind transport (aeolian transport—relocation by wind) can also occur and is particularly relevant when dust from inorganic solids (even coke dust that contains deposited inorganic chemicals) and inorganic catalyst dust are also considered. Dust becomes airborne when winds traversing arid land with little vegetation cover pick up small particles such as catalyst dust, coke dust, and other inorganic debris and send them skyward after which the movement of pollutants in the atmosphere is caused by transport, dispersion, and deposition. Dispersion results from local turbulence, that is, motions that last less than the time used to average the transport. Deposition processes, including precipitation, scavenging, and sedimentation, cause downward movement of pollutants in the atmosphere, which ultimately move the pollutants back to the ground surface but not necessary in the locals from which the pollutants originated.

6.3.2 Dissolution

Dissolution is the process whereby gases, liquids, or solids dissolve into or are dissolved into a liquid or other solvent by which these original states (gas, liquid, or solid) become a solute (dissolved component), forming a solution of the gas, liquid, or solid in the original solvent. Solid solutions are the result of dissolution of one solid into another (such as metal alloys) and occur where the formation is governed and described by the relevant phase diagram. In the case of a crystalline solid dissolving in a liquid, the crystalline structure must be disintegrated such that the separate atoms, ions, or molecules are released. For liquids and gases, the molecules must be able to form noncovalent intermolecular interactions with those of the solvent for a solution to form. Dissolution is of fundamental importance in all inorganic chemical processes especially the solubility of inorganic chemicals in water, which can be enhanced by the occurrent of acid rain.

Generally, most inorganic chemicals are polar and are usually soluble in water but insoluble in organic solvents (such as diethyl ether, dichloromethane, chloroform, and hexane). The solubility (dissolution) characteristics of inorganic molecules in water are complex and very much dependent upon the structure and properties of the inorganic chemical(s). The inorganic chemical may be a gas, a liquid, or a solid (the solute) that dissolves in the water; the solubility depends on the physical and chemical properties of the chemicals and on temperature, pressure, and the pH (acidity or alkalinity of the water).

Furthermore, the extent of solubility of the chemical can range from infinitely soluble (without limit) to poorly soluble (such as some lead salts); the term *insoluble* is often applied to poorly or very poorly soluble compounds, and (in the world of inorganic chemistry) a common threshold to describe an inorganic chemical as insoluble is a solubility <0.1 g/100 mL of water. However, solubility of an inorganic chemical in water should not be confused with the ability of water to dissolve the chemical because apparent solubility of the chemical might also occur because of a chemical reaction (*reactive solubility*).

Solubility of an inorganic chemical in water applies not only to environmental inorganic chemistry but also to areas of chemistry, such as (alphabetically) biochemistry, geochemistry, inorganic chemistry, and physical chemistry. In all cases, the solubility of the inorganic chemical depends on the physical conditions (temperature, pressure, and concentration of the chemical) and the enthalpy and entropy directly relating to the water and the chemicals concerned. Water is, by far, the most common solvent in inorganic chemistry and is a solvent for a wide range of inorganic chemicals, especially those chemicals that readily form ions. This is a crucial factor in which acidity and alkalinity play a role as in much of the area known as environmental inorganic chemistry.

In addition, the term *dissolved inorganic chemicals* (*or total dissolved solids*) is used as a broad classification for inorganic chemicals of varied origin and composition within aquatic systems (the aquasphere). The source of the dissolved inorganic chemical in freshwater systems and in marine systems depends on the body of water. When water contacts highly ionizable inorganic chemicals, these components can drain into rivers and lakes as a dissolved chemical. Whatever the source of the dissolved inorganic chemical, it is also extremely important in the transport of metals in aquatic systems; certain metals can form extremely strong metallic complexes with already dissolved inorganics, which enhances the solubility of the metal in aqueous systems while also reducing the bioavailability of the metal.

In terms of inorganic chemicals that are not naturally occurring, knowledge of the structure and properties can be used to examine relationships between the solubility properties of the inorganic chemical and its structure and vice versa. In fact, structure dictates function, which means that by knowing the structure of an inorganic chemical, it may be (but not always) possible to predict the properties of the chemical such as its solubility, acidity or basicity, stability, and reactivity. In the context of environmental distribution of a chemical, predicting the solubility of an inorganic molecule is a useful component of knowledge.

Dissolved inorganic compounds are generally associated with the various characteristics of drinking water (such as taste, smell, color, and odor), and the quality of drinking water, as perceived by the senses, largely determines the acceptability of water. For health-related contaminants—a contaminant that is unsafe for one is unsafe for all—the general characteristics (taste, smell, and odor) are subject to social, economic, and cultural considerations.

On the other hand, the presence of nondissolved suspended solids in water gives rise to turbidity. Suspended solids may consist of clay, silt, airborne particulates, colloidal organic particles, plankton, and other microscopic organisms. The presence of particulate matter in water not only may be offensive to the general senses but also can, in a more serious aspect, protect bacteria and viruses from the action of disinfectants. In addition, the adsorptive capacity of some suspended particulates can lead to entrapment of undesirable inorganic and organic compounds present in the water, and in this way, turbidity can bear an indirect relationship to the health aspects of water quality.

6.3.3 Emulsification

An emulsion is a mixture of two or more liquids that are usually immiscible. Examples include crude oil and water that can form an *oil-in-water emulsion*, wherein the oil is the dispersed phase and water is the dispersion medium, or a *water-in-oil emulsion*, wherein the water is the dispersed phase and the oil is the dispersion medium. The physical structure of an emulsion is based on the droplet size; most emulsions contain droplets with a mean diameter of more than around 1 μm (1 μm, 1×10^{-6} m); however, miniemulsions and nanoemulsions can be formed with droplet sizes in the 100–500 nm range (1 nm, 1×10^{-9} m), and with proper formulation, highly stable micro emulsion can be prepared having droplets as small as a few nanometers.

The stability of an emulsion refers to the ability of the emulsion to resist change in form and properties over time. There are three types of instability in emulsions: (i) flocculation, (ii) creaming, and (iii) coalescence. Flocculation occurs when there is an attractive force between the droplets, so they form a flocculant mass (flocs) in the fluid through precipitation or aggregation of suspended particles. Creaming occurs when the droplets rise to the top of the emulsion under the influence of buoyancy. Coalescence occurs when droplets collide and combine to form a larger droplet, so the average droplet size increases over time. Use of a surface-active agent (surfactant) can increase the stability of an emulsion so that the size of the droplets does not change significantly with time and the emulsion is then defined as a *stable emulsion.*

This demulsification process can be slowed with the help of emulsifiers that are surface-active substances with strong hydrophilic properties that are used to eliminate the prolonged effects of spills in which emulsions are formed. Emulsifiers help to stabilize emulsions and promote dispersing the dispersed phase oil to form microscopic (invisible) droplets, which accelerates may the decomposition of the pollutant in the water.

The formation of an emulsion based on water and inorganic chemicals is less likely that in the case of water and organic chemicals. Nevertheless, the properties of inorganic chemicals as a function of the solubility in water is important because of the potential use of the inorganic chemicals to break, for example, an oil-in-water emulsion that forms after a crude oil spill into the aquasphere and the potential to aid in the cleanup process.

6.3.4 Evaporation

Evaporation is the process by which water changes from a liquid to a gas or vapor and is the primary pathway that water moves from the liquid state back into the water cycle as atmospheric water vapor. Evaporation is the opposite of condensation, and sublimation is the phenomenon that occurs when a solid become a gas without the conversion of the solid to a liquid and thence the liquid to a gas.

$$\text{Evaporation : Liquid} \rightarrow \text{gas}$$

$$\text{Sublimation : Solid} \rightarrow \text{gas}$$

Both phenomena are important parts of the environmental inorganic chemical cycle since both processes involve disappearance of and inorganic contaminant in the water or on the land, *but* the contaminant does appear in the atmosphere. Of the two processes, the evaporation is the most common process since inorganic chemicals that go through the sublimation processes are not as obvious or as common as chemicals that evaporate. In the field of environmental inorganic chemistry, evaporation is more typical when water evaporated from an aqueous solution of the chemical rather than evaporation (volatilization) of the inorganic chemical.

Many factors affect the evaporation process. For example, if the air is already saturated with other chemicals (or the humidity is high when the water content is high), there is typically little chance of a liquid evaporating as quickly when the air is not saturated (or humidity is low). In addition, the air pressure also affects the evaporation process since, under high air pressure, an inorganic chemical is much more difficult to evaporate. Temperature also affects the ability of an inorganic chemical to evaporate.

6.3.5 Leaching

Leaching is a natural process by which water-soluble substances (such as water-soluble inorganic chemicals or hydrophilic inorganic chemicals) are washed out from soil or waste-disposal areas (such as a landfill). These leached out chemicals (leachates) can cause pollution of surface waters (ponds, lakes, rivers, and sea) and subsurface groundwater aquifers.

In the environment, leaching is the process by which inorganic chemicals or radionuclides are released from the solid phase into the water phase under the influence of mineral dissolution, desorption, and complexation processes as affected by pH, redox chemistry, and biological activity. The process itself is universal, as any material exposed to contact with water will leach components from its surface or its interior depending on the porosity of the material. In terms of the effect of pH (acidity or alkalinity), the leachability of a chemical (i.e., the ability of the chemical to be leached form the source) can be enhanced by the occurrence of acid rain, which can enhance the solubility of the chemical in the (now acidified) water.

Thus, leaching is the process by which (in the current context) inorganic contaminants are released from the solid phase into the water phase under the influence of dissolution, desorption, or complexation processes as affected by acidity or alkalinity (the pH value). The process itself is universal, as any material exposed to contact with water will leach components from its surface or its interior depending on the porosity of the material under consideration. Leaching often occurs naturally with soil contaminants such as inorganic chemicals with the result that the chemicals end up in potable waters.

Many inorganic chemicals occur in a mixture of different components in a solid (such as the metallic deposits on petroleum coke or metals in spent catalysts). In order to separate the desired solute constituent or remove an undesirable solute component from the solid phase, the solid is brought into contact with a liquid during which time the solid and liquid are in contact and the solute or solutes can diffuse from the solid into the solvent, resulting in separation of the components originally in the solid (*leaching, liquid-solid leaching*). In addition, leaching may also be referred to as *extraction* because the inorganic chemical is being extracted from the solid or the process may also be referred to as *washing* because a chemical is removed from a solid with water.

Leaching is affected by (i) the texture of the solid that holds the chemical to be leached, (ii) structure of the chemical to be leached, and (iii) water content of the solid that holds the chemical. If the solid holding the chemical is soil, then in terms of soil texture, for example, the proportions of sand, silt, and clay affect the movement of water through the soil. Coarse-textured soil contains more sand particles that have large pores, and the soil is highly permeable, which allow the water to move rapidly through the pore system. On the downside, inorganic chemicals (such as phosphate pesticides) carried by water through coarse-textured soil are more likely to reach and contaminate groundwater. On the other hand, clay-textured soils have low permeability and tend to retain more water and adsorb more inorganic chemicals from the water. This slows the downward movement of chemicals, helps increase the chance of degradation and adsorption to soil particles, and reduces the chance of groundwater contamination.

In terms of soil structure, loosely packed soil particles allow speedy movement of water through the soil, while tightly compacted soil holds water back and does not allow the water to move freely through it. Plant roots penetrate soil, creating excellent water channels when they die and rot away. These openings and channels may permit relatively rapid water movement even through clay-containing soil. On the other hand, the amount of water already in the soil has a direct bearing on whether rain or irrigation results in the recharging of groundwater and possible leaching of inorganic chemicals into an aquifer. Soluble chemicals are more likely to reach groundwater when the water content of the soil approaches or is at saturation. Saturation is typical in the spring when rain and snowmelt occurs, but when soil is dry, the added water just fills the pores in the soil near the soil surface, making it unlikely that the water will reach the groundwater supply.

Leaching processes introduce inorganic chemicals into the water phase by solubility and entrainment. In addition, the leaching processes that move inorganic chemicals from soil can have a variety of potential scenarios; the spill of inorganic chemicals onto soil may cause partitioning in water that has been in contact with the contamination.

6.3.6 Sedimentation or Adsorption

In the current context, sedimentation is the tendency for inorganic chemicals in suspension in water to settle out of the water and come to rest against a barrier. In geology, sedimentation is often used as the opposite of erosion, i.e., the terminal end of sediment transport. Settling is the falling of suspended particles through the liquid, whereas sedimentation is the termination of the settling process. Sedimentation may pertain to objects of various sizes, ranging from large rocks in flowing water to suspensions of dust that produce significant sedimentation. The term is typically used in geology to describe the deposition of sediment that results in the formation of sedimentary rock.

Thus, sedimentation occurs when particles in suspension to settle out of the fluid in which they are entrained and come to rest against a barrier, which is typically the basement of a waterway. Settling is due to the motion of the particles through the fluid in response to the forces acting on them, which in an ecosystem, can be due to gravity. The term *sedimentation* is often used as the opposite of erosion, i.e., the terminal end of sediment transport. Settling occurs when suspended particles fall through the liquid, whereas sedimentation is the termination of the settling process.

Sedimentation can be generally classified into three different types that are all applicable to inorganic chemicals. Type 1 sedimentation is characterized by particles that settle discretely at a constant settling velocity and typically these particles settle as individual particles and do not flocculate or stick to other during settling. Type 2 sedimentation is characterized by particles that flocculate during sedimentation and because of this the particle size is constantly changing and therefore their settling velocity is changing. Type 3 sedimentation (also known as zone sedimentation) involves particles that are at a high concentration (e.g., >1000 mg/L) such that the particles tend to settle as a mass and a distinct clear zone and sludge zone are present. Zone settling occurs in active sludge sedimentation and sedimentation of sludge thickeners.

In terms of the sedimentation of inorganic chemicals, some of the chemicals may be adsorbed on the suspended material (especially if the suspended material is clay or another highly adsorptive mineral) and deposited to the bottom. This mainly happens in shallow waters where particulate matter is abundant and water is subjected to intense mixing, usually through turbulence. Simultaneously, the process of biosedimentation can also occur when plankton and other organisms absorb the inorganic chemical. The suspended forms of the inorganic chemicals undergo intense chemical and biological (microbial) decomposition in the water. However, this situation radically changes when the suspended inorganic chemical reaches the lake bed, river bed, or seabed and the decomposition rate of the inorganic chemical(s) buried on the bottom abruptly drops. The oxidation processes slow down, especially under anaerobic conditions in the bottom environment and the inorganic chemical(s) accumulated inside the sediments can be preserved for many months and even years.

Although related to sedimentation, adsorption (a surface phenomenon) has some differences insofar as this process is the adhesion of an inorganic chemical from a gas, liquid, or dissolved solid to a surface; it might be considered as a sedimentation process that is encourage or enhanced by the properties of the chemical and the adsorbing properties of the surface to which the chemical is adsorbed. The (adsorption) process creates a film of the *adsorbate* on the surface of the *adsorbent.* This process differs from absorption in which a fluid (the *absorbate*) is dissolved by or permeates a liquid or solid (the *absorbent*), respectively. Adsorption is a surface-based process, while absorption involves the whole volume of the material. The term *sorption* encompasses both processes, while *desorption* is the reverse of it.

6.3.7 Spreading

Spreading is a form of dispersion in which the movement of IOCs through the subsurface is complex and is difficult to predict since different types of chemicals react differently with soils, sediments, and other geologic materials and commonly travel along different flow paths and at different velocity. Inorganic chemicals released from a source area (typically on the surface) infiltrate (through solubility in water) into the subsurface and migrate downward by gravity as an aqueous solution through the vadose zone (the zone that extends from the top of the ground surface to the water table). When low-permeability soil units are encountered, the contaminants in aqueous solution can also spread laterally along the permeability contrast.

Most IOCs are introduced into the subsurface by percolation through soil strata, and the interactions between the soil and the chemical are important for assessing the fate and transport of the contaminant in the subsurface, especially in the groundwater system. Contaminants that are highly soluble, such as salts and any potentially ionic derivatives, can move readily from surface soil to saturated materials below the water table and often occur during and after rainfall events. Those contaminants that may not be highly soluble may have considerably longer residence times in the surface strata (the soil zone). Some chemicals adsorb readily onto soil particles and slowly dissolve during precipitation events, resulting in migration. Once below the water table, IOCs are also subject to dispersion (mechanical mixing with uncontaminated water) and diffusion (dilution by concentration gradients). Many inorganic contaminants begin to spread immediately after entry into the environment.

6.3.8 Sublimation

Sublimation is the transition of a substance directly from the solid phase to the gas phase without passing through the intermediate liquid phase (Table 6.8 and Fig. 6.2). Sublimation is an endothermic phase transition that occurs at temperatures and pressures below the triple point of a chemical in the phase diagram. The reverse process of sublimation is the process of *deposition* in which some chemicals pass directly from the gas phase to the solid phase, again without passing through the intermediate liquid phase.

The term *sublimation* refers to a *physical* change of state and is not used to describe transformation of a solid to a gas in a chemical reaction. For example, the dissociation on heating of solid ammonium chloride (NH_4Cl) into

Table 6.8 Phase Transformations From Gas to Liquid to Solid and the Reverse

	To		
From	**Gas**	**Liquid**	**Solid**
Gas	N/A	Condensation	Deposition
Liquid	Evaporation[a]	N/A	Freezing
Solid	Sublimation	Melting	Transformation[b]

[a]Also, boiling.
[b]For example, change in crystal structure.

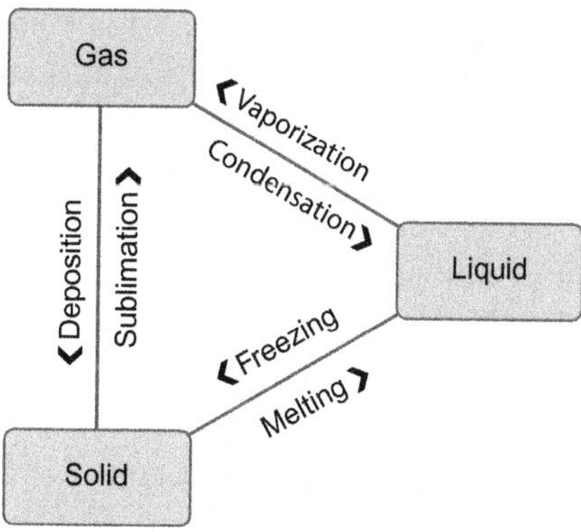

Fig. 6.2 Representation of phase changes.

ammonia (NH_3) and hydrogen chloride (HCl) is *not* sublimation but a chemical reaction:

$$NH_4Cl \rightarrow NH_3 + HCl$$

Sublimation requires additional energy and is an endothermic change, and the enthalpy of sublimation (also referred to as the *heat of sublimation*) can be calculated by adding the enthalpy of fusion and the enthalpy of vaporization.

6.4 TYPES OF CHEMICALS

As described above, inorganic chemicals can enter the air, water, and soil when they are produced, used, or disposed. The impact of these chemicals on the environment is determined by the amount of the chemical that is released, the type and concentration of the chemical, and where it is found. Some chemicals can be harmful if released to the environment even when there is not an immediate, visible impact. On the other hand, some chemicals are of concern as they can work their way into the food chain and accumulate and/or persist in the environment for many years.

Inorganic chemicals, in fact all chemicals, that enter the environment should be categorized and ranked using hazard assessment criteria. This not only would ensure that truly pressing environmental issues are identified and prioritized but also would maximize the use of limited resources. In the case of soluble inorganic chemicals, surrogate data such as persistence and bioaccumulation have been used, in combination with toxicity, for the purpose of hazard categorization. However, for insoluble or sparingly soluble inorganic chemicals such as metals and metal compounds, persistence and bioaccumulation are neither appropriate nor useful. Unfortunately, this is not always recognized by regulators or even by scientists.

The use of persistent, bioaccumulative, and toxic (PBT) criteria for chemicals was developed to address the hazards posed by synthetic organic chemicals. In fact, the criteria and test methods to evaluate persistence (i.e., the lack of degradability of a chemical) and bioaccumulation (the dispersion of a chemical through knowledge of the water-octanol partition coefficient) were developed to be used in combination with toxicity in order to reduce the importance given to the use of toxicity data alone. These test methods were based on an understanding of the chemistry of chemicals of concern at the time and of the biological interactions that the chemicals would have with the surrounding biota. Specifically, it was realized that if some chemicals exerted high intrinsic toxicity under standardized laboratory test conditions but did not persist or bioaccumulate, the environmental hazard of such chemicals would be lower.

As mentioned above, persistence is measured by determining the lack of degradability of a substance from a form that is biologically available and active to a form that is less available. This applies to many inorganic substances; inorganic metals and metal compounds tend to be in forms that are not bioavailable. Only under specific conditions would inorganic metals or metal compounds transform into a bioavailable form. Thus, rather than persistence, the key criterion for classifying inorganic metals and metal compounds should be their capacity to transform into bioavailable form(s). Furthermore, although bioavailability is a necessary precursor to toxicity, it does not inevitably lead to toxicity. Although inorganic metals and metal compounds stay in the environment for long periods of time, the risk they may pose generally decreases over time. For example, metals introduced into the aquatic environment are subject to removal/immobilization processes (e.g., precipitation, complexation, and absorption).

Similarly, the use of bioaccumulation has significant limitations for predicting hazard for inorganic metals and metal compounds. Generally, either bioconcentration factors (BCFs) or bioaccumulation factors (BAFs) are used for this purpose. A bioconcentration factors is the ratio of the concentration of a substance in an organism, following direct uptake from the surrounding environment (water), to the concentration of the same substance in the surrounding environment. A bioaccumulation factor considers uptake from food as well. In contrast to organic compounds, uptake of inorganic metals is not based on lipid partitioning. Further, organisms have internal mechanisms (homeostasis) that allow them to regulate (bioregulate) the uptake of essential metals and to control the presence of other metals. Thus, if the concentration of an essential metal in the surrounding environment is low and the organism requires more, it will actively accumulate that metal. This will result in an elevated bioconcentration factors (or bioaccumulation factor) value that, while of concern in the case of organic substances, is not an appropriate measure in the case of metals.

The primary determining factor of hazard for inorganic metals and metal compounds is therefore toxicity, which requires consideration of dose (indeed, the fundamental tenet of toxicology is *the dose makes the poison*). Historically, it has been the practice to measure the toxicity of soluble metal salts or indeed the toxicity of the free metal ion. However, in different media, metal ions compete with different types or forms of organic matter (e.g., fish gills, suspended solids, and soil particulate material) to reduce the total amount of metals present in bioavailable form. Toxicity of the bioavailable fraction (i.e., as determined through transformation processes) is the most appropriate and technically defensible method for categorizing and ranking the hazard of inorganic metals and metal compounds.

The relative proportion of hazardous constituents present in any collection of inorganic chemicals (crude-oil-derived products included) is variable and rarely consistent because of site differences. Therefore, the extent of the contamination will vary from one site to another, and in addition, the farther a contaminant progresses from low molecular weight to high molecular weight, the greater the occurrence of polynuclear aromatic hydrocarbons, complex ring systems (not necessity aromatic ring systems), and increase in the composition of the semivolatile inorganic chemicals or the nonvolatile inorganic chemicals. These latter inorganic chemical constituents (many of which are not as immediately toxic as the volatiles) can result in long-term chronic impacts to the flora and fauna of the environment. Thus, any complex mixture of inorganic chemicals should be analyzed for the semivolatile compounds, which may pose the greatest long-term risk to the environment.

Bioconcentration is the accumulation of a chemical in or on an organism and is also the process by which a chemical concentration in an organism exceeds that in the surrounding environment as a result of the exposure of the organism to the chemical. Bioconcentration can be measured and assessed and these include (i) octanol-water partition coefficient (K_{OW}), (ii) the bioconcentration factor, BCF, (iii) the bioaccumulation factor, BAF, and (iv) the biota-sediment accumulation factor, BSAF. Each of these factors can be calculated using either empirical data or measurements and from mathematical models. The bioconcentration factor can also be expressed as the ratio of the concentration of a chemical in an organism to the concentration of the chemical in the surrounding environment and is a measure of the extent of chemical sharing between an organism and the surrounding environment. Thus,

$$\mathrm{BCF} = (\text{concentration in biota})/(\text{concentration in ecosystem})$$

The bioconcentration factor can also be related to the octanol-water partition coefficient (K_{ow}), which is correlated with the potential for a chemical to bioaccumulate in floral and faunal organisms. The bioconcentration factor can be predicted from the octanol-water partition coefficient:

$$\text{Log}(\text{bioconcentration factor}) = m \log K_{\mathrm{OW}} + b$$

$$K_{\mathrm{OW}} = (\text{concentration in octanol})/(\text{concentration in water})$$
$$= C_{\mathrm{O}}/C_{\mathrm{W}} \text{ at equilibrium}$$

Heavy metals are common inorganic chemical pollutants. The most common heavy metals found at contaminated sites, in order of abundance are Pb, Cr, As, Zn, Cd, Cu, and Hg. Those metals are important since they are capable of decreasing crop production due to the risk of bioaccumulation and biomagnification in the food chain. There is also the risk of superficial and groundwater contamination. Knowledge of the basic chemistry, environment, and associated health effects of these heavy metals is necessary in understanding their speciation, bioavailability, and remedial options. The fate and transport of a heavy metal in soil depends significantly on the chemical form and speciation of the metal. Once in the soil, heavy metals are adsorbed by initial fast reactions (minutes, hours), followed by slow adsorption reactions (days, years) and are, therefore, redistributed into different chemical forms with varying bioavailability, mobility, and toxicity (Shiowatana et al., 2001). This distribution is believed to be controlled by reactions of heavy metals in soils such as (i) mineral precipitation and dissolution; (ii) ion exchange, adsorption, and desorption; (iii) aqueous complexation; (iv) biological immobilization and mobilization; and (v) plant uptake (Levy et al., 1992). The toxicity of metals varies greatly with pH, water hardness, dissolved oxygen levels, salinity, temperature, and other parameters. Physiological impacts occur at small concentrations.

The specific type of metal contamination found in a contaminated soil is directly related to the operation that occurred at the site. The range of contaminant concentrations and the physical and chemical forms of contaminants will also depend on activities and disposal patterns for contaminated wastes on the site. Other factors that may influence the form, concentration, and distribution of metal contaminants include soil and groundwater chemistry and local transport mechanisms.

6.5 PHYSICAL PROPERTIES AND DISTRIBUTION IN THE ENVIRONMENT

Pollution by inorganic chemicals and the distribution of these contaminants in the environment are a major concern. However, the movement of inorganic chemicals through the environment is complex and is difficult to predict. Different types of chemicals react differently with soils, sediments, and other geologic materials and commonly travel along different flow paths and at different velocities. For example, using soil as the example, contaminants that are highly soluble, such as salts (e.g., sodium chloride, NaCl), move readily from surface soils to saturated materials below the water table. This often occurs during and after rainfall events. Other contaminants that are not highly soluble may have considerably longer residence times in the soil zone. Some contaminants adsorb readily onto soil particles and slowly dissolve during precipitation events, resulting in dissolve fraction concentrations of contaminants migrating to groundwater. Once below the water table, inorganic chemicals are also subject to dispersion (mechanical mixing with uncontaminated water) and diffusion (dilution by concentration gradients).

Typically, inorganic chemicals released from various sources are ultimately dispersed among and can at times accumulate in various environmental compartments (such as soil and various floral and faunal species). Some contaminants may contribute primarily to environmental compartments on a local scale, but other contaminants that are more persistent in the environment can be distributed over much greater distances—even up to a regional scale, a national scale, or an international scale.

Thus, understanding the potential environmental impact of dispersed chemicals requires an understanding of the relative contribution of the various sources of the pollutants and the types of the (Chapter 5) potential for a pollutant to undergo chemical (or physical) transformation in the environment (Chapter 7). Therefore, an investigation of a potential contaminant must account for transport of the contaminant through ecosystems. The required characterization of concentrations of contaminants in an environmental medium, such as air, involves accounting for the gains (or inputs to) and losses from that medium and transport through it.

Some examples of atmospheric pollutants include nitrogen dioxide (NO_2), sulfur dioxide (SO_2), and carbon monoxide (CO). The first two pollutants combine with water to form acids, which not only irritate the lungs but also contribute to the long-term destruction of the environment due to the generation of acid rain (Chapter 7):

$$2[C]_{\text{fossil fuel}} + O_2 \rightarrow 2CO$$
$$[C]_{\text{fossil fuel}} + O_2 \rightarrow CO_2$$
$$2[N]_{\text{fossil fuel}} + O_2 \rightarrow 2NO$$
$$[N]_{\text{fossil fuel}} + O_2 \rightarrow NO_2$$
$$[S]_{\text{fossil fuel}} + O_2 \rightarrow SO_2$$
$$2SO_2 + O_2 \rightarrow 2SO_3$$

Carbon monoxide, generated by the incomplete combustion of hydrocarbons, displaces and prevents oxygen from binding to hemoglobin and causes asphyxiation. Also, it binds with metallic pollutants and causes them to be more mobile in air and water.

Fresh, clean, and drinkable water is a necessary but limited resource on the planet. Industrial, agricultural, and domestic wastes can contribute to the pollution of this valuable resource, and water pollutants can damage human and animal health. Three important classes of water pollutants are heavy metals, inorganic pollutants, and organic pollutants. Heavy metals include transition metals such as cadmium, mercury, and lead, all of which can contribute to brain damage. Inorganic pollutants like hydrochloric acid, sodium chloride, and sodium carbonate change the acidity, salinity, or alkalinity of the water, making it undrinkable or unsuitable for the support of floral and faunal species.

The use of pesticides (such as inorganic phosphates) in agriculture contributes to environmental pollution. Pesticides are used to control the growth of insects, weeds, and fungi, which compete with humans in the consumption of crops. This use not only increases crop yields and decreases grocery prices but also controls diseases such as malaria and encephalitis. However, the spraying of crops and the water runoff from irrigation transports these harmful chemicals to the habitats of nontarget animals. Chemicals can build up in the tissues of these animals, and when humans consume the animals, the increased potency of the pesticides is manifested as health problems and in some cases death. Chemists have recently developed naturally occurring pesticides that are toxic only to their particular targets and are benign to birds and mammals. The most significant pesticide of the 20th century was DDT (Chapter 1), which was highly effective as an insecticide but did not break down in the environment and led to the death of birds, fish, and some humans.

To effectively monitor changes in the environmental behavior of inorganic chemicals that are of most concern, it is extremely important to understand how these chemicals typically behave in natural systems and in specific ecosystems. Equally important is an understanding of how these chemicals might respond to specific best management practices. Some of the discharged inorganic chemicals only become problems at high concentrations that impair the beneficial uses of these ecosystems. An effective monitoring program explicitly considers how these chemicals may change as they move from a source into the groundwater, surface water, or into the soil. This includes an understanding of how a specific inorganic chemical may be introduced or mobilized within an ecosystem and how the chemical moves or through an ecosystem the transformations that may occur during this process.

As an example, groundwater will normally look clear and clean because the ground naturally filters out particulate matter. But, natural and human-induced chemicals can be found in groundwater (Tables 6.9 and 6.10). As groundwater flows through the ground, metals such as iron and manganese are dissolved and may later be found in high concentrations in the water. Industrial discharges, urban activities, agriculture, groundwater pumpage, and disposal of waste all can affect groundwater quality. Contaminants can be human-induced, as from leaking fuel tanks or toxic chemical spills. Pesticides and fertilizers applied to lawns and crops can accumulate and migrate to the water table. Leakage from septic tanks and/or waste-disposal sites also can introduce bacteria to the water, and pesticides and fertilizers that seep into farmed soil can eventually end up in water drawn from a well, or a well might have been placed in land that was once used for something like a garbage or chemical dump site.

However, the entry of chemicals into the environment and the distribution of these chemicals within the environment are often complex, and there have been many occasions when a significant amount of a chemical (or a mixture of chemicals) has entered an ecosystem and the effects of contamination are well defined. Generally, the assumption is that the inorganic chemical (or a mixture thereof) does not rapidly diffuse away but remains in the immediate vicinity at a noticeably high concentration or perhaps

Table 6.9 Physical Properties of Groundwater

Contaminant	Sources to groundwater	Potential health and other effects
Turbidity	Caused by the presence of suspended matter such as clay, silt, and fine particles of organic and inorganic matter, plankton, and other microscopic organisms. A measure how much light can filter through the water sample	Objectionable for esthetic reasons. Indicative of clay or other inert suspended particles in drinking water. May not adversely affect health but may cause need for additional treatment. Following rainfall, variations in groundwater turbidity may be an indicator of surface contamination
Color	Can be caused by decaying leaves, plants, organic matter, copper, iron, and manganese, which may be objectionable. Indicative of large amounts of organic chemicals, inadequate treatment, and high disinfection demand. Potential for the production of excess amounts of disinfection by-products	Suggests that treatment is needed. No health concerns. Esthetically unpleasing
pH	Indicates, by numerical expression, the degree to which water is alkaline or acidic. Represented on a scale of 0–14 where 0 is the most acidic, 14 is the most alkaline, and 7 is neutral	High pH causes a bitter taste; water pipes and water-using appliances become encrusted; depresses the effectiveness of the disinfection of chlorine, thereby causing the need for additional chlorine when pH is high. Low-pH water will corrode or dissolve metals and other substances
Odor	Certain odors may be indicative of organic or nonorganic contaminants that originate from municipal or industrial waste discharges or from natural sources	
Taste	Some substances such as certain organic salts produce a taste without an odor and can be evaluated by a taste test. Many other sensations ascribed to the sense of taste actually are odors, even though the sensation is not noticed until the material is taken into the mouth	

Table 6.10 Inorganic Contaminants Found in Groundwater

Contaminant	Sources	Effects
Aluminum	Occurs naturally in some rocks and drainage from mines	Can precipitate out of water after treatment, causing increased turbidity or discolored water
Antimony	Enters environment from natural weathering, industrial production, municipal waste disposal, and manufacturing of flame retardants, ceramics, glass, batteries, fireworks, and explosives	Decreases longevity, alters blood levels of glucose and cholesterol in laboratory animals exposed at high levels over their lifetime
Arsenic	Enters environment from natural processes, industrial activities, pesticides, and industrial waste, smelting of copper, lead, and zinc ore	Causes acute and chronic toxicity and liver and kidney damage; decreases blood hemoglobin. A carcinogen
Barium	Occurs naturally in some limestones, sandstones, and soils in the eastern United States	Can cause a variety of cardiac, gastrointestinal, and neuromuscular effects. Associated with hypertension and cardiotoxicity in animals
Beryllium	Occurs naturally in soils, groundwater, and surface water. Often used in electric industry equipment and components, nuclear power, and space industry. Enters the environment from mining operations, processing plants, and improper waste disposal. Found in low concentrations in rocks, coal, and petroleum and enters the ground	Causes acute and chronic toxicity; can cause damage to the lungs and bones. Possible carcinogen
Cadmium	Found in low concentrations in rocks, coal, and petroleum and enters the groundwater and surface water when dissolved by acidic waters. May enter the environment from industrial discharge, mining waste, metal plating, water pipes, batteries, paints and pigments, plastic stabilizers, and landfill leachate	Replaces zinc biochemically in the body and causes high blood pressure, liver and kidney damage, and anemia. Destroys testicular tissue and red blood cells. Toxic to aquatic biota

Continued

Table 6.10 Inorganic Contaminants Found in Groundwater—cont'd

Contaminant	Sources	Effects
Chloride	May be associated with the presence of sodium in drinking water when present in high concentrations. Often from saltwater intrusion, mineral dissolution, industrial, and domestic waste	Deteriorates plumbing, water heaters, and municipal waterworks equipment at high levels. Above secondary maximum contaminant level, taste becomes noticeable
Chromium	Enters environment from old mining operations runoff and leaching into groundwater, fossil-fuel combustion, cement-plant emissions, mineral leaching, and waste incineration. Used in metal plating and as a cooling-tower water additive	Chromium III is a nutritionally essential element. Chromium VI is much more toxic than chromium III and causes liver and kidney damage, internal hemorrhaging, respiratory damage, dermatitis, and ulcers on the skin at high concentrations
Copper	Enters environment from metal plating, industrial and domestic waste, mining, and mineral leaching	Can cause stomach and intestinal distress, liver and kidney damage, and anemia in high doses. Imparts an adverse taste and significant staining to clothes and fixtures. Essential trace element but toxic to plants and algae at moderate levels
Cyanide	Often used in electroplating, steel processing, plastics, synthetic fabrics, and fertilizer production; also from improper waste disposal	Poisoning is the result of damage to the spleen, brain, and liver
Dissolved solids	Occur naturally but also enters environment from man-made sources such as landfill leachate, feedlots, or sewage. A measure of the dissolved "salts" or minerals in the water. May also include some dissolved organic compounds	May have an influence on the acceptability of water in general. May be indicative of the presence of excess concentrations of specific substances not included in the Safe Water Drinking Act, which would make water objectionable. High concentrations of dissolved solids shorten the life of hot-water heaters

Table 6.10 Inorganic Contaminants Found in Groundwater—cont'd

Contaminant	Sources	Effects
Fluoride	Occurs naturally or as an additive to municipal water supplies; widely used in industry	Decreases incidence of tooth decay but high levels can stain or mottle teeth. Causes crippling bone disorder (calcification of the bones and joints) at very high levels
Hardness	Result of metallic ions dissolved in the water; reported as concentration of calcium carbonate. Calcium carbonate is derived from dissolved limestone or discharges from operating or abandoned mines	Decreases the lather formation of soap and increases scale formation in hot-water heaters and low-pressure boilers at high levels
Iron	Occurs naturally as a mineral from sediment and rocks or from mining, industrial waste, and corroding metal	Imparts a bitter astringent taste to water and a brownish color to laundered clothing and plumbing fixtures
Lead	Enters environment from industry, mining, plumbing, gasoline, coal, and as a water additive	Affects red blood cell chemistry; delays normal physical and mental development in babies and young children. Causes slight deficits in attention span, hearing, and learning in children. Can cause slight increase in blood pressure in some adults. Probable carcinogen
Manganese	Occurs naturally as a mineral from sediment and rocks or from mining and industrial waste	Causes esthetic and economic damage and imparts brownish stains to laundry. Affects taste of water and causes dark brown or black stains on plumbing fixtures. Relatively nontoxic to animals but toxic to plants at high levels
Mercury	Occurs as an inorganic salt and as organic mercury compounds. Enters the environment from industrial waste, mining, pesticides, coal, electric equipment (batteries, lamps, and switches), smelting, and fossil-fuel combustion	Causes acute and chronic toxicity. Targets the kidneys and can cause nervous system disorders

Continued

Table 6.10 Inorganic Contaminants Found in Groundwater—cont'd

Contaminant	Sources	Effects
Nickel	Occurs naturally in soils, groundwater, and surface water. Often used in electroplating, stainless steel and alloy products, mining, and refining	Damages the heart and liver of laboratory animals exposed to large amounts over their lifetime
Nitrate (as nitrogen)	Occurs naturally in mineral deposits, soils, seawater, freshwater systems, the atmosphere, and biota. More stable form of combined nitrogen in oxygenated water. Found in the highest levels in groundwater under extensively developed areas. Enters the environment from fertilizer, feedlots, and sewage	Toxicity results from the body's natural breakdown of nitrate to nitrite. Causes "blue-baby disease" or methemoglobinemia, which threatens oxygen-carrying capacity of the blood
Nitrite (combined nitrate/nitrite)	Enters environment from fertilizer, sewage, and human or farm-animal waste	Toxicity results from the body's natural breakdown of nitrate to nitrite. Causes "blue-baby disease" or methemoglobinemia, which threatens oxygen-carrying capacity of the blood
Selenium	Enters environment from naturally occurring geologic sources, sulfur, and coal	Causes acute and chronic toxic effects in animals—"blind staggers" in cattle. Nutritionally essential element at low doses but toxic at high doses
Silver	Enters environment from ore mining and processing, product fabrication, and disposal. Often used in photography, electric and electronic equipment, sterling and electroplating, alloy, and solder. Because of great economic value of silver, recovery practices are typically used to minimize loss	Can cause argyria, a blue-gray coloration of the skin, mucous membranes, eyes, and organs in humans and animals with chronic exposure

Table 6.10 Inorganic Contaminants Found in Groundwater—cont'd

Contaminant	Sources	Effects
Sodium	Derived geologically from leaching of surface and underground deposits of salt and decomposition of various minerals. Human activities contribute through deicing and washing products	Can be a health risk factor for those individuals on a low-sodium diet
Sulfate	Elevated concentrations may result from saltwater intrusion, mineral dissolution, and domestic or industrial waste	Forms hard scales on boilers and heat exchangers; can change the taste of water and has a laxative effect in high doses
Thallium	Enters environment from soils; used in electronics, pharmaceuticals manufacturing, glass, and alloys	Damages the kidneys, liver, brain, and intestines in laboratory animals when given in high doses over their lifetime
Zinc	Found naturally in water, most frequently in areas where it is mined. Enters environment from industrial waste, metal plating, and plumbing, and is a major component of sludge	Aids in the healing of wounds. Causes no ill-health effects except in very high doses. Imparts an undesirable taste to water. Toxic to plants at high levels

moves but in such a way that concentration levels of the chemical remain high as it moves. Such cases would normally occur when large quantities of a substance were being stored, transported, or otherwise handled in concentrated form. Thus, due to leakages, spills, improper disposal, and accidents during transport, inorganic compounds have become subsurface contaminants that threaten various ecosystems.

Background knowledge of the sources, chemistry, and potential risks of toxic heavy metals in contaminated soils is necessary for the selection of appropriate remedial options. Remediation of soil contaminated by heavy metals is necessary to reduce the associated risks, make the land resource available for agricultural production, enhance food security, and scale down land tenure problems. Immobilization, soil washing, and phytoremediation are frequently listed among the best available technologies for cleaning up heavy-metal-contaminated soils. These technologies are recommended for field applicability and commercialization in the developing countries also where agriculture, urbanization, and industrialization are leaving a legacy of environmental degradation (Wuana and Okieimen, 2011).

REFERENCES

Bahadori, A., 2014. Natural Gas Processing: Technology and Engineering Design. Gulf Professional Publishing, Elsevier, Amsterdam.

Kidnay, A.J., Parrish, W.R., McCartney, D.G., 2011. Fundamentals of Natural Gas Processing, second ed. CRC Press, Taylor & Francis Group, Boca Raton, FL.

Levy, D.B., Barbarick, K.A., Siemer, E.G., Sommers, L.E., 1992. Distribution and partitioning of trace metals in contaminated soils near Leadville, Colorado. J. Environ. Qual. 21 (2), 185–195.

Miller, D.R., 1984. In: Sheehan, P.J., Miller, D.R., Butler, G.C., Bourdeau, P. (Eds.), Effects of Pollutants at the Ecosystem Level: SCOPE 22. Scientific Committee on Problems of the Environment (SCOPE) of the International Council of Scientific Unions (ICSU). John Wiley & Sons Inc., Hoboken, NJ. http://dge.stanford.edu/SCOPE/SCOPE_22/SCOPE_22_front%20material.pdf.

Mokhatab, S., Poe, W.A., Speight, J.G., 2006. Handbook of Natural Gas Transmission and Processing. Elsevier, Amsterdam.

Shiowatana, J., McLaren, R.G., Chanmekha, N., Samphao, A., 2001. Fractionation of arsenic in soil by a continuous flow sequential extraction method. J. Environ. Qual. 30 (6), 1940–1949.

Speight, J.G., 2007. Natural Gas: A Basic Handbook. GPC Books, Gulf Publishing Company, Houston, TX.

Speight, J.G., 2014. The Chemistry and Technology of Petroleum, fifth ed. CRC Press, Taylor & Francis Group, Boca Raton, FL.

Speight, J.G., 2017. Handbook of Petroleum Refining. CRC Press, Taylor & Francis Group, Boca Raton, FL.

Wuana, R., Okieimen, F.E., 2011. Heavy metals in contaminated soils: a review of sources, chemistry, risks and best available strategies for remediation. ISRN Ecol.. 2011(2011) 402647. http://dx.doi.org/10.5402/2011/402647https://www.hindawi.com/journals/isrn/2011/402647/.

CHAPTER SEVEN

Transformation of Inorganic Chemicals in the Environment

7.1 INTRODUCTION

Inorganic chemicals can be emitted directly into the environment or formed through chemical reactions of so-called precursor species. The inorganic pollutants can undergo chemical reactions converting highly toxic substances to less toxic or inert products and convert nontoxic chemicals or chemicals of low toxicity to more chemicals of high toxicity. Thus, exposure to hazardous inorganic chemicals can occur on timescales that range from minutes to days or even weeks, and a variety of chemistry and physical processes can play an important role in defining the effect of the exposure.

Within the major groups of inorganic pollutants, the principal focus to understand the reactions for the degradation or transformation of these chemicals requires (i) the knowledge of the type of pollutants and (ii) the reactivity of the pollutants so that the design of the relevant method to remove the pollutants from the environment can be achieved. Although much emphasis has been and continues to be placed on biotic reactions carried out by the various biota (such as bacteria), important transformation reactions of the pollutants in the environment must not be omitted. Some of these chemical reactions will be beneficial—in terms of reduced toxicity and pollutant removal, while other chemical reactions may have adverse effects on pollutant removal by the opposite effect in which the pollutant is converted to a product that is more capable of remaining in the environment and may even prove to be persistent inorganic chemical. Thus, emphasis must be placed on the occurrence of partial degradation of full degradation of the inorganic chemical and the role of any intermediate products that are toxic to floral and faunal organisms, inhibit further degradation, or have adverse effects on the environment.

The chemical transformation of an inorganic chemical in the environment is an issue that needs to be given serious consideration because of

Environmental Inorganic Chemistry for Engineers
http://dx.doi.org/10.1016/B978-0-12-849891-0.00007-2

the chemical and physical changes (often nonbenign) that can occur to the chemical. It would be unusual of the chemical transformation that did not show some effect on the properties of the discharged chemical. Thus, the chemical transformations of inorganic chemicals released into the environment are, in the context of this book, considered to be the transformation of the released chemical into a product that is still of environmental concern in terms of toxicity. Furthermore, knowledge of the relative amounts of each species present is critical because of the potential for differences in behavior and toxicity (including the possibility of enhanced toxicity), which are of concern when examining the potential fate of such chemicals.

As used here, the term *fate* refers to the ultimate disposition of the inorganic chemical in the ecosystem, either by chemical or biological transformation to a new form that (hopefully) is nontoxic (degradation) or, in the case of an ultimately persistent inorganic pollutants, by conversion to a less offensive chemicals or even by sequestration in a sediment or other location that is expected to remain undisturbed. This latter option—the sequestration in a sediment or other location—is not a viable option as for safety reasons that the chemical must be dealt with at some stage of its environmental life cycle. Using the old adage *bad pennies always turn up* can also be applied to a hidden chemical, and it is likely to manifest its presence (usually by an adverse action) at some future date. In summary, hiding the chemical away on paper (a note in a file giving the written but unproven or hypothetical reason why the chemical is considered to be of limited danger) is not an effective way of protecting the environment. However, for inorganic chemicals that are effectively degraded in nature, whether by hydrolysis, photolysis, microbial degradation, or other chemical transformation in the ecosystem, it is necessary (even essential) to collect, tabulate, and store (for ready retrieval) any information related to the chemical reaction parameters that can serve as indicators of the processes and the rates at which transformation (i.e., degradation) would occur.

Moreover, inorganic chemicals are subject to two processes that determine the fate of the chemical in the environment: (i) the potential for transportation of the chemical and (ii) the chemical changes that can occur once the chemical has been released to the environment and that depend upon a variety of physical and chemical properties (Table 7.1) some of which may lead to corrosion (Table 7.2). Thus, release into the environment of a persistent inorganic pollutant leads to an exposure level that ultimately depends on the length of time the chemical remains in circulation and how many times it is recirculated in some sense, before ultimate termination of the

Table 7.1 Various Physical and Chemical Properties that Influence the Behavior of an Inorganic Chemical in the Environment

Chemical transport processes	***Chemical fate processes***
• Erosion • Leaching • Movement in water system streams, rivers, or groundwater • Runoff • Wind	• Biological processes • Sorption • Transport • Transformation/degradation • Volatilization
Transformation and degradation processes	
• Biological transformations due to microorganism **(1)** Aerobic processes **(2)** Anaerobic processes	
Physical properties	***Chemical properties***
Boiling point Color Density Electrical conductivity Hardness Melting point Solubility Sublimation point State (gas, liquid, solid) Vapor pressure	Ability to act as an oxidizing agent Ability to act as a reducing agent Absorption into a liquid or porous solid Adsorption to a surface Can cause corrosion Decomposition into lower-molecular-weight chemicals Reaction with acids Reaction with alkalis (bases) Reaction with other chemicals Reaction with oxygen (oxidation) Reaction with oxygen (combustion) Decomposition into lower-molecular-weight chemicals

environmental life cycle of the chemical—the same rationale applied to product formed from the pollutant by any form of chemical transformation. In addition, the potential for transportation and chemical change (either before or after transportation) raises the potential for the chemical to behave in an unpredictable manner.

A particular question that needs to be addressed more often for persistent inorganic pollutants relates to the fraction that remains in circulation (until the end of the life cycle) and the means by which the environmental existence of the inorganic chemical can be terminated as expeditiously as possible and without further harm to the environment. The findings may not

Table 7.2 Typical Corrosive Inorganic Chemicals

Acidic corrosives
Inorganic acids
Hydrochloric acid (HCl)
Nitric acid (HNO_3)
Sulfuric acid (H_2SO_4)
Alkaline or basic corrosives
Sodium hydroxide (NaOH)
Potassium hydroxide (KOH)
Corrosive dehydrating agents
Phosphorous pentoxide (P_2O_5)
Calcium oxide (CaO)
Corrosive oxidizing agents
Halogen gases (F_2, Cl_2, Br_2)
Hydrogen peroxide (H_2O_2, concentrated)
Perchloric acid ($HClO_4$)

always be positive but must be given serious consideration in terms of an as-near-as-possible complete removal of the inorganic chemical and any products of chemical transformation.

Furthermore, in the present context, inorganic chemicals in the environment are of concern because of the high potential for toxicity to a wide variety of floral and faunal species. Many inorganic chemicals are well known for their adverse effects on flora and fauna at high levels of exposure. These chemicals typically have no known essential role in the human body, and these nonessential chemicals are tolerated, at very low exposure, with little, if any, adverse effect, but at higher exposure, their toxicity is exerted, and health consequences become obvious. In between the low exposure and high exposure limits, there may be an acceptable range of exposure within which the floral and faunal species are able to regulate an optimum level of the element; the level will be species-dependent. Generally, it is safer to assume that inorganic chemicals (other than those prescribed by a physician in regulated dosages) are harmful to floral and faunal species and should be treated/handled accordingly.

When an inorganic chemical (or a mixture of inorganic chemicals) is released into the environment, the issues that need to be considered are (i) the toxicity of the inorganic chemical, (ii) the concentration of the released inorganic chemical, (iii) the concentration of the toxic inorganic chemical in the mixture if a mixture is released, (iv) the potential of the inorganic chemical to migrate to other sites, (v) the potential of the inorganic chemical to produce a toxic degradation product, (vi) whether or not the toxicity is lower or higher than the toxicity of the released chemical, (vii) the potential of the toxic degradation product or products to migrate to other sites, (viii) the persistence of the inorganic chemical in an ecosystem, (ix) the persistence of any toxic degradation product in an ecosystem, (x) the potential for the toxic degradation product to degrade even further into harmful or nonharmful constituents and the rate of degradation, and (xi) the degree to which the chemical or any degradation product of the chemical can accumulate in an ecosystem. Other factors that may be appropriate because of site specificity may also be considered; this list *is not meant to be complete* but does serve to indicate the types of issues that must be given serious consideration preferably before a spill or discharge of an inorganic chemical into the environment.

Thus, in order to complete such a list and monitor the behavior and effects of inorganic chemicals in an ecosystem, an understanding of chemical transformation processes in which a disposed or discharged chemical might particulate is valuable to any study of the effects on the environment. Typically, chemical transformation processes change the chemical composition and structure of the discharged chemical, which can change the properties (and possibly the toxicity) of the chemical and influence behavior and life cycle of the chemical in the environment.

Examples of chemical transformation of inorganic chemicals in the environment where weathering processes (such as oxidative processes and physical-change processes) are ever-present include phenomena such as (i) evaporation and sublimation, in which part of the chemical or part of a mixture can vaporize into the atmosphere; (ii) leaching, in which part of the chemical or part of a mixture is transferred to the aqueous phase through dissolution; (iii) entrainment, in which part of the chemical or part of a mixture is subject to physical transport along with the aqueous phase; (iv) chemical oxidation, in which part of the chemical or part of a mixture is oxidized to a derivative or to completely different chemical; and (v) microbial degradation, in which part of the chemical or part of a mixture is converted by bacteria to a derivative or to completely different chemical. Furthermore, the rate of transformation of the chemical is highly dependent on environmental conditions. Unfortunately, the database on such transformations and the available on the composition of spilled chemicals that have been transformed in the environment are limited.

However, the various chemical transformation processes, which influence the presence and the analysis of inorganic chemicals at a particular site, although often represented by simple (and convenient) chemical equations, can be very complex, and the true nature of the chemical transformation process is difficult to elucidate. The extent of transformation is dependent on many factors including (i) the properties of the chemical; (ii) the geology of the site; (iii) the climatic conditions, such as temperature, oxygen levels, and moisture; (iv) the type of microorganisms present at the site; and (v) any other environmental conditions that can influence the life cycle of the chemical. In fact, the primary factor controlling the extent of chemical transformation is the molecular composition of the inorganic chemical contaminant.

However, it must be reemphasized that an inorganic chemical deposited into the environment has a potential to undergo transformation (unless the chemical is a persistent inorganic pollutant) to another chemical form, which is still of concern in terms of toxicity. Moreover, when released into the environment, the fate of inorganic chemicals depends on the physical and chemical properties of the compound(s) and the ability of these chemicals to undergo transformation to products—that is, the reactivity of the chemical. In addition, it is not only the structure of the chemical deposited into the environment but also the chemical forms that can result from the chemical transformations that are the result of the chemicals undergoing weathering (oxidation) and other environmental effects that cause change to the chemical structure. Thus, inorganic chemicals that are not directly toxic to environmental floral species and faunal species (including humans) at current environmental concentrations can become capable of causing environmental damage after chemical transformation has occurred.

Thus, the focus in this chapter is to present a fundamental understanding of the nature of these chemical processes, so that activities that have an effect on the environment and chemistry can be presented.

7.2 CHEMICAL AND PHYSICAL TRANSFORMATION

In the chemical sciences, a transformation is the chemical or physical conversion of a chemical (the substrate) to another chemical (the product). A typical chemical process generates products and wastes from raw materials such as substrates and excess reagents. If most of the reagents and the solvent can be recycled, the mass flow looks quite different, and the prevention of waste (to be disposed into the environment) can be achieved if most of the reagents can be recycled. Furthermore, there has been the suggestion that an efficiency factor (E-factor, i.e., the mass efficiency in terms of mass of the reactants) of a chemical process can be used to assess the transformation of a chemical in the environment:

$$\mathrm{E-factor} = (\text{mass of original waste})/(\text{mass of transformed product})$$

Alternatively,

$$\mathrm{E-factor} = (\text{mass of raw feedstock} - \text{mass of product})/\text{mass of product}$$

Typically, E-factors used in the manufacturing processes for bulk and fine chemicals give reliable data, but in those processes, the conditions are controlled, and any examination of amounts and properties can be achieved relatively conventionally (Chapter 6). The examination of chemical properties and chemical transformation in an ecosystem is much more difficult, and the E-factors may, more than likely, not be as meaningful. In addition, any such E-factor and any related factors do not account for any type of toxicity of the chemical waste. Efforts (to determine an E-factor) are at a very preliminary stage, and the parameters for the calculation of these factors need much more consideration and development before being applied to chemical wastes in the environment.

Reliable estimation of the behavior of inorganic chemicals in the environment is not a simple task and may only experience rare success. While acknowledging that the use of E-factors is a step in the right direction, it appears that there are too many unknowns and the unreliability of the environmental reaction parameters needs to be finely tuned to provide a much needed tool for estimating the behavior of chemicals under such variable conditions. In view of the lack of any estimation of satisfactory error limits for the produced data, an arbitrary outer difference limit may not be sufficiently accurate to define whether or not the data are satisfactory and of any use.

7.2.1 Chemical Transformation

An inorganic chemical transformation involves changing an inorganic chemical (or chemicals) from a beginning chemical (the reactant or reactants) mass to a different resulting (the product or products) chemical substance. The transformation produces new chemicals that have different physical and chemical properties when compared with the physical and chemical properties of the starting material(s). Such changes are usually irreversible in environment because the newly formed chemical(s) cannot easily change back into the original chemical(s). In the laboratory, chemical changes can easily be identified with the help of change in color, odor, energy level, and physical state. Any chemical change can easily be represented with the help of chemical equations, which is a simple representation of a chemical reaction and involves the molecular or atomic formulas of the reactants and products. On the basis of cleavage and formation of chemical bonds, most inorganic chemical reactions can be classified into four broad categories: (i) combination reactions, (ii) decomposition reactions, (iii) single-displacement reactions, and (iv) double-displacement reactions. Thus,

Combination reaction:

$$A + B \rightarrow AB$$
$$2Na(s) + Cl_2(g) \rightarrow 2NaCl(s)$$
$$8Fe + S8 \rightarrow 8FeS$$
$$S + O_2 \rightarrow SO_2$$

Decomposition reaction:

$$AB \rightarrow A + B$$
$$2HgO \rightarrow 2Hg + O_2$$

Single-displacement reaction:

$$A + BC \rightarrow AC + B$$
$$Mg + 2H_2O \rightarrow Mg(OH)_2 + H_2$$
$$Cu(s) + 2AgNO_3(aq) \rightarrow 2Ag(s) + Cu(NO_3)_2(aq)$$
$$Zn(s) + CuSO_4(aq) \rightarrow Cu(s) + ZnSO_4(aq)$$

Double-displacement reaction:

$$AB + CD \rightarrow AD + CB$$
$$Pb(NO_3)_2 + 2KI \rightarrow PbI_2 + 2KNO_3$$
$$CaCl_2(aq) + 2AgNO_3(aq) \rightarrow Ca(NO_3)_2(aq) + 2AgCl(s)$$

Double-displacement reaction—acid-base reaction:

$$HA + BOH \rightarrow H_2O + BA$$
$$HBr(aq) + NaOH(aq) \rightarrow NaBr(aq) + H_2O(l)$$
$$HCl(aq) + NaOH(aq) \rightarrow NaCl(aq) + H_2O(l)$$

These reactions are equally likely to occur in the environment as in the laboratory and can occur on a regular basis once an inorganic chemical is released into an ecosystem. However, the rate of the reaction does depend upon the chemical and the conditions that exist in the ecosystem.

One reaction that is often omitted from the above is the *combustion reaction*. While this type of reaction is typically assigned plied to the effect of organic chemicals on the environment, the combustion reaction does (using butane as the example) produce inorganic gases that are of serious environmental concern, such as carbon dioxide (complete combustion) and carbon monoxide, with some particulate matter such as soot or carbon (incomplete combustion). Thus,

$$2C_4H_{10} + 13O_2 \rightarrow 8CO_2 + 10H_2O \quad \text{(complete combustion)}$$
$$C_4H_{10} + 3O_2 \rightarrow CO + \; + 3C + 5H_2O \quad \text{(incomplete combustion)}$$

In the environment, a chemical transformation is typically based on the same principle as in the laboratory or in any of the chemical process industries—the transformation of a substrate to a product—but whether or not the product is benign and less likely to harm the environment (relative to the substrate) or is more detrimental by exerting a greater impact on the environment depends upon the origin, properties, and reactivity pathways of the starting substrate. Thus, a chemical transformation requires a chemical reaction to lead to the transformation of one chemical substance to another (Habashi, 1994). Typically, chemical reactions encompass changes that only involve the positions of electrons in the forming and breaking of chemical bonds between atoms, with no change to the nuclei (no change to the elements present), and can often be described by a relatively simple chemical equation; thus,

$$A + B \rightarrow C$$

However, although the various chemical transformation processes that occur and influence the presence and the analysis of inorganic chemicals at a particular site are often represented by simple (and convenient) chemical equations, the chemical transformation reactions can be much more complex than the equation indicated.

The inorganic chemical (substrate and reactant) initially involved in a chemical reaction (the reactant) is usually characterized by a chemical change and yields one or more products, which usually have properties different from the original reactant. Reactions often consist of a sequence of individual (and often complex) substeps, and the information on the precise course of action that is part of the reaction mechanism is not always clear. Chemical reactions typically occur under a specific set of parameters (temperature, pressure, chemical concentration, time, and ratios of reactants) and (under these parameters) at a characteristic reaction rate. Typically, reaction rates increase with increasing temperature because there is more thermal energy available to reach the activation energy necessary for breaking the bonds between the constituent atoms. The general rule of thumb is that for every 10°C (18°F) increase in temperature, the rate of an inorganic chemical reaction is doubled.

In addition, a chemical reaction may proceed in the forward direction and processed to completion and in the reverse direction until the reactants and products reach an equilibrium state:

$$A + B \rightarrow C + D$$

$$C + D \rightarrow A + B$$

Thus,

$$A + B \leftrightarrow C + D$$

Reactions that proceed in the forward direction to approach equilibrium are often described as spontaneous, requiring no input of free energy to go forward. On the other hand, nonspontaneous reactions require input of free energy to go forward (e.g., application of heat for the reaction to proceed). In inorganic chemical synthesis, different chemical reactions are used in the combinations during the reaction in order to obtain a desired product. Also, in inorganic chemistry, a consecutive series of chemical reactions (where the product of one reaction is the reactant of the next reaction) are often encourage to proceed by a variety of catalysts that increase the rates of biochemical reactions without themselves (the catalysts) being consumed in the reaction, so that syntheses and decompositions impossible under ordinary conditions can occur at the temperatures, pressures, and reactant concentrations present within a reactor and, by inference, in the current content within the environment:

$$A + B \rightarrow C$$

$$C \rightarrow D + E$$

This simplified equation does not truly illustrate the potential complexity of an inorganic chemical reaction, but such complexity must be anticipated when an inorganic chemical is transformed in an environmental ecosystem. Mother Nature, a tricky old lady, can often encourage or instigate complexity!

A physical conversion that is often ignored (or not recognized as such) is the change in composition that occurs when precipitation and dissolution reactions exert a major effect on the concentrations of inorganic ions in solution. The precipitation of a solid phase in environmental systems rarely results in a pure mineral phase. Minor elements, such as cadmium, can coprecipitate with major elements such as calcium to form a solid solution. In addition, the equilibrium activity of the component of a solid solution does not correspond to the solubility calculated from the solubility products of pure minerals. Furthermore, it is often impossible to distinguish chemically between a minor component coprecipitated in a solid and the adsorption of that minor component onto the solid surface.

Also, complexation is an important process that will determine in some cases if mineral solubility limits are reached, the amount of adsorption that occurs and the redox state that exists in the water. Inorganic chemicals can also form stable, soluble complexes with organic ligands. Ligands in leachates could include synthetic chelating agents (such as ethylenediamine), partially oxidized biodegradation products (such as organic acids), or natural humic materials. In fact, the movement of metals in the subsurface is strongly influenced by the concentration and chemical properties of ligands. However, the types and concentrations of ligands in most waste leachates are largely unknown.

To predict the persistency of a chemical in the environment, the physical-chemical properties and reactivity of the inorganic chemical in the environment need to be known or at least estimated. The chemicals that react in the environment are not likely to persist in the environment, while those that did not show any observable reactivity may persist for a very long time.

Finally, it must be recognized that there are certain limitations of chemical equations. While the phenomena described in the following phrases may be identified in the laboratory, they are more difficult to identify as occurring in the environment. For example, (i) the chemical equation does not usually clarify the state of the substances, and (s) for solid, (l) for liquid, (g) for gas, and (v) for vapor may have to be added; (ii) the reaction may or may not be concluded, but the equation does not reveal it; (iii) the chemical equation does not give any information about the rate (speed) of the reaction; (iv) the chemical equation does not give the concentration of the substances, and in some cases, the terms like diluted and concentrated are used; (v) the chemical equation does not give the general parameter of the reaction such as temperature, pressure, and whether or not a catalyst is necessary; (vi) the chemical equation will not give any idea about color changes during the exchange, which has to be mentioned separately; and (vii) the chemical equation will never give any indication regarding the production or absorption of heat. This is mentioned separately.

7.2.2 Physical Transformation

On the other hand, many processes arising in chemical science and technology involve physical transformations in addition to or even without chemical reactions, such as phase change and adsorption onto a solid surface. Most mechanistic work has focused on chemical reactions in solution or extremely simple processes in the gas phase. There is increasing interest in reactions in solids or on solid surfaces, such as the surfaces of solid catalysts in contact with reacting gases. Some such catalysts act inside pores of defined size, such as those in zeolites. In these cases, only certain molecules can penetrate the pores to get to the reactive surface, and they are held in defined positions when they react. Perhaps, the saving grace is that in the environment there is the potential that physical changes can be reversed, whereas chemical changes cannot be reversed with the substance changed back without extraordinary means, if at all. However, but reversibility is not always a certain criterion for classification between chemical and physical changes.

Physical changes are changes affecting the form of an inorganic chemical but not always the chemical composition. Physical changes are used to separate mixtures into their component compounds, such as the use of liquid-solid chromatography, but cannot usually be used to separate compounds into chemical elements or simpler compounds. Thus, physical changes occur when objects or substances undergo a change that does not change their chemical composition. This contrasts with the concept of chemical change in which the composition of a substance changes or one or more substances combine or break up to form new substances.

A physical change involves a change in physical properties such as melting point or melting range, transition to another phase, change in crystal form, and changes in color, volume, and density. Many physical changes also involve the rearrangement of atoms most noticeably in the formation of crystals. Although chemical changes may be recognized by an indication such as odor, color change, or production of a gas, everyone of these indicators can result from physical change.

Physical transformations between states of matter, not necessarily involving chemical reaction, can occur on changes in temperature or pressure or application of external forces or fields. Such phase transitions have been central to quantitative research in chemical sciences. Suitable choices or changes in temperature, pressure, and other controllable properties can produce abrupt changes (first-order transitions) in the state of a substance: for example, boiling or freezing or the formation of systems having two phases coexisting with distinct phase boundaries between the phases.

The importance of physical transformations in causing changes to spills or chemical disposal implies that behavior of the spilled material can change and requires an understanding of the behavior of the various constituents of the mixture. For example, absorption of part of the spillage into water or adsorption onto a mineral (clay) surface can cause part of the spilled material to be retained in the water or on the soil. Control over such forces is difficult, and the selective absorption or adsorption of any part of a spilled chemical mixture can cause changes in the effect of the nonabsorbed or nonadsorbed chemical on the environment. The converse is also true insofar as the selective absorption or adsorption of any part of a spilled chemical mixture can cause changes in the effect absorbed or adsorbed chemical on the environment. And in either cases, the environmental effects can be considerable.

A common example that we will treat as a physical transformation is the settling of suspended sediment particles. Although settling does not actually transform the sediment into something else, it does remove sediment from our control volume by depositing it on the riverbed. This process can be expressed mathematically by heterogeneous transformation equations at the riverbed; hence, we will discuss it as a transformation.

Another example of a physical change comes from the field of nuclear physics and is radioactive decay. Radioactive decay is the process by which an atomic nucleus emits particles or electromagnetic radiation to become either a different isotope of the same element or an atom of a different element. The three radioactive decay paths are alpha decay (the emission of a helium nucleus), beta decay (the emission of an electron or positron), and gamma decay (the emission of a photon). Gamma decay alone does not result in transformation, but it is generally accompanied by beta emission, which does.

7.3 INORGANIC REACTIONS

Inorganic chemicals, such as metals, compounds, and radionuclides, are not readily digested and destroyed by the flora and fauna in an ecosystem and exhibit variable behavior in the environment (Summers and Silver, 1978). Thus, with inorganic chemicals, other natural processes, such as chemical transformation, dilution, sorption, and radioactive decay, are responsible for natural attenuation, as broadly defined by US Environmental Protection Agency (www.epa.gov). The health and environmental risks posed by those contaminants may be reduced by changing the amount of exposure, the exposure pathway, or the toxicity of the chemical (Rahm et al., 2005). At this point, several comments on the rate of a reaction are warranted.

The energy requirements of a reaction fall within the realm of *thermodynamics*, but in the present context, the rate of the reaction falls within the realm of *kinetics* (Chapter 4). It is essential that the rate of the reaction can be controlled not only for commercial but also for safety reasons. For example, if a reaction takes too long to reach equilibrium or completion, the rate at which a product is manufactured would not be viable. On the other hand, in a runaway reaction—a reaction that progresses too fast and becomes uncontrollable—there are inherent dangers such as fires and/or explosions.

The rate at which a reaction takes place can be affected by the concentration of reactants, pressure, temperature, the size of particles of solid reactants, or the presence of catalysts (i.e., additives that alter the speed of reactions without being consumed during the reaction) as well as impurities. Catalysts tend to be specific to a particular reaction or family of reactions. For example, nickel is used to facilitate hydrogenation reactions (e.g., add hydrogen to carbon-carbon double bonds in a refinery), whereas platinum is used to catalyze specific oxidation reactions. In most cases, chemical reactions require that the reactants are pure since impurities can act as unwanted catalysts; alternatively, catalysts can be deactivated by poisoning—the deposition of metals or coke-like products onto the catalyst during the reaction.

For reactions that progress slowly at room temperature, it may be necessary to heat the mixture or add a catalyst for the reaction to occur at an economically viable rate. For very rapid reactions, the reactant mixture may need to be cooled or solvent added to dilute the reactants and hence reduce the speed of reaction to a manageable (a controllable and nondangerous) rate. In general, the rate of a chemical speed of reaction (i) doubles for every 10°C (18°F) rise in temperature, (ii) is proportional to the concentration of reactants in solution, (iii) increases with decreased particle size for reactions involving a solid, and (iv) increases with pressure for gas-phase reactions.

7.3.1 Hydrolysis Reactions

Hydrolysis is a double decomposition reaction (Chapter 2) with water as one of the reactants. Put simply, if an inorganic chemical is represented by the formula AB in which A and B are atoms or groups and water is represented by the formula HOH, the hydrolysis reaction may be represented by the reversible chemical equation:

$$AB + HOH \rightleftharpoons AH + BOH$$

The reactants other than water and the products of hydrolysis may be neutral molecules (as in most hydrolysis reactions involving organic compounds) or ionic molecules (charged species), as in hydrolysis reactions of salts, acids, and bases.

In the current context, hydrolysis typically means the cleavage of inorganic chemical bonds by the addition of water or a base that supplied the hydroxyl ion ($—OH^-$) in which a chemical bond is cleaved and two new bonds are formed, each one having either the hydrogen component (H) or the hydroxyl component (OH) of the water molecule. Many inorganic compounds can be altered by a direct reaction of the chemical with water. The rate of a hydrolysis reaction is typically rates that are expressed in terms of the acid-catalyzed, neutral-catalyzed, and base-catalyzed hydrolysis rate constants.

Typically, the hydroxyl replaces another chemical group on the inorganic molecule, and hydrolysis reactions are usually catalyzed by hydrogen ions or hydroxyl ions. This produces the strong dependence on the acidity or alkalinity (pH) of the solution often observed, but in some cases, hydrolysis can occur in a neutral (pH 7) environment. Adsorption onto a mineral sediment (such as a clay sediment that has strong adsorptive powers) generally reduces the rates of hydrolysis for acid- or base-catalyzed reactions. Neutral reactions appear to be unaffected by adsorption although there is always the possibility that the mineral sediment can cause catalyzed chemical transformation reactions.

In inorganic chemistry, a common type of hydrolysis occurs when a salt of a weak acid or weak base (or both) is dissolved in water. The water spontaneously ionizes into hydronium cations (H_3O^+, for simplicity usually represented as H^+) and hydroxide anions ($—OH^-$). The salt, for example, using sodium acetate (CH_3COONa), dissociates into the constituent cations (Na^+) and anions (CH_3COO^-). The sodium ions tend to remain in the ionic form (Na^+) and react very little with the hydroxide ions (OH^-), whereas the acetate ions combine with hydronium ions to produce acetic acid (CH_3COOH). In this case, the net result is a relative excess of hydroxide ions, and the solution has basic properties. On the other hand, strong acids also undergo hydrolysis, and when sulfuric acid (H_2SO_4) is dissolved in (mixed with) water, the dissolution is accompanied by hydrolysis to produce hydronium ion (H_3O^+) and a bisulfate ion (HSO_4^-), which is the conjugate base of sulfuric acid.

There are four possible general rules involving the reactions of salts in the aqueous environment: (i) If the salt is formed from a *strong* base and *strong* acid, then the salt solution is neutral, indicating that the bonds in the salt solution will not break apart (indicating no hydrolysis occurred), and is basic; (ii) if the salt is formed from a *strong* acid and *weak* base, the bonds in the salt solution will break apart and becomes acidic; (iii) if the salt is formed from a *strong* base and *weak* acid, the salt solution is basic and hydrolyzes; and (iv) if the salt is formed from a weak base and weak acid, it will hydrolyze but the acidity or basicity depends on the equilibrium constants of K_a and K_b. If the K_a value is greater than the K_b value, the resulting solution will be acidic, and if the K_b value is greater than the K_a value, the resulting solution will be basic (Petrucci et al., 2010).

7.3.2 Photolysis Reactions

Photolysis (also called photodissociation and photodecomposition) is a chemical reaction in which an inorganic chemical (or an organic chemical) is broken down by photons and is the interaction of one or more photons with one target molecule. The photolysis reaction is not limited to the effects of visible light, but any photon with sufficient energy can cause the chemical transformation of the inorganic bonds of a chemical. Since the energy of a photon is inversely proportional to the wavelength, electromagnetic waves with the energy of visible light or higher, such as ultraviolet light, X-rays, and gamma rays, can also initiate photolysis reactions.

Photolysis should not be confused with photosynthesis that is a two-part process in which natural chemicals (typically organic chemicals) are synthesized by a living organism as part of the life cycle (life chemistry) of the organism.

The primary step of a photolysis reaction is

$$X + h\nu \rightarrow X^*$$

where X^* is an electronically excited state of molecule X and subsequently undergo either physical or chemical processes:

Physical processes:

Fluorescence:

$$X^* \rightarrow X + h\nu$$

Collisional deactivation:

$$X^* + M \rightarrow X + M$$

Chemical processes:
Dissociation:

$$X^* \rightarrow Y + Z$$

Isomerization:

$$X^* \rightarrow X'$$

Direct reaction:

$$X^* + Y \rightarrow Z_1 + Z_2$$

Intramolecular rearrangement:

$$X^* \rightarrow Y \quad \text{Ionization} \quad X^* \rightarrow X+ \; + e$$

The general form of photolysis reactions is

$$X + h\nu \rightarrow Y + Z$$

The rate of reaction is

$$-d/dt[X] = d/dt[Y] = d/dt[Z] = k[X]$$

k is the photolysis rate constant for this reaction in units of 1/s.

Thus, photolysis is a chemical process by which chemical bonds are broken as the result of transfer of light energy (direct photolysis) or radiant energy (indirect photolysis) to these bonds. The rate of photolysis depends upon numerous chemical and environmental factors including the light adsorption properties and reactivity of the chemical and the intensity of solar radiation. In the process, the photochemical mechanism of photolysis is divided into three stages: (i) the adsorption of light that excites electrons in the molecule, (ii) the primary photochemical processes that transform or deexcite the excited molecule, and (ii) the secondary (dark) thermal reactions that transform the intermediates produced in the previous step (step ii).

Indirect photolysis or sensitized photolysis occurs when the light energy captured (absorbed) by one molecule is transferred to the inorganic molecule of concern. The donor species (the sensitizer) undergoes no net reaction in the process but has an essentially catalytic effect. Moreover, the probability of a sensitized molecule donating its energy to an acceptor molecule is proportional to the concentration of both chemical species. Thus, complex mixtures may, in some cases, produce enhancement of photolysis rates of individual constituents through sensitized reactions.

Photolysis occurs in the atmosphere as part of a series of reactions by which primary pollutant nitrogen oxides react to form secondary pollutants such as peroxyacyl nitrate derivatives. The two most important photolysis reactions in the troposphere are

$$O_3 + h\nu \rightarrow O_2 + O^*$$

The excited oxygen atom (O^*) can react with water to give the hydroxyl radical:

$$O^* + H_2O \rightarrow 2OH\cdot$$

The hydroxyl radical is central to atmospheric chemistry because it can initiate the oxidation of hydrocarbons in the atmosphere.

Another reaction involves the photolysis of nitrogen dioxide to produce nitric oxide and an oxygen radical that is a key reaction in the formation of tropospheric ozone:

$$NO_2 + h\nu \rightarrow NO + O$$

The formation of the ozone layer is also caused by photolysis. Ozone in the stratosphere is created by the ultraviolet light striking oxygen molecules (O_2) and splitting the oxygen molecules into individual oxygen atoms (atomic oxygen). The atomic oxygen then combines with unbroken oxygen molecule to create ozone (O_3):

$$O_2 \rightarrow 2O$$
$$O_2 + O \rightarrow O_3$$

In addition, the absorption of radiation in the atmosphere can cause photodissociation of nitrogen (as one of the several possible reactions) that can lead to the formation of nitric oxide (NO) and nitrogen dioxide (NO_2) that can act as a catalyst to destroy ozone:

$$N_2 \rightarrow 2N$$
$$O_2 \rightarrow 2O$$
$$CO_2 \rightarrow C + 2O$$

$$H_2O \rightarrow 2H + O$$
$$2NH_3 \rightarrow 3H_2 + N_2$$
$$N + 2O \rightarrow NO_2$$

As shown in the above equations, most photochemical transformations occur through a series of simple steps (Wayne and Wayne, 2005).

7.3.3 Radioactive Decay

Radionuclides are inorganic chemicals that emit radiation because their atoms are unstable and they disintegrate or decay as they release energy in the form of radioactive particles or waves. There are many different types of radioactive decay (Table 7.3). A decay or loss of energy from the nucleus results when an atom with an initial type of nucleus (the *parent radionuclide* or *parent radioisotope*) transforms into a *daughter nuclide*. The transformation produces an atom in a different state (a nucleus containing a different number of protons and neutrons). In some decay processes, the parent and the daughter nuclides are different chemical elements, and thus, the decay process results in the creation of an atom of a different element (nuclear transmutation).

Table 7.3 Common Forms of Radioactive Decay

Mode of Decay	Participating Particles
Decays with emission of nucleons	
Alpha decay	An alpha particle emitted from the nucleus
Proton emission	A proton ejected from the nucleus
Neutron emission	A neutron ejected from the nucleus
Spontaneous fission	Nucleus disintegrates into two or more smaller nuclei and other particles
Different modes of beta decay	
β^- decay	A nucleus emits an electron and an electron antineutrino
β^+ decay	A nucleus emits a positron and an electron neutrino
Electron capture	A nucleus captures an orbiting electron and emits a neutrino
Transitions between states of the same nucleus	
Isomeric transition	Excited nucleus releases a high-energy photon (gamma ray)
Internal conversion	Excited nucleus transfers energy to an orbital electron, which is subsequently ejected from the atom

Another type of radioactive decay results in products that are not defined, but appear in a range of molecular fragments of the original nucleus (spontaneous fission), which occurs when a large unstable nucleus spontaneously splits into two (and occasionally three) smaller daughter nuclei, and generally leads to the emission of gamma rays, neutrons, or other particles from those products.

In the environment, radioactive decay is a natural process that happens spontaneously, and as a radionuclide decays, radioactive isotopes transform into other, often less radioactive isotopes of the same element, or even sometimes into isotopes of other elements. The identity and chemical properties of an element are determined by the number of protons in its nucleus, the atomic number. Many elements, however, occur as different isotopes, which are defined by the number of neutrons in the nucleus. For example, the common isotope of hydrogen has one proton and no neutrons in its nucleus. Tritium, the radioactive isotope of hydrogen, has one proton and two neutrons. Uranium-235 and Uranium-238, two isotopes of the element Uranium, both have 92 protons, but they have 143 and 146 neutrons, respectively.

Unlike other inorganic elements or chemicals, radionuclides degrade naturally, but the degradation process itself is hazardous to living things. The risk posed by radionuclides is a function of the type and amount of radiation and exposure pathways. The half-life of a radioactive element is the time required to reduce the concentration of a chemical to 50% of the initial concentration of a radioactive element.

If the flora and fauna of an ecosystem are isolated from the radionuclide, the risk of contamination is low. However, active remediation does not affect the decay rate but is designed to stabilize, prevent the migration of, or isolate the substance. In comparing the natural attenuation of radionuclides to active remediation, sorption and dilution are often considered as the mechanisms for isolating the material and decreasing the concentration. Some radioactive isotopes decay to form other, more hazardous radionuclides. As with the degradation of organic contaminants, therefore, it is essential to consider the hazard posed by the daughter products and radiation from the original hazard when evaluating the effects of natural radioactive decay. Radioactive chemicals range in half-life; for example, $Uranium^{238}$ (also written as ^{238}Uranium or Uranium-238) has a half-life of 4.5 billion years, while tritium (^{3}H) has a half-life of 12.33 years.

Though nonradioactive metals do not break down in the environment, sometimes chemical changes in the subsurface environment reduce their toxicity. Copper in certain forms is considered so safe that we use it to pipe drinking water, yet trace concentrations can be detrimental to aquatic life. Chromium in one form (hexavalent or Cr^{6+}) forms compounds that are highly toxic and very soluble, yet under certain conditions, it transforms into trivalent chromium (Cr^{3+}), which is less toxic and insoluble. The potential for such a change differs for each substance, but it can be estimated by conducting a chemical and biological analysis of the subsurface environment.

Finally, the radioactive half-life for a given radioisotope is a measure of the tendency of the nucleus to decay (disintegrate) and is based upon that probability. The half-life is independent of the physical state (solid, liquid, or gas), temperature, pressure, chemical compound in which the nucleus exists, and essentially any other outside influence. The half-life is also independent of the chemistry of the atomic surface and independent of the ordinary physical factors of the environment in which the chemical exists.

7.3.4 Rearrangement Reactions

A rearrangement reaction falls into a broad class of (inorganic and organic) reactions where a molecule is rearranged to produce a structural isomer of the original molecule.

Inorganic reactions typically yield products that are in accordance with the generally accepted mechanism of the reactions. However, in some instances, inorganic reactions do not give exclusively and solely the anticipated products but may lead to other product that arise from unexpected and mechanistically different reaction paths. This type of unexpected product is often referred to as a rearranged product, and while such a product may not be the expected product, it may be the major product of the reaction. Thus, the reaction has involved a rearrangement of the expected product to an unexpected product; a rearrangement reaction has occurred. More than likely, this may have resulted from a plausible rearrangement occurring during the mechanistic course of the reaction to fulfill the principle of the minimum energy state of the whole system, that is, of the transition state that assumed another configuration to maintain a minimum energy balance to the system. In many cases, the rearrangement affords products of an isomerization, coupled with some stereochemical changes. An energetic requirement is also observed in order for a rearrangement to take place; that is, the rearrangement usually involves an evolution of energy (typically in the form of heat, i.e., the reaction is overall an exothermic reaction) to be able to yield a more stable compound.

7.3.5 Redox Reactions

Redox reactions (reduction-oxidation reactions) are reactions in which one of the reactants is reduced and another reactant is oxidized. Therefore, the oxidation state of the species involved must change. The word *reduction* originally referred to the loss in weight upon heating a metallic ore such as a metal oxide to extract the metal; the ore was *reduced* to the metal. However, the meaning of *reduction* has become generalized to include all processes involving gain of electrons. Thus, in redox reactions, one species is oxidized, while another is reduced by the net transfer of electron from one to the other. As may be expected, the change in the oxidation states of the oxidized species must be balanced by any changes in the reduced species. For example, the production of iron from the iron oxide ore is

$$Fe_2O_3 + 3CO \rightarrow 2Fe + 3CO_2$$

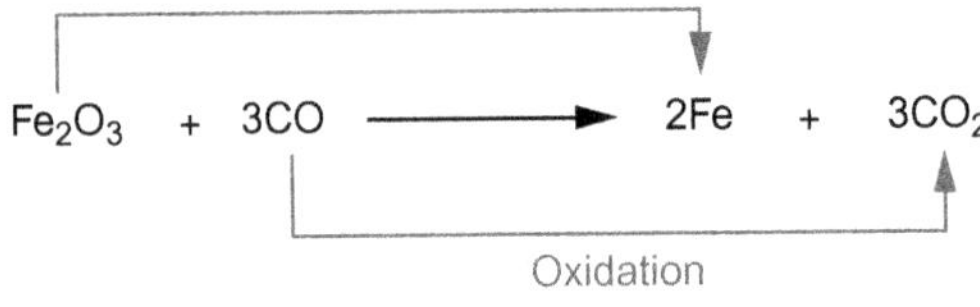

To complicate matters even further, the oxidizing and reducing agents can be the same element or compound, as in the case when disproportionation of the reactive species occurs. For example,

$$2A \rightarrow (A + n) + (A - n)$$

In this equation, n is the number of electrons transferred. Disproportionation reactions do not need to commence with a neutral molecule and can involve more than two species with differing oxidation states.

Within redox reactions, the pair of reactions must always occur, i.e., a reduction reaction must be accompanied by an oxidation process, as electrons are transferred from one species to another. Each of the singular reactions in this pair is called a half reaction, in which the electrons lost or gained are included explicitly, allowing electron balance to be accounted as well. The two sides of the reaction, given by the half reactions, should be balanced accordingly. The additional terminology comes from the definition that, within redox processes, a reductant transfers electrons to an oxidant; hence, the reductant (reducing agent) loses electrons, so is oxidized, while the oxidant (oxidizing agent) gains electrons, so is reduced.

Redox reactions are important for a number of applications, including energy storage devices (batteries), photographic processing, and energy production and utilization in living systems including humans. For example, a *reduction reaction* is a reaction in which an atom gains an electron and therefore decreases (or reduces its oxidation number). The result is that the positive character of the species is reduced. On the other hand, an oxidation reaction is a reaction in which an atom loses an electron and therefore increases its oxidation number. The result is that the positive character of the species is increased.

Although oxidation reactions are commonly associated with the formation of oxides from oxygen molecules, these are only specific examples of a more general concept of reactions involving electron transfer. Redox reactions are a matched set, that is, there cannot be an oxidation reaction without a reduction reaction happening simultaneously. The oxidation reaction and the reduction reaction always occur together to form a whole reaction. Although oxidation and reduction properly refer to *a change in the oxidation state*, the actual transfer of electrons may never occur. The oxidation state of an atom is the fictitious charge that an atom would have if all bonds between atoms of different elements were 100% ionic. Thus, oxidation is best defined as an *increase in oxidation state* and reduction as a *decrease in oxidation state*. In practice, the transfer of electrons will always cause a change in oxidation state, but there are many reactions that are classed as redox reactions even though no electron transfer occurs (such as those involving covalent bonds).

The key to identifying oxidation-reduction reactions is recognizing when a chemical reaction leads to a change in the oxidation number of one or more atoms.

7.4 CATALYSTS

Catalysts are substances that increase the rate of a reaction by providing a low-energy alternate route from reactants to products. In some cases, reactions occur so slowly that, without a catalyst, they are of little value.

The study of environmental interfaces and environmental catalysis is central to finding more effective solutions to air pollution and in understanding of how pollution impacts the natural environment. Surface catalysis of airborne particles—including ice, trace atmospheric gases, aerosolized soot nanoparticles, and mineral dust surfaces—and particles in contact with ground water and their role in surface adsorption, surface catalysis, hydrolysis, dissolution, precipitation, oxidation, and ozone decomposition must be continuously investigated. With increasing ground water pollution and increasing particulates in the atmosphere, there is an increasing need to remove pollutants from industrial and automotive sources.

In catalysis reactions, the reaction does not proceed directly, but through reaction with a third substance (the catalyst) and although the catalyst takes part in the reaction, it is (in theory) returned to its original state by the end of the reaction and so is not consumed. However, the catalyst is not immune to being inhibited, deactivated, or destroyed by secondary processes.

Industrial catalysts are often metals, as most metals have many electrons that allow the metals to use these electrons to help out in reactions before resuming the normal electronic configuration when the reaction is over. Examples are iron-based catalysts used for making ammonia (the Haber-Bosch process):

$$N_2 + 3H_2 \rightarrow 2NH_3$$

In the absence of a catalyst, a high temperature on the order of 3000°C (5430°F) would be required to cause the reactant molecules to collide with enough force as to break chemical bonds.

Catalysts can be used in a different phase (heterogeneous catalysis) or in the same phase (homogeneous catalysis) as the reactants. In heterogeneous catalysis, typical secondary processes include coking (coke production from inorganic starting materials) where the catalyst becomes covered by ill-defined high-molecular weight by-side products. Heterogeneous catalysis is used in automobile exhaust systems to decrease nitrogen oxide, carbon monoxide, and unburned hydrocarbon emissions. The exhaust gas is vented through a high-surface-area chamber lined with platinum, palladium, and rhodium. For example, the carbon monoxide is catalytically converted to carbon dioxide by reaction with oxygen.

Additionally, heterogeneous catalysts can dissolve into the solution in a solid-liquid system or evaporate in a solid-gas system. Catalysts can only speed up the reaction; chemicals that slow down the reaction are called inhibitors, and there are chemicals that increase the activity of catalysts (catalyst promoters) and chemicals that deactivate catalysts (catalytic poisons). With a catalyst, a reaction that is kinetically inhibited by a high activation energy can take place in circumvention of this activation energy. Heterogeneous catalysts are usually solids, powdered in order to maximize their surface area. Of particular importance in heterogeneous catalysis are the platinum metals and other transition metals, which are used in crude oil refining processes such as hydrogenation and catalytic reforming.

Homogeneous catalysis involvers a reaction in which the soluble catalyst is in solution—as long as the catalyst is in the same phase as the reactants. Although the term is used almost exclusively to describe reactions (and catalysts) in solution, it often implies catalysis by organometallic compounds but can also apply to phase reactions and solid phase reactions. Homogeneous catalysis differs from heterogeneous catalysis insofar as the catalyst is in a different phase than the reactants. The advantage of homogeneous catalysts is the ease of mixing them with the reactants, but they may also be difficult to separate from the products. Therefore, heterogeneous catalysts are preferred in many industrial processes for the production and transformation (conversion) of the starting compound(s). However, heterogeneous catalysis offers the advantage that products are readily separated from the catalyst and heterogeneous catalysts are often more stable and degrade much slower than homogeneous catalysts. However, heterogeneous catalysts are difficult to study, so their reaction mechanisms are often unknown.

Environmental catalysis has continuously grown in importance over the latter half of the 20th century, and the development of innovative catalysts for protection of the environment is a crucial factor toward the objective of developing a new sustainable industrial chemistry. In the last decade, considerable expansion of the traditional area of environmental catalysis has occurred, mainly removal of nitrogen oxides from stationary and mobile sources and conversion of volatile organic compounds (VOCs). New areas include (i) catalytic technologies for liquid or solid waste reduction or purification; (ii) use of catalysts in energy-efficient catalytic technologies and processes; (iii) reduction of the environmental impact in the use or disposal of catalysts; (iv) new eco-compatible refinery, chemical, or nonchemical catalytic processes; (v) catalysis for greenhouse gas control; (vi) use of catalysts for user-friendly technologies and reduction of indoor pollution; (vii) catalytic processes for sustainable chemistry; and (viii) reduction of the environmental impact of transport (Centi et al., 2002; Grassian, 2005).

7.5 SORPTION AND DILUTION

Inorganic chemicals interact with the environment in different ways, and once a chemical is released into the environment, there are two physical effects that can influence the distribution of the chemicals: (i) sorption and (ii) dilution.

Sorption occurs when an inorganic chemical attaches to underground particles, immobilizing and limiting the availability and movement of the chemical. The most common host materials for sorption are iron hydroxides, clay minerals, and carbonate minerals. Sorption is most effective for a chemical that is present in low concentration and that do not overwhelm the potential sires for sorption. With sorption, the contaminants are mobilized, but not destroyed, and while sorption is always reversible, it is conventionally termed irreversible if the sorbent (such as soil) encloses (traps) the chemical. When this happens, the sorbed chemical (the sorbate) is less likely to be affected by changes in the environment that might desorb or dissolve the chemical, such as changes in acidity (or alkalinity) or exposure of the chemical to an additional mutually reactive chemical.

As part of the sorption process, adhesion must also be recognized and considered. Adhesion is the tendency of similar and/or dissimilar particles and/or surfaces to cling to one another; on the other hand, cohesion is the tendency of similar or identical particles/surfaces to cling to one another. The forces that cause adhesion and cohesion can be divided into several types: (i) chemical adhesion, (ii) dispersive adhesion, and (iii) diffusive adhesion.

There are times when the sorption (or adhesion) of an inorganic chemical can be enhanced through a chemical transformation reaction or chemical stabilization. That is, before the metals attach to soil particles, they form solid compounds with chemicals, such as oxygen and sulfur, that are found in the soil. This has two advantages: (i) The chemical will not travel as easily with the flow of groundwater, and (ii) the new chemical formed by reaction may be more likely to adhere to the soil, especially the clay minerals in the soil. However, over a period of time that is dependent upon the properties of the chemical and the environment into which it is released, changes in acidity or the concentration of charged particles and reactive chemicals in the soil can destabilize the sorptive capacity of the soil and the ability of the chemical to be sorbed.

Unfortunately, sorption is more difficult to measure than the degradation of inorganic chemicals. With inorganic degradation, the by-products of chemical reactions and of biological reactions can be measured in order to estimate how the extent of any degradation. In contrast, inorganic chemicals are sorbed by displacing ions that are abundant in the soil, and measuring relatively small variations in the concentration of those ions is very difficult, and the results are often less reliable.

In evaluating sorption as a natural attenuation technique, the following three factors must be considered: (i) based upon available soil data, such as the sorptive capacity of the soil for the specific contaminant and whether all of the inorganic chemical be sorbed by soil particles; (ii) If sorption of the chemical can reduce the contamination to acceptable levels, it is necessary to determine the length of time it will take and whether or not any contamination reach receptors before sorption is complete, and (iii) the stability of the sorbed contaminants and whether or not any changes in the soil chemistry will cause the contaminant to desorb.

Sorption, like other natural attenuation processes, occurs whether or not it has been approved as a remedy. In fact, sometimes, *irreversible sorption (fixed sorption)* makes active removal of the contamination now bound to the soil much more difficult.

Furthermore, sorption can be subdivided into two categories: (i) adsorption and (ii) absorption; both are important phenomena with differences in the outcomes (Table 7.4). The structures (physical and electronic) of the inorganic molecules play a role in both phenomena and such properties as water solubility, and (in the case of mixtures) the composition is particularly important. Evaluation of adsorption or absorption can be obtained either through laboratory measurements or by the use of several property correlations. In addition, any deductions from laboratory measurement must also take into account the potential for the transformation of the inorganic chemicals in the environment and degradation of the chemicals.

7.5.1 Adsorption

Adsorption is the physical accumulation of material (usually a gas or liquid) on the surface of a solid adsorbent and is a *surface phenomenon*. Typically, adsorption processes remove solutes from liquids based on their mass transfer from liquids to porous solids. Ion exchange is the exchange of dissolved ions for ions on solid media. The process can be used to remove water hardness and toxic metals during wastewater treatment. Disinfection is the removal or inactivation of pathogenic organisms in wastewater prior to be discharged to the receiving body of water.

Table 7.4 Comparison of Adsorption and Absorption

	Adsorption	Absorption
Definition	Accumulation of the molecular species at the surface rather than in the bulk of the solid or liquid	Assimilation of molecular species throughout the bulk of the solid or liquid
Characteristic	A surface phenomenon	A bulk phenomenon
Reaction type	Exothermic process	Endothermic process
Temperature	Unaffected by temperature	Not affected by temperature
Reaction rate	Increases to equilibrium	Occurs at a uniform rate
Concentration	Different at surface to bulk	Same throughout

The adsorption process creates a film of the *adsorbate* on the surface of the *adsorbent*, and the process differs from the absorption process in which a fluid (the *absorbate*) is dissolved by a liquid or permeates into a solid (the *absorbent*), respectively. Thus, adsorption is a surface-based process, while absorption involves the whole volume of the material. The term *sorption* encompasses both processes, while *desorption* is the reverse of sorption. In the environment, inorganic compounds will be collected on the surfaces of particles, such as soil or suspended sediment. Most of these particles are covered with a layer of inorganic material; thus, the adsorption results from the attraction of two inorganic materials for one another.

In nature, a variety of potential natural adsorbents exit in the soil; adsorption occurs in many natural, physical, biological, and chemical systems (especially in the environment) where inorganic molecules can adsorb onto minerals (such as clay) or onto charred wood that remains after a forest fire. In fact, clay minerals are particularly good adsorbents and have a high adsorption capacity for inorganic chemicals that have been released into the environment.

A natural clay mineral is not composed of one clay mineral only. Impurities such as calcite ($CaCO_3$), quartz (SiO_2), feldspar ($KAlSi_3O_8/NaAlSi_3O_8/CaAl_2Si_2O_8$), iron oxides, ($FeO$ and Fe_2O_3), and humic acids (degradation products of organic materials) are the most common components in addition to the pure clay mineral. Calcite, iron oxides, and humic acids can be removed by chemical treatments. Quartz and feldspar can be removed by sedimentation if the particle size is bigger than that of the clay minerals, but traces of quartz are often found in the purified samples. Physically, clay minerals are typically ultrafine grained (normally considered to be less than 2 μm (<2 μm, $<2\times10^{-6}$ m) in size on standard particle-size classifications). In the present context, clay minerals, which can be classified into various chemical groups, such as the silicate clay mineral groups (Table 7.5) are an important part of many soils, thus rendering the soil capable of having a high adsorption capacity for inorganic chemicals. Generally, no two clay minerals are the same, and the adsorption capacity will vary accordingly.

Adsorption of an inorganic chemical onto a solid adsorbent, such as a clay mineral or any other mineral, is measured by a partition coefficient, which is the ratio of the concentration the inorganic chemical on the solid to the concentration of the chemical in the fluid (usually water) surrounding the solid:

$$K_d = C_{solid}/C_{water}$$

Table 7.5 Illustration of Various Clay Mineral Groups

Group	Layer Type	Layer Charge (x)	Type of Chemical Formula
Kaolinite	1:1	<0.01	$[Si_4]Al_4O_{10}(OH)_8 \cdot nH_2O$ ($n=0$ or 4)
Illite	2:1	1.4–2.0	$M_x[Si_{6.8}Al_{1.2}]Al_3Fe.025\ Mg_{0.7}5O_20(OH)_4$
Vermiculite	2:1	1.2–1.8	$M_x[Si_7Al]AlFe.05\ Mg0.5O_20(OH)_4$
Smectite	2:1	0.5–1.2	$M_x[Si_8]Al_{3.2}Fe_{0.2}Mg_{0.6}O_20(OH)_4$
Chlorite	2:1:1	Variable	$(Al(OH)_{2.55})4[Si_{6.8}AlO_{1.2}Al_{3.4}Mg_{O.6})20(OH)_4$

The concentration on the solid has units of mol/kg, and the concentration in the water is mol/L and thus, the adsorption coefficient (K_d) has units of L/kg. Assuming a solid density of 1 kg/L, these units are often ignored. The adsorption coefficient will often depend on how much of the total mass of the particle is inorganic material. Thus, the adsorption coefficient can be corrected by the fraction of inorganic material (f_{om}) in the particles:

$$K_{om} = K_d / f_{om}$$

Adsorbed molecules are those that are resistant to washing with the same solvent medium in the case of adsorption from solutions. The washing conditions can thus modify the measurement results, particularly when the interaction energy is low. The exact nature of the bonding depends on the details of the chemical species involved, but the adsorption process is generally classified as physisorption (which is characteristic of weak van der Waals forces) or chemisorption (which is characteristic of covalent bonding). It may also occur due to electrostatic attraction.

The interactions involved in the sorption of heavy metal cations and anions to the surfaces of inorganic materials such as soil are complex. Specific adsorption/surface precipitation onto various mineral phases present on composite inorganic materials explains the capacity of such materials to adsorb metals. However, differences in operational parameters make a comparison of the adsorption capacity between materials difficult to occur and are not fully defined or explained. The ease of desorption of the inorganic is also an important consideration, because in the treatment of wastewaters, materials are used primarily as ion exchangers, while for in situ immobilization, the metals need to be irreversibly bound to the added adsorbent (Zhou and Haynes, 2010).

7.5.2 Absorption

Absorption is another phenomenon that can be a beneficial or adverse influence of the environment and involves the uptake of one substance into the inner structure of another, most typically a gas into a liquid solvent. Furthermore, absorption is a physical or chemical phenomenon or a process in which atoms, molecules, or ions enter some bulk phase—gas, liquid, or solid material. This is a different process from *adsorption*, since molecules undergoing absorption are taken up by the volume, not by the surface (as in the case for adsorption). A more general term is *sorption*, which covers absorption and adsorption; the former (absorption) is a condition in which something takes in another substance. In many processes important in technology, the chemical absorption is used in place of the physical process. It is possible to extract from one liquid phase to another a solute without a chemical reaction. The process of absorption means that a substance captures and transforms energy and the absorbent distributes the material it captures throughout whole and adsorbent only distributes it through the surface.

In chemical absorption (sometimes referred to in the shortened word form as *chemisorption*), the absorbed material is generally converted to a product different to the starting material. Thus, chemical absorption or *reactive absorption* involves a chemical reaction between the absorbent (the absorbing substance) and the absorbate (the absorbed substance) and may be combined with the physical absorption phenomenon. This type of absorption depends upon the stoichiometry of the reaction and the concentration of the potential reactants.

Physical absorption or nonreactive absorption is made between two phases of matter: A liquid absorbs a gas, or a solid absorbs a liquid. When a liquid solvent absorbs a gas mixture or part of it, a mass of gas moves into the liquid. For example, water may absorb oxygen from the air. This mass transfer takes place at the interface between the liquid and the gas, at a rate depending on both the gas and the liquid. This type of absorption depends on the solubility of gases, the pressure, and the temperature. The rate and amount of absorption also depend on the surface area of the interface and its duration in time. For example, when the water is finely divided and mixed with air, as may happen in a waterfall or a strong ocean surf, the water absorbs more oxygen. When a solid absorbs a liquid mixture or part of it, a mass of liquid moves into the solid. This mass transfer takes place at the interface between the solid and the liquid, at a rate depending on both the solid and the liquid. Absorption is essentially a molecule attaching them to a substance and will not be attracted from other molecules.

On the other hand, chemical absorption or reactive absorption is a chemical reaction between the absorbed and the absorbing substances. Sometimes, it combines with physical absorption. This type of absorption depends upon the stoichiometry of the reaction and the concentration of its reactants.

7.5.3 Dilution

Dilution is the process in which an inorganic chemical in an ecosystem becomes less concentrated, and there is a decrease in the concentration of a solute in solution, usually simply by mixing with more solvent (such as water). Dilution is also a reduction in the acidity or alkalinity of an inorganic chemical (gas, vapor, or solution). To dilute a solution means to add more solvent without the addition of more solute. The resulting solution is thoroughly mixed to ensure that all parts of the solution are identical. Mathematically, dilution can be represented by a simple equation:

$$C_1 \times V_1 = C_2 \times V_2$$

In this equation, C_1 is the initial concentration of the solute, V_1 is the initial volume of the solution, C_2 is the final concentration of the solute, and V_1 is the final volume of the solution. However, this type of simple equation does not consider any solute-solvent interactions, solute-solute interactions, and solvent-solvent interactions.

Dilution of the chemical may reduce the risk to the floral and faunal species because the potential individual receptors are likely to be exposed to lower, less toxic concentrations of the hazard. However, the chronic effects associated with the dilution of a toxic inorganic chemical (or any chemical for that matter) to a less concentrated form are extremely difficult to measure. By itself, however, dilution does not reduce the mass of the chemical but, rather, spreads the area of potential exposure to the chemical. In addition, some inorganic contaminants are believed to be hazardous (such as the always lethal potassium cyanide, KCN) even at levels that may be too dilute to be detected with standard field characterization equipment and techniques.

7.6 BIODEGRADATION

Biodegradation (biotic degradation, biotic decomposition) is the chemical degradation of contaminants by bacteria or other biological means. Organic material can be degraded aerobically (in the presence of oxygen) or anaerobically (in the absence of oxygen). Most biodegradation systems operate run under aerobic conditions, but a system under anaerobic conditions may permit microbial organisms to degrade inorganic chemical species that are otherwise nonresponsive to aerobic treatment and vice versa. Thus, biodegradation is a natural process (or a series of processes) by which spilled inorganic chemicals or other inorganic waste material can be broken down (degraded) into nutrients that can be used by other organisms. The ability of a chemical to be biodegraded is an indispensable element in the understanding the risk posed by that chemical on the environment.

Biodegradation is a key process in the natural attenuation (reduction or disposal) of inorganic chemical compounds at hazardous waste sites. The contaminants of concern must be amenable to degradation under appropriate conditions, but the success of the process depends on the ability to determine these conditions and establish them in the contaminated environment. Important site factors required for success include (i) the presence of metabolically capable and sustainable microbial populations; (ii) suitable environmental growth conditions, such as the presence of oxygen; (iii) temperature, which is an important variable—keeping a substance frozen or below the optimal operating temperature for microbial species, can prevent biodegradation; most biodegradation occurs at temperatures between 10°C and 35°C (50°F and 95°F); (iv) the presence of water; (v) appropriate levels of nutrients and contaminants; and (vi) favorable acidity or alkalinity (Table 7.6). In regard to the last parameter, soil pH is extremely important because most microbial species can survive only within a certain pH range; for example, the biodegradation of chemicals might be optimal at a pH 7 (neutral), but the *acceptable* (or optimal) pH range might be on the order of 6–8. Furthermore, soil (or water) pH can affect the availability of nutrients. Thus, through biodegradation processes, living microorganisms (primarily not only bacteria but also yeasts, molds, and filamentous fungi) can alter and/or metabolize various classes of inorganic chemicals.

Temperature influences the rate of biodegradation by controlling the rate of enzymatic reactions within microorganisms. Generally, the rate of an enzymatic reaction approximately doubles for each 10°C (18°F) rise in

Table 7.6 Essential Factors for Microbial Bioremediation

Factor	Optimal Conditions
Microbial population	Suitable kinds of organisms that can biodegrade all of the contaminants
Oxygen	Enough to support aerobic biodegradation (about 2% oxygen in the gas phase or 0.4 mg/L in the soil water)
Water	Soil moisture should be from 50% to 70% of the water holding capacity of the soil
Nutrients	Nitrogen, phosphorus, sulfur, and other nutrients to support good microbial growth
Temperature	Appropriate temperatures for microbial growth (0–40°C, 32–104°F)
pH	Optimal range of 6.5–7.5

temperature (Nester et al., 2001). However, there is an upper limit to the temperature that microorganisms can withstand. Most bacteria found in soil, including many bacteria that degrade petroleum hydrocarbons, are mesophile organisms that have an optimum working temperature range on the order of 25–45°C (77–113°F) (Nester et al., 2001). Thermophilic bacteria (those that survive and thrive at relatively high temperatures), which are normally found in hot springs and compost heaps, exist indigenously in cool soil environments and can be activated to degrade chemicals with an increase in temperature to 60°C (140°F). This indicates the potential for natural attenuation in cool soils through thermally enhanced bioremediation techniques.

In the absence of oxygen, some microorganisms obtain energy from fermentation and anaerobic oxidation of inorganic carbon. Many anaerobes use nitrate, sulfate, and salts of iron (Fe^{3+}) as practical alternates to oxygen acceptor. The anaerobic reduction process of nitrates, sulfates, and salts of iron is an example:

$$2NO_3^- + 10e^- + 12H^+ \rightarrow N_2 + 6H_2O$$

$$SO_4^{2-} + 8e^- + 10H^+ \rightarrow H_2S + 4H_2O$$

$$Fe(OH)_3 + e^- + 3H^+ \rightarrow Fe^{2+} + 3H_2O$$

Anaerobic biodegradation is a multistep process performed by different bacterial groups. It involves hydrolysis of polymeric substances like proteins or carbohydrates to monomers and the subsequent decomposition to soluble acids, alcohols, molecular hydrogen, and carbon dioxide. Depending on the prevailing environmental conditions, the final steps of ultimate anaerobic biodegradation are performed by denitrifying, sulfate-reducing, or methanogenic bacteria.

In contrast to the strictly anaerobic sulfate-reducing and methanogenic bacteria, the nitrate-reducing microorganisms and many other decomposing bacteria are mostly facultative anaerobic insofar as these microorganisms able to grow and to degrade inorganic substances under aerobic and anaerobic conditions. Thus, aerobic and anaerobic environments represent the two extremes of a continuous spectrum of environmental habitats that are populated by a wide variety of microorganisms with specific biodegradation abilities.

7.7 CHEMISTRY IN THE ENVIRONMENT

In terms of inorganic chemicals, a chemical transformation is the conversion of a substrate (or reactant) to a product. In more general terms, a chemical transformation involves (or is) a chemical reaction that is characterized by a chemical change and yields one or more products, which usually have properties substantially different from the properties of the individual reactants. Reactions often consist of a sequence of individual substeps that can be described by means of chemical equations, which symbolically present the starting materials, end products, and sometimes intermediate products and reaction conditions.

Chemical reactions occur at a characteristic rate (the reaction rate) at a given temperature and chemical concentration. Typically, reaction rates increase with increasing temperature because there is more thermal energy available to reach the activation energy necessary for breaking bonds between atoms. The general rule of thumb (see above) is that for every 10°C (18°F) increase in temperature the rate of an inorganic chemical reaction is doubled and there is no reason to doubt that this would not be the case for inorganic chemicals discharged into the environment.

Chemicals can enter the environment (air, water, and soil) when they are produced, used, or disposed, and the impact on the environment is determined by the amount of the chemical that is released, the type and concentration of the chemical, and where it is found, as well as through any chemical transformation that occur after the chemical has entered the environment whether it is in the atmosphere, the aquasphere, or the terrestrial biosphere. Some chemicals can be harmful if released to the environment even when there is not an immediate, visible impact. Some chemicals are of concern as they can work their way into the food chain and accumulate and/or persist in the environment for prolonged periods, including years that is in direct contradiction of the earlier *conventional wisdom* (or unbridled optimism) that assumed that inorganic chemicals would either (i) degrade into the harmless by-products as a result of microbial or chemical reactions, (ii) immobilize completely by binding to soil solids, or (iii) volatilize to the atmosphere where dilution to harmless levels was assured. This false assurance led to years of agricultural chemical use and chemical waste disposal with no monitoring of atmosphere or groundwater (the aquasphere) or soil (the terrestrial biosphere) in the vicinity of discharge. Thus, the volatility of an inorganic chemical is of concern predominantly for surface-located chemicals and is affected by (i) the temperature of the soil; (ii) the water content of the soil; (iii) the adsorptive interaction of the chemical and the soil; (iv) the concentration of the chemical in the soil; (v) the vapor pressure of the chemical; and (vi) the solubility of the chemical in water, which is the predominant liquid in the soil.

7.7.1 Chemistry in the Atmosphere

The study of chemistry in the atmosphere (atmospheric chemistry) is a branch of science in which the chemistry of the atmosphere of the Earth is the prime focus (Warneck, 2000; Wayne, 2000; Seinfeld and Pandis, 2006). Studying the atmosphere includes studying the interactions between the atmosphere and living organisms. The composition of the atmosphere changes as result of natural processes such as emission from volcanoes, lightning, and bombardment by solar particles from the corona of the Sun. It has also been changed by human activity, examples of which include acid rain, ozone depletion, photochemical smog, greenhouse gas emissions, and climate change, although there are other geologic factors involved such as the Earth being in an interglacial period (Speight and Islam, 2016).

Atmospheric chemistry is an important discipline for understanding air pollution and its impacts. However, atmospheric composition and chemistry is not always simple and is complex, being controlled by the emission and photochemistry. The transport of many trace gases is often represented by convenient chemical equations. Understanding the timescale and the chemical and spatial patterns of perturbations to trace gases is needed to evaluate possible environmental damage (e.g., stratospheric ozone depletion or climate change) caused by anthropogenic emissions (Baulch et al., 1980; Finlayson-Pitts, 2010).

Inorganic chemicals can be emitted directly into the atmosphere or formed by chemical conversion through chemical reactions of precursor species. In these reaction, highly toxic inorganic chemicals can be converted into less toxic products, but the result of the reactions can also be products having a higher toxicity than the starting chemicals. In order to understand these reactions, it is also necessary to understand the chemical composition of the natural atmosphere, the way gases, liquids, and solids in the atmosphere interact with each other and with the Earth's surface and associated biota and how human activities may be changing the chemical and physical characteristics of the atmosphere (Baulch et al., 1980; Gaffney and Marley, 2003; Prather, 2007; Finlayson-Pitts, 2010).

In addition, the atmosphere is composed of a number of important regions that are defined by thermal structure as a function of altitude (Chapter 1) (Finlayson-Pitts and Pitts, 2000) that contribute to various physical effects. For example, as altitude increases, air pressure drops off rapidly, and furthermore, the temperature structure of the atmosphere is somewhat complex because of exothermic chemical reactions in the upper atmosphere that are caused by the absorption of high-energy photons (ultraviolet-C radiation and ultraviolet-B radiation) from the incoming solar radiation.

The rupture of molecular bonds produces atomic species and results in the recombination reactions that form ozone (O_3). In this process, an oxygen molecule is split (photolyzed) by higher-frequency ultraviolet (UV) light into two oxygen atoms after which each oxygen atom then combines with an oxygen molecule to form an ozone molecule:

$$O_2 + h\nu \rightarrow 2O\bullet$$

$$O\bullet + O_2 \rightarrow O_3$$

Other important gases may also be formed and that can filter the incoming ultraviolet and act as a protective optical shield against harmful solar radiation. Most of this shielding occurs in the stratosphere, where a great deal of recent interest has been focused because of the depletion of the stratospheric ozone layer by anthropogenic use of various chemicals. As altitude increases from sea level, temperature in the troposphere decreases up to the tropopause, where temperature begins to increase in the stratosphere.

Thus, there are a number of critical environmental issues associated with a changing atmosphere, including photochemical smog, global climate change, toxic air pollutants, acidic deposition, and stratospheric ozone depletion. In fact, photochemical smog has been observed in almost every major city on the planet, and the release of nitrogen oxides (as well as hydrocarbon derivatives) into the air has been found to lead to the photochemical production of ozone and a variety of secondary species including aerosols (Gaffney and Marley, 2003). Much of this anthropogenic (human) impact on the atmosphere has been associated with the increasing use of fossil fuels as an energy source—for things such as heating, transportation, and electric power production. Photochemical smog and the generation of tropospheric ozone are serious environmental problems that have been associated with the use of fossil fuels. In fact, the combustion of fossil fuels (which are in fact, inorganic chemicals) is one of the most common sequences of chemistry that causes pollution in the atmosphere. This phenomenon may not be classed as direct pollution (in the sense of inorganic chemistry) but is certainly and indirect form of pollution (again, in the sense of inorganic chemistry).

Over the past decade, it has become clear that aqueous chemical processes occurring in cloud droplets and wet atmospheric particles are an important sources of organic atmospheric particulate matter (McNeill, 2015). Reactions of water-soluble VOCs or semivolatile organic gases (SVOCs) in these aqueous media lead to the formation of highly oxidized organic particulate matter (secondary organic aerosol) and key tracer species, such as organosulfate derivatives. These processes are often driven by a combination of anthropogenic and biogenic emissions, and therefore, their accurate representation in models is important for effective air quality management. One of the results of these various processes is the formation and deposition of acid rain.

Acid rain is formed when sulfur dioxide and nitrogen oxides react with water vapor and other chemicals in the presence of sunlight to form various acidic compounds in the air. The principle source of acid-rain-causing pollutants, sulfur dioxide and nitrogen oxides, are from fossil fuel combustion and from the combustion of fossil-fuel-derived fuels:

$$2[\mathrm{C}]_{\text{fossil fuel}} + \mathrm{O_2} \rightarrow 2\mathrm{CO}$$

$$[\mathrm{C}]_{\text{fossil fuel}} + \mathrm{O_2} \rightarrow \mathrm{CO_2}$$

$$2[\mathrm{N}]_{\text{fossil fuel}} + \mathrm{O_2} \rightarrow 2\mathrm{NO}$$

$$[\mathrm{N}]_{\text{fossil fuel}} + \mathrm{O_2} \rightarrow \mathrm{NO_2}$$

$$[\mathrm{S}]_{\text{fossil fuel}} + \mathrm{O_2} \rightarrow \mathrm{SO_2}$$

$$2\mathrm{SO_2} + \mathrm{O_2} \rightarrow 2\mathrm{SO_3}$$

Hydrogen sulfide and ammonia are produced from processing sulfur-containing and nitrogen-containing feedstocks:

$$[S]_{\text{fossil fuel}} + H_2 \rightarrow H_2S + \text{hydrocarbons}$$
$$2[N]_{\text{fossil fuel}} + 3H_2 \rightarrow 2NH_3 + \text{hydrocarbons}$$
$$SO_2 + H_2O \rightarrow H_2SO_3(\text{sulfurous acid})$$
$$SO_3 + H_2O \rightarrow H_2SO_4(\text{sulfuric acid})$$
$$NO + H_2O \rightarrow HNO_2(\text{nitrous acid})$$
$$3NO_2 + 2H_2O \rightarrow HNO_3(\text{nitric acid})$$

Two of the pollutants that are emitted are hydrocarbons (e.g., unburned fuel) and nitric oxide (NO). When these pollutants build up to sufficiently high levels, a chain reaction occurs from their interaction with sunlight in which the NO is converted to nitrogen dioxide (NO_2)—a brown gas—and at sufficiently high levels can contribute to urban haze. However, a more serious problem is that nitrogen dioxide (NO_2) can absorb sunlight and break apart to produce oxygen atoms that combine with the oxygen in the air to produce ozone (O_3), a powerful oxidizing agent and a toxic gas.

7.7.2 Chemistry in the Aquasphere

Water is the most abundant molecule on the Earth's surface and one of the most important molecules to study in chemistry and plays an important role as a chemical substance. Water has a simple molecular structure; it is composed of one oxygen atom and two hydrogen atoms (H_2O). Each hydrogen atom is covalently bonded to the oxygen via a shared pair of electrons. Oxygen also has two unshared pairs of electrons, and thus, oxygen is an electronegative atom compared with hydrogen. Also, water is a polar molecule because of the uneven distribution of electron density; water has a partial negative charge ($\delta -$) near the oxygen atom due to the unshared pairs of electrons and partial positive charges ($\delta +$) near the hydrogen atoms resulting in the formation of hydrogen bonds. In fact, many other unique properties of water are due to the hydrogen bonds. For example, the unique physical properties, including a high heat of vaporization, strong surface tension, high specific heat, and nearly universal solvent properties of water, are also due to hydrogen bonding. The hydrophobic effect or the exclusion of compounds containing carbon and hydrogen (nonpolar compounds) is another unique property of water caused by the hydrogen bonds. The hydrophobic effect is particularly important in the formation of cell membranes.

Thus, the many important functions of water include (i) being a good solvent for dissolving many solids, (ii) serving as an excellent coolant both mechanically and biologically, and (iii) acting as a reactant in many chemical reactions. For example, the ability of acids to react with bases depends on the tendency of hydrogen ions to combine with hydroxide ions to form water. This tendency is very great, so the reaction is practically *complete*:

$$H^{+}(aq) + OH^{-}(aq) \rightarrow H_2O$$

No reaction, however, is really 100% complete; at *equilibrium* (when there is no further net change in amounts of substances), there will be at least a minute concentration of the reactants in the solution. Another way of expressing this is to say that any reaction is at least slightly *reversible*. In pure water, the following reaction will proceed to a very slight extent. Thus,

$$H_2O \rightarrow H^{+}(aq) + OH^{-}(aq)$$

Water pollution has become a widespread phenomenon and has been known for centuries, particularly the pollution of rivers and groundwater. By way of example, in ancient time up to the early part of the 20th century, many cities deposited waste into the nearby river or even into the ocean. It is only very recently (because of serious concerns for the condition of the environment) that an understanding of the behavior and fate of toxic chemicals, which are discharged to the aquatic environment (Table 7.7) because of these activities, is essential to the control of water pollution. In rivers, the basic physical movement of pollutant molecules is the result of advection, but superimposed upon this are the effects of dispersion and mixing with tributaries and other discharges. Some of the chemicals discharged are relatively inert, so their concentration changes only due to advection, dispersion, and mixing. However, many inorganic chemicals are not conservative in their behavior and undergo changes due to chemical processes or biochemical processes, such as oxidation. The dispersibility of an inorganic chemical is a measure or indication of the potential of the chemical to spread in the air, in the water, or on the land. In some spill situations and under appropriate conditions, dispersants are employed to interfere with the tendency of the chemical to adhere to or adsorb on mineral surfaces. This may be an effective countermeasure for minimizing contamination of the air, shorelines, and/or land-based sites.

In addition, there are many indications that the inorganic chemical materials in the aquasphere (also called, when referring to the sea, the marine

Table 7.7 Changes in Physicochemical Properties to Favor Reduced Aquatic Toxicity (NRC, 2014)

Physicochemical Property	Changes
Molecular size and weight	Generally, as molecular weight increases, aquatic bioavailability and toxicity decrease. At MW > 1000, bioavailability is negligible. Caution must be taken, however, to consider possible breakdown products that may have MW < 1000 and exert toxicity
Octanol-water partition coefficient (log *P*) and octanol-water distribution coefficient at biological pH (log $D_{7.4}$)	log *P* usually correlates exponentially with acute aquatic toxicity by narcosis for nonionic organic chemicals up to a value of about 5–7. Chemicals with log $P<2$ have a higher probability of having low acute and chronic aquatic toxicity. For ionizable organic chemicals, log $D_{7.4}$ is a more appropriate measure; ionizable compounds with log $D_{7.4}<1.7$ have been shown to have increased probability of being safe to freshwater fish than those with log $D_{7.4}>1.7$
Water solubility	Generally, compounds with higher log *P* have lower water solubility. Very poorly water-soluble chemicals (<1 ppb) generally have low bioavailability and are less toxic

aquasphere) are subject to intense chemical transformations and physical recycling processes imply that a total inorganic-carbon approach is not sufficient to resolve the numerous processes occurring.

The effects of an inorganic chemical released into the marine environment (or any part of the aquasphere) depend on several factors such as (i) the toxicity of the chemical; (ii) the quantity of the chemical; (iii) the resulting concentration of the chemical in the water column; (iv) the length of time that floral and faunal organisms are exposed to that concentration; and (v) the level of tolerance of the organisms, which varies greatly among different species and during the life cycle of the organism. Even if the concentration of the chemical is below what would be considered as the lethal concentration, a sublethal concentration of an inorganic chemical can still lead to a long-term impact within the aqueous marine environment. For example, chemically induced stress can reduce the overall ability of an organism to reproduce, grow, feed, or otherwise function normally within a few generations. In addition, the characteristics of some inorganic chemicals can result in an accumulation of the chemical within an organism (*bioaccumulation*), and the organism may be particularly vulnerable to this problem. Furthermore, subsequent biomagnification may also occur if the inorganic chemical (or a toxic product produced by one or more transformation reactions) can be passed on, following the food chain up to higher flora and fauna.

In terms of the marine environment and a spill of crude oil, complex processes of crude oil transformation start developing almost as soon as the oil contacts the water although the progress, duration, and result of the transformations depend on the properties and composition of the oil itself, parameters of the actual oil spill, and environmental conditions. The major operative processes are (i) physical transport, (ii) dissolution, (iii) emulsification, (iv) oxidation, (v) sedimentation, (vi) microbial degradation, (vii) aggregation, and (viii) self-purification.

In terms of *physical transport*, the distribution of an inorganic chemical oil spilled into the aquasphere is subject to transportation and changes due to the presence of oxygen in the water. However, the major changes take place under the combined impact of meteorologic and hydrologic factors and depend mainly on the power and direction of wind, waves, and currents. A considerable part of the chemical will disperse in the water as ion droplets that can be transported over large distances away from the place of the entry point of the chemical into the water. This is analogous to dilution but should not be construed as being in order because of the high volume of water. Water like any other solvent can only tolerate so much in the way of the amount of solute before the properties of the water have an adverse effect on the floral and faunal species in the water spill. *Oxidation*, a complex process even in water, can also occur as a series of chemical reactions involving the hydroxyl radical (•OH).

As these processes occur, some of the crude oil constituents are adsorbed on any suspended material and deposited on the ocean floor (sedimentation), the rate of which is dependent upon the ocean depth; in deeper areas remote from the shore, sedimentation of oil (except for the heavy fractions) is a slow process. Simultaneously, the process of *biosedimentation* occurs; in this process, plankton and other organisms absorb the emulsified oil, and the crude oil constituents are sent to the bottom of the ocean as sediment with the metabolites of the plankton and other organisms. However, this situation radically changes when the suspended oil reaches the sea bottom; the decomposition rate of the oil on the ocean bottom abruptly ceases—especially under the prevailing anaerobic conditions, and any crude oil constituents accumulated inside the sediments can be preserved for many months and even years. These products can be swept to the edge of the ocean (the beach) by turbulent condition at some later time.

The fate of most of the constituents of crude oil in the marine environment is ultimately defined by their transformation and degradation due to *microbial degradation*. The degree and rates of biodegradation depend upon the structure of the chemical, and the constituents of the mixture biodegrade faster than aromatic constituents and naphthenic constituents, and with increasing complexity of molecular structure and with increasing molecular weight, the rate of microbial decomposition usually decreases. Besides, this rate depends on the physical state of the oil, including the degree of its dispersion and environmental factors such as temperature, availability of oxygen, and the abundance of oil-degrading microorganisms.

In the chemical sense, *aggregation* (especially *particle agglomeration*) refers to the formation of assemblages in a suspension and represents a mechanism leading to destabilization of colloidal systems. During this process, particles dispersed in the liquid phase stick to each other and spontaneously form irregular particle clusters, flocs, or aggregates. This phenomenon (also referred to as *coagulation* or *flocculation*) and such a suspension is also called *unstable*. Particle aggregation can be induced by adding salts or another chemical referred to as coagulant or flocculant.

Particle aggregation is normally an irreversible process, and once particle aggregates have formed, they will not easily disrupt. In the course of aggregation, the aggregates will grow in size, and as a consequence, they may settle to the bottom of the container, which is referred to as sedimentation. Alternatively, a colloidal gel may form in concentrated suspensions that change the rheological properties. The reverse process whereby particle aggregates are disrupted and dispersed as individual particles, referred to as peptization, hardly occurs spontaneously but may occur under stirring or shear. Colloidal particles may also remain dispersed in liquids for long periods of time (days to years). This phenomenon is referred to as *colloidal stability* and such a suspension is said to be *stable*. Stable suspensions are often obtained at low salt concentrations or by the addition of chemicals referred to as *stabilizers* or *stabilizing agents*. Similar aggregation processes occur in other dispersed systems. For example, in an emulsion, they may also be coupled to droplet coalescence and lead not only to sedimentation but also to creaming. In aerosols, airborne particles may equally aggregate and form larger clusters. In the chemical sense, creaming is the migration of the dispersed phase of an emulsion, under the influence of buoyancy. The particles float upwards or sink, depending on how large they are and how much less dense or more dense they may be than the continuous phase and also how viscous or how thixotropic (a time-dependent shear thinning property) the continuous phase might be. For as long as the particles remain separated, the process is called *creaming*.

Self-purification is a result of the processes previously described above in which crude oil in the marine environment rapidly loses its original properties and disintegrates into various fractions. These fractions have different chemical composition and structure and exist in different migrational forms, and they undergo chemical transformations that slow after reaching thermodynamic equilibrium with the environmental parameters. Eventually, the original and intermediate compounds disappear, and carbon dioxide and water form. This form of self-purification inevitably happens in water ecosystems if the toxic load does not exceed acceptable limits.

While acid-base reactions are not the only chemical reactions important in aquatic systems, they do present a valuable starting point for understanding the basic concepts of chemical equilibriums in such systems. Carbon dioxide (CO_2), a gaseous inorganic chemical substance of vital importance to a variety of environmental processes, including growth and decomposition of biological systems, climate regulation, and mineral weathering, has acid-base properties that are critical to an understanding of its chemical behavior in the environment. The phenomenon of acid rain is another example of the importance of acid-base equilibriums in natural aquatic systems.

Furthermore, the almost unique physical and chemical properties of water as a solvent are of fundamental concern to aquatic chemical processes. For example, in the liquid state, water has unusually high boiling-point and melting-point temperatures compared with its hydride analogues from the periodic table such as ammonia (NH_3), hydrogen fluoride (HF), and hydrogen sulfide (H_2S) (Table 7.8). Hydrogen bonding between water molecules means that there are strong intermolecular forces making it relatively difficult to melt or vaporize.

In addition, pure water has a maximum density at 4°C (39°F), higher in temperature than its freezing point (0°C, 32°F), and that ice is substantially less dense than liquid water and is important (and fortunate) in several contexts. Thus, as water in a lake is cooled at its surface by loss of heat to the atmosphere, the ice structures formed will float. Furthermore, a dynamically

Table 7.8 Melting- Point and Boiling Point Temperatures (at Normal Pressure) for Hydrides of the Groups 15–17 Elements in the Periodic Table That Are Analogous to Water

Melting Points (°C)		
NH_3: −78	H_2O: 0	HF: −83
PH_3: −133	H_2S: −86	HCl: −115
AsH_3: −116	H_2Se: −60	HBr: −89
SbH_3: −-88	H_2Te: −49	HI: −51
Boiling points (°C)		
NH_3: −33	H_2O: 100	HF: 20
PH_3: −88	H_2S: −61	HCl: −85
AsH_3: −55	H_2Se: −42	HBr: −67
SbH_3: −17	H_2Te: −2	HI: −35

stable water layer near 4°C (39°F) will tend to accumulate at the bottom of the lake, and the overlying, less dense water able to continue cooling down to the freezing point. This means that ice will eventually coalesce at the surface, forming an insulating layer that greatly reduces the rate of freezing of the underlying water. This situation is obviously important for plants and animals that inhabit lake waters.

In addition, the presence of salt components means that the temperature of maximum density for seawater is shifted to lower temperatures; in fact, the density of seawater continues to increase right down to the freezing point. The high concentration of electrolytes in seawater assists in breaking up the open, hydrogen-bonded icelike structure of water near its freezing point. Because the salt components tend to be excluded from the ice formed by freezing seawater, sea ice is relatively fresh and still floats on water. Much of the salt it contains is not truly part of the ice structure but contained in brines that are physically entrained by small pockets and fissures in the ice.

The dielectric constant of water (78.2 at 25°C, 77°F) is high compared with most liquids. Of the common liquids, few have comparable values at this temperature, e.g., hydrogen cyanide (HCN, 106.8), hydrogen fluoride (HF, 83.6), and sulfuric acid (H_2SO_4, 101). By contrast, most nonpolar liquids have dielectric constants on the order of 2. The high dielectric constant helps liquid water to solvate ions, making it a good solvent for ionic substances, and arises because of the polar nature of the water molecule and the tetrahedrally coordinated structure in the liquid phase.

In many electrolyte solutions of interest, the presence of ions can alter the nature of the water structure. Ions tend to orient water molecules that are near to them. For example, cations attract the negative oxygen end of the water dipole toward them. This reorientation tends to disrupt the icelike structure further away. This can be seen by comparing the entropy change on transferring ions from the gas phase to water with a similar species that does not form ions.

7.7.3 Chemistry in the Terrestrial Biosphere

The biosphere consists of the parts of the Earth where life exists and extends from the deepest root systems of trees to the dark environment of ocean trenches, to the rain forests, and to the high mountaintops. Often, the Earth is described in terms of spheres—(i) the solid surface layer of the Earth is the lithosphere; (ii) the water on the surface of the Earth, in the ground and in the air, constitutes the hydrosphere; and (iii) the atmosphere is the layer of air that stretches above the lithosphere.

Since life exists on the ground, in the air, and in the water, the biosphere overlaps all these spheres. Although the biosphere measures about 20 km (12 miles) from top to bottom, almost all life exists between about 500 m (1640 ft) below the ocean's surface and about 6 km (3.75 miles) above sea level. The addition of oxygen to the biosphere allowed more complex life forms to evolve. Millions of different plants and other photosynthetic species developed. Animals, which consume plants (and other animals), evolved. Bacteria and other organisms evolved to decompose or break down dead animals and plants. The biosphere benefits from this food web; the remains of dead plants and animals release nutrients into the soil and into the ocean. These nutrients are reabsorbed by growing plants. This exchange of food and energy makes the biosphere a self-supporting and self-regulating system. The biosphere is sometimes considered to be one large ecosystem—that is, in fact many ecosystems that exist as a complex community of living and nonliving things functioning as a single unit. This delicate balance can be upset by the discharge of toxic inorganic chemicals into the biosphere.

In a general context, the terrestrial biosphere is predominantly the soil that is on the surface of the Earth and that can house many different types of inorganic chemicals and properties (Table 7.9). Soil is a matrix of solids including sand, silt, clay, and organic matter particles and aggregates of various sizes formed from them and pore space, which may be filled with air or water. However, soils are chemically different from the rocks and minerals from which they are formed in that soils contain less of the water-soluble weathering products, calcium, magnesium, sodium, and potassium, and

Table 7.9 General Characteristics of Soil

Property	Characteristic	Comments
Composition	Porous body	Surface horizons have approximately equal volumes of solids and pores containing water or air
	Weathering	Weatherable minerals decompose and more resistant minerals form
	Transported chemicals	Small particles may be transported by water, wind, or gravity
Physical properties	Texture	Surface area per unit volume and particle size; influence on properties and reactivity
	Structure	Influence on porosity and large and small pore ratio; affects adsorption-absorption
	Consistency	Physical behavior of soil changes as moisture content changes
Chemical properties	Cation exchange	Ability of soil clay minerals to reversibly adsorb and exchange cations
	pH	A measure of the acidity or alkalinity can change soil chemistry
Environment	Water	Can change soil chemistry, hydrolysis
	Temperature	Can change soil chemistry, decomposition
	Aeration	Can change soil chemistry, redox reactions

more of the relatively insoluble elements such as iron and aluminum. Older, highly weathered soil typically has a high concentration of aluminum oxide and iron oxide.

An ecosystem is a community of living organisms in conjunction with the nonliving components of their environment (such as air, water, and mineral soil), interacting as a system. Ecosystems are controlled both by (i) external factors and (ii) internal factors. External factors such as climate, the parent material that forms the soil, and topography control the overall structure of an ecosystem and the way things work within it, but are not themselves influenced by the ecosystem. Internal factors not only control ecosystem processes but also are controlled by them and are often subject to feedback loops. Other internal factors include disturbance, succession, and the types of species present. Although humans exist and operate within ecosystems, their cumulative effects are large enough to influence external factors like climate. An important aspect of the chemistry of the biosphere is the chemistry that occurs in the soil (Strawn et al., 2015).

The soil itself is a complex mixture of inorganic chemicals—even before pollution occurs. Eight chemical elements comprise the majority of the mineral matter in soils. Of these eight elements, oxygen, a negatively charged ion (anion) in crystal structures, is the most prevalent on both weight and volume basis. The next most common elements, all positively charged ions that, in decreasing order, are silicon, aluminum, iron, magnesium, calcium, sodium, and potassium. Ions of these elements combine in various ratios to form different minerals.

The most chemically active fraction of soils consists of colloidal clays and organic matter. Colloidal particles are so small (<0.0002 mm) that they remain suspended in water and exhibit a very large surface area per unit weight. These materials also generally exhibit net negative charge and high adsorptive capacity. Several different silicate clay minerals exist in soils, but all have a layered structure. It is these clay minerals that act as adsorbents for inorganic pollutants (as well as organic pollutants).

The soil transports and moves water, provides refuge for bacteria and other fauna, and has many different arrangements of weathered rock and minerals. When soil and minerals weather over time, the chemical composition of soil also changes. However, many of the problems of soil chemistry are concerned with environmental sciences such as the accidental or deliberate disposal of chemicals. This can involve several physical-chemical interactions that can cause changes to the soil: (1) ion exchange, (ii) soil pH, (iii) sorption and precipitation, and (iv) oxidation-reduction reactions.

Ion exchange involves the movement of cations (positively charged elements such as calcium, magnesium, and sodium) and anions (negatively charged elements such as chloride and compounds like nitrate) through the soils. More specifically, *cation exchange* is the interchanging between a cation in the solution of water around the soil particle and another cation that is on the surface of the clay mineral.

A cation is a positively charged ion, and most inorganic contaminants are cations, such as ammonium (NH_4^+), calcium (Ca^{2+}), copper (Cu^{2+}), magnesium (Mg^{2+}), manganese (Mn^{2+}), potassium (K^+), and zinc (Zn^{2+}). These cations are in the soil solution and are in dynamic equilibrium with the cations adsorbed on the surface of clay and organic matter. It is the cation-exchange interaction of ions with clay minerals in the soil that gave spoil the ability of soil to adsorb and exchange cations with those in soil solution (water in soil pore space). The adsorption capacity of clay minerals can cause a mixture of inorganic pollutants to physically transform when one or more constituents of the mixture adsorbs onto a clay mineral. In addition, the total amount of positive charges that the clay can absorb (the *cation-exchange capacity*, CEC) impacts the rate of movement of the pollutants through the soil and the physical and chemical changes that can occur to the pollutants. For example, soil (i.e., clay) with a low cation-exchange capacity is much less likely to retain pollutants, and furthermore, the soil (i.e., the clay) is less able to retain inorganic chemicals with the potential that the chemicals are released into the groundwater.

Soil pH is a commonly measured soil chemical property and is also one of the more informative properties since the data imply certain characteristics that might be associated with a soil. For example, the pH of soil is a measure of the acidity or alkalinity: values from 0 to 7 indicate acidity, while values from 7 to 14 indicate alkalinity. Typically, the pH value for soil is on the order of soils usually range from 4 to 10. The pH is one of the most important properties involved in plant growth and understanding how rapidly reactions occur in the soil. For example, the element iron becomes less available to plants at higher alkalinity ($pH > 7$) that creates iron deficiency problems. Crops usually have optimal growth at pH values between 5.5 and 8, but the value depends on the crop. The pH of soil also affects the ability of organisms to feed on contaminants. Thus, any chemical spillage onto soil can seriously affect the overall chemistry of the soil and the ability of flora and fauna to survive.

The other principal variables affecting life in soil in the environment include moisture, temperature, and aeration, and the balance of these factors controls the abundance and activities of the floral and faunal inhabitants in the soil, which in turn have a marked influence on the critical processes of soil aggregation retention of pollutants.

Soil water is usually derived from rainfall or some other form of overland flow. The amount of water that enters the soil is a function of soil structure that in turn is a function of texture and higher-order structure, i.e., aggregation (e.g., well-aggregated, porous soils allow greater infiltration). Larger pores created by root channels and animal burrows (ant tunnels and wormholes) and other types of macropores also greatly facilitate movement of water into and through the soil profile. This is why these soil animals are usually regarded as highly beneficial in soil and why some scientists often use the abundance of earthworms as an indicator of a healthy soil, though it should be remembered that highly fertile, productive soils do not always contain large numbers of earthworms and other soil faunal species.

Sorption is the process in which one substance takes up or holds another. In the current context, soil that has a high sorption capacity can hold a high level of environmental contaminants by sorption of the contaminants onto the soil particles, which have the ability to capture different nutrients and ions. *Precipitation* occurs during chemical reactions when a nutrient or chemical in the soil solution (water around the soil particles) transforms into a solid. The presence or absence of oxygen determines the manner by which soil can react chemically. *Oxidation* is the loss of electrons, and *reduction* is a gain of electrons at the soil surface.

Although soil is an inorganic matrix of metals and minerals, the main reason why the soil becomes contaminated is due to the presence of anthropogenic waste. This waste is typically (i) chemicals in one form or another that are not indigenous to the soil or (ii) chemicals that are indigenous chemicals that are deposited into the soil in amounts that exceed the natural abundance. The typical actions that lead to pollution of the soil by inorganic chemicals are (i) industrial activities, (ii) agricultural activities, (iii) waste disposal, and (iv) acid rain.

Industrial activities have been the biggest contributor to the soil pollution since the time of the *industrial revolution*, especially since the amount of mining and manufacturing increased almost logarithmically in the 160 years immediately following 1800. Most industries are dependent on extracting minerals from the Earth, and the by-products have not been sent for disposal in a manner that can be considered safe. Thus, the industrial waste lingers in the soil surface for a long time and makes it unsuitable for use. At the same time, physical changes and chemical changes to the inorganic pollutant also occur for the better or for worse.

Agricultural activities have caused a rise in the use of inorganic chemicals since technology provided the means of producing inorganic pesticides and inorganic fertilizers. They are full of chemicals that are not produced in nature and cannot be broken down by it. These chemicals seep into the ground after they mix with water and slowly change the character or of the soil by adsorption and by changing the acidity/alkalinity of the water in the soil. Other chemicals damage the composition of the soil and make it easier to erode by water and air.

Waste disposal has also been a major cause for concern in how we dispose of our waste. While industrial waste is sure to cause contamination, this type of waste typically contains toxins and chemicals that are now seeping into the land and causing pollution of soil. Finally, acid rain is caused when pollutants present in the air mix up with the rain and fall back to the soil. The polluted (acidified) water has the ability to dissolve away some of the soil constituents and, been more drastic, change the structure of the soil. This latter action could release pollutants that were adsorbed (and may have even remained adsorbed) and once released can find their way into the groundwater.

REFERENCES

Baulch, D.L., Cox, R.A., Hampson Jr., R.F., Kerr, J.A., Roe, J., Watson, R.T., 1980. Evaluation of kinetic and photochemical data for atmospheric chemistry. J. Phys. Chem. Ref. Data 9 (2), 295–471.

Centi, G., Ciambelli, P., Perathoner, S., Russo, P., 2002. Environmental catalysis: trends and outlook. Catal. Today 75 (1–4), 3015.

Finlayson-Pitts, B.J., 2010. Atmospheric chemistry. Proc. Natl. Acad. Sci. U. S. A. 107 (15), 6566–6567.

Finlayson-Pitts, B.J., Pitts Jr., J.N., 2000. Chemistry of the Upper and Lower Atmosphere: Theory, Experiments, and Applications. Academic Press, San Diego, CA.

Gaffney, J.S., Marley, N.A., 2003. Atmospheric chemistry and air pollution. Sci. World J. 3, 199–234.

Grassian, V.H., 2005. Environmental Catalysis. CRC Press/Taylor & Francis Group, Boca Raton, FL.

Habashi, F., 1994. Conversion reactions in inorganic chemistry. J. Chem. Educ. 71 (2), 130.

McNeill, V.F., 2015. Aqueous organic chemistry in the atmosphere: sources and chemical processing of organic aerosols. Environ. Sci. Technol. 49 (3), 1237–1244.

Nester, E.W., Anderson, D.G., Roberts Jr., C.E., Pearsall, N.N., Nester, M.T., 2001. Microbiology: A Human Perspective, third ed. McGraw-Hill, New York, NY.

NRC, 2014. Physicochemical Properties and Environmental Fate: A Framework to Guide Selection of Chemical Alternatives. National Research Council, Washington, DC.

Petrucci, R.H., Herning, G.E., Madura, J., Bissonnette, C., 2010. General Chemistry: Principles and Modern Application, 11th ed. Prentice Hall, Upper Saddle River, NJ.

Prather, M.J., 2007. Lifetimes and time-scales in atmospheric chemistry. Phil. Trans. R. Soc. A 365, 1705–1726.

Rahm, S., Green, N., Norrgran, J., Bergman, Å., 2005. Hydrolysis of environmental contaminants as an experimental tool for indication of their persistency. Environ. Sci. Technol. 39 (9), 3128–3133.

Seinfeld, J.H., Pandis, S.N., 2006. Atmospheric Chemistry and Physics: From Air Pollution to Climate Change. In: second ed. John Wiley & Sons, Hoboken, NJ.

Speight, J.G., Islam, M.R., 2016. Peak Energy—Myth or Reality. Scrivener Publishing, Salem, MA.

Strawn, D.G., Bohn, H.L., O'Connor, G.A., 2015. Spoil Chemistry, fourth ed. John Wiley & Sons Inc., Hoboken, NJ.

Summers, A.O., Silver, S., 1978. Microbial transformation of metals. Annu. Rev. Microbiol. 32, 1–709.

Warneck, P., 2000. Chemistry of the Natural Atmosphere, second ed. Academic Press Inc., New York, NY.

Wayne, R.P., 2000. Chemistry of Atmospheres, third ed. Oxford University Press, Oxford.

Wayne, C.E., Wayne, R.P., 2005. Photochemistry. Oxford University Press, Oxford.

Zhou, Y.-F., Haynes, R.J., 2010. Sorption of heavy metals by inorganic and organic components of solid wastes: significance to use of wastes as low-cost adsorbents and immobilizing agents. Crit. Rev. Environ. Sci. Technol. 40 (11), 909–977.

CHAPTER EIGHT

Environmental Regulations

8.1 INTRODUCTION

The latter part of the 20th century started with the important realization that all chemicals, especially inorganic chemicals, can act as environmental pollutants depending upon the ecosystem into which the chemicals are discharged and the amount of chemical discharged. In addition, there came the realization, in the current context, that emissions of inorganic compounds such as carbon dioxide (CO_2), methane (CH_4), and nitrous oxide (N_2O) to the atmosphere had either a direct impact or even an indirect impact on the global climate and on depletion of the ozone layer. As a result, unprecedented efforts were then made to reduce all global emissions of all chemicals in order to maintain a *green perspective* though the evolution of environmental-oriented thinking was followed by the formulation and passage of environmental regulations (Table 8.1).

Inorganic chemicals that are produced by various industrial processes and used in a range of products are an intrinsic part of modern lifestyles. Many products that have contributed to making lifestyles more comfortable, including domestic appliances, detergents, pharmaceuticals, and personal computers, involve the use of chemicals to varying degrees. Chemicals are also an essential component of industrial production and are used in sectors ranging from agriculture and mining to manufacturing. Besides the benefits chemicals bring for the economy, trade, and employment, the rapid increase in the use and build up of chemicals in the environment also comes at a price, including the loss of for the flora and fauna (and any effects on including human health) if not managed effectively. Potential adverse impacts on the environmental flora and fauna include acute poisoning and even long-term effects in humans such as cancerous growths, neurological disorders, and a variety of birth defects. In addition, harmful chemicals can trigger eutrophication of water bodies, ozone depletion, and also pose a threat to sensitive ecosystems and biodiversity. On this note, the Endangered Species Act (ESA) was put into law to prevent extinction of endangered

Environmental Inorganic Chemistry for Engineers
http://dx.doi.org/10.1016/B978-0-12-849891-0.00008-4

Table 8.1 Chronology of Environmental Events and Regulations in the United States (Not Necessarily Related to the Inorganic Chemicals Industry or the Crude Oil-Natural Gas Industry)

1906	The Pure Food and Drug Act established the Food and Drug Administration (FDA) that now oversees the manufacture and use of all foods, food additives, and drugs; Amendments (1938, 1958, and 1962) strengthened the law considerably
1924	The Oil Pollution Act
1935	The Chemical Manufacturers Association (CMA), a private group of people working in the chemical industry and especially involved in the manufacture and selling of chemicals, established a Water Resources Committee to study the effects of their products on water quality
1948	The Chemical Manufacturers Association established an Air Quality Committee to study methods of improving the air that could be implemented by chemical manufacturers
1953	The Delaney Amendment to the Food and Drug Act defined and controlled food additives; any additives showing an increase in cancer tumors in rats, even if extremely large doses were used in the animal studies, had to be outlawed in foods; recent debates have focused on a number of additives, including the artificial sweetener cyclamate
1959	Just before Thanksgiving the government announced that it had destroyed cranberries contaminated with a chemical, aminotriazole that produced cancer in rats; the cranberries were from a lot frozen from 2 years earlier when the chemical was still an approved weed killer
1960	Diethylstilbestrol (DES), taken in the late 1950s and early 1960s to prevent miscarriages and also used as an animal fattener, was reported to cause vaginal cancer in the daughters of these women as well as premature deliveries, miscarriages, and infertility
1962	Thalidomide, a prescription drug used as a tranquilizer and flu medicine for pregnant women in Europe to replace dangerous barbiturates that cause 2000–3000 deaths per year by overdoses, was found to cause birth defects. Thalidomide had been kept off the market in America because of the insistence that more safety data be produced for the drug
1962	The Kefauver-Harris Amendment to the Food and Drug Act began required that drugs be proven safe before put on the market
1962	The publication of Silent Spring (authored by Rachel Carson) that outlined many environmental problems associated with chlorinated pesticides, especially DDT and its use was banned in 972
1965	Nonlinear, nonbiodegradable synthetic detergents made from propylene tetramer were banned after these materials were found in large amounts in rivers, so much as to cause soapy foam in many locations. Phosphates in detergents were banned in detergents by many states in the 1970s
1965	Mercury poisoning from concentration in the food chain recognized
1966	Polychlorinated biphenyls (PCBs) were first found in the environment and in contaminated fish; banned in 1978 except in closed systems
1968	TCDD (a dioxin derivative) tested positive as a teratogen in rats

Table 8.1 Chronology of Environmental Events and Regulations in the United States (Not Necessarily Related to the Inorganic Chemicals Industry or the Crude Oil-Natural Gas Industry)—cont'd

1969	The artificial sweetener cyclamate was banned because of its link to bladder cancer in rats fed with large doses; many 20 subsequent studies have failed to confirm this result but cyclamate remains banned
1972	Federal Water Pollution Control Act
1974	Safe Drinking Water Act
1977	Saccharin found to cause cancer in rats; banned by the FDA temporarily but Congress placed a moratorium on this ban because of public pressure; saccharin is still available
1970	Earth Day recognized because of concern about with the effects of many substances on the environment
1970	The Clean Air
1971	TCDD (see above) outlawed by the Environmental Protection Agency
1971	The Chemical Manufacturers Association established the Chemical Emergency Transportation System (CHEMTREC) to provide immediate information on chemical transportation emergencies
1972	The Clean Water Act
1974	Vinyl chloride investigated as a possible carcinogen
1976	The Toxic Substances Control Act (TSCA or TOSCA); Environmental Protection Agency developed rules to limit manufacture and use of PCBs
1976	The Resource Conservation and Recovery Act (RCRA)
1977	Dibromochloropropane (DBCP) investigated for causes leading to sterility; now banned
1977	Benzene was linked to an abnormally high rate of leukemia; increased concern with benzene use in industry
1978	Ban on chlorofluorocarbons (CFCs) as aerosol propellants; react with ozone in the stratosphere causing an increase the penetration of ultraviolet sunlight and increase the risk of skin cancer
1978	Love Canal, Niagara Falls, New York
1980	CHEMTREC (see above) recognized by the Department of Transportation as the central service to provide immediate information on chemical transportation emergencies
1980	The Comprehensive Environmental Response, Compensation, Liability Act
1986	The Safe Drinking Water Act Amendments
1986	The Emergency Planning and Community-Right-to-Know Act; companies must also report inventories of specific chemicals kept in the workplace and annual release of hazardous materials into the environment
1986	The Superfund Amendments and Reauthorization Act
1989	Pasadena, TX: explosion caused by leakage of exploded when ethylene and iso-butane leaked from a pipeline

Continued

Table 8.1 Chronology of Environmental Events and Regulations in the United States (Not Necessarily Related to the Inorganic Chemicals Industry or the Crude Oil-Natural Gas Industry)—cont'd

1990	Channelview, TX: explosion in a petrochemical treatment tank of wastewater and chemicals
1991	Sterlington, LA: explosion at a nitro-paraffin plant
1991	Charleston, SC: explosion at a plant manufacturing Antiblaze 19, a phosphonate ester and flame retardant used in textiles and polyurethane foam; manufactured from trimethyl phosphite, dimethyl methylphosphonate, and trimethyl phosphate

plants and animals and to recover these populations by preventing threats to their survival.

The evolution of industrial processes and the rising consumption of inorganic chemicals have also led to a rapid increase in generation of hazardous wastes. These wastes not only pose risks and hazards because of their chemical properties and behavior in an ecosystem. Furthermore, if (advertent or inadvertent) mixing is allowed to occur, the potential to contaminate large quantities of otherwise nonhazardous waste chemicals exists. Thus, it has been recognized that the sound management of chemicals throughout their lifecycle and the proper segregation, treatment, and disposal of chemical wastes (hazardous and nonhazardous) are critical to the protection of vulnerable ecosystems, the resident floral and faunal species, including the livelihood and health of human communities (Wagner, 1999).

The Code of Federal Regulations (CFR) is the codification of the general and permanent rules published in the Federal Register by the executive departments and agencies of the Federal Government. It is divided into 50 titles that represent broad areas subject to Federal regulation. Each volume of the CFR is updated once each calendar year and is issued on a quarterly basis.

Notwithstanding early efforts at environmental protection and control, the concept of laws and regulations that were oriented to the protection of the environment but as a separate and distinct from the typical civil laws is a 20th century development (Lazarus, 2004). However, the collective recognition of the public and lawmakers alike that the environment is a fragile organism that is need of legal protection did not occur until late in the 1960s (Chapter 1). At that time, numerous influences including (i) a growing awareness of the unity and fragility of the various ecosystems, (ii) increased public concern over the impact of industrial activity on natural resources and the various flora and fauna including human health, (iii) the increasing strength of the western government and the regulatory ability of these governments, and (iv) the success of various civil movements to protect the environment led to a collections of laws in a relatively short period of time (Tables 8.1 and 8.2). While the modern history of environmental law is one of continuing (political) discussion and evolution, environmental laws had been established by the end of the 20th century as a component of the legal landscape in all of the developed and industrialized counties of the world, which was followed by similar regulatory movements in many developing countries.

By way of introduction to the term, *environmental law* includes regulation of pollutants and natural resource conservation and allocation. The regulations refer to energy development and use, agriculture, real estate, and land

Table 8.2 Federal Regulations Relevant to Chemicals, Including Inorganic Chemicals[a]

Atomic Energy Act (AEA)
Beaches Environmental Assessment and Coastal Health (BEACH) Act
Chemical Safety Information, Site Security, and Fuels Regulatory Relief Act
Clean Air Act (CAA)
Clean Water Act (CWA); original title: Federal Water Pollution Control Amendments of 1972
Comprehensive Environmental Response, Compensation and Liability Act (CERCLA, also known as Superfund or the Superfund Act); also contains the Superfund Amendments and Reauthorization Act (SARA)
Emergency Planning and Community Right-to-Know Act (EPCRA)
Endangered Species Act (ESA)
Energy Independence and Security Act (EISA)
Energy Policy Act
Federal Food, Drug, and Cosmetic Act (FFDCA)
Federal Insecticide, Fungicide, and Rodenticide Act (FIFRA)
Food Quality Protection Act (FQPA)
Marine Protection, Research, and Sanctuaries Act (MPRSA, also known as the Ocean Dumping Act)
National Environmental Policy Act (NEPA)
National Technology Transfer and Advancement Act (NTTAA)
Nuclear Waste Policy Act (NWPA)
Occupational Safety and Health (OSHA)
Oil Pollution Act (OPA)
Pesticide Registration Improvement Act (PRIA)
Pollution Prevention Act (PPA)
Resource Conservation and Recovery Act (RCRA)
Safe Drinking Water Act (SDWA)
Shore Protection Act (SPA)
Toxic Substances Control Act (TSCA)

[a]The various Acts are listed alphabetically and not in the order of importance to the chemical and refining industries. Executive orders signed by the president are not included in this list.

use and have been expanded to include international environmental governance, international trade, environmental justice, and climate change. The practice and application of environmental law typically requires extensive knowledge of administrative law and aspects of tort law, property, legislation, constitutional law, and land-use law.

In concert with the evolution and institution of laws that are designed to the protect the environment, commercial processes have been designed to reduce the direct emissions of various chemicals (including inorganic chemicals) and the related products and emissions of inorganic chemical by-products into the air (*the atmosphere*), in the water (*the aquasphere*), and on to the land (*the terrestrial biosphere*). There has also been, and continues to be, the movement to recycle and to reuse as much of these chemicals and chemical waste as possible. Advanced technologies for the rapid, economical, and effective elimination of industrial and domestic chemical wastes have been developed and employed on a large scale. In fact, advanced technologies for the control and monitoring of chemical pollutants continue to be developed and implemented on regional, national, and global levels. To ensure that the movement to a clean environment continues, satellite-based instruments are employed to detect, to quantify, and to monitor areas where there is a potential to discharge a wide range of chemical pollutants into the various ecosystems. In addition, through an increased knowledge and behavior of the properties of various chemicals (Chapter 4), there is an accompanying increased understanding of the fate and consequences of the discharge of these chemicals into the environment (Chapters 6 and 7). Studies that are focused on chemical properties and chemical behavior in the environment has increased dramatically, and there are now available means of predicting, with much greater precision, many of the environmental, ecological, and biochemical consequences of the inadvertent introduction of inorganic chemicals into the environment.

Furthermore, because of the evolving environmental awareness, inorganic chemical refinery operators face more stringent regulation of the treatment, storage, and disposal of hazardous wastes. Waste management laws govern the transport, treatment, storage, and disposal of all manner of chemical waste, including municipal solid waste, hazardous waste, and nuclear waste. Waste laws are generally designed to minimize or eliminate the uncontrolled dispersal of waste materials into the environment in a manner that may cause ecological or biological harm and include laws designed to reduce the generation of waste and promote or mandate waste recycling. Regulatory efforts include identifying and categorizing waste types and mandating transport, treatment, storage, and disposal practices. Thus, under recent regulations, a larger number of inorganic chemical compounds have been, and are continuing to be, studied. Long-time acceptable methods of disposal of these chemicals, such as land farming of chemical waste, are being phased out, even forbidden in some countries. New regulations are becoming even more stringent, and these regulations encompass a broader range of chemical constituents and processes.

To address concerns related to harmful inorganic chemicals and waste, it is necessary to promote chemical safety and provide ready access to information about toxic inorganic chemicals (in fact, about all chemicals) and also to promote chemical safety by providing technical guidance on the regulations that govern the manufacture, use, and disposal of inorganic chemicals. Although it may be felt that the effect and activity of inorganic chemicals in the environment must focus on the chemical and physical properties of these chemicals, knowledge of the various environmental regulations (Table 8.2) is always helpful in determining the analyses that must be performed. These laws are designed to help protect human health and the environment; the Environmental Protection Agency is charged with administering all or a part of each law.

However, it is not the purpose of this chapter here to enter into any political discussion and the levy of fines for infringement of the environmental laws. The purpose of this chapter is to introduce the reader to an overview of a selection of the many and varied regulations instituted in the United States that regulate the disposition of inorganic chemicals into the environment. Thus, this chapter provides a general overview of the processes involved and some of the potential environmental hazards associated with inorganic chemicals (Chapters 3, 4, and 7).

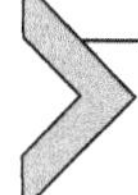

8.2 ENVIRONMENTAL IMPACT OF PRODUCTION PROCESSES

Any form of pollution that can be traced to an immediate source to industrial practices is known as industrial pollution. Most of the pollutants can be traced back to one or more industrial operations. In fact, the issue of industrial pollution has taken on grave importance for agencies trying to fight against environmental degradation. Countries facing sudden and rapid growth of such industries are finding it to be a serious problem that has to be brought under control immediately. In fact, the issue of industrial pollution is (or should be) a concern for every nation. As a result, many steps have been taken to seek permanent solutions to the problem. Better technology is being developed for disposal of waste and recycling as much polluted water in the industries as possible. Organic methods are being used to clean the water and soil, such as using microbes that naturally use heavy metals and waste as feed. Policies are being set into place to prevent further misuse of land. However, industrial pollution still occurs in some countries and may take many years to be brought under control.

Inorganic chemicals can enter the air, water, and soil during production, usage, or disposal. The impact of these chemicals on the environment is determined by (i) the type of the chemical, (ii) the concentration of the chemical, which is related to the amount of the chemical that is released, and (iii) the location where the chemical is introduced into the environment. In addition, the adverse effects of the chemical on an ecosystem may not be evident for some time after the introduction of the chemical. Nevertheless, the inadvertent or deliberate disposal of chemicals into the environment must also be of concern as any inorganic chemical has the ability to become part of the floral and faunal food chain (including humans) after which the chemical can also accumulate and/or persist in the environment for many years, even for decades.

In order to combat such effects, a series of environmental regulations have been enacted in many countries. The broad category of *environmental regulations* may be broken down into a number of more specific regulatory subjects, and while there is no single agreed-upon division or subdivision into specific categories, the core environmental regulations that are of interest in the context of this book can be employed address the environmental impact of the production, use, and disposal of inorganic chemicals on the environment.

In a more localized context, the inorganic chemicals industry is one of the largest industries in the United States, and potential environmental hazards have caused the increased need for environmental. Briefly, production of inorganic chemicals involves a series of steps that includes separation and blending of inorganic chemicals. Production facilities are generally considered a major source of pollutants in areas where they are located and are regulated by a number of environmental laws related to air, land, and water (Table 8.2). Thus, the inorganic chemical industry is (correctly or incorrectly, without any form of condemnation here) has been considered to be a major source of pollutants in areas where they are located and are regulated by a number of environmental laws related to air (the atmosphere), water (the aquasphere), and land (the terrestrial biosphere).

8.2.1 Air Pollution

Air pollution can come from several sources within the inorganic chemicals industry including (i) equipment leaks from valves or other devices, (ii) high-temperature combustion processes in the actual burning of fuels for electricity generation, (iii) the heating of steam and process fluids, and (iv) the transfer of products. These pollutants are typically emitted into the environment over the course of a year through normal emissions, fugitive releases, accidental releases, or plant upsets. The combination of oxides of nitrogen, oxides of sulfur, and volatile hydrocarbons also contribute to ozone formation, one of the most important air pollution problems.

Air quality laws govern the emission of air pollutants into the atmosphere, and air quality laws are designed specifically to limit or completely eliminate the concentration of airborne pollutants and the complete elimination of some airborne pollutants. Other regulations are designed to address broader ecological problems, such as the limitation on chemicals that affect the ozone layer and emissions trading programs to address acid rain or to address climate change. Regulatory efforts include (i) identifying and categorizing air pollutants; (ii) setting limits on acceptable emissions levels, if any; and (iii) dictating or suggesting or recommending the necessary or appropriate mitigation technologies.

Inorganic chemicals are a source of hazardous and toxic air pollutants; they are also a major source of criteria air pollutants: particulate matter (PM), nitrogen oxides (NO_x), carbon monoxide (CO), hydrogen sulfide (H_2S), and sulfur oxides (SO_x). However, it is worthy of note that, at the current time and because of improved process for gas-cleaning operations, inorganic chemical production (including refinery operations) releases less toxic inorganic chemicals than in prior decades.

8.2.2 Water Pollution

Inorganic chemicals are found in water supplies and may be natural in the geology or caused by activities of man through mining, industry, or agriculture. It is common to have trace amounts of many inorganic chemicals in water supplies. However, in terms of pollution, inorganic chemicals (as either ionic species or nondissociated compounds) are the greatest proportion of chemical contaminants in drinking water and are usually present in natural waters at much higher concentrations than their organic counterparts. The inorganics are present in greatest quantity because of natural processes, but several important contaminants are present as a result of anthropogenic activities such as arising from the plumbing material through which water is transported to the consumer. Therefore, inorganic contaminants are the most important determinants of acceptability to the consumer, affecting taste, and color as well as scale deposition on pipes and fittings.

Many of the inorganic chemicals are naturally occurring—such as calcium carbonate, $CaCO_3$, and calcium bicarbonate, $Ca(HCO_3)_2$, in hard water—and should be considered as an integral part of that type of water rather than as contaminants. However, there are many inorganic components usually present in much lower concentrations that could be considered as contaminants and are of greater interest in terms of their effect on water quality and health than the major components.

Sometimes, the most startling differences in the inorganic content of drinking water arise as a consequence of the difference between groundwater and surface water. Such differences are usually a reflection of the solution of minerals as water percolates through the ground. However, some are due to relatively low groundwater flows compared with surface water, and the subsequent buildup of pollutants such as nitrate. Moreover, an assessment of the potential health effects of inorganic contaminants in drinking water may be complicated by the limited database on the toxicity of these chemicals by the oral route and the fact that many of these elements are essential for human nutrition.

Water quality laws govern the release of the various types of pollutants into water resources, including surface water, groundwater, and stored drinking water. Some water-quality laws, such as drinking water regulations, may be designed solely with particular reference to human health. Many other regulations, including restrictions on the disposal of inorganic chemicals into water resources, may also reflect broader efforts to protect the aquasphere (i.e., aquatic ecosystems).

Typically, the regulatory efforts include (i) identifying and categorizing water pollutants; (ii) specifying acceptable pollutant concentrations in water resources; and, above all, (iii) limiting pollutant discharges from effluent sources. Regulatory areas include disposal of industrial waste, sewage treatment and disposal, management of agricultural chemical waste and agricultural wastewater, and control of surface runoff from construction sites and other urban environments.

Production facilities are also potential contributors to groundwater and surface water contamination. Some companies have continued to use (with the appropriate permits) deep injection wells to dispose of wastewater generated by the various processes, but this method of disposal may allow some of these wastes end up in aquifers and groundwater. Industrial wastewater may be highly contaminated and may arise from various processes (such as wastewaters from desalting, water from cooling towers, storm water, distillation, or cracking). This water is recycled through many stages during the production process and goes through several treatment processes, including a wastewater treatment plant, before being released (through government-issued permits) into surface waters.

Many wastewater issues (such as chemicals in waste process waters) face the inorganic chemicals industry (which also includes the fossil fuel industries and other process industries). However, efforts by the industry are being continued to eliminate any water contamination that may occur, whether it be from inadvertent leakage of petroleum or petroleum products or leakage of contaminated water from one or more processes. In addition to monitoring the more complex salts (ionized chemicals) in the water, metals concentration must be continually monitored since heavy metals tend to concentrate in the tissues of floral and, in particular, faunal species (such as fish and animals) and increase in concentration as they go higher in the food chain. In addition, general sewage problems related to the disposal of sewage face every municipal sewage treatment facility, regardless of size.

Primary treatment (solid settling and removal) is required and secondary treatment (use of bacteria and aeration to enhance inorganic degradation) is becoming more routine, tertiary treatment (filtration through activated carbon, applications of ozone, and chlorination) have been, or are being, implemented by many chemical production companies. Wastewater pretreaters that discharge wastewater into sewer systems have more stringent requirements and a variety of pollutant standards for sewage sludge have been enacted. Toxic inorganic chemicals in the water must be identified and plans must be developed and accepted by the various levels of governmental authority to alleviate any potential problems. In addition, regulators have established and continue to establish, water-quality standards for priority toxic pollutants.

Many of the wastes are regulated under the Safe Drinking Water Act (SDWA), while the wastes discharged into surface waters are subject to state discharge regulations and are regulated under the Clean Water Act (CWA). As examples of the regulations, the discharge guidelines typically limit the amounts of sulfides, ammonia, suspended solids (particulate matter or sediment) and other inorganic constituents that may be present in the wastewater. Although these guidelines are in place, contamination from past unmanaged discharge of inorganic chemicals may be persistent in surface water bodies.

8.2.3 Soil Pollution

Contamination of soils from the various processes is also an issue. Past (pre-1970) production practices may have led to spills on company property that now need to be cleaned up. In some cases, the natural bacteria that may use the inorganic chemicals as food are often effective at cleaning up inorganic chemicals spills and leaks compared with many other pollutant chemicals. Many waste materials are produced during the production of inorganic chemicals, and some of them are recycled through other stages in the process. Other wastes are collected and disposed of in landfills, or they may be recovered by other facilities. Soil contamination including some hazardous wastes, spent catalysts or coke dust, tank bottoms, and sludge from the treatment processes can occur from leaks and accidents or spills on- or off-site during the transport process.

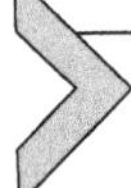

8.3 ENVIRONMENTAL REGULATIONS IN THE UNITED STATES

The term *environmental law* (or *environmental regulations*) is a term that is commonly used to signify a collection of laws that work together and often overlap in areas and the terms covers the laws that regulate the discharge or inorganic chemicals into the environment. Moreover, the broad category of *environmental law* may be broken down into a number of more specific regulatory subjects (Table 8.2). While there is no single agreed-upon classification of these regulations, the core environmental regulation addresses environmental pollution, and many do specify pollution by chemical or by inorganic chemicals such as pesticides. A related but distinct set of regulatory regulations focus on the management of specific natural resources, such as forests, minerals, or fisheries. Other areas, such as environmental impact assessment, may not fit neatly into either category, but are nonetheless important components of the protection of the environmental.

In the United States, there are also *common law protections* that allow a landowner whose land is being polluted to seek judgment (sue) the polluter. A landowner may sue under a theory of trespass (a physical invasion of the property) or nuisance (an interference with the landowner's enjoyment of his property). However, each of these theories must include an element of reasonableness, and there can be no recovery (of financial damages) if the neighbor is making a reasonable use of the land. Additionally, the degree of *reasonableness* depends on the facts of the specific case. Also, an action may be brought under public nuisance where the suit is brought by a public entity if it is the public that is harmed (rather than a uniquely harmed individual).

State laws also reflect the same concerns and common law actions that allow adversely affected property owners to seek a judicial remedy for environmental harms. Although laws on the state level vary from state to state, many of use the federal laws as a base thereby allowing an additional forum for aggrieved landowners to be heard. Finally, state laws may require a higher level of protection then federal law.

The toxic chemicals found within the inorganic chemicals industry are not necessarily unique, and although general air pollution, water pollution, and land pollution controls are affected by the inorganic chemicals defining, these problems and solutions are not unique to the industry. In fact, because the issues are so diverse, the inorganic chemicals industry (and because a segment of the industry—the refining industry—is an industrial complex consisting of many integrated unit processes) may be looked upon as a series of complex pollution-prevention issues, each one unique to the unit processes from which the effluent originates. Therefore, there may be many examples of laws and controls that have been enacted by governments with input from the producers of chemicals that address pollution prevention and control and discharge of hazardous chemicals into the environment.

Thus, air quality laws govern the emission of chemical pollutants into the atmosphere (Chapter 6). Other initiatives are designed to address broader ecological problems, such as limitations on chemicals that affect the ozone layer, and emissions trading programs that address the formation and disposition of acid rain. Regulatory efforts include identifying and categorizing air pollutants, setting limits on acceptable emissions levels, and dictating necessary or appropriate mitigation technologies.

Similarly, water-quality laws govern the release of chemical pollutants into water resources, including surface water, groundwater, and drinking water (Chapter 6). Some of the water-quality laws, such as regulations that influence the quality of drinking water regulations, may be designed solely with reference to human health but many other water-quality laws, including restrictions on the alteration of the chemical, physical, radiological, and biological characteristics of water resources are designed to protect the aquatic ecosystems and are often broader in their respective application to ecosystems. Regulatory efforts may also include identifying and categorizing water pollutants, dictating acceptable pollutant concentrations in water resources, and limiting pollutant discharges from effluent sources. Regulatory areas include sewage treatment and disposal, industrial wastewater management and agricultural wastewater management, and control of surface runoff from construction sites and urban environments, especially where there is a high likelihood that chemicals are dissolved in the wastewater.

Waste management regulations govern the transport, treatment, storage, and disposal of all manner of waste, including municipal solid waste, hazardous waste, and nuclear waste, among many other types of chemical-containing waste. These laws are typically designed to minimize or eliminate the uncontrolled dispersal of waste (chemical) materials into the environment in a manner that may cause ecological or biological harm, and the regulations also include laws that are designed to reduce the generation of waste and promote or mandate waste recycling. Regulatory efforts include identifying and categorizing waste types and mandating transport, treatment, storage, and disposal practices.

Environmental cleanup laws govern the removal of (chemical) pollutants or (chemical) contaminants from environmental ecosystems such as soil, sediment in aqueous ecosystems, surface water, or groundwater. Unlike pollution control laws (which are designed to be followed before the fact), cleanup laws are designed to respond after the fact to environmental contamination and consequently must often define not only the necessary response actions but also the parties who may be responsible for undertaking the actual cleanup and bearing the cost of the cleanup actions. Regulatory requirements may include rules for emergency response, liability allocation, site assessment, remedial investigation, feasibility studies, remedial action, postremedial monitoring, and site reuse.

Thus, pollution prevention and control of hazardous chemical materials is an issue not only for the inorganic chemicals industry but also for many industries and has been an issue for decades (Table 8.1) (Noyes, 1993). In this context, there are specific definitions for terms such as *hazardous substances*, *toxic substances*, and *hazardous waste* (Chapter 1). These are all terms of art and must be fully understood in the context of their statutory or regulatory meanings and not merely limited to their plain English or dictionary meanings. It is absolutely imperative from a legal sense that each statute or regulation promulgated be read in conjunction with terms defined in that specific statute or regulation (Majumdar, 1993).

In order to combat any threat to the environment, it is necessary to understand the nature and magnitude of the problems involved. It is in such situations that environmental technology has a major role to play.

Environmental issues even arise when outdated laws are taken to task. Thus, the concept of what seemed to be a good idea at the time the action occurred may no longer be satisfactory when the law influences the environment.

Finally, it is worth of note that regulatory disincentives to voluntary reductions of emissions from inorganic chemicals also exist. Many environmental statutes define a baseline period and measure progress in pollution reductions from that baseline. Any reduction in emissions before it is required could lower a facility's baseline emissions. Consequently, future regulations requiring a specified reduction from the baseline could be more difficult (and, consequently, have a much greater effect on the economic bottom line) to achieve because the most easily applied and, hence, the most cost-effective reductions would already have been made and establishing the environmental base case may by no longer realistic. With no credit given for voluntary reductions, those facilities that do the minimum may be in fact be rewarded when emissions reductions are required.

As a start to the passage of various federal laws in the United States, the National Environmental Policy Act (NEPA) was passed in 1970 along with the Environmental Quality Improvement Act, the National Environmental Education Act, and the Environmental Protection Agency (EPA). The National Environmental Policy Act has been described as one of most relevant pieces of environmental legislation passed by the Congress of the United States. The basic purpose of National Environmental Policy Act is to ensure that governmental agencies to consider the effects of their decisions on the environment. Thus, the main objective of these federal enactments was to assure that the environment be protected against both public and private actions that failed to take account of costs or harms inflicted on the ecosystem. It is the duty of the United States Environmental Protection Agency to monitor and analyze the environment, conduct research, and work closely with state and local governments to devise pollution control policies.

Thus, over time, the Congress of the United States (as has also happened in many other countries) has enacted a variety of protections with the goal of protecting the environment. These laws are summarized in the following sections.

8.3.1 Clean Air Act

The formulators of the Clean Air Act recognized that virtually all metals are present in the atmosphere at low levels. Particulate matter emitted from combustion of fossil fuels contains trace metals that were present in the original fuel sample. The greatest health hazard is from aerosols that are smaller than 2.5 micrometers (μm) ion diameter and contain lead, beryllium, mercury, cadmium, and chromium. On the upside, these particles can settle out of the atmosphere over time or by precipitation events. On the downside, the particles must go somewhere, usually into the oceans and on to the land. Therefore, removal from the atmosphere by natural processes is not the answer to pollution. Furthermore, there is also the issues of mercury as a pollutant; mercury is the only metal that can exist as a gas (it is liquid at room temperature) and is therefore the only metal that can exist in a steady-state concentration in air. All other metals are emitted to the atmosphere from natural or anthropogenic sources and then removed by settling or precipitation events.

However, gaseous inorganic chemicals in various forms (Table 8.3) remain in the atmosphere for long residence times and would continue to build up much higher levels than observed currently if it was not for the fact that the inorganic gases undergo reactions in the gas-phase reactions that convert these chemicals substances to other chemical soluble species, such as acid rain (Chapter 7).

Furthermore, although ozone (O_3) is a desirable substance in the stratosphere, it is a major environmental hazard at ground level. Ozone is a by-product of photochemical smog and reacts with hydrocarbons to form peroxy-nitrate derivatives (Chapter 5) that can cause severe environmental damage. Excessive ozone levels in the troposphere have been blamed for causing floral destruction through reactions with chlorophyll. Ozone is formed naturally when oxygen molecules are photochemically dissociated into oxygen atoms that can then react with a second oxygen molecule to make ozone. The presence of nitrogen oxides (NO and NO_2) leads to higher than normal background levels of ozone through several well-understood photochemical reactions.

Carbon monoxide is a product of the incomplete combustion of fossil fuels (Chapter 5) and as much as 20% (v/v) of the carbon monoxide released

Table 8.3 Gaseous Inorganic Pollutants

Element	Atmospheric Forms
Oxygen	O_3
Carbon	CO, CO_2
Nitrogen	NH_3, N_2O, NO, NO_2, N_2O_5
Sulfur	H_2S, SO_2, SO_3

into the atmosphere each year comes from natural sources, but the greatest health problem is in metropolitan areas near high densities of vehicular traffic. It has been estimated that carbon monoxide has a 4-month lifetime in the atmosphere, where it reacts with the hydroxyl radical to form carbon dioxide:

$$CO + HO\cdot \rightarrow CO_2 + H\cdot$$
$$H\cdot + O_2 \rightarrow HOO\cdot$$
$$2HOO\cdot \rightarrow H_2O_2 + O_2$$
$$H_2O_2 + h\nu \rightarrow 2HO\cdot$$

Atmospheric carbon dioxide levels are determined by a long-term equilibrium between carbon dioxide in the air and (i) carbon dioxide dissolved in the aquasphere, (ii) releases of carbon dioxide from natural and anthropogenic sources, and (iii) losses by plant growth. Elevated levels of atmospheric carbon dioxide may have a major impact on the climate of the Earth.

Also, sulfur dioxide that is generated by coal-fired electric power generating plants is also an issue. Gas cleaning technologies (scrubbing technologies) have been used since the early 1970s to remove sulfur dioxide from power plant emissions (*flue gas desulfurization*, FGD) (Mokhatab et al., 2006; Speight, 2007, 2013, 2014, 2017; Kidnay et al., 2011; Bahadori, 2014). All current technologies involve exposing the combustion gases to a substance that will absorb most of the sulfur dioxide. The flue gas desulfurization technologies are categorized into *wet technologies* that expose the flue gases to an aqueous solution and *dry technologies* that expose the flue gas to solid absorbents. Thus,

Lime scrubbing:

$$Ca(OH)_2 + SO_2 \rightarrow CaSO_3 + H_2O$$

Limestone slurry scrubbing:

$$CaCO_3 + SO_2 \rightarrow CaSO_3 + CO_2(g)$$

Magnesium oxide scrubbing:

$$Mg(OH)_2 + SO_2 \rightarrow MgSO_3 + H_2O$$

Sodium base scrubbing:

$$Na_2SO_3 + H_2O + SO_2 \rightarrow 2NaHSO_3$$
$$2NaHSO_3 \rightarrow Na_2SO_3 + H_2O + SO_2(\text{regeneration})$$

Double alkali scrubbing:

$$2NaOH + SO_2 \rightarrow Na_2SO_3 + H_2O$$

$$Ca(OH)_2 + Na_2SO_3 \rightarrow 2NaOH + CaSO_3(s)(\text{regeneration})$$

There is also air pollution caused by the presence of particulate matter (PM), which is a term used for a mixture of solid particles and liquid droplets found in the air. Some particles, such as dust, dirt, soot, or smoke, are large or dark enough to be seen with the naked eye. Others are so small; they can only be detected using an electron microscope. Pollution by such material includes (i) PM_{10}, which refers to inhalable particles with a diameter on the order of 10 micrometers (10 μm and smaller, and (ii) $PM_{2.5}$, which refers to fine inhalable particles with a diameter on the order of 2.5 micrometers (2.5 μm) and smaller. These particles have various sizes and shapes and can be made up of hundreds of different chemicals. Some are emitted directly from a source, such as construction sites, unpaved roads, fields, smokestacks, or fires. Most particles form in the atmosphere as a result of complex reactions of chemicals such as sulfur dioxide (SO_2) and nitrogen oxides (NO_x), which are pollutants emitted from power plants, industries, and automobiles.

Because of these air pollution issues, the Clean Air Act (CAA) was enacted in 1970 and is designed to protect air quality by regulating stationary and mobile sources of pollution. The Amendments (CAAA) of 1990 have made significant changes in the basic Clean Air Act. The Clean Air allowed the establishment of air quality standards and provisions for their implementation and enforcement. This law was strengthened in 1977, and the Clean Air Act Amendments of 1990 imposed many new standards that included controls for industrial pollutants.

The Clean Air Act of 1970 and the 1977 Amendments that followed consist of three titles (i) Title I deals with stationary air emission sources, (ii) Title II deals with mobile air emission sources, and (iii) Title III includes definitions of appropriate terms, provisions for citizen suits, and applicable standards for judicial review. However, in contrast to the previous clean air statutes, the 1990 Amendments contained extensive provisions for control of the accidental release of air toxics from storage or transportation and the formation of acid rain (Chapter 7). At the same time, the 1990 Amendments provided new and added requirements for such original ideas as state implementation plans for attainment of the national ambient air quality standards and permitting requirements for the attainment and nonattainment areas. Title III now calls for a vastly expanded program to regulate *hazardous air pollutants* (HAPs) or the so-called *air toxics*.

Under the Clean Air Act Amendments of 1990, the mandate is to establish, during the first phase, technology-based maximum achievable control technology (MACT) emission standards that apply to the major categories or subcategories of sources of the listed hazardous air pollutants. In addition, Title III provides for health-based standards that address the issue of residual risks due to air toxic emissions from the sources equipped with MACT and to determine whether the MACT standards can protect health with an *ample margin of safety*.

Section 112 of the original Clean Air Act that dealt with hazardous air pollutants has been greatly expanded by the 1990 Amendments. The list of hazardous air pollutants has been increased many fold. In addition, the standards for emission control have been tightened and raised to a very high level, referred to as the *best of the best*, in order to reduce the risk of exposure to various hazardous air pollutants.

Thus, the 1990 Clean Air Act Amendments aimed to encourage voluntary reductions above the regulatory requirements by allowing facilities to obtain emission credits for voluntary reductions in emissions. These credits would serve as offsets against any potential future facility modifications resulting in an increase in emissions. Other regulations established by the amendments, however, will require the construction of major new units within existing chemicals producers to reduce emissions even further, and these new operations will require emission offsets in order to be permitted. This will consume many of the credits available for existing facility modifications. A shortage of credits for facility modifications will make it difficult to receive credits for emission reductions through pollution prevention projects.

Thus, under this Clean Air Act, the Environmental Protection Agency sets limits on how much of a pollutant can be in the air anywhere in the United States. The law does allow individual states to have stronger pollution controls, but states are not allowed to have weaker pollution controls than those set for the whole country. The law recognizes that it makes sense for states to take the lead in carrying out the Clean Air Act, because pollution control problems often require special understanding of local industries and geography as well as housing developments near to industrial sites.

In addition, Title IV of the 1990 Amendments to the Clean Air Act (CAA) (Title IV—Acid Deposition Control) mandates requirements for the control of acid deposition (acid rain). The purpose of this title is to reduce the adverse effects of acid deposition through reductions in annual emissions of sulfur dioxide and also, in combination with other provisions of this Act, through reductions of nitrogen oxides emissions in the 48 contiguous states and the district of Columbia. It is the intent of this title to effectuate such reductions by requiring compliance by affected sources with prescribed emission limitations by specified deadlines, which limitations may be met through alternative methods of compliance provided by an emission allocation and transfer system. It is also the purpose of this title to encourage energy conservation, use of renewable and clean alternative technologies, and pollution prevention as a long-range strategy, consistent with the provisions of this title, for reducing air pollution and other adverse impacts of energy production and use. Furthermore, individual states are required to develop a state implementation plans (SIP) that is a collection of the regulations a state will use to clean up polluted areas. The states must involve the public, through hearings and opportunities to comment, in the development of each state implementation plan. The Environmental Protection Agency must approve each plan and if a state implementation is not acceptable, the Environmental Protection Agency can take over enforcing the Clean Air Act in that state.

Air pollution often travels from its source in one state to another state. In many metropolitan areas, people live in one state and work or shop in another; air pollution from cars and trucks may spread throughout the interstate area. The 1990 Clean Air Act Amendments provide for interstate commissions on air pollution control, which are to develop regional strategies for cleaning up air pollution. The 1990 Amendments also cover pollution that originates in nearby countries, such as Mexico and Canada, and drifts into the United States and pollution from the United States that reaches Canada and Mexico.

In the current context, the 1990 Amendments provide economic incentives for cleaning up pollution. For instance, producers of inorganic chemicals can get credits if they produce cleaner products than required, and use the credits when the product falls short of the requirements. Furthermore, inorganic chemicals (like many industrial products) can be extremely toxic to the environment. As an example, to combat such effects refiners have started to reformulate gasoline sold in the formerly smog-prone areas. This gasoline contains less volatile inorganic chemicals such as benzene (which is also a hazardous air pollutant that causes cancer and aplastic anemia, a potentially fatal blood disease). The reformulated gasoline also contains detergents, which, by preventing buildup of engine deposits, keep engines working smoothly and burning fuel cleanly.

8.3.2 Clean Water Act

The Clean Water Act came into law because of the contamination of the aquasphere (groundwater, aquifers, lakes, rivers, and oceans) that occurs when pollutants are directly or indirectly discharged into water bodies without prior adequate treatment to remove the pollutants. Water pollution affects the entire biosphere—floral and faunal species that live in the various bodies of water. In almost all cases, the effect is damaging not only to the natural biological communities, including users (human users) of the water.

Water is typically referred to as polluted when it is impaired by contaminants from anthropogenic sources and (1) either does not support a human use, such as drinking water or (ii) undergoes a marked shift in its ability to support its constituent biotic communities, such as fish. Natural phenomena such as volcanoes, algal blooms, storms, and earthquakes also cause major changes in water quality and the ecological status of water.

The specific contaminants leading to pollution in water include a wide spectrum of inorganic chemicals and physical changes such as elevated temperature and discoloration. While many of the chemicals and substances that are regulated may be naturally occurring (such as derivatives of calcium, sodium, iron, and manganese), the concentration (relative to the indigenous concentration) is often the key in determining what a natural component of water is and what a contaminant is. High concentrations of naturally occurring substances can have negative impacts on aquatic flora and fauna. It is safest to assume that many of the anthropogenic inorganic chemicals are toxic, unless proven otherwise. Even then if the chemicals are nontoxic, there is no reason why anthropogenic inorganic chemicals should be in the water body. The inorganic chemicals (nontoxic or toxic) can cause alteration of physical chemistry of the water body, which includes acidity (a change in the pH), electric conductivity, temperature, and eutrophication. Eutrophication is an increase in the concentration of chemical nutrients in an ecosystem to an extent that increases in the primary productivity of the ecosystem. Depending on the degree of eutrophication, subsequent negative environmental effects such as anoxia (oxygen depletion) and severe reduction in water quality may occur, affecting the floral population and the animal population of the water body.

Thus, the Clean Water Act is designed to protect water by preventing discharge of pollutants into navigable waters from point sources. The Clean Water Act started life as a regulatory act The Federal Water Pollution Control Act of 1948 was the first major US law to address water pollution.

Growing public awareness and concern for controlling water pollution led to amendments in 1972, and as amended in 1972, the law became commonly known as the Clean Water Act (CWA). The 1972 Amendments (i) established the basic structure for regulating pollutant discharges into the waters of the United States; (ii) gave the Environmental Protection Agency the authority to implement pollution control programs such as setting wastewater standards for industry; (iii) maintained existing requirements to set water-quality standards for all contaminants in surface waters; (iv) made it unlawful for any person to discharge any pollutant from a point source into navigable waters, unless a permit was obtained under its provisions; (v) funded the construction of sewage treatment plants under the construction grants program; and (vi) recognized the need for planning to address the critical problems posed by nonpoint source pollution.

Subsequent amendments modified some of the earlier provisions of the Clean Water Act. For example, revisions in 1981 streamlined the municipal construction grants process, improving the capabilities of treatment plants built under the program. Further changes to the Act in 1987 phased out the construction grants program, replacing it with the State Water Pollution Control Revolving Fund (the Clean Water State Revolving Fund), which addressed water-quality needs by building on EPA-state partnerships.

Over the years, many other laws have changed parts of the Clean Water Act. Title I of the Great Lakes Critical Programs Act of 1990, for example, put into place parts of the Great Lakes Water Quality Agreement of 1978, signed by the United States and Canada, where the two nations agreed to reduce certain toxic pollutants in the Great Lakes. That law required the Environmental Protection Agency to establish water-quality criteria for the Great Lakes addressing 29 toxic pollutants with maximum levels that are safe for humans, wildlife, and aquatic life. It also required the Environmental Protection Agency to help the states implement the criteria on a specific schedule.

Thus, the Clean Water Act (CWA or the Water Pollution Control Act) is the cornerstone of surface water-quality protection in the United States and employs a variety of regulatory and nonregulatory tools to sharply reduce direct pollutant discharges into waterways and manage polluted runoff. The objective of the Clean Water Act is to restore and maintain the chemical, physical, and biological integrity of water systems. The Act established the basic structure for regulating discharges of pollutants into the waters of the United States and regulating quality standards for discharge of pollutants into the waters of the United States and gave the

Environmental Protection Agency the authority to implement pollution control programs such as setting wastewater standards for industry. The Clean Water Act also continued requirements to set water-quality standards for all contaminants in surface waters. The Act is credited with the first comprehensive program for controlling and abating water pollution. In addition, the Act made it unlawful to discharge any pollutant from a point source into navigable waters, unless a permit was obtained. Point sources are discrete conveyances such as pipes or man-made ditches. Individual homes that are connected to a municipal system, use a septic system, or do not have a surface discharge do not need an NPDES permit; however, industrial, municipal, and other facilities must obtain permits if their discharges go directly to surface waters.

The statute makes a distinction between conventional and toxic pollutants. As a result, two standards of treatment are required prior to their discharge into the navigable waters of the nation. For conventional pollutants that generally include degradable nontoxic inorganic chemicals, the applicable treatment standard is best conventional technology (BCT). For toxic pollutants, on the other hand, the required treatment standard is best available technology (BAT), which is a higher standard than BCT.

The statutory provisions of Clean Water Act have five major sections that deal with specific issues: (i) nationwide water-quality standards, (ii) effluent standards for the inorganic chemicals industry, (iii) permit programs for discharges into receiving water bodies based on the National Pollutant Discharge Elimination System (NPDES), (iv) discharge of toxic chemicals, including oil spills, and (v) construction grant program for publicly owned treatment works (POTW). In addition, Section 311 of Clean Water Act includes elaborate provisions for regulating intentional or accidental discharges of oil and hazardous substances. Included there are response actions required for oil spills and the release or discharge of toxic and hazardous substances. Pursuant to this, certain elements and compounds are designated as hazardous substances, and an appropriate list has been developed (40 CFR 116.4). The person in charge of a vessel or an onshore or offshore facility from which any designated hazardous substances is discharge, in quantities equal to or exceeding its reportable quality, must notify the appropriate federal agency as soon as such knowledge is obtained. Such notice should be provided in accordance with the designated procedures (33 CFR 153.203).

Under the Clean Water Act, discharge of waterborne pollutants is limited by National Pollutant Discharge Elimination System (NPDES) permits.

Chemical-producing companies that easily meet their permit requirements may find that the permit limits will be changed to lower values and be less stringent. However, because occasional system upsets do occur resulting in significant excursions above the normal performance values, many companies may feel that they must maintain a large operating margin below the permit limits to ensure continuous compliance. Those companies that can significantly reduce waterborne emissions may find the risk of having their permit limits lowered to be a substantial disincentive.

8.3.3 Comprehensive Environmental Response, Compensation, and Liability Act

The cleanup of environmental pollution involves a variety of techniques, ranging from simple biological processes to advanced engineering technologies. Cleanup activities may address a wide range of contaminants, from common industrial chemicals such as agricultural chemicals and metals to radionuclides. Cleanup technologies may be specific to the contaminant (or contaminant class) and to the site (Chapter 9).

Cleanup costs can vary dramatically depending on the contaminants, the *media* affected, and the size of the contaminated area. Much of the remediation has been in response to such historical chemical management practices as dumping, poor storage, and uncontrolled release or spillage of chemicals. Greater effort in recent years has been directed toward pollution prevention, which is more cost-effective than remediation.

However, in most cases, it is financially or physically impractical to completely remove all traces of contamination. In such cases, it is necessary to set an acceptable level of residual contamination. As a result, evolution of cleanup technologies has yielded four general categories of remediation approaches: (i) physical removal, with or without treatment; (ii) in situ conversion by physical or chemical means to less toxic or less mobile forms; (iii) containment; and (iv) passive cleanup or natural attenuation (Chapter 9). Combinations of the four technology types may be used at some contaminated sites.

Thus, the Comprehensive Environmental Response, Compensation, and Liability Act (CERCLA), commonly known as Superfund, 1980, is aimed at cleaning up already polluted areas. This statute assigns liability to almost anyone associated with the improper disposal of hazardous waste and is designed to provide funding for clean up. To achieve this goal, a tax was created on the chemical and inorganic chemicals industries and provided broad federal authority to respond directly to releases or threatened releases of hazardous substances that may endanger public health or the environment.

The Act was amended by the Superfund Amendments and Reauthorization Act (SARA) in 1986 and stressed the importance of permanent remedies and innovative treatment technologies in cleaning up hazardous waste sites. Thus, the Act provides a Federal *superfund* to clean up uncontrolled or abandoned hazardous-waste sites as well as accidents, spills, and other emergency releases of pollutants and contaminants into the environment. Through CERCLA, the Environmental Protection Agency was given power to seek out those parties responsible for any release and assure their cooperation in the cleanup.

A CERCLA response or liability will be triggered by an actual release or the threat of a *hazardous substance or pollutant or contaminant* being released into the environment. A hazardous substance [CERCLA 101(14)] is any substance requiring special consideration due to its toxic nature under the Clean Air Act, the Clean Water Act, or the Toxic Substances Control Act (TSCA) and as defined under RCRA. Additionally, a pollutant or contaminant can be any other substance not necessarily designated or listed, but that "will or may reasonably" be anticipated to cause any adverse effect in organisms and/or their offspring [CERCLA 101(33)].

The central purpose of CERCLA is to provide a response mechanism for cleanup of any hazardous substance released, such as an accidental spill, or of a threatened release of a hazardous substance (Nordin et al., 1995). Section 102 of CERCLA is a catchall provision because it requires regulations to establish *that quantity of any hazardous substance the release of which shall be reported pursuant to Section 103* of CERCLA. Thus, under CERCLA, the list of potentially responsible parties (PRPs) can include all direct and indirect culpable parties who have either released a hazardous substance or violated any statutory provision. In addition, responsible private parties are liable for cleanup actions and/or costs and for reporting requirements for an actual or potential release of a hazardous substance, pollutant, or contaminant.

CERCLA (Superfund) legislation deals with actual or potential releases of hazardous materials that have the potential to endanger people or the surrounding environment at uncontrolled or abandoned hazardous waste sites. The Act requires responsible parties or the government to clean up waste sites. Among CERCLA's major purposes are the following: (i) site identification, (ii) evaluation of danger from sites where waste chemicals have been deposited, (iii) evaluation of damages to natural resources, (iv) monitoring of release of hazardous substances from sites, and (v) removal or cleanup of wastes by responsible parties or government.

The Superfund Amendments and Reauthorization Act (SARA) addresses closed hazardous waste disposal sites that may release hazardous substances into any environmental medium. Title III of SARA also requires regular review of emergency systems for monitoring, detecting, and preventing releases of extremely hazardous substances at facilities that produce, use, or store such substances.

The most revolutionary part of SARA is the Emergency Planning and Community Right-to-Know Act (EPCRA) that is covered under Title III of SARA. EPCRA includes three subtitles and four major parts: emergency planning, emergency release notification, hazardous chemical reporting, and toxic chemical release reporting. Subtitle A is a framework for emergency planning and release notification. Subtitle B deals with various reporting requirements for *hazardous chemicals* and *toxic chemicals*. Subtitle C provides various dimensions of civil, criminal, and administrative penalties for violations of specific statutory requirements.

Other provisions of SARA basically reinforce and/or broaden the basic statutory program dealing with the releases of hazardous substances (CERCLA Section 313). It requires owners and operators of certain facilities that manufacture, process, or otherwise use one of the listed chemicals and chemical categories to report all environmental releases of these chemicals annually. This information about total annual releases of chemicals from the industrial facilities can be made available to the public.

The Act also requires (under Section 4) testing of chemicals by manufacturers, importers, and processors where risks or exposures of concern are found and (under Section 5) issuance of significant new use rules (SNURs) when a significant new use is identified for a chemical that that could result in exposures to, or releases of, a substance of concern. In addition, companies or persons importing or exporting chemicals are required to comply with certification reporting and/or other requirements that also require (under Section 8) reporting and record-keeping by persons who manufacture, import, process, and/or distribute chemical substances in commerce and that any person who manufactures (including imports), processes, or distributes in commerce a chemical substance or mixture and who obtains information that reasonably supports the conclusion that such substance or mixture presents a substantial risk of injury to health or the environment to immediately inform the Environmental Protection Agency, except where the Agency has been adequately informed of such information.

The Toxic Substances Control Acts is also authorized (under Section 8) to maintain the TSCA Inventory that contains more than 83,000 chemicals, and as new chemicals are commercially manufactured or imported, they are placed on the list.

8.3.4 Hazardous Materials Transportation Act

A *hazardous material* is any material that is a hazardous material and a *hazardous waste* is any material that is subject to the hazardous waste manifest requirements of the United States Environmental Protection Agency. It was recognized that the transportation of such materials can cause environmental problems when (for whatever reason) spillage occurs and there was the need to establish a means of protecting the environment against the risks to the environment that are inherent in the transportation of hazardous material in intrastate, interstate, and foreign commerce.

In the 1960s and 1970s, the high cost of disposal of hazardous material and hazardous waste led to increased dumping of materials that were increasingly being deemed hazardous by the public and by government. Illegal dumping took place on vacant lots, along highways, or on the actual highways themselves. At the same time, increased accidents and incidents with hazardous materials during transportation was a growing problem, causing damage to the environment. The increasing frequency of illegal *midnight dumping* and spills, along with the already existing inconsistent regulations and fragmented enforcement, led to the passing of legislation to mitigate such activities.

Thus, the Hazardous Materials Transportation Act, passed in 1975, is the law governing transportation of chemicals and hazardous materials. It is the principal federal law in the United States that regulates the transportation of hazardous materials (such as, in the current context, inorganic chemicals). The purpose of the Act is to *protect against the risks to life, property, and the environment that are inherent in the transportation of hazardous material in intrastate, interstate, and foreign commerce* under the authority of the US Secretary of Transportation.

The Act was passed as a means to improve the uniformity of existing regulations for transporting hazardous materials and to prevent spills and illegal dumping endangering the public and the environment, a problem exacerbated by uncoordinated and fragmented regulations. Regulations are enforced through four key provisions that encompass federal standards under Title 49 of the US Code (a code that regards the role of transportation in the United States): (i) procedures and policies, (ii) material designations and labeling, (iii) packaging requirements, and (iv) operational rules. Violation of the Act regulations can result in civil or criminal penalties, unless a special permit is granted under the discretion of the secretary of transportation.

Thus, the basic purpose of Hazardous Materials Transportation Act is to ensure safe transportation of hazardous materials through the nation's highways, railways, and waterways. The basic theme of the Act is to prevent any person from offering or accepting a hazardous material for transportation anywhere within this nation if that material is not properly classified, described, packaged, marked, labeled, and properly authorized for shipment pursuant to the regulatory requirements. In addition, the Act includes a comprehensive assessment of the regulations, information systems, container safety, and training for emergency response and enforcement. The regulations apply to *any person who transports, or causes to be transported or shipped, a hazardous material; or who manufactures, fabricates, marks, maintains, reconditions, repairs, or tests a package or container which is represented, marked, certified, or sold by such person for use in the transportation in commerce of certain hazardous materials.*

Under this statutory authority, the Secretary of Transportation has broad authority to determine what a hazardous material is, using the dual tools of quantity and type. By this two-part approach, any material that may pose an unreasonable risk to human health or the environment may be declared a hazardous material. Such a designated hazardous material obviously includes both the quantity and the form that make the material hazardous. Furthermore, under the Department of Transportation (DOT) regulations, a hazardous material is *any substance or material, including a hazardous substance and hazardous waste that is capable of posing an unreasonable risk to health, safety, and property when transported in commerce.* DOT thus has broad authority to regulate the transportation of hazardous materials that, by definition, include hazardous substances and hazardous wastes.

8.3.5 Occupational Safety and Health Act

All employees have the right to a workplace that is reasonably free of safety and health hazards, and there was a need to assure the safety and health of workers in the United States by setting and enforcing workplace safety standards.

Thus, the objective of the OSHA Hazard Communication Standard is to inform workers of potentially dangerous substances in the work place and to train them on how to protect themselves against potential dangers. The Act is formulated to *to assure safe and healthful working conditions for working men and women; by authorizing enforcement of the standards developed under the Act; by assisting and encouraging the States in their efforts to assure safe and healthful working conditions; by providing for research, information, education, and training in the field of occupational safety and health; and for other purposes.*

The goal of OSHA is to ensure that *no employee will suffer material impairment of health or functional capacity* due to a lifetime occupational exposure to chemicals and hazardous substances. The statute imposes a duty on the employers to provide employees with a safe workplace environment, free of known hazards that may cause death or serious bodily injury. Thus, the Act is entrusted with the major responsibility for workplace safety and worker's health (Wang, 1994). It is responsible for the means by which chemicals are contained through the inspection of workplaces to ensure compliance and enforcement of applicable standards under OSHA. It is also the means by which guidelines have evolved for the destruction of chemicals used in chemical laboratories.

The statute covers all employers and their employees in all the states and federal territories with certain exceptions (Lunn and Sanstone, 1994). Generally, the statute does not cover self-employed persons, farms solely employing family members, and those workplaces covered under other federal statutes. Chemicals producers must evaluate whether the chemicals that the company manufactures and sell are hazardous. Under the General Duty Clause of OSHA, employers are required to provide an environment that is free from recognized hazards that could cause physical harm or death.

All employers are required to develop, implement, and maintain at the work place a written hazard communication program. The program must include the following components: (i) a list of hazardous chemicals in the work place; (ii) the methods the employer will use to inform employees of the hazards associated with these chemicals; and (iii) a description of how the labeling, material safety data sheet (MSDS), and employee training requirements will be met.

The following information must be included in the program for employers who produce, use, or store hazardous chemicals in the workplace: (i) the means by which manufacturer's safety data sheets (MSDS) will be made available to the outside contractor for each hazardous chemical, (ii) the means by which the employer will inform the outside contractor of precautions necessary to protect the contractor's employees both during normal operating conditions and in foreseeable emergencies, and (iii) the methods that the employer will use to inform contractors of the labeling system used in the workplace.

8.3.6 Resource Conservation and Recovery Act

As the 20th century evolved, there was an increased need to the environment from (i) the potential hazards of waste disposal, (ii) to conserve energy and natural resources, (iii) to reduce the amount of waste generated, and (iv) to ensure that wastes were managed in an environmentally sound manner. As a result, there was the need for regulations that required corrective actions to address the investigation and cleanup of releases of hazardous chemicals in accordance with state and federal requirements. The degree of investigation and subsequent corrective action necessary to protect the environment varies significantly among facilities. Cleanup progress at these facilities is measured, in part, by interim cleanup milestones known as environmental indicators. The hazardous waste regulatory program began with the Resource Conservation and Recovery Act (RCRA) in 1976 addressed these issues and provided a *cradle-to-grave* system of preventing pollution by use of a manifest system to ensure that waste is properly disposed of, and thus not dumped into the environment.

Since the enactment of Resource Conservation and Recovery Act, the Act has been amended several times, to promote safer solid and hazardous waste management programs (Dennison, 1993). The Used Oil Recycling Act of 1980 and the Hazardous and Solid Waste Amendments of 1984 (HSWA) were the major amendments to the original law. The 1984 Amendments also brought the owners and operators of underground storage tanks under the Resource Conservation Recovery Act umbrella. This can have a significant effect on refineries that store inorganic chemicals in underground tanks. Now, in addition to the hazardous waste being controlled, the Resource Conservation Recovery Act Subtitle I regulates the handling and storage of inorganic chemicals.

The Resource Conservation Recovery Act controls disposal of solid waste and requires that all wastes destined for land disposal be evaluated for their potential hazard to the environment. Solid waste includes liquids, solids, and containerized gases and is divided into nonhazardous waste and hazardous waste. The various amendments are aimed at preventing the disposal problems that lead to a need for the Comprehensive Environmental

Response Compensation and Liability Act (CERCLA), or Superfund, as it is known.

Subtitle C of the original Resource Conservation Recovery Act lists the requirements for the management of hazardous waste. This includes criteria for identifying hazardous waste, and the standards for generators, transporters, and companies that treat, store, or dispose of the waste. The Resource Conservation Recovery Act regulations also provide standards for design and operation of such facilities. However, before any action under the Act is planned, it is essential to understand what constitutes a solid waste and what constitutes a hazardous waste. The first step to be taken by a generator of waste is to determine whether that waste is hazardous. Waste may be hazardous by being listed in the regulations, or by meeting any of the four characteristics: ignitability, corrosivity, reactivity, and extraction procedure (EP) toxicity.

Section 1004(27) of the Resource Conservation Recovery Act defines *solid waste* as garbage, refuse, sludge, from a waste treatment plant, water supply treatment plant, or air pollution control facility and other discarded material, including solid, liquid, semisolid, or contained gaseous material resulting from industrial, commercial, mining and agricultural operations and from community activities, but not including solid or dissolved materials in domestic sewage, or solid or dissolved materials in irrigation return flows or industrial discharges that are point sources subject to permits under Section 402 of the Federal Water Pollution Control Act, as amended (86 Stat. 880), or source, special nuclear, or by-product materials as defined by the Atomic Energy Act of 1954, as amended (68 Stat. 923).

This statutory definition of solid waste is pursuant to the regulations of the Environmental Protection Agency insofar as a solid waste is a hazardous waste if it exhibits any one of four specific characteristics: (i) ignitability, (ii) reactivity, (iii) corrosivity, and (iv) toxicity. However, a waste chemical listed solely for the characteristic of ignitability, reactivity, and/or corrosivity is excluded from regulation as a hazardous waste once it no longer exhibits a characteristic of hazardous waste [Section 261.3(g)(1)].

In terms of *ignitability*, a waste is an *ignitable hazardous waste* if it has a flash point of less than 140°F (40 CFR 261.21) as determined by the Pensky-Martens closed cup flash point test, readily causes fires and burns so vigorously as to create a hazard, or is an ignitable compressed gas or an oxidizer (as defined by the Department of Transport regulations). A simple method of determining the flash point of a waste is to review the material. Ignitable wastes carry the waste code D1001. Naphtha is an example of an ignitable hazardous waste.

On the other hand, a *corrosive waste* is a liquid waste that has a pH of less than or equal to 2 (a highly acidic waste) or greater than or equal to 12.5 (a highly alkaline waste) is considered to be a corrosive hazardous waste (40 CFR 261.22). Also, a corrosive waste may be a liquid and corrodes steel (SAE 1020) at a rate greater than 6.35 mm (0.250 in.) per year at a test temperature of 55°C (130°F). Corrosivity testing is conducted using the standard test method as formulated by the National Association of Corrosion Engineers (NACE)—Standard TM-01-69 or, in place of this method, an EPA-approved equivalent test method.

For example, sodium hydroxide (caustic soda), with a high pH, is often used by the inorganic chemicals production industry (especially by the refining industry and the natural gas industry) in the form of a caustic wash to remove sulfur compounds or acid gases. When these caustic solutions become contaminated and must be disposed of, the waste would be a corrosive hazardous waste. Corrosive wastes carry the waste code D002. Acid solutions also fall under this category.

A chemical waste material is considered to be a *reactive* hazardous waste if it is normally unstable, reacts violently with water, generates toxic gases when exposed to water or corrosive materials, or if it is capable of detonation or explosion when exposed to heat or a flame (40 CFR 261.23). Materials that are defined as forbidden explosives or Class A or B explosives by the DOT are also considered reactive hazardous waste.

Typically, reactive wastes are solid wastes that exhibit any of the following properties as defined at 40 CFR 61.23(a): (i) It is normally unstable and readily undergoes violent change without detonating; (ii) it reacts violently with water; (iii) it forms potentially explosive mixtures with water; (iv) when mixed with water, it generates toxic gases, vapors, or fumes in a quantity sufficient to present a danger to human health or the environment; (v) it is a cyanide or sulfide-bearing waste that, when exposed to pH conditions between 2 and 12.5, can generate toxic gases, vapors or fumes in a quantity sufficient to present a danger to human health or the environment; (vi) it is capable of detonation or explosive reaction if it is subjected to a strong initiating source or if heated under confinement; (vii) it is readily capable of detonation or explosive decomposition or reaction at standard temperature and pressure; and (viii) it is a forbidden explosive as defined in 49 CFR 173.51, or a Class A explosive as defined in 49 CFR 173.53 or a Class B explosive as defined in 49 CFR 173.88.

The Environmental Protection Agency has assigned the Hazardous Waste Number D003 to reactive characteristic waste.

The fourth characteristic that could make a waste a hazardous waste is *toxicity* (40 CFR 261.24). To determine if a waste is a toxic hazardous waste, a representative sample of the material must be subjected to a test conducted in a certified laboratory using a test procedure (toxicity characteristic leaching procedure, TCLP). Under federal rules (40 CFR 261), all generators are required to use the toxicity characteristic leaching procedure test when evaluating wastes.

Wastes that fail a toxicity characteristic test are considered hazardous under the Resource Conservation and Recovery Act. There is less incentive for a company to attempt to reduce the toxicity of such waste below the toxicity characteristic levels because, even though such toxicity reductions may render the waste nonhazardous, it may still have to comply with new land disposal treatment standards under subtitle C of the Resource Conservation and Recovery Act before disposal. Similarly, there is little positive incentive to reduce the toxicity of listed hazardous wastes because, once listed, the waste is subject to subtitle C regulations without regard to how much the toxicity levels are reduced.

Besides the four characteristics of hazardous wastes, the Environmental Protection Agency has established three hazardous waste lists: (i) hazardous wastes from nonspecific sources, such as spent nonhalogenated solvents; (ii) hazardous wastes from specific sources, such as bottom sediment sludge from the treatment of wastewaters from wood preserving; and (iii) discarded commercial chemical products and off-specification species, containers, and spill residues. However, under regulations of the Environmental Protection Agency, certain types of solid wastes (e.g., household waste) are not considered to be hazardous wastes irrespective of their characteristics. Additionally, the Environmental Protection Agency has provided certain regulatory exemptions based on very specific criteria. For example, hazardous waste generated in a product or raw material storage tank, transport vehicle, or manufacturing processes and samples collected for monitoring and testing purposes are exempt from the regulations.

Finally, in terms of waste classification, the Environmental Protection Agency has also designated certain wastes as *incompatible wastes*, which are hazardous wastes that, if placed together, could result in potentially dangerous consequences. As defined at 40 CFR 260.10, an *incompatible waste* is a hazardous waste that is unsuitable for (i) placement in a particular device or facility because it may cause corrosion or decay of containment materials, such as container inner liners or tank walls, or (ii) comingling with another waste or material under uncontrolled conditions because the commingling might produce heat or pressure, fire or explosion, violent reaction, toxic dusts, mists, fumes, or gases, or flammable fumes or gases.

Once the physical and chemical properties of a hazardous waste have been adequately characterized, hazardous waste compatibility charts can be consulted to identify other types of wastes with which it is potentially incompatible. For example, Appendix V to both 40 CFR 264 and 265 presents examples of potentially incompatible wastes and the potential consequences of their mixture.

Under the Resource Conservation Recovery Act, the hazardous waste management program is based on a *cradle-to-grave* concept so that all hazardous wastes can be traced and fully accounted for. Section 3010(a) of the Act requires all generators and transporters of hazardous wastes as well as owners and operators of all TSD facilities to file a notification with the Environmental Protection Agency within 90 days after the promulgation of the regulations. The notification should state the location of the facility and include a general description of the activities and the identified and listed hazardous wastes being handled.

Submission of the part A permits application for existing facilities prior to Nov. 19, 1980, qualified a refinery for interim status. This meant that the refinery was allowed to continue operation according to certain regulations during the permitting process. The Hazardous and Solid Waste Amendments (HSWA) represented a strong bias against land disposal of hazardous waste. Some of the provisions that affect the companies (especially refineries) involved in the production of inorganic chemicals are the following:

- A ban on the disposal of bulk or noncontainerized liquids in landfills. The prohibition also bans solidification of liquids using absorbent material including absorbents used for spill cleanup.
- Five hazardous wastes from specific sources come under scheduled for disposal prohibition and/or treatment standards. These five are dissolved air flotation (DAF) float, slop oil emulsion solids, and heat exchanger bundle cleaning sludge. API separator sludge and leaded tank bottoms (EPA waste numbers: K047–K051).
- Producers of inorganic chemicals (such as refineries) must retrofit surface impoundments that are used for hazardous waste management. Retrofitting must involve the use of double liners and leak detection systems.

Under the Resource Conservation Recovery Act, the Environmental Protection Agency has the authority to require a company to clean up releases of hazardous waste or waste constituents. The regulation provides for cleanup of hazardous waste released from active treatment, storage, and disposal facilities. Superfund was expected to handle contamination that had occurred before that date.

8.3.7 Safe Drinking Water Act

In terms of drinking water, the term *contaminant* means any physical, chemical, biological, or radiological substance or matter in water that is, in actual fact, broadly defined as any molecular species other than water molecules. Drinking water may reasonably be expected to contain at least small amounts of some contaminants but some drinking water contaminants may be harmful if consumed at certain levels in drinking water, while others may be harmless.

There four general categories of drinking water contaminants and examples of each are the following: (i) physical contaminants, (ii) chemical contaminants, (iii) radiological contaminants, and (iv) biological contaminants.

Physical contaminants primarily impact the physical appearance or other physical properties of water. Inorganic examples of physical contaminants are sediment or material suspended in the water of lakes, rivers, and streams from soil erosion. *Chemical contaminants* are elements or compounds and are often of anthropogenic origin; examples of inorganic chemical contaminants include nitrate ($—NO_3$) derivatives, bleach (NaOCl), salts, inorganic pesticides, and metals. *Radiological contaminants* are chemical elements with an unbalanced number of protons and neutrons resulting in unstable atoms that can emit ionizing radiation (Chapter 4); examples of radiological contaminants include cesium (Cs), plutonium (Pu), and uranium (U). Finally, although out of the current context of this text but still worthy of mention because of possible interactions with the other three contaminants, *biological contaminants* are organisms in the water and are also referred to as microbes or microbiological contaminants. Examples of biological or microbial contaminants include bacteria, viruses, parasites, and protozoans. The latter organisms (protozoans) are single-celled eukaryotes (organisms whose cells have nuclei) that commonly show characteristics usually associated with animals, most notably mobility and ate heterotrophic insofar as the organism that cannot fix carbon from inorganic sources (such as carbon dioxide) but uses organic carbon for sustenance and growth.

The Safe Drinking Water Act (SDWA) was enacted in 1974 to assure high-quality water supplies through public water system. In the United States, it is the federal law that protects public drinking water supplies throughout the nation. Under the Safe Drinking Water Act, the

Environmental Protection Agency is authorized to set the standards for drinking water quality and with its partners implements various technical and financial programs to ensure drinking water safety.

The Act is truly the first federal intervention to set the limits of contaminants in drinking water. The 1986 Amendments came 2 years after the sage of the Hazardous and Solid Waste Amendments (HSWA) or the so-called Resource Conservation Recovery Act Amendments of 1984. As a result, certain statutory provisions were added to these 1986 Amendments to reflect the changes made in the underground injection control (UIC) systems. In addition, the Superfund Amendments and Reauthorization Act (SARA) of 1986 set the groundwater standards the same as the drinking water standards for the purpose of necessary cleanup and remediation of an inactive hazardous waste disposal site.

The law was amended in 1986 and 1996 and requires many actions to protect drinking water and its sources—rivers, lakes, reservoirs, springs, and ground water wells. (The Safe Drinking Water Act does not regulate private wells that serve fewer than 25 individuals.) For example, the 1986 Amendments of Safe Drinking Water Act included additional elements to establish maximum contaminant level goals (MCLGs) and national primary drinking water standards. The maximum contaminant level goals must be set at a level at which no known or anticipated adverse effects on human health occur, thus providing an *adequate margin of safety*. Establishment of a specific maximum contaminant level goal depends on the evidence of carcinogenicity in drinking water or a reference dose that is individually calculated for each specific contaminant. The maximum contaminant level goals, an enforceable standard, however, must be set to operate as the nation primary drinking water standard (NPDWS). The 1996 Amendments greatly enhanced the existing law by recognizing source water protection, operator training, funding for water system improvements, and public information as important components of safe drinking water. This approach ensures the quality of drinking water by protecting it from source to tap.

The Safe Drinking Water Act calls for regulations that (i) apply to public water systems, (ii) specify contaminants that may have any adverse effect on the health of persons, and (iii) specify contaminant levels. The difference between primary and secondary drinking water regulations are defined, as well as other applicable terms. Information concerning national drinking water regulations and protection of underground sources of drinking water is given.

In the context of the inorganic chemicals industry, the priority list of drinking water contaminants is very important since it includes the contaminants known for their adverse effect on public health. Furthermore, most if not all are known or suspected to have hazardous or toxic characteristics that can compromise human health.

8.3.8 Toxic Substances Control Act

A toxic substance is (in the current context since there are toxic substances other than inorganic chemicals) any chemical that may cause harm to an individual (floral or faunal) organism if it enters the organism. Toxic chemicals may enter the organism in different ways—referred to as routes of exposure (Table 8.4). An inorganic chemical can be toxic, or hazardous, or both. In fact, any inorganic chemical can be toxic or harmful under certain conditions. Some chemicals are hazardous because of their physical properties; they can explode, burn, or react easily with other chemicals. Some substances are more toxic than others. The toxicity of a substance is described by the types of effects it causes and its potency (Table 8.4).

The degree of hazard associated with any toxic chemical is related to (i) the individual chemical, (ii) concentration of the material, (iii) the route into the organism, and (iv) the amount absorbed by the organism—the dose. Individual susceptibility of the organism also plays a role. Once a chemical enter an organism, the effects may occur immediately, or the effects may be delayed—acute effects or chronic effects (Table 8.4). Generally, acute effects are caused by a single, relatively high exposure, while chronic effects tend to occur over a longer period of time and involve lower

Table 8.4 General Terminology Related to Toxic Chemicals

Types of Effects	Dependent Upon the Chemical; May Be Quick Reacting or Slow Reacting
Potency	A measure of the toxicity of a chemical
Exposure	A chemical can cause health effects only when it contacts or enters the body
Routes of exposure	Inhalation, ingestion or direct contact
Dose	The amount of a substance that enters or contacts a organism
Exposure medium	Exposure to chemicals occurs when exposed to any medium (air, water, or land) that contains chemicals
Length of exposure	Acute (short-term) exposure or chronic (long-term exposure) may cause serious effects

exposures (e.g., exposure to a smaller amount over time). Some toxic chemicals can have both acute and chronic health effects.

The Toxic Substances Control Act (TSCA), enacted in 1976, was designed to understand the use or development of chemicals and to provide controls, if necessary, for those chemicals that may threaten human health or the environment (Ingle, 1983; Sittig, 1991). The Act provides the Environmental Protection Agency with the authority to require reporting, record-keeping and testing requirements, and restrictions relating to chemical substances and/or mixtures.

This Act has probably had more effect on the producers of chemicals, and the reining industry, than any other Act. It has caused many changes in the industry and may even create further modifications in the future. The basic purpose of the Act is (i) to develop data on the effects of chemicals on our health and environment, (ii) to grant authority to the Environmental Protection Agency to regulate substances presenting an unreasonable risk, and (iii) to assure that this authority is exercised so as not to impede technological innovation. In short, the Act calls for regulation of *chemical substances* and *chemical mixtures* that present an unreasonable risk or injury to health or the environment. Furthermore, the introduction and evolution of this Act has led to a central bank of information on existing commercial chemical substances and chemical mixtures, procedures for further testing of hazardous chemicals, and detailed permit requirements for submission of proposed new commercial chemical substances and chemical mixtures.

As used in the Act, the term *chemical substance* means any inorganic chemical of a particular molecular identity, including any combination of such substances occurring in whole or in part as a result of a chemical reaction or occurring in nature, and any element or uncombined radical. Items not considered *chemical substances* are listed in the definition section of the act. The term *mixture* means any combination of two or more chemical substances if the combination does not occur in nature and is not, in whole or in part, the result of a chemical reaction; except that such term does include any combination that occurs, in whole or in part, as a result of a chemical reaction if none of the chemical substances comprising the combination is a new chemical substance and if the combination could have been manufactured for commercial purposes without a chemical reaction at the time the chemical substances comprising the combination were combined.

For many, familiarity with the Toxic Substances Control Act generally stems from its specific reference to polychlorinated biphenyls, which raise a vivid, deadly characterization of the harm caused by them. But the Act is not a statute that deals with a single chemical or chemical mixture or product. In fact, under the Toxic Substances Control Act, the Environmental Protection Agency is authorized to institute testing programs for various chemical substances that may enter the environment. Under Tosca's broad authorization, data on the production and use of various chemical substances and mixtures may be obtained to protect public health and the environment from the effects of harmful chemicals. In actuality, the Act supplements the appropriate sections dealing with toxic substances in other federal statutes such as the Clean Water Act (Section 307) and the Occupational Safety and Health Act (Section 6).

At the heart of the Toxic Substances Control Act is a premanufacture notification (PMN) requirement under which a manufacturer must notify the Environmental Protection Agency at least 90 days prior to the production of a new chemical. In this context, a *new chemical* is a chemical that is not listed in the Act-based Inventory of Chemical Substances or is an unlisted reaction product of two or more chemicals. For chemicals already on this list, a notification is required if there is a new use that could significantly increase human or environmental exposure. No notification is required for chemicals that are manufactured in small quantities solely for scientific research and experimentation.

The Chemical Substances Inventory of the Toxic Substances Control Act is a comprehensive list of the names of all existing chemical substances and currently contains over 70,000 existing chemicals. Information in the inventory is updated every 4 years. A facility must submit a premanufacture notice (PMN) prior to manufacturing or importation for any chemical substances not on the list and not excluded by the Act. Examples regulated chemicals include lubricants, paints, inks, fuels, plastics, and solvents.

8.4 OUTLOOK

The chemicals production industry will increasingly feel the effects of the land bans on their hazardous waste management practices. Current practices of land disposal must change along with management attitudes for waste handling. The way companies handle the waste products in the future depends largely on the ever-changing regulations. Waste management is the focus and reuse/recycle options must be explored to maintain a balanced waste management program. This requires that a waste be recognized as either *nonhazardous* or *hazardous*.

A good deal is already known about the effects of chemical substances on man during their production, transport and use, from the study of occupational medicine, toxicology, pharmacology, etc., and to a lesser extent their effects on resource organisms, from veterinary science and plant pathology. Much less is known about the effects of a chemical upon wildlife species, following its disposal in the environment. This last area of knowledge needs improving because ecological cycles and food chains may deliver the potentially hazardous chemical from affected wildlife back to humans. Moreover, wildlife can often be used as indicators of environmental states and trends for a potentially harmful substance, giving an early warning of future risks to man. We are also frequently ignorant of how far wildlife may be supportive to human well-being: as a food base for an important resource species, e.g., fisheries or grazing animals; as a key species maintaining the stability of economically valuable ecosystems; as predators of crop or livestock pests; as a species involved in mineral recycling or biodegradation; and as an important amenity.

Failure to recognize the mutually interactive roles of man, resource species, wildlife organisms and climate in the biosphere and their different tolerances to chemical substances has hindered the development of a unitary environmental management policy embracing all four biosphere components. Although a good deal is already known about the influence of molecular structure on the toxicity to human beings of drugs and certain other chemicals, much less is known about the influence of molecular structure on the environmental persistence of a chemical. For wildlife, persistence is probably the most important criterion for predicting potential harm because there is inevitably some wild species or others that are sensitive to any compound and any persistent chemical, apparently harmless to a limited number of toxicity test organisms, which will eventually be delivered by biogeochemical cycles to a sensitive target species in nature. This means that highly toxic, readily biodegradable substances may pose much less of an environmental problem, than a relatively harmless persistent chemical, which may well damage a critical wild species. The study of chemical effects in the environment resolves itself into a study of (a) the levels of a substance accumulating in air, water, soils (including sediments) and biota (including man) and (b), when the threshold action level has been reached, effects produced in biota that constitute a significant adverse response (i.e., environmental dose-response curve). In order to predict trends in levels of a chemical, much more information is needed about rates of injection, flow and partitioning between air, water, soils, and biota, and loss via degradation (environmental balance sheets).

These dynamic phenomena are governed by the physicochemical properties of the molecule. Fluid mechanics and meteorology may in future provide the conceptual and technical tools for producing predictive models of such systems. Most of our knowledge of effects derives from acute toxicology and medical studies on man, but since environmental effects are usually associated with chronic exposure, studies are being increasingly made of long-term continuous exposure to minute amounts of a chemical. The well-known difficulty of recognizing such effects when they occur in the field is aggravated by the fact that many of the effects are nonspecific and are frequently swamped by similar effects deriving from exposure to such natural phenomena as famines, droughts, and cold spells. Even when a genuine effect is recognized, a candidate causal agent must be found and correlated with it. This process must be followed by experimental studies, unequivocally linking chemical cause and adverse biological effect. All three stages are difficult and costly, and it is not surprising that long delays are often experienced between the recognition of a significant adverse effect and a generally agreed chemical cause. There is often ample uncertainty to allow underreaction and overreaction to potential hazards, both backed up by "scientific" evidence.

In the mid-20th century, solid waste management issues rose to new heights of public concern in many areas of the United States because of increasing solid waste generation, shrinking disposal capacity, rising disposal costs, and public opposition to the siting of new disposal facilities. These solid waste management challenges continue as many communities are struggling to develop cost-effective, environmentally protective solutions. The growing amount of waste generated has made it increasingly important for solid waste management officials to develop strategies to manage wastes safely and cost effectively.

8.4.1 Hazardous Waste Regulations

By way of introduction, a hazardous waste is a waste with properties that make it dangerous or capable of having a harmful effect on floral and faunal species (including human species) or the environment. Inorganic hazardous waste is generated from many sources, ranging from industrial manufacturing process wastes to batteries and may come in many forms, including liquids, solids gases, and sludge. The hazardous waste regulatory program, as is currently practiced, began with the Resource Conservation and Recovery Act (RCRA) in 1976. The Used Oil Recycling Act of 1980 and Hazardous and Solid Waste Amendments of 1984 (HSWA) were the major amendments to the original law.

The Resource Conservation Recovery Act provides for the tracking of hazardous waste from the time it is generated, through storage and transportation, to the treatment or disposal sites. The Act and the various amendments are aimed at preventing the disposal problems that lead to a need for the Comprehensive Environmental Response Compensation and Liability Act (CERCLA), or Superfund, as it is known. Subtitle C of the original Resource Conservation Recovery Act lists the requirements for the management of hazardous waste. This includes the Environmental Protection Agency criteria for identifying hazardous waste, and the standards for generators, transporters, and companies that treat, store, or dispose of the waste. The Resource Conservation Recovery Act regulations also provide standards for design and operation of such facilities.

New regulations are becoming even more stringent, and they encompass a broader range of chemical constituents and processes. Continued pressure from the US Congress has led to more explicit laws allowing little leeway for industry, the United States Environmental Protection Agency (Environmental Protection Agency), or state agencies. A summary of the current regulations and what they mean is given in the following.

8.4.2 Requirements

The first step to be taken by a generator of waste is to determine whether that waste is hazardous. Waste may be hazardous by being listed in the regulations, or by meeting any of the four characteristics: ignitability, corrosivity, reactivity, and extraction procedure (EP) toxicity.

Generally, (i) if the material has a flash point less than 140°F, it is considered ignitable; (ii) if the waste has a pH less than 2.0 or above 12.5, it is considered corrosive—it may also be considered corrosive if it corrodes stainless steel at a certain rate; (iii) a waste is considered reactive if it is unstable and produces toxic materials, or it is a cyanide or sulfide-bearing waste that generates toxic gases or fumes; and (iv) a waste that is analyzed for EP toxicity and fails is also considered a hazardous waste. This procedure subjects a sample of the waste to an acidic environment. After an appropriate time has elapsed, the liquid portion of the sample (or the sample itself if the waste is liquid) is analyzed for certain metals and pesticides. Limits for allowable concentrations are given in the regulations. The specific analytical parameters and procedures for these tests are referred to in 40CRF 261.

The 1984 Amendments also brought the owners and operators of underground storage tanks under the umbrella of the Resource Conservation Recovery Act. This can have a significant effect on chemicals production companies (such as refineries) that store products in underground tanks. In addition, inorganic chemicals are also regulated by the Resource Conservation Recovery Act, Subtitle I.

As the regulatory programs evolved, the United States Environmental Protection Agency has developed a regulatory definition and process that identifies specific substances known to be hazardous and provides objective criteria for including other materials in the regulated hazardous waste universe (Fig. 8.1). In order for a material to be classified as a hazardous waste, it must first be a solid waste. Therefore, the first step in the hazardous waste identification process is determining if a material is a solid waste. The second step in this process examines whether or not the waste is specifically excluded from regulation as a solid or hazardous waste. Once a generator determines that their waste meets the definition of a solid waste, they investigate whether or not the waste is a listed or characteristic hazardous waste. Finally, a chemical facility can petition the United States Environmental Protection Agency to delist a waste from the Resource Conservation and Recovery Act (RCRA) Subtitle C regulation.

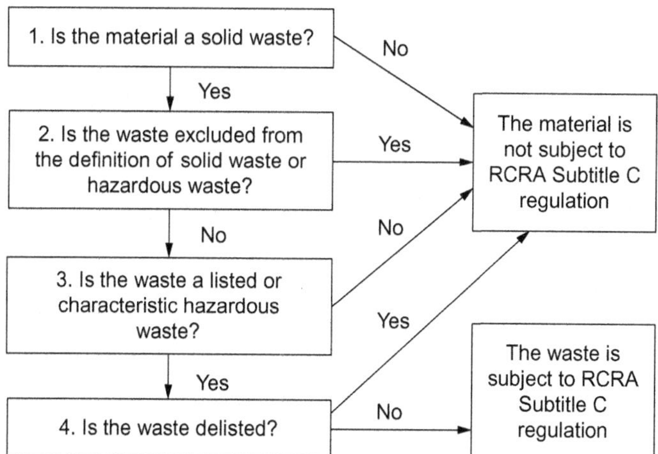

Fig. 8.1 The US EPA hazardous waste identification process.

Finally, operators of inorganic chemicals production companies face stringent regulation of the treatment, storage, and disposal of hazardous wastes. Under recent regulations, a larger number of compounds have been, and are being, studied and long-time methods of disposal, such as land farming of chemical waste, are being phased out. Thus, many companies are changing their waste management practices.

REFERENCES

Bahadori, A., 2014. Natural Gas Processing: Technology and Engineering Design. Gulf Professional Publishing, Elsevier, Amsterdam.

Dennison, M.S., 1993. RCRA Regulatory Compliance Guide. Noyes Data Corp, Park Ridge, NJ.

Ingle, G.W. (Ed.), 1983. TSCA' s Impact on Society and the Chemical Industry. ACS Symposium SeriesAmerican Chemical Society, Washington, DC.

Kidnay, A.J., Parrish, W.R., McCartney, D.G., 2011. Fundamentals of Natural Gas Processing, second ed. CRC Press, Taylor & Francis Group, Boca Raton, FL.

Lazarus, R., 2004. The Making of Environmental Law. Cambridge University Press, Cambridge.

Lunn, G., Sanstone, E.B., 1994. Destruction of Hazardous Chemicals in the Laboratory, second ed. McGraw-Hill, New York, NY.

Majumdar, S.B., 1993. Regulatory Requirements for Hazardous Materials. McGraw-Hill, New York, NY.

Mokhatab, S., Poe, W.A., Speight, J.G., 2006. Handbook of Natural Gas Transmission and Processing. Elsevier, Amsterdam.

Nordin, J.S., Sheesley, D.C., King, S.B., Routh, T.K., 1995. Environ. Solut. 8 (4), 49.

Noyes, R. (Ed.), 1993. Pollution Prevention Technology Handbook. Noyes Data Corp, Park Ridge, NJ.

Sittig, M., 1991. Handbook of Toxic and Hazardous Chemicals and Carcinogens, third ed. Noyes Data Corp, Park Ridge, NJ.

Speight, J.G., 2007. Natural Gas: A Basic Handbook. GPC Books, Houston, TX.

Speight, 2013. The Chemistry and Technology of Coal, third ed. CRC Press, Taylor & Francis Group, Boca Raton, FL.

Speight, J.G., 2014. The Chemistry and Technology of Petroleum, fifth ed. CRC Press, Taylor & Francis Group, Boca Raton, FL.

Speight, J.G., 2017. Handbook of Petroleum Refining. CRC Press, Taylor & Francis Group, Boca Raton, FL.

Wagner, T.P., 1999. The Complete Guide to Hazardous Waste Regulations. John Wiley & Sons Inc., New York, NY

Wang, C.C.K., 1994. OSHA Compliance and Management Handbook. Noyes Data Corp, Park Ridge, NJ.

FURTHER READING

CFR, 2004. Code of Federal Regulations. United States Government, Washington, DC.

EPA, 2004. Environmental Protection Agency, Washington, DC. http://www.epa.gov.

CHAPTER NINE

Removal of Inorganic Compounds From the Environment

9.1 INTRODUCTION

In the early days of the 20th century, few people were aware or were even concerned about the potentially negative effects of the unmanaged disposal of chemical into the environment and the effects of these chemicals on the environment and recognized only the need and despite for creating new useful materials and products. However, in the book *Silent Spring* (Carson, 1962), the author (Rachel Carson) awakened the chemical industry (and other pertinent industries) and the public to the dangers to all natural floral and faunal ecosystems from the misuse of pesticides (specifically, chemicals such as dichlorodiphenyltrichloroethane, DDT). This is not surprising that over the last seven decades, industry and the public have become aware of the adverse effects on the environment of the past practices of unmanaged disposal of chemicals and the need to protect the environment from such actions. Since that time, there has been focus on the unmanaged disposal of chemicals and the lifetime effects of chemicals on the environment.

Over the entire life of an inorganic chemical product (often referred to as the *cradle-to-grave* cycle), there is a potential for a negative impact on man and the environment. The cradle-to-grave analysis (also known as life-cycle analysis, life-cycle assessment, and ecobalance) is a technique to assess environmental impacts associated with all the stages of the life of a chemical from the extraction of the raw material from the Earth through material processing, manufacture, distribution, use, and disposal or recycling. This involves (i) compiling an inventory of relevant energy and material inputs and environmental releases, (ii) evaluating the potential impacts associated with identified inputs and releases, and (iii) interpreting the results to help make a more informed decision.

Environmental Inorganic Chemistry for Engineers
http://dx.doi.org/10.1016/B978-0-12-849891-0.00009-6

Phase:	Scope and definition	Inventory analysis	Impact assessment
Need:	*Interpretation*	*Interpretation*	Interpretation

This investigation also includes (i) identification of the type of chemical, (ii) specific examples of the types of chemical spilled, and (iii) the compatibility/incompatibly of the spill chemical with the chemical used in the cleanup process (Table 9.1).

As a user of raw materials, the inorganic chemical industry can impact on the supply of nonrenewable resources. And, as these materials are in general based on the use of hydrocarbon fuels, combustion of these fuels can lead to the production and emission of pollutant inorganic chemicals such as carbon dioxide (CO_2, a greenhouse gas) and nitrogen oxides (NOx), which contribute to the formation of tropospheric ozone (O_3) and photochemical smog (often referred to simply as *smog*). Because there are many potential sources of inorganic pollutants and sources of these pollutants (Tables 9.2 and 9.3), in fact, the life cycle of each potential pollutants (from extraction, material processing, manufacture, distribution, use, and disposal or recycling) can vary widely and is chemical-dependent. Thus, processing the raw materials to produce useable chemicals can result in the release of hazardous pollutants to the environment as can their actual use, by either other industries or consumers. In addition, and perhaps even more important, hazardous waste can be generated by the chemical industry as a

Table 9.1 General Guidelines for Segregation of Spilled Chemicals

Class of Chemicals	Common Examples	Incompatible Chemicals
Toxic chemicals	Cyanides, heavy metal compounds	Flammable liquids, acids, bases
Corrosive acids	Hydrochloric acid, sulfuric acid, phosphoric acid	Bases, oxidizers, cyanides, sulfides
	Chromic acid, nitric acid	Bases, oxidizers, cyanides, sulfides
Corrosive bases	Ammonium hydroxide, potassium hydroxide	Acids, oxidizers, hypochlorite
	Sodium hydroxide	Acids, oxidizers, hypochlorite
Oxidizers	Perchlorates, permanganates, nitrates	Water-reactive chemicals
Water-reactive chemicals	Sodium, lithium, and potassium metals	Water, aqueous solutions, oxidizers

Table 9.2 Examples of the Types and Forms of Inorganic Chemical Pollutants

Industrial Source	Type of Pollutant	Form of Pollutant
Chloralkali	Brine	Aqueous solution
	Calcium chloride	Aqueous solution
	Mercury	Liquid
Desalination	Brine	Aqueous solution
Electric/electronics	Mercury	Liquid
	Copper	Solid
	Precious metals	Solid
	Selenium	Solid
Explosives	Barium	Solid
	Lead	Solid
	Manganese	Solid, particulate matter
	Mercury	Liquid
	Nitric acid	Aqueous solution
	Phosphorus	Solid
Fertilizer	Ammonia	Gas, aqueous solution
	Oxides of nitrogen (NOx)	Gas
	Nitrates	Solid
	Phosphates	Solid
Glass and ceramic	Arsenic	Aqueous solution
	Barium	Solid
	Manganese	Solid
	Selenium	Solid
Hydrofluoric acid	Calcium sulfate	Aqueous solution, particulate matter
Nitric acid	Oxides of nitrogen	Gas
Nuclear fuel and power	Radioactive materials	Solid
	Radioisotopes	Solid
Phosphoric acid	Calcium sulfate	Aqueous solution, particulate matter
	Hydrofluoric acid	Aqueous solution
Petroleum refining	Alkali	Aqueous solution
	Mineral acids	Liquid
	Spent catalysts	Solid
	Cobalt	Solid
	Iron	Solid
	Manganese	Solid
	Nickel	Solid
	Platinum	Solid
	Vanadium	Solid

Continued

Table 9.2 Examples of the Types and Forms of Inorganic Chemical Pollutants—cont'd

Industrial Source	Type of Pollutant	Form of Pollutant
Petrochemical production	Fluorine	Aqueous solution
	Hydrochloric acid	Aqueous solution
	Hydrofluoric acid	Aqueous solution
	Sulfuric acid	Aqueous solution
	Sulfur oxides	Gas
Power generation	Nitrogen oxides	Gas
	Cooling water	Liquid
	Fly ash	Solid, particulate matter
	Bottom ash	Solid
	Slag (clinker)	Solid

Table 9.3 Examples of Sources of Inorganic Pollutants

Chemical	Examples	Industrial Source
Mineral acids	Hydrochloric acid	Pickling of metals
	Nitric acid	Chemical reagent
	Sulfuric acid	By-products, petrochemicals
Alkali	Sodium hydroxide	Electroplating
	Lime	Beverage production
		Photography
		Vegetable and fruit processing
Ammonia		Nitric acid production
		Urea and ammonium nitrate works
Arsenic	Arsine	Pigment and dye
	Arsenous acid and salts	Pesticide and herbicide production
		Metallurgical processing of other metals
		Glass and ceramic industries
		Tanneries
Asbestos		Obsolete building products
		Insulation removal operations
Carbon dioxide		Combustion
		Fermentation
Carbon monoxide		Coke ovens
		Incomplete combustion generally
		Smelting
		Vehicle exhausts
		Metal extraction and refining

Table 9.3 Examples of Sources of Inorganic Pollutants—cont'd

Chemical	Examples	Industrial Source
Chlorine and chlorides		Chloralkali
		Paper and pulp processing
		Petrochemicals
		Metal extraction and refining
Chromium and compounds		Anodizing
		Cement
		Dyes
		Electroplating
		Paint
		Tanneries
Cobalt and compounds		Refinery catalysts
		Paint
		Paper and pulp processing
		Tungsten carbide binder
Copper and compounds		Electroplating
		Electric and electronics
		Etching
Cyanide		Heat treatment of metal
		Electroplating
		Photographic
		Synthetic fiber
Fluorides		Aluminum
Iron and compounds		Aluminum refining
		Steelworks
		Electronics
		Electroplating
Lead and compounds		Batteries
		Explosives
		Paint
		Petrochemicals
		Printing
Manganese and compounds		Batteries
		Catalyst
Mercury		Herbicides
		Batteries
		Catalysts
		Chloralkali process
		Combustion of coal and oil
		Electric and electronic
		Explosives
Nitrogen oxides		Combustion processes
		Explosives
		Electricity generation
		Forage tower silos
		Nitric acid

Continued

Table 9.3 Examples of Sources of Inorganic Pollutants—cont'd

Chemical	Examples	Industrial Source
Phosphorus and compounds	Phosphoric acid	Corrosion protection
		Metal finishing
Platinum and compounds		Catalysts
Silicates		Cement
		Metal extraction and refining
Sulfur oxides	Sulfur dioxide	Combustion of coal and heavy fuel oil
	Sulfur trioxide	Electricity generation
Tin and compounds		Tinplating
Titanium and compounds	Titanium dioxide	Astronautics
		Paint
		Paper
Vanadium and compounds	Catalysts	Petroleum refining
Zinc and compounds		Electroplating
		Galvanizing

by-product of manufacturing and from products that work their way through the supply chain and are eventually disposed of after final use.

Thus, one of the major and continuing environmental problems is contamination resulting from the activities related to industrial processes and products. Contamination of the air, land, and water causes extensive damage of local ecosystems since accumulation of pollutants in animals and plant tissue may cause death or mutations. Thus, once a spill has occurred, every effort must be made to rid the environment of the toxins. The chemicals of known toxicity range in degree of toxicity from low to high represent considerable danger to human health and must be removed. Many of these chemical substances come in contact with and are sequestered by soil or water systems. While conventional methods to remove, reduce, or mitigate the effects of toxic chemical in nature are available including (i) pump and treat systems, (ii) soil vapor extraction, (iii) incineration, and (iv) containment, each of these conventional methods of treatment of contaminated soil and/or water suffers from recognizable drawbacks and may involve some level of risk. In short, these methods, depending upon the chemical constituents of the spilled material, may limit effectiveness and can be expensive.

Although the effects of bacteria (microbes) on chemicals, especially organic chemicals such as petroleum-derived hydrocarbons, have been known for decades, this technology biodegradation (also known as *bioremediation* in the sense of applied cleanup of a site by other than natural means) has shown promise and, in some cases, high degrees of effectiveness for the treatment of sites contaminated by inorganic chemicals since it is cost-effective and will lead to complete mineralization. The concept of biodegradation may also refer to complete *mineralization* of the inorganic contaminants into carbon dioxide, water, and inorganic compounds to other simpler inorganic compounds that are not detrimental to the environment. In fact, unless they are overwhelmed by the amount of the spilled material or it is toxic, many indigenous microorganisms in soil and/or water are capable of degrading chemical contaminants.

The capabilities of microorganisms and plants to degrade and transform contaminants provide benefits in the cleanup of petroleum-derived pollutants from spill sites and storage sites (Speight and Arjoon, 2012) and offer considerable potential for use after a spill of inorganic chemicals. These remediation ideas have provided the foundation for many ex situ waste treatment processes (including sewage treatment) and a host of in situ biodegradation methods that are currently in practice.

The US Environmental Protection Agency (US EPA) uses biodegradation because it takes advantage of natural processes and relies on microbes that occur naturally or can be laboratory cultivated; these consist of bacteria, fungi, actinomycetes, cyanobacteria, and, to a lesser extent, plants. These microorganisms either consume and convert the contaminants or assimilate within them all harmful compounds from the surrounding area, thereby rendering the region virtually contaminant-free. Generally, the substances that are consumed as an energy source are inorganic compounds, while those, which are assimilated within the organism, are heavy metals. Biodegradation harnesses this natural process by promoting the growth and/or rapid multiplication of these organisms that can effectively degrade specific contaminants and convert them to nontoxic by-products.

In the current context, biodegradation at contaminated sites is the natural or stimulated cleanup of spills of inorganic chemicals using microbes to break down the inorganic contaminants of the spill into less harmful (usually lower molecular weight) and easier-to-remove products (biodegradation). The microbes transform the contaminants through metabolic or enzymatic processes, which vary greatly, but the final product is usually harmless and includes carbon dioxide, water, and cell biomass. Thus, the emerging science and technology of biodegradation offers an alternative method to detoxify soil and water from chemical contaminants. Furthermore, by means of clarification, *biodegradation (biotic degradation* and *biotic decomposition) is the chemical degradation of contaminants by bacteria or other biological means*. Inorganic material can be degraded aerobically (in the presence of oxygen) or anaerobically (in the absence of oxygen). Most biodegradation reactions operate under aerobic conditions, but a system under anaerobic conditions may permit microbial organisms to degrade chemical species that are otherwise nonresponsive to aerobic treatment and vice versa.

Thus, biodegradation is a natural process (or a series of processes) by which spilled inorganic chemicals are broken down (degraded) into nutrients that can be used by other organisms. Thus, the ability of a chemical to be biodegraded is an indispensable element in understanding the risk posed by that chemical on the environment.

Biodegradation is a key process in the natural attenuation (reduction or disposal) of chemical compounds at hazardous waste sites, but the success of the process depends on the ability to determine these conditions and establish them in the contaminated environment. Thus, important site factors required for success include (i) the presence of metabolically capable and sustainable microbial populations; (ii) suitable environmental growth conditions, such as the presence of oxygen; (iii) temperature, which is an important variable—keeping a substance frozen or below the optimal operating temperature for microbial species can prevent biodegradation—most biodegradation occurs at temperatures between 10°C and 35°C (50°F and 95°F); (iv) the presence of water; (v) appropriate levels of nutrients and contaminants; and (vi) favorable acidity or alkalinity. In regards to the last parameter, soil pH is extremely important because most microbial species can survive only within a certain pH range; as a basis for comparison, generally, the biodegradation of hydrocarbons is optimal at a pH 7 (neutral), and the *acceptable* (or optimal) pH range is on the order of 6–8 (Speight and Arjoon, 2012; Speight, 2017b). Furthermore, soil (or water) pH can affect availability of nutrients and activity of the microbes.

Thus, through biodegradation processes, living microorganisms (primarily not only bacteria but also yeasts, molds, and filamentous fungi) can alter and/or metabolize various classes of chemical compounds. Furthermore, biodegradation also alters subsurface accumulations of chemicals.

Temperature influences rate of biodegradation by controlling rate of enzymatic reactions within microorganisms. Generally, the rate of an enzymatic reaction approximately doubles for each 10°C (18°F) rise in temperature.

However, there is an upper limit to the temperature that microorganisms can withstand. Most bacteria found in soil are mesophile organisms, which are organisms that experience optimal growth in moderate temperature conditions that are typically between 20°C and 45°C (68°F and 113°F) but do not survive at all well in temperature regimes exceeding 50°C (122°F).

In order to enhance and make favorable the parameters presented above to ensure microbial activity, there are two other enhanced biodegradation methods that offer useful options for cleanup of spills of chemicals, (i) fertilization and (ii) seeding. *Fertilization* (*nutrient enrichment*) is the method of adding nutrients such as phosphorus and nitrogen to a contaminated environment to stimulate the growth of the microorganisms capable of biodegradation. Limited supplies of these nutrients in nature usually control the growth of native microorganism populations. When more nutrients are added, the native microorganism population can grow rapidly, potentially increasing the rate of biodegradation. *Seeding* is the addition of microorganisms to the existing native degrading population. Some species of bacteria that do not naturally exist in an area will be added to the native population. As with fertilization, the purpose of seeding is to increase the population of microorganisms that can biodegrade the spilled chemical. Thus, biodegradation is an environmentally acceptable naturally occurring process that takes place when all of the nutrients and physical conditions involved are suitable for growth. The process allows for the breakdown of a compound to fully either oxidized or reduced simple molecules such as carbon dioxide/methane, nitrate/ammonium, and water. However, in some cases, where the process is not complete, the products of biodegradation can be more harmful than the substance degraded.

Intrinsic biodegradation is the combined effect of natural destructive and nondestructive processes to reduce the mobility, mass, and associated risk of a contaminant. Nondestructive mechanisms include sorption, dilution, and volatilization. Destructive processes are aerobic and anaerobic biodegradation. *Intrinsic aerobic biodegradation* is well documented as a means of remediating soil and groundwater contaminated with fuel hydrocarbons but has not seen much application to remediation of spills of inorganic chemicals.

9.2 CLEANUP

The expression environmental cleanup (also known as *environmental remediation*) relates (in the current context) to the removal of inorganic contaminants from environmental media such as air, water (including groundwater and surface water), and soil (including sediment). This would mean that once requested by the government or a land remediation authority, immediate action should be taken as this can impact negatively on human health and the environment. Remedial action is generally subject to a collection of regulatory requirements and can be based on assessments of the condition of the flora (plant life) and fauna (animal life) within an ecosystem and any ecological risks where no legislated standards exist or where standards are advisory rather than regulatory.

There are two approaches through which pollution can be reduced: (i) reducing consumption or usage of a polluting product and (ii) treatment of wastes, discharges, and disposals of a pollutant. Yet, waste treatment can only be effective if pollution is coming from a defined and accessible source (point source).

More generally, the cleanup of environmental pollutants involves a variety of techniques, ranging from simple biological processes to advanced engineering technologies. Cleanup activities may address a wide range of contaminants, from common industrial chemicals, agricultural chemicals, and metals, as well as radionuclides. Cleanup technologies may be specific to the contaminant (or contaminant class) and to the site. However, before cleanup, the sources and types of pollutants (Chapter 5) must be identified, and although elemental analysis is not a focus of this book, it is necessary to acknowledge the need for accurate analytic data at this point.

However, before environmental cleanup efforts are initiated, there should be a full analysis of the charter of the base soil and the number and amounts of contaminants in the soil. Though many nonradioactive metals do not break down in the environment, sometimes chemical changes in the subsurface environment can reduce the toxicity of these compounds (Chapter 7). For example, copper in certain forms is considered so safe that we use it to pipe drinking water, yet trace concentrations can be detrimental to aquatic life. Chromium in one form (hexavalent, Cr^{6+}) forms compounds that are highly toxic and very soluble, yet under certain conditions, it transforms into trivalent chromium (Cr^{3+}), which is less toxic and insoluble. The potential for such a change differs for each substance, but it can be estimated by conducting a chemical and biological analysis of the subsurface environment.

The precise and accurate analyses of environmental pollutants and indigenous materials form the basis for most environmental studies. Analytic programs must include both laboratory-based determinations of the stable or stabilized analytes and the in-field determination of the nonstable analytes. Difficulties in providing such analyses include the (i) large range of analyte and matrix element concentrations, (ii) phase associations of the analyte elements, (iii) sample size, (iv) sample homogeneity, and (v) analyte volatility or stability. Furthermore, the growth in environmental analytic chemistry has also given rise to increased awareness of the need for a wide range of appropriate standard reference materials, both for total analyses and for operationally defined extraction procedures.

Once the pollutants have been identified, the cleanup of environmental pollution by inorganic chemicals involves a variety of techniques, ranging from simple biological processes to advanced engineering technologies. Cleanup activities may address a wide range of contaminants, from common industrial chemicals, such as inorganic agricultural chemicals and metals, to radionuclides for which an accurate analysis is required. Cleanup technologies may be specific to the contaminant (or contaminant class) and to the site (sire-specific).

In some cases, it may be physically impractical to completely remove all traces of contamination, and in such cases, it is necessary to decide upon an acceptable level of residual contamination. This decision will be based upon the criteria such as evaluation of toxicity data and extrapolation to potential exposure scenarios, and the result of these evaluations is an estimate of risk for given adverse outcome. Risk-based target levels typically determine when cleanup is complete. As a result, evolution of cleanup technologies has yielded four general categories of remediation approaches: (i) physical removal with or without treatment, (ii) in situ conversion by physical or chemical means to less toxic or less mobile forms, (iii) containment, and (iv) passive cleanup, also known as natural attenuation. A fifth option involves the combination of two or more of these technologies may be used at some sites (Tedder and Pohland, 2000).

9.2.1 The Chemistry of Cleaning

The chemistry of environmental remediation is variable and subject to (i) the area (the ecosystem) to be cleaned, (ii) the floral and faunal species within the ecosystem, (iii) the chemical properties of the contaminants, and (iv) the chemical properties of the contaminants. The answer to these and other questions lies within words like surfactant, solvent, chelating agent, and builder. Understanding the basic elements of the effectiveness of a remedial operation agent against different types of air, soil, and water is essential to understanding the chemistry of the remediation process.

9.2.1.1 Soil Types

Soil can be broken down into three broad categories: (i) inorganic-contaminated soil, (ii) organic-contaminated soil, and (iii) a combination of the first two categories. However, there are several classification systems that are applicable to soil.

In the present context, inorganic-contaminated soil includes rust, scale, hard water deposits, and minerals such as sand, silt, and clay. In some cases, acids are used to remove inorganic deposits such as rust and scale. Combination soils often present the toughest challenge for a cleaner since the soil contains both organic and inorganic components, but careful investigation of the soil and identification of the inorganic pollutants is critical. Most combination soils are removed with a very concentrated, highly built cleaner that also contains solvent. On the other hand, organic-contaminated soils encompass a broad range and include food soils such as fat, grease, protein, and carbohydrate; living matter such as mold, yeast, and bacteria; and petroleum soils such as motor oil, axle grease, and cutting oils. Most of the time, contaminants in organic-contaminated soil are best removed using alkaline cleaners or solvents.

In addition, when investigating the potential of soil cleanup, consideration must be given to the characteristics of the base (noncontaminated) soil. The general types are (i) sandy soil, (ii) silty soil, clay soil, (iv) peaty soil, and (v) saline soil. *Sandy soil* has the largest particles among the different soil types and has large spaces between the particles. As a result, sandy soil cannot retain water for any length of time and is unlikely to retain inorganic contaminants because of the potential for wash—though during periods of rain (and acid rain). On the other hand, *silty soil* has much smaller particles than sandy soil and retains water (and contaminants) longer. *Clay soil* has the smallest particles among the three, so it has good water storage qualities. Due to the extremely small size of the clay particles, this type of soil retains inorganic contaminants by adsorption, which may not be easily reversible. *Peaty soil* typically has a high-water content, and the water is acidic, which may have an adverse effect on the ability of this soil to retain inorganic contaminants. *Saline soil* is usually brackish because of its high salt content, and the reactivity of the soil to inorganic contaminants will be dependent upon the salt content of the soil.

When using soil texture as a means of soil classification, soil type usually is more commonly related to the different sizes of mineral particles in the soil. It is the finely ground rock particles that are grouped by size and give the soil the designation sandy soil, silty soil, and clayey soil (as described above).

Each component of the soil plays an important role in the retention of inorganic contaminants. For example, the largest particles determine the aeration and drainage characteristics, while the tiniest, submicroscopic clay particles are chemically active and bind readily with inorganic cations and anions. Generally, clayey soils decrease in permeability as the clay content increases. In addition to the mineral composition of soil, humus (organic material) also plays an important role in soil characteristics and the ability of the soil to retain inorganic cations and anions.

9.2.1.2 Surfactants

A surfactant (*surface active agent*) can be the most important part of any cleaning operation. In general, they are chemicals that, when dissolved in water or another solvent, orient themselves at the interface (boundary) between the liquid and a solid (the dirt that is being removed) and modify the properties of the interface. All surfactants have a common molecular similarity. One end of the molecule has a long nonpolar chain that is attracted to oil, grease, and dirt (the hydrophobe). Another part of the molecule is attracted to water (the hydrophile). The surfactant lines up at the interface as diagrammed below. The hydrophobic end of the molecule gets away from the water, and the hydrophilic end stays next to the water. When dirt or grease is present (hydrophobic in nature), the surfactants surround it until it is dislodged from the boundary.

It should be noted that a surfactant can be either a soap or a synthetic detergent. Soaps have been used for centuries because they are made from natural materials such as animal fat and lye. Synthetics have only become available over the last 40 years. Soaps are still commonly used in personal hygiene products because of their mildness. Synthetic detergents are the surfactants of choice for almost all other cleaning agents.

9.2.1.3 Chelating Agents

Soil removal is a complex process that is much more involved than just adding soap or surfactant to water. One of the major concerns in dealing with cleaning compounds is water hardness. Water is made *hard* by the presence of calcium, magnesium, iron, and manganese metal ions. These metal ions interfere with the cleaning ability of detergents. The metal ions act like dirt and interact with the surfactants, making them unavailable to act on the surface that is to be cleaned. A chelating agent combines itself with these disruptive metal ions in the water. The metal ions are surrounded by the clawlike chelating agent that alters the electronic charge of the metal ions from positive to negative. This makes it impossible for the metal ions to be precipitated with the surfactants. Thus, chelated metal ions remain tied up in solution in a harmless state where they will not use up the surfactants. Some common chelating agents used in industrial cleaning compounds include phosphates, ethylenediaminetetraacetic acid (EDTA), sodium citrate, and zeolite compounds.

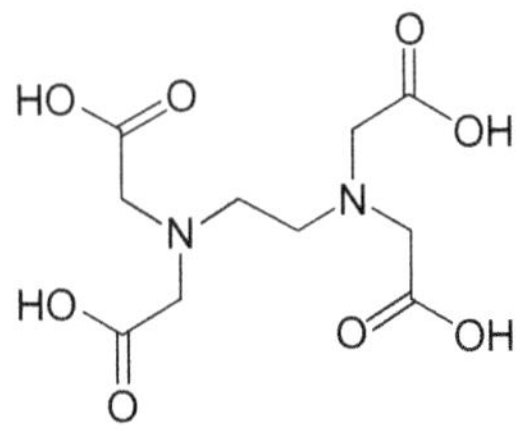

EDTA

Household cleaning agents, such as laundry soap, used phosphate type chelating agents heavily in the 1950s and 1960s. In the 1970s, phosphate bans were imposed because phosphates entered back into the environment unchanged through sewerage works and caused oxygen depletion in waterways. Alternative chelating agents such as EDTA have been developed as phosphate substitutes.

9.2.1.4 Builders

Detergents consist of surfactants and chelating agents; the surfactant removes dirt from a soiled surface, and the chelating agent is used to surround unwanted metal ions found in cleaning solutions. The chelating process, though very effective, is not always necessary and adds to the cost of formulating detergents. Builders are often a good alternative.

Builders are added to a cleaning compound to upgrade and protect the cleaning efficiency of the surfactant(s). Builders have several functions including softening, buffering, and emulsifying. Builders soften water by deactivating hardness minerals (metal ions like calcium and magnesium). They do this through one of the two ways: (i) sequestration, which involves holding metal ions in solution, and (ii) precipitation, which involves removing metal ions from solution as insoluble materials.

Builders, in addition to softening, provide a desirable level of alkalinity (increase pH), which aids in cleaning. They also act as buffers to maintain proper alkalinity in wash water. Finally, the chemicals that constitute builders help emulsify oil-contaminated and grease-contaminated soil by breaking it up into tiny globules. Many builders will peptize or suspend loosened dirt and keep it from settling back on the cleaned surface. Below are three of the most common builders used in today's heavy-duty detergents: (i) phosphate derivatives, (ii) sodium carbonate, and (iii) sodium silicate.

Phosphates, usually sodium tripolyphosphate (STPP), have been used as builders extensively in heavy-duty industrial detergents. They combine with hardness minerals to form a soluble complex that is removed with the wash water. They also sequester dissolved iron and manganese, which can interfere with detergency. *Sodium carbonate* (*soda ash*) is used as a builder but can only soften water through precipitation. Precipitated calcium and magnesium particles can build up on surfaces, especially clothing, and therefore, sodium carbonate is not used in laundry detergents. Sodium silicate serves as a builder in some detergents when used in high concentrations. When used in lower concentrations, it inhibits corrosion and adds crispness to detergent granules.

9.2.1.5 Solvents

As stated above, detergents consist of surfactants, chelating agents, and builders. Remember that surfactants are designed to remove dirt from a soiled surface. Chelating agents and builders are added to the formula to keep water hardness from interfering with the cleaning process. Water makes up a large percentage of most liquid cleaner formulas. It is not uncommon for water-based detergents to contain 50% water or more. Some ready-to-use formulations may contain as much as 90%–95% water! With this much water present in a cleaner, why do they work so well? Water can be considered an active ingredient that adds to the detergency of cleaners. It performs several very important functions in liquid cleaners. Most importantly, it adds to the "detergency" of a cleaner. Water acts as a solvent that breaks up soil particles after the surfactants reduce the surface tension and allow the water to penetrate soil (water is commonly referred to as "the universal solvent").

Water is necessary for the detergent to work correctly, and the water also aids in the suspension and antiredeposition of soils. Once the soil has been dissolved and emulsified away from the surface, it must be prevented from being redeposited. Water keeps the soil suspended away from the clean surface so that it can be carried away easily during the rinsing process. Without this water, the cleaning formulas would be much less effective.

In addition to water, other chemical solvents are often added to cleaners to boost performance. Compounds such as 2-butoxyethanol (butyl), isopropyl alcohol (rubbing alcohol), and D-limonene are all considered solvents. Their main function is to liquefy grease and oils or dissolve solid soil into very small particles so surfactants can more readily perform their function.

9.2.1.6 Preservatives

A preservative is nothing more than a substance that protects soaps and detergents against the natural effects of aging such as decay, discoloration, oxidation, and bacterial degradation. Synthetic detergents are preserved differently from soaps. In soaps, preservatives are used to forestall the natural tendency to develop rancidity and oxidize upon aging. Butylated hydroxytoluene and stannic chloride are commonly used in this application; EDTA is also, used in small amounts. In detergents, preservatives are used to prevent bacteria from spoiling the solution. Methylparaben and propylparaben are very common for this application. Detergents would not be preserved if they were not biodegradable. Bacteria found in air, in waste treatment systems, and in soil decompose the surfactants and other ingredients found in our cleaners once they enter the environment.

Furthermore, the cleanup of environmental pollution involves a variety of techniques, ranging from simple biological processes to advanced engineering technologies. Cleanup activities may address a wide range of contaminants, including from common industrial inorganic chemicals, agricultural chemicals, metals, and radionuclides (Page, 1997; Tedder and Pohland, 2000; Testa and Winegardner, 2000). Cleanup technologies may be specific to the contaminant (or contaminant class) and to the site.

Cleanup costs can vary dramatically depending on the contaminants, the media affected, and the size of the contaminated area. Much of the remediation has been in response to such historical chemical management practices as dumping, poor storage, and uncontrolled release or spillage. However, greater efforts have been directed toward pollution prevention, which is more cost effective than remediation. In most cases, it is financially or physically impractical to completely remove all traces of contamination. In such cases, it is necessary to set an acceptable level of residual contamination. Evaluating experimental toxicity data and then extrapolating to potential exposure scenarios form the basis for such decisions. The result of these evaluations is an estimate of risk for given adverse outcome (e.g., cancer or death). Risk-based target levels typically determine when cleanup is complete. Thus, evolution of cleanup technologies has yielded four general categories of remediation approaches: (i) physical removal with or without treatment, (ii) in situ conversion by physical or chemical means to less toxic or less mobile forms, (iii) containment, (iv) passive cleanup or natural attenuation, and (v) combinations of any of the aforementioned technologies.

9.2.2 Physical Removal

The physical removal processes invariably refer to contaminant removal from the ground, that is, from the soil. Removal from the atmosphere or from the aquasphere also involves treating process but from a different perspective.

The physical removal of contaminated soil and groundwater has been, and continues to be, a common cleanup practice. However, physical removal does not eliminate the contamination but rather transfers it to another location such as a facility that is specially designed to contain the contamination for a sufficient period of time. In this way, proper removal reduces risk by reducing or removing the potential for exposure to the contamination. Removal options vary dramatically for soil and groundwater,

The properties of the inorganic pollutant, including the potential of the pollutant to adhere to the soil, may require the physical removal of contaminated soil and groundwater that has been and continues to be a common cleanup practice. However, physical removal does not eliminate the contamination but, instead, transfers the contamination to an alternate location. Thus, cleanup cannot be accomplished by the transfer process (the unacceptable out of sight, out of mind option), but it is essential that the alternate location will be a facility that is specially designed to contain the contamination for a sufficient period. There may also be treatment of the soil to convert the inorganic contaminate to a less virulent form after which cleanup, say, by washing may be accomplished. In this way, proper removal can reduce the risk imposed by the contaminant by removing the potential for exposure to the contamination.

The excavation of chemical-contaminated soils works well for limited areas of contamination that are close to the ground surface. Under ideal conditions and by following the necessary local or state or regional or national regulations, the disposal location is a designed, regulated, and controlled disposal facility (e.g., a landfill or incineration facility). Alternatively, chemical-contaminated soil may be excavated and consolidated in a prepared facility on-site. Prepared disposal facilities range from simple excavations with impermeable covers (caps) to sophisticated containment structures such as those used in modern landfills.

On the other hand, engineering for landfill construction now includes the use of liners, leachate controls, and management practices to prevent groundwater contamination or other forms of cross pollution. The liners in a landfill may typically consist of multiple layers of impermeable materials that are often combinations of synthetic (plastic) liners and compacted layers of dense clays, piping to collect and transport liquids generated within the landfill (leachate), and systems of sensors within and surrounding the landfill to detect leaks. Although caution must be exercised since many inorganic chemical can (alone or in the presence of oxygen) degrade the liners to the point where an impermeable liner becomes permeable. When contaminated soil is excavated, transported, and disposed of properly, physical removal can be an effective and economical cleanup option. However, the landfill treatment option can be troublesome since landfill space decreases and public opposition to the deposition of environmental contaminants in a landfill is very evident.

Treatment of excavated soil, to either destroy the contaminant or to reduce its toxicity or mobility often is associated with physical removal. Treatment following removal will differ with the chemical of concern. Many inorganic chemicals may be incinerated or landfilled effectively; the landfill category typically excludes the persistent inorganic pollutants may be incinerated. In some cases, metals require conversion to compounds that will not react with other substances before being transferred to a landfill. Effective air pollution controls must be available in order to manage and control incinerator emissions, although opposition to incineration is still active in some areas. Once the environmental issues have been agreed upon for incinerator use, incineration can serve to convert inorganic contaminant to less harmful metal oxides that may serve a useful purpose as feedstock to the inorganic chemical industry:

$$2MX_2 + O_2 \rightarrow \underset{\text{Metal oxide}}{2MO} + 2X_2$$

Typically, if the emitted chemical (X_2) is a gas or volatile liquid, it can be recovered by any one of a variety of gas cleaning operations (Mokhatab et al., 2006; Speight, 2007, 2014a, 2017a; Kidnay et al., 2011; Bahadori, 2014).

In addition to excavation, more selective removal technologies have been developed for contaminants in soil, including soil washing, which uses processing equipment and chemical solvents to wash (leach) contaminants from soil. However, this form of treatment must be done with little delay.

Any inorganic chemicals that will each use a solvent typically water in which many inorganic chemicals have a high degree of solubility (Chapter 4) will (sooner or later) be subject to rainfall, and when in place, the chemicals may be leached into groundwater aquifers. In addition, the occurrence of acid rain may enhance the solubility of the chemicals in the water and cause further problems.

In practice, soil washing often is complicated and expensive. Phytoextraction (which is the use of plants to remove soil contaminants) has achieved some success and favor but not in all applications. Selected plant species may remove and concentrate inorganic contaminants such as heavy metals and radionuclides in the aboveground or belowground tissues. If phytoextraction is successful, the resulting plant tissue will have high levels of the soil contaminant and be classified as hazardous waste, requiring appropriate treatment or disposal options (see previous section).

Sediments are the inorganic materials (such as clay, silt, and sand) and organic materials (plant and animal remains) that settle to the bottom of water bodies. Aquatic sediments often become contaminated by a wide variety of man-made chemicals including (i) agricultural inorganic chemicals such as pesticides—for example nitrate derivatives, sulfate derivatives, and phosphate derivatives—that are washed into water bodies, (ii) industrial chemicals that are released into water bodies or that leak from containment structures, and (iii) many inorganic chemicals that are transported by water. Contamination in aquatic sediments may affect the organisms that live within the sediments or may bioaccumulate through the food chain as larger species feed on organisms that have absorbed the contamination. Remediating such contamination requires choosing between the risks associated with leaving the contamination in place and the risks associated with excavating the sediments (and resuspending them in the water) followed by transportation and disposal of the chemicals.

Inorganic chemicals that are sent for disposal by burial or by direct release on to the ground surface can migrate down into the soil structure (especially when it rains and the rain may be acid rain) and come in contact with groundwater. Final disposition of these chemicals depends on the water solubility of the chemicals. Aqueous-phase chemicals (i.e., chemicals that are soluble in water and water-reactive chemicals) (Chapter 4) dissolve in and move with groundwater. Nonaqueous phase chemicals such as certain salts and radionuclides that do not dissolve (or have a low solubility) in water can have a significant impact on the detection and remediation of inorganic-organic contamination. When contamination is detected in groundwater, one common cleanup approach is to drill wells and then pump out the water followed by contaminant removal by a variety of methods. However, this *pump-and-treat* approach addresses only the dissolved, aqueous phase of contamination.

To remove sources of groundwater contamination, technologies are needed to accurately detect and measure the amounts of these chemicals. Well drilling is commonly used to investigate or remediate contaminated sites, and it brings up contaminated soil that must be disposed of properly. *Direct push* technologies use large vehicles equipped with hydraulic rams or percussion equipment to push metal tubes into the ground. Sensors on the advancing tip of these tubes provide information on the nature of the sediments being penetrated. Recent advances in this technology allow special chemical sensors to be deployed on the end of the tube providing information on the presence and concentration of chemicals in the ground. The hollow tube also can be used to collect soil and groundwater samples. When sampling is complete, the rods typically are removed from the ground, and the hole is sealed. While depth and geology limit a direct push technology, it is generally faster than well drilling, and it does not contaminate the soil.

9.2.3 Conversion

Inorganic chemistry—the study of the synthesis, reactions, structures, and properties of compounds of the elements—encompasses the chemistry of the nonorganic compounds and overlaps with organic chemistry in the area of organometallic chemistry, in which metals are bonded to carbon-containing ligands and molecules (Chapter 2). Inorganic chemistry is fundamental to many practical technologies including catalysis and materials, energy conversion and storage, and electronics. Inorganic compounds are also found in biological systems where they are essential to life processes. The problem arises when inorganic compounds are released to the environment and require cleanup, and this is where conversion of the inorganic chemical can play a major role.

In the case of pollution resulting from a buildup of (toxic) inorganic chemicals, reduction of the accessibility of the chemicals to the environment must be ensured to rebuild ecosystem services in a polluted area. Although physical or chemical methods such as change in acidity or absorption into the soil can help decrease the availability of chemicals, additional monitoring and securing is necessary to make sure that the pollutant is not brought back into the environment. Ideally, the system should convert or degrade the inorganic chemical(s) by microbes or fungi, as this will irreversibly destroy the toxicant.

Many inorganic chemicals take a long time to biodegrade in the environment, which means that their buildup rate is almost proportional to the total rate of pollution at any given time. These are also often some of the most potent and generally poisonous materials and thus strongly toxic even in low concentrations. Influential inorganic pollutants include nonmetals like ammonia and cyanide and heavy metals such as copper (Cu), mercury (Hg), and Cd, which are all toxic in various degrees. Many inorganic discharges are point sources, so proper treatment of material is generally possible through biological degradation with microbes and fungi or electrokinetic treatment (the use of electricity to reduce heavy metal ions and turn them into elemental precipitates). Also, most heavy metals are much less toxic in alkaline environments, a fact that can be used in treatment plans. Some combination of these three techniques should be established to lower emissions for point-source metal pollution.

Thus, conversion uses chemical reactions to change contaminants into less toxic or less mobile forms. These chemical reactions may be produced by the introduction of reactive chemicals to the contaminated area or by the action of living organisms such as bacteria. Conversion of the inorganic pollutant is a process that is typically characterized by a chemical reaction that results in chemical change. In this type of reaction, the products are different from the starting materials (reactants). Chemical reactions involve the breaking and formation of chemical bonds. There are several different types of chemical reactions and more than one way of classifying them, but for simplicity, the conversion reaction can take any one or more of the general types of inorganic reactions: (i) combination reaction, (ii) decomposition reactions, (iii) single-displacement reactions, and (iv) double-displacement reactions (Chapter 2).

By way of a refresher, *combination reactions* occur when two or more reactants form one product in a combination reaction. Thus,

$$S(s) + O_2(g) \rightarrow SO_2(g)$$

Decomposition reactions occur when an inorganic compound breaks down into two or more substances, which usually results from heating. Thus,

$$2HgO(s) + \text{heat} \rightarrow 2Hg(l) + O_2(g)$$

A *single-displacement reaction* occurs when an atom or ion of a single compound replacing an atom of another element. Thus,

$$Zn(s) + CuSO_4(aq) \rightarrow Cu(s) + ZnSO_4(aq)$$

A double-displacement reaction occurs when the elements from two compounds displace each other to form new compounds. These reactions may occur when one product is removed from the solution as a gas or precipitate or when two species combine to form a weak electrolyte that remains undissociated in solution. Thus,

$$CaCl_2(aq) + 2AgNO_3(aq) \rightarrow Ca(NO_3)_2(aq) + 2AgCl(s)$$

A neutralization reaction is an example of a specific type of double-displacement reaction that occurs when an acid reacts with a base, producing a solution of a metal salt and water. Thus,

$$\underset{\text{Acid}}{HCl(aq)} + \underset{\text{Base}}{NaOH(aq)} \rightarrow \underset{\text{Salt}}{NaCl(aq)} + \underset{\text{Water}}{H_2O(l)}$$

However, in spite of this general differentiation of the type of inorganic reactions, a reaction can fit into more than one category.

In addition, there are also more specific categories in which a chemical is combusted in air (oxygen) or a reaction in which a product is precipitated from a solution. Thus,

$$Pb(NO_3)_2 + H_2SO_4 \rightarrow \underset{\text{Precipitate}}{PbSO_4 \downarrow} + 2HNO_3$$

Knowledge of these general reactions will assist in the choice and design of a cleanup operation in which the inorganic chemical pollutant is converted to a more manageable and less polluting product.

In water, where inorganic chemicals are likely to react and undergo conversion, there are several types of reactions that can occur in an aqueous ecosystem. For example, three important types of reactions in water are (i) precipitation reaction, (ii) acid-base reaction, and (iii) oxidation-reduction (redox) reaction.

As shown above, in a precipitation reaction, an anion and a cation contact each other and form an insoluble ionic compound that precipitates out of solution. Thus,

$$Ag^+(aq) + Cl^-(aq) \rightarrow \underset{\text{Precipitate}}{AgCl(s)}$$

In an acid-base reaction, an acid (HCl) and a base (NaOH) combine to form salt and water. Thus,

$$HCl(aq) + NaOH(aq) \rightarrow NaCl + H_2O$$

The hydrochloric acid acts as an acid by donating H^+ ions or protons, and the sodium hydroxide acts as a base and provides the hydroxyl ions.

In an oxidation-reduction (redox reaction), there is an exchange of electrons between two reactants—the species that loses electrons is oxidized and the species that gains electrons is reduced. An example of a redox reaction occurs between hydrochloric acid and zinc metal, where the zinc atoms lose electrons and are oxidized to form Zn^{2+} ions:

$$Zn(s) \rightarrow Zn^{2+}(aq) + 2e^-$$

The H^+ ions of the hydrochloric acid gain electrons and are reduced to hydrogen atoms that combine to form the hydrogen molecule:

$$2H^+(aq) + 2e^- \rightarrow H_2(g)$$

Thus:

$$Zn(s) + 2H^+(aq) \rightarrow Zn^{2+}(aq) + H_2(g)$$

As simple as these reactions may seem, there is, nevertheless, an important principle applies: the equations must be balanced according to the principle of stoichiometry (Chapter 2). The total charge must be the same on both sides of a balanced equation and can (even in ionic reactions) be zero as long as the charge is the same on both the reactants and products sides of the equation.

Thus, cleanup by conversion uses known chemical reactions (Chapters 4 and 7) to change inorganic contaminants into less toxic or less mobile forms (e.g., see Habashi, 1994). These chemical reactions may be produced by the introduction of *reactive chemicals* to the contaminated area or by the action of living organisms such as bacteria. The use of biological systems to clean up contamination (bioremediation) includes all cleanup technologies that take advantage of biological processes to remove inorganic contaminants from soil and groundwater; the most common technique is microbial metabolism, if the colony of microbes can metabolize (feed upon) the inorganic chemical and survive. For decades, scientists have known that microbes can degrade some (but not all) certain inorganic contaminants, and in cases of historical contamination, microbial communities often adapt to take advantage of the energy released when these chemicals are degraded (i.e., metabolized). By assessing the existing conditions in the ecosystem, other chemicals (such as nutrients or oxygen) that the microbes require to break down the inorganic contaminants may be added to enhance the biodegradation process.

This is the concept behind a technology to remove contamination from soils that also contain low levels of radioactive materials. The combination of hazardous materials and radiation places this soil in the regulatory category of *mixed waste*, for which disposal is extremely difficult. By using biodegradation to remove any other inorganic contaminants, the remaining soil can be classified as low-level radioactive waste, which may have an accepted (but regulated) disposal mechanism.

Heavy metals are a common target for conversion approaches, although removal may not be practical when such metals contaminate large areas of surface soil. In these cases, chemical approaches often are sought to convert the metals to a less toxic and less mobile form. Such conversions often involve the use of reactive agents such as sulfur to create immobile sulfide salts of metals. Reducing the mobility of soil contaminants often refers to reducing the water solubility of the compounds. Reducing water solubility lowers the potential for contaminants to become dissolved in and move with water in the subsurface.

9.2.4 Containment

In some cases, a situation may exist in which technology is not available or practical to remove or convert contaminants. In those situations, it is often possible to contain the contamination as a final solution or as an interim measure until appropriate technologies become available. Environmental containment protects against liquid or solid spillage that could pollute the ground and ultimately preserves the environment. Thus, the purpose of a containment system is to enclose a chemical waste site (or a hazardous waste site), such as a lagoon or impoundment, and to prevent vertical and horizontal infiltration and migration of contaminated water (leachate or groundwater). This can be accomplished with a system consisting of vertical barriers to provide lateral containment such as (i) slurry walls, (ii) barrier walls, (iii) permeable reactive barrier walls also known as PRB walls, (iv) cofferdams, (v) cutoff walls and trenches, and sheet pile barrier walls.

Briefly and by way of definition, a *permeable reactive barrier* is a subsurface emplacement of reactive materials through which a dissolved contaminant plume must move as it flows, typically under natural gradient. Treated water exits the other side of the permeable reactive barrier. This in situ method for remediating inorganic dissolved-phase contaminants in groundwater combines a passive chemical or biological treatment zone with subsurface fluid flow management. A *cofferdam* (also called a *coffer*) is a temporary enclosure built within, or in pairs across, a body of water and constructed to allow the enclosed area to be pumped out. This pumping creates a dry work environment for the major work to proceed.

Situations may exist in which a cleanup technology is not available or it is not practical to remove or convert inorganic contaminants. In those situations, it may be possible to contain the contaminant as a final solution or as an interim measure until appropriate technologies become available. Containment has been used with varying degrees of success in cleaning up oil spills, but spills of inorganic contaminants are affected by the solubility of the inorganic chemicals in the water. However, containment of inorganic chemicals is often difficult because of the solubility of these materials in water—natural water (and its already dissolved salts) or water modified by the occurrence of acid rain water.

Radionuclides from historical weapon production sites and nuclear testing sites and from industrial uses of radioactive materials appear to be a good match for the use of containment technologies. Removing radioactive contamination from soil is problematic from a worker-safety standpoint, and it may create further contamination of equipment, containers, and surrounding areas.

Groundwater is not generally suitable for use of a containment technology because of the relative freedom of movement required by groundwater. However, between containment and conversion is a technology known as reactive barriers. Reactive barriers intercept contaminated groundwater plumes and are constructed of chemically reactive materials (e.g., iron) that bind or convert dissolved contaminants. Reactions between the contaminant and the iron either immobilize or degrade the contaminant by altering its chemical form (redox manipulation).

9.2.5 Passive Cleanup

Passive remediation technologies are increasingly common in some applications and take advantage of naturally occurring chemical or biological processes that degrade contaminants to less toxic forms. The accepted term for this group of technologies is monitored natural attenuation (MNA), which is the result of regulatory recognition that natural biological processes are capable of degrading certain contaminants under specific conditions and that dispersion may aid in achieving objectives. MNA is employed for the cleanup of a variety of contaminants in situations where the longer time frame associated with MNA does not increase the risks posed by the contamination. MNA recognizes that, while these processes are possible, they must be monitored to ensure that the expected progression occurs.

Cleaning inorganic contaminants from the ocean is an unenviable task that may be impossible to bring to any form of conclusion. The ocean is a complex solution of many inorganic compounds (Table 9.4), and in contrast to the behavior of most oceanic substances, the concentrations of the principal inorganic constituents of the oceans are remarkably constant. Calculations indicate that for the main constituents of seawater, the time required for thorough oceanic mixing is quite short compared with the time that would be required for input or removal processes to significantly change the concentration of an oceanic constituent. The concentrations of the principal constituents of the oceans vary primarily in response to a comparatively rapid exchange of water (precipitation and evaporation), with relative concentrations remaining nearly constant.

In addition, a variety of elements essential to the growth of marine organisms and some elements that have no known biological function exhibit nutrient-like behavior broadly similar to nitrate and phosphate. Silicate $\left(SiO_3{}^{2-}\right)$ is incorporated into the structural parts of certain types of marine organisms (diatoms) that are abundant in the upper ocean. Also, the concentration of zinc, a metal essential to a variety of biological functions, is observed to generally parallel silicate distributions. Cadmium, though having no

Table 9.4 Inorganic Constituents of Seawater

Ionic Constituent	g/kg of Seawater	Relative Concentration
Chloride	19.162	1
Sodium	10.679	0.8593
Magnesium	1.278	0.0974
Sulfate	2.680	0.0517
Calcium	0.4096	0.0189
Potassium	0.3953	0.0187
Carbon (inorganic)	0.0276	0.0043
Bromide	0.0663	0.00154
Boron	0.0044	0.00075
Strontium	0.0079	0.000165
Fluoride	0.0013	0.000125

known biological function, generally exhibits distributions that are covariant with phosphate and concentrations that are even lower than those of zinc. Attempts to clean up inorganic spills in the ocean may serve little purpose except to disturb the balance of nature.

In terms of passive cleanup of the oceans, it would take insurmountable expenditures to clean up the oceans, and as a result, ocean cleanup could be accomplished by a passive method of removing marine debris in or near the ocean gyre (any large system of circulating ocean currents, particularly those involved with large wind movements) by means of exceptionally long networks of floating barriers, anchored to the ocean floor. These V-shaped barrier networks are designed to interact with natural ocean currents, funneling plastic debris toward a central point where the plastic can be extracted by a platform and stored for transportation and recycling.

9.3 BIOREMEDIATION

Toxic inorganic chemicals are major contributors to environment contamination, and prevention of future contamination from these compounds presents an immense technical challenge. While various physical processes and chemical processes have been developed for treating these pollutants, these approaches are often prohibitively expensive, nonspecific, or have the potential for introducing secondary contamination. Thus, there has been an increased interest in eco-friendly bio-based treatments commonly known as bioremediation (Brar et al., 2006; Singh et al., 2008). Bioremediation (the actual chemical process is sometimes referred to as biodegradation) is often considered a cost-effective and environmental friendly method and is gradually making inroads for environmental cleanup applications.

Bioremediation relies on improved detoxification and degradation of toxic pollutants either through intracellular accumulation or via enzymatic transformation to less or nontoxic compounds. Many microorganisms naturally possess the ability to degrade, transform, or chelate various toxic chemicals. However, these natural transformations are limited by the relative slow rates. Development of new genetic tools and a better understanding of the natural transformation ability of a microorganism at the genetic level are essential to accelerate the progress of designer microbes for improved hazardous waste removal. Several attempts have been made recently to enhance biotransformation and bioaccumulation of toxic wastes by microorganisms.

Extensive environmental pollution by heavy metals and radionuclides (Table 9.5) primarily by anthropogenic origin is adversely affecting human

Table 9.5 Examples of Inorganic Chemical and Radionuclide Contaminants

Contaminant	Sources of Contaminant
Arsenic	Erosion of natural deposits; runoff from
	Runoff from glass and electronic production wastes
Cadmium	Corrosion of galvanized pipes
	Erosion of natural deposits
	Discharge from metal refineries
	Runoff from waste batteries and paints
Lead	Corrosion of household plumbing systems
	Erosion of natural deposits
Mercury	Erosion of natural deposits
	Discharge from factories and refineries
	Runoff from landfills
Radium	Erosion of natural deposits
Uranium	Erosion of natural deposits

health and environment and has led to stricter regulatory limits (Singh et al., 2008). Although traditional methods are adequate for treating high concentrations of contamination, they are not cost effective at reducing the levels to regulatory limits. Recently, pathway-engineering techniques have been explored for selective and high-capacity bioremediation of heavy metals and radionuclides. In fact, in addition to heavy metals, radionuclide contamination either through nuclear plant leaks or by nuclear weapons is a major environmental issue. Naturally occurring bacteria highly resistant to radiation are ideal metabolic engineering candidates for enhanced radionuclide cleanup (Singh et al., 2008).

Bioremediation is looked upon as an environmentally friendly technique used to restore soil and water to its original state by using indigenous microbes to break down and eliminate contaminants. Biological technologies are often used as a substitute to chemical or physical cleanup of chemical spills because biodegradation does not require as much equipment or labor as other methods; therefore, it is usually cheaper. It also allows cleanup workers to avoid contact with polluted soil and water.

Biodegradation technology exploits various naturally occurring mitigation processes: (i) *natural attenuation*, (ii) *biostimulation*, and (iii) *bioaugmentation*. Biodegradation that occurs without human intervention other than monitoring is often called *natural attenuation*. This natural attenuation relies on natural conditions and behavior of soil microorganisms that are indigenous to soil. *Biostimulation* also utilizes indigenous microbial populations to remediate contaminated soils and consists of adding nutrients and other substances to soil to catalyze natural attenuation processes. *Bioaugmentation* involves introduction of exogenic microorganisms (sourced from outside the soil environment) capable of detoxifying a contaminant, sometimes employing genetically altered microorganisms.

9.4 REMEDIATION OF HEAVY METAL-CONTAMINATED SITES

The overall objective of any approach to removing environmental contamination (Table 9.6) is to create a final solution that is protective of human health and the environment (Wuana and Okieimen, 2011). Remediation is generally subject to an array of regulatory requirements and can also be based on assessments of human health and ecological risks where no legislated standards exist or where standards are advisory. The regulatory authorities will normally accept remediation strategies that center on reducing metal bioavailability only if reduced bioavailability is equated with reduced risk and if the bioavailability reductions are demonstrated to be long term. For heavy-metal-contaminated soils, the physical and chemical form of the heavy metal contaminant in soil strongly influences the selection of the appropriate remediation treatment approach (Table 9.7). Information about the physical characteristics of the site and the type and level of contamination at the site must be obtained to enable accurate assessment of site contamination and remedial alternatives. The contamination in the soil should be characterized to establish the type, amount, and distribution of

Table 9.6 General Cleanup Methods for Spills of Inorganic Compounds

Acids, Inorganic Adsorb With	**Apply Sodium Bicarbonate/Calcium Oxide or Sodium Carbonate/Calcium Oxide**
Bases (caustic alkalis)	Neutralize with acid, citric acid, or commercial chemical neutralizers
Halides	Apply sodium bicarbonate
Hydrazine	Apply slaked lime, $Ca(OH)_2$
Hydrofluoric acid	Absorb with calcium carbonate (limestone, $CaCO_3$) or lime (calcium oxide, CaO)
Inorganic salt solutions	Apply soda ash
Oxidizing agents	Apply sodium bisulfite ($NaHSO_3$)
Reducing substances	Apply soda ash or sodium bicarbonate

Table 9.7 Technologies for Remediation of Heavy Metal Contaminated Soil

Electrokinetics	Removes contaminants from soil by application of an electric field
Encapsulation	Physical isolation and containment of the contaminated material. The impacted soils are isolated by low permeability caps or walls to limit the infiltration of precipitation
Isolation	Capping, subsurface barriers
Immobilization	Solidification/stabilization, vitrification, chemical treatment
Mobility reduction	Chemical treatment, permeable treatment, biological treatment bioaccumulation, phytoremediation (phytoextraction, phytostabilization, and rhizofiltration), bioleaching, biochemical processes
Physical separation	Extraction, soil washing, pyrometallurgical extraction, in situ soil flushing, and electrokinetic treatment
Soil flushing	Flood-contaminated soils with a solution that moves the contaminant to an area where they can be removed. Soil flushing is accomplished by passing an extraction fluid through soils using an injection or infiltration process. Recovered fluids with the absorbed contaminants may need further treatment
Soil vapor extraction	Involves the installation of wells in the contaminated area. Vacuum is applied through the wells to evaporate the volatile constituents of the contaminated mass, which are subsequently withdrawn through an extraction well. Afterward, the extracted vapors are adequately treated
Soil washing	The process separates coarse soil (sand and gravel) from fine soil (silt and clay), where contaminants tend to bind and sorb. This soil fraction must be further treated with other technologies
Solidification	Encapsulates the waste materials in a monolithic solid of high structural integrity
Stabilization/ immobilization	Reduces the risk posed by a waste by converting the contaminant into a less soluble, immobile, and toxic form
Thermal desorption	Contaminated soil is excavated, screened, and heated to temperatures such that the boiling point of the contaminants is reached, and they are released from the soil. The vaporized contaminants are often collected and treated by other means
Vitrification	Uses a powerful source of energy to "melt" soil at extremely high temperatures (1600–2000°C, 2910–3630°F), immobilizing most inorganics into a chemically inert, stable glass product and destroying organic pollutants by pyrolysis

heavy metals in the soil. Once the site has been characterized, the desired level of each metal in soil must be determined. This is done by comparison of observed heavy metal concentrations with soil quality standards for a regulatory domain or by performance of a site-specific risk assessment. Remediation goals for heavy metals may be set as total metal concentration or as leachable metal in soil or as some combination of these.

Several technologies exist for the remediation of metal-contaminated soil, which can be classified into three categories of hazard-alleviating measures: (i) gentle in situ remediation measures, (ii) in situ harsh soil restrictive measures, and (iii) in situ or ex situ harsh soil destructive measures. The goal of the last two harsh alleviating measures is to avert hazards to either man, plant, or animal, while the main goal of gentle in situ remediation is to restore the soil fertility, which allows a safe use of the soil. At present, a variety of approaches have been suggested for remediating contaminated soils. The US EPA has broadly classified remediation technologies for contaminated soils into (i) source control and (ii) containment remedies. Source control involves in situ and ex situ treatment technologies for sources of contamination.

In situ or in place means that the contaminated soil is treated in its original place, unmoved and unexcavated, remaining at the site or in the subsurface. In situ treatment technologies treat or remove the contaminant from soil without excavation or removal of the soil. Ex situ means that the contaminated soil is moved, excavated, or removed from the site or subsurface. Implementation of ex situ remedies requires excavation or removal of the contaminated soil. Containment remedies involve the construction of vertical engineered barriers, caps, and liners used to prevent the migration of contaminants.

Another classification places remediation technologies for heavy-metal-contaminated soils under five categories of general approaches to remediation: isolation, immobilization, toxicity reduction, physical separation, and extraction. In practice, it may be more convenient to employ a hybrid of two or more of these approaches for more cost-effectiveness. The key factors that may influence the applicability and selection of any of the available remediation technologies are (i) cost, (ii) long-term effectiveness/permanence, (iii) commercial availability, (iv) general acceptance, (v) applicability to high metal concentrations, (vi) applicability to mixed wastes (heavy metals and organics), (vii) toxicity reduction, (viii) mobility reduction, and (ix) volume reduction.

Background knowledge of the sources, chemistry, and potential risks of toxic heavy metals in contaminated soils is necessary for the selection of appropriate remedial options. Remediation of soil contaminated by heavy metals is necessary to reduce the associated risks, make the land resource available for agricultural production, enhance food security, and scale down land tenure problems. Immobilization, soil washing, and phytoremediation are frequently listed among the best available technologies for cleaning up heavy-metal-contaminated soils but have been mostly demonstrated in developed countries. These technologies are recommended for field applicability and commercialization in the developing countries also where agriculture, urbanization, and industrialization are leaving a legacy of environmental degradation.

9.4.1 Immobilization Techniques

Ex situ and in situ immobilization techniques are practical approaches to remediation of metal-contaminated soils. The ex situ technique is applied in areas where highly contaminated soil must be removed from its place of origin, and its storage is connected with a high ecological risk (e.g., in the case of radio nuclides). The method's advantages are (i) fast and easy applicability and (ii) relatively low costs of investment and operation. The disadvantages of this method include (i) high invasivity to the environment; (ii) generation of a significant amount of solid wastes (twice as large as volume after processing); (iii) the by-product must be stored on a special landfill site; (iv) in the case of changing of the physicochemical condition in the side product or its surroundings, there is serious danger of the release of additional contaminants to the environment; and (v) permanent control of the stored wastes is required.

In the in situ technique, the fixing agent amendments are applied on the unexcavated soil. The technique's advantages are (i) its low invasivity, (ii) simplicity and rapidity, (iii) relatively inexpensive, (iv) small amounts of wastes are produced, (v) high public acceptability, and (vi) covers a broad spectrum of inorganic pollutants. The disadvantages of in situ immobilization are (i) it is only a temporary solution (contaminants are still in the environment); (ii) the activation of pollutants may occur when soil physicochemical properties change; (iii) the reclamation process is applied only to the surface layer of soil—12–20 in.; and (iv) permanent monitoring is necessary.

Immobilization technology often uses organic and inorganic amendment to accelerate the attenuation of metal mobility and toxicity in soils. The primary role of immobilizing amendments is to alter the original soil metals to more geochemically stable phases via sorption, precipitation, and complexation processes. The mostly applied amendments include clay, cement, zeolites, minerals, phosphates, organic composts, and microbes. Recent studies have indicated the potential of low-cost industrial residues such as red mud and terminate in immobilization of heavy metals in contaminated soils. Due to the complexity of soil matrix and the limitations of current analytic techniques, the exact immobilization mechanisms have not been clarified, which could include precipitation, chemical adsorption and ion exchange, surface precipitation, formation of stable complexes with organic ligands, and redox reaction. Most immobilization technologies can be performed ex situ or in situ. In situ processes are preferred due to the lower labor and energy requirements, but implementation of in situ will depend on specific site conditions.

9.4.1.1 Solidification/Stabilization

Solidification involves the addition of binding agents to a contaminated material to impart physical/dimensional stability to contain contaminants in a solid product and reduce access by external agents through a combination of chemical reaction, encapsulation, and reduced permeability/surface area. Stabilization (also referred to as fixation) involves the addition of reagents to the contaminated soil to produce more chemically stable constituents. Conventional stabilization is an established remediation technology for contaminated soils and treatment technology for hazardous wastes in many countries in the world.

The general approach for solidification/stabilization treatment processes involves mixing or injecting treatment agents to the contaminated soils. Inorganic binders (Table 9.8), such as clay (bentonite and kaolinite), cement, fly ash, blast furnace slag, calcium carbonate, Fe/Mn oxides, charcoal, and

Table 9.8 Inorganic Amendments for Heavy Metal Immobilization

Material	Heavy Metal Immobilized
Lime (from lime factory)	Cd, Cu, Ni, Pb, Zn
Phosphate salt (from fertilizer plant)	Pb, Zn, Cu, Cd
Hydroxyapatite (from phosphorite)	Zn, Pb, Cu, Cd
Fly ash (from thermal power plant)	Cd, Pb, Cu, Zn, Cr
Slag (from thermal power plant)	Cd, Pb, Zn, Cr
Ca-montmorillonite (mineral)	Zn, Pb
Portland cement (from cement plant)	Cr, Cu, Zn, Pb
Bentonite	Pb

zeolite; and organic stabilizers such as bitumen, composts, and manures; or a combination of organic-inorganic amendments may be used (Wuana and Okieimen, 2011). The dominant mechanism by which metals are immobilized is by precipitation of hydroxides within the solid matrix. Solidification/stabilization technologies are not useful for some forms of metal contamination, such as species that exist as oxyanions (e.g., $Cr_2O_7^{2-}$, AsO_3^-) or metals that do not have low-solubility hydroxides (e.g., Hg). Solidification/stabilization may not be applicable at sites containing wastes that include organic forms of contamination, especially if volatile organics are present. Mixing and heating associated with binder hydration may release organic vapors. Pretreatment, such as air stripping or incineration, may be used to remove the organics and prepare the waste for metal stabilization/solidification. The application of solidification-stabilization technologies will also be affected by the chemical composition of the contaminated matrix, the amount of water present, and the ambient temperature. These factors can interfere with the solidification/stabilization process by inhibiting bonding of the waste to the binding material, retarding the setting of the mixtures, decreasing the stability of the matrix, or reducing the strength of the solidified area.

Cement-based binders and stabilizers are common materials used for implementation of solidification-stabilization technologies. Portland cement, a mixture of calcium silicates, aluminates, aluminoferrites, and sulfates, is an important cement-based material. Pozzolanic materials, which consist of small spherical particles formed by coal combustion (such as fly ash) and in lime and cement kilns, are also commonly used for solidification-stabilization. Pozzolanas exhibit cement-like properties, especially if the silica content is high. Portland cement and pozzolanas can be used alone or together to obtain optimal properties for a particular site. Organic binders may also be used to treat metals through polymer microencapsulation. This process uses organic materials such as bitumen, polyethylene, paraffins, waxes, and other polyolefins as thermoplastic or thermosetting resins. For polymer encapsulation, the organic materials are heated and mixed with the contaminated matrix at elevated temperatures (120–200°C). The organic materials polymerize and agglomerate the waste, and the waste matrix is encapsulated. Organics are volatilized and collected, and the treated material is extruded for disposal or possible reuse (e.g., as paving material). The contaminated material may require pretreatment to separate rocks and debris and dry the feed material. Polymer encapsulation requires more energy and more complex equipment than cement-based S/S operations.

Asphalt (called *bitumen* in some countries) is the cheapest and most common thermoplastic binder. Solidification-stabilization is achieved by mixing the contaminated material with appropriate amounts of binder/stabilizer and water. The mixture sets and cures to form a solidified matrix and contain the waste. The cure time and pour characteristics of the mixture and the final properties of the hardened cement depend upon the composition (amount of cement, pozzolana, and water) of the binder/stabilizer.

Ex situ solidification-stabilization can be easily applied to excavated soils because methods are available to provide the vigorous mixing needed to combine the binder/stabilizer with the contaminated material. Pretreatment of the waste may be necessary to screen and crush large rocks and debris. Mixing can be performed via in-drum, in-plant, or area-mixing processes. In-drum mixing may be preferred for treatment of small volumes of waste or for toxic wastes. In-plant processes utilize rotary drum mixers for batch processes or pug mill mixers for continuous treatment. Larger volumes of waste may be excavated and moved to a contained area for area mixing. This process involves layering the contaminated material with the stabilizer/binder and subsequent mixing with a backhoe or similar equipment. Mobile and fixed treatment plants are available for ex situ S/S treatment. Smaller pilot-scale plants can treat up to 100 tons of contaminated soil per day, while larger portable plants typically process 500–1000 tons per day. Stabilization/stabilization techniques are available to provide mixing of the binder/stabilizer with the contaminated soil in situ. In situ S/S is less labor- and energy-intensive than ex situ process that requires excavation, transport, and disposal of the treated material. In situ solidification-stabilization is also preferred if volatile or semivolatile organics are present because excavation would expose these contaminants to the air. However, the presence of bedrock, large boulder cohesive soils, oily sands, and clays may preclude the application of in situ solidification-stabilization at some sites. It is also more difficult to provide uniform and complete mixing through in situ processes. Mixing of the binder and contaminated matrix may be achieved using in-place mixing, vertical auger mixing, or injection grouting. In-place mixing is similar to ex situ area mixing except that the soil is not excavated prior to treatment. The in situ process is useful for treating surface or shallow contamination and involves spreading and mixing the binders with the waste using conventional excavation equipment such as draglines, backhoes, or clamshell buckets. Vertical auger mixing uses a system of augers to inject and mix the binding reagents with the waste. Larger (6–12 ft diameter) augers are used for shallow (10–40 ft) drilling and can treat 500–1000 yd^3 per day. Deep stabilization/solidification (up to 150 ft) can be achieved by using ganged augers (up to 3 ft in diameter each) that can treat 150–400 yd^3 per day. Finally, injection grouting may be performed to inject the binder containing suspended or dissolved reagents into the treatment area under pressure. The binder permeates the surrounding soil and cures in place.

9.4.1.2 Vitrification

The mobility of metal contaminants can be decreased by high-temperature treatment of the contaminated area that results in the formation of vitreous material, usually an oxide solid. During this process, the increased temperature may also volatilize and/or destroy organic contaminants or volatile metal species (such as Hg) that must be collected for treatment or disposal. Most soils can be treated by vitrification, and a wide variety of inorganic and organic contaminants can be targeted. Vitrification may be performed ex situ or in situ although in situ processes are preferred due to the lower energy requirements and cost. Typical stages in ex situ vitrification processes may include excavation, pretreatment, mixing, feeding, melting and vitrification, off-gas collection and treatment, and forming or casting of the melted product. The energy requirement for melting is the primary factor influencing the cost of ex situ vitrification. Different sources of energy can be used for this purpose, depending on local energy costs. Process heat losses and water content of the feed should be controlled to minimize energy requirements.

Vitrified material with certain characteristics may be obtained by using additives such as sand, clay, and/or native soil. The vitrified waste may be recycled and used as clean fill, aggregate, or other reusable materials. In situ vitrification (ISV) involves passing electric current through the soil using an array of electrodes inserted vertically into the contaminated region. Each setting of four electrodes is referred to as a melt. If the soil is too dry, it may not provide sufficient conductance, and a trench containing flaked graphite and glass frit (ground glass particles) must be placed between the electrodes to provide an initial flow path for the current. Resistance heating in the starter path melts the soil. The melt grows outward and down as the molten soil usually provides additional conductance for the current. A single melt can treat up to 1000 t of contaminated soil to depths of 20 ft, at a typical treatment rate of 3–6 t/h. Larger areas are treated by fusing together multiple individual vitrification zones. The main requirement for ISV is the ability of the soil melt to carry current and solidify as it cools. If the alkali content (as Na_2O and K_2O) of the soil is too high (1.4%, w/w), the molten soil may not provide enough conductance to carry the current.

Vitrification is not a classical immobilization technique. The advantages include (i) easily applied for reclamation of heavily contaminated soils (Pb), cadmium (Cd), chromium (Cr), asbestos, and materials containing asbestos; (ii) during applying, this method qualification of wastes could be changed, say from a hazardous waste to a neutral waste.

An example of disposal of vitrified material is the disposal of either bottom ash or boiler slag, which is subject to local environmental regulations. These materials are either landfilled or sluiced to storage lagoons. When sluiced to storage lagoons, the bottom ash or boiler slag is usually combined with fly ash. This blended fly ash and bottom ash or boiler slag are referred to as ponded ash. Approximately 30% of all coal ash is handled wet and disposed of as ponded ash. Ponded ash is potentially useable but variable in its characteristics because of its manner of disposal. Because of differences in the unit weight of fly ash and bottom ash or boiler slag, the coarser bottom ash or boiler slag particles settle first, and the finer fly ash remains in suspension longer. Ponded ash can be reclaimed and stockpiled, during which time it can be dewatered. Under favorable drying conditions, ponded ash may be dewatered into a range of moisture that will be within the vicinity of its optimum moisture content. The higher the percentage of bottom ash or boiler slag there is in ponded ash, the easier it is to dewater and the greater its potential for reuse. Reclaimed ponded ash has been used in stabilized base or subbase mixes and in embankment construction and can also be used as fine aggregate or filler material in flowable fill.

Both bottom ash and boiler slag have been used as fine aggregate substitute in hot mix asphalt wearing surfaces and base courses and emulsified asphalt cold mix wearing surfaces and base courses. Because of the *popcorn*, clinker-like low-durability nature of some bottom ash particles, bottom ash has been used more frequently in base courses than wearing surfaces. Boiler slag has been used in wearing surfaces, base courses, and asphalt surface treatment or seal coat applications.

Screening of oversized particles and blending with other aggregates will typically be required to use bottom ash and boiler slag in paving applications. Pyrite (FeS_2) that may be present in the bottom ash should also be removed (with electromagnets) prior to use. Pyrite (iron sulfide, FeS_2) is volumetrically unstable and expansive and produces a reddish stain when exposed to water over an extended time period.

Both bottom ash and boiler slag (Chapter 5) have occasionally been used as unbound aggregate or granular base material for pavement construction. Bottom ash and boiler slag are considered fine aggregates in this use. To meet required specifications, the bottom ash or slag may need to be blended with other natural aggregates prior to its use as a base or subbase material. Screening or grinding may also be necessary prior to use, particularly for the bottom ash, where large particle sizes, typically greater than 19 mm (0.75 in.), are present in the ash.

Bottom ash and boiler slag have been used in stabilized base applications. Stabilized base or subbase mixtures contain a blend of aggregate and cementitious materials that bind the aggregates, providing the mixture with greater bearing strength. Types of cementitious materials typically used include Portland cement, cement kiln dust, or pozzolanas with activators, such as lime, cement kiln dusts, and lime kiln dusts. When constructing a stabilized base using either bottom ash or boiler slag, both moisture control and proper sizing are required. Deleterious materials such as pyrites should also be removed.

Bottom ash and ponded ash have been used as structural fill materials for the construction of highway embankments and/or the backfilling of abutments, retaining walls, or trenches. These materials may also be used as pipe bedding in lieu of sand or pea gravel. To be suitable for these applications, the bottom ash or ponded ash must be at or reasonably close to its optimum moisture content, free of pyrites and/or "popcorn"-like particles, and must be noncorrosive. Reclaimed ponded ash must be stockpiled and adequately dewatered prior to use. Bottom ash may require screening or grinding to remove or reduce oversize materials greater than 19 mm (3/4 in.) in size.

Bottom ash has been used as an aggregate material in flowable fill mixes. Ponded ash also has the potential for being reclaimed and used in flowable fill. Since most flowable fill mixes involve the development of comparatively low compressive strength (to be able to be excavated at a later time, if necessary), no advance processing of bottom ash or ponded ash is needed. Neither bottom ash nor ponded ash needs to be at any moisture content to be used in flowable fill mixes because the amount of water in the mix can be adjusted in order to provide the desired flowability.

9.4.1.3 Assessment of Efficiency and Capacity of Immobilization

The efficiency (E) and capacity (P) of different additives for immobilization and field applications can be evaluated using the expressions

$$E(\%) = (Mo - Me)/Mo \times 100$$
$$P = [(Mo - Me)V]/m$$

E is efficiency of immobilization agent, P is the capacity of immobilization agent, Me is the equilibrium extractable concentration of single metal in the immobilized soil (mg/L), *Mo* is the initial extractable concentration of single metal in preimmobilized soil (mg/L), *V* is the volume of metal salt solution (mg/L), and *m* is the weight of immobilization agent (g). High values of *E* and *P* represent the perfect efficiency and capacity of an additive that can be used in field studies of metal immobilization (Wuana and Okieimen, 2011).

After screening out the best efficient additive, another experiment could be conducted to determine the best ratio (soil/additive) for the field-fixing treatment. After the fixing treatment of contaminated soils, a lot of methods including biological and physiochemical experiments could be used to assess the remediation efficiency. Environmental risk could also be estimated after confirming the immobilized efficiency and possible release.

9.4.2 Soil Washing

Soil washing is essentially a volume reduction/waste minimization treatment process. It is done on the excavated (physically removed) soil (ex situ) or on-site (in situ). Soil washing as discussed in this review refers to ex situ techniques that employ physical and/or chemical procedures to extract metal contaminants from soils. During soil washing, (i) those soil particles that host the majority of the contamination are separated from the bulk soil fractions such as by physical separation; (ii) contaminants are removed from the soil by aqueous chemicals and recovered from solution on a solid substrate, such as chemical extraction, or (iii) a combination of both. In all cases, the separated contaminants then go to hazardous waste landfill (or occasionally are further treated by chemical, thermal, or biological processes). By removing most of the contamination from the soil, the bulk fraction that remains can be (i) recycled on the site being remediated as relatively inert backfill, (ii) used on another site as fill, or (iii) disposed of relatively cheaply as nonhazardous material.

Ex situ soil washing is particularly frequently used in soil remediation because it (i) completely removes the contaminants and hence ensures the rapid cleanup of a contaminated site, (ii) meets specific criteria, (iii) reduces or eliminates long-term liability, (iv) may be the most cost-effective solution, and (v) may produce recyclable material or energy. The disadvantages include the fact that the contaminants are simply moved to a different place, where they must be monitored, the risk of spreading contaminated soil and dust particles during removal and transport of contaminated soil and the relatively high cost. Excavation can be the most expensive option when large amounts of soil must be removed or disposal as hazardous or toxic waste is required.

Acid and chelator soil washing are the two most prevalent removal methods. Soil washing currently involves soil flushing an in situ process in which the washing solution is forced through the in-place soil matrix, ex situ extraction of heavy metals from the soil slurry in reactors, and soil heap leaching. Another heavy metal removal technology is electroremediation, which mostly involves electrokinetic movement of charged particles suspended in the soil solution, initiated by an electric gradient. The metals can be removed by precipitation at the electrodes. Removal of most of the contaminants from the soil does not mean that the contaminant-depleted bulk is totally contaminant-free. Thus, for soil washing to be successful, the level of contamination in the treated bulk must be below a site-specific action limit (e.g., based on risk assessment). Cost-effectiveness with soil washing is achieved by offsetting processing costs against the ability to significantly reduce the amount of material requiring costly disposal at a hazardous waste landfill.

Typically, the cleaned fractions from the soil-washing process should be >70%–80% of the original mass of the soil, but where the contaminants have a very high associated disposal cost and/or where transport distances to the nearest hazardous waste landfill are substantial, a 50% reduction might still be cost effective. There is also a generally held opinion that soil washing based on physical separation processes is only cost effective for sandy and granular soils where the clay and silt content (particles less than 0.063 mm) is less than 30%–35% of the soil. Soil washing by chemical dissolution of the contaminants is not constrained by the proportion of clay as this fraction can also be leached by the chemical agent. However, clay-rich soils pose other problems such as difficulties with materials handling and solid-liquid separation. Full-scale soil-washing plants exist as fixed centralized treatment centers or as mobile/transportable units. With fixed centralized facilities, contaminated soil is brought to the plant, whereas with mobile/transportable facilities, the plant is transported to a contaminated site, and soil is processed on the site. Where mobile/transportable plant is used, the cost of mobilization and demobilization can be significant. However, where large volumes of soil are to be treated, this cost can be more than offset by reusing clean material on the site (therefore avoiding the cost of transport to an off-site centralized treatment facility and avoiding the cost of importing clean fill).

9.4.2.1 Principles of Soil Washing

Soil washing is a volume reduction/waste minimization treatment technology based on physical and/or chemical processes. With physical soil washing, differences between particle grain size, settling velocity, specific gravity, surface chemical behavior, and rarely magnetic properties are used to separate those particles that host the majority of the contamination from the bulk, which are contaminant-depleted. The equipment used is standard mineral processing equipment, which is more generally used in the mining industry. Mineral processing techniques as applied to soil remediation have been reviewed in literature.

With chemical soil washing, soil particles are cleaned by selectively transferring the contaminants on the soil into solution. Since heavy metals are sparingly soluble and occur predominantly in a sorbed state, washing the soils with water alone would be expected to remove too low several cations in the leachates; chemical agents have to be added to the washing water. This is achieved by mixing the soil with aqueous solutions of acids, alkalies, complexing agents, other solvents, and surfactants. The resulting cleaned particles are then separated from the resulting aqueous solution. This solution is then treated to remove the contaminants (e.g., by sorption on activated carbon or ion exchange).

The effectiveness of washing is closely related to the ability of the extracting solution to dissolve the metal contaminants in soils. However, the strong bonds between the soil and metals make the cleaning process difficult. Therefore, only extractants capable of dissolving large quantities of metals would be suitable for cleaning purposes. The realization that the goal of soil remediation is to remove the metal and preserve the natural soil properties limits the choice of extractants that can be used in the cleaning process.

9.4.2.2 Chemical Extractants for Soil Washing

Owing to the different nature of heavy metals, extracting solutions that can optimally remove them must be carefully sought during soil washing. Several classes of chemicals used for soil washing include surfactants, cosolvents, cyclodextrins, chelating agents, and organic acids. All of the soil-washing extractants have been developed on a case-by-case basis depending on the contaminant type at a particular site. A few studies have indicated that the solubilization/exchange/extraction of heavy metals by washing solutions differs considerably for different soil types. Strong acids attack and degrade the soil crystalline structure at extended contact times. For less damaging washes, organic acids and chelating agents are often suggested as alternatives to mineral acid use.

Natural, low-molecular-weight organic acids (LMWOAs) including oxalic, citric, formic, acetic, malic, succinic, malonic, maleic, lactic, aconitic, and fumaric acids are natural products of root exudates, microbial secretions, and plant and animal residue decomposition in soils. Thus, metal dissolution by organic acids is likely to be more representative of a mobile metal fraction that is available to biota. The chelating organic acids are able to dislodge the exchangeable, carbonate, and reducible fractions of heavy metals by washing procedures. Although many chelating compounds including citric acid (2-hydroxypropane-1,2,3-tricarboxylic acid)

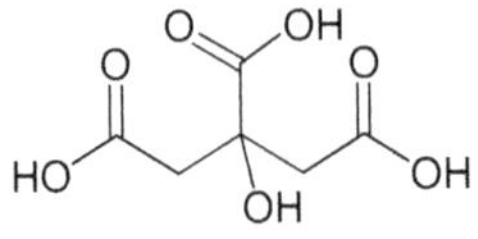

Citric acid

tartaric acid (2,3-dihydroxysuccinic acid)

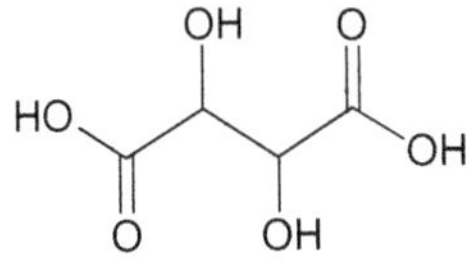

Tartaric acid

and EDTA (Wuana and Okieimen, 2011) for mobilizing heavy metals have been evaluated, there are still uncertainties as to the optimal choice for full-scale application.

The identification and quantification of coexisting solid metal species in the soil before and after treatment are essential to design and assess the efficiency of soil-washing technology. Furthermore, changes in nickel (Ni), copper (Cu), zinc (Zn), cadmium (Cd), and lead (Pb) speciation and uptake before and after washing with three chelating organic acids indicated that EDTA and citric acid appeared to offer greater potentials as chelating agents for remediating the permeable soil. Tartaric acid was, however, recommended in events of moderate contamination.

Soil washing using water may continue to be a process that is used to remove contaminates from soil. On the downside, soil washing does not destroy or immobilize the contaminants. Consequently, the resulting concentrated soil must be disposed of carefully. Furthermore, soil contaminated with both metals and organic compounds make formulating a single suitable washing solution difficult. In this case, sequential washing, using different wash formulations may be required. Some washing formulation may have a tendency to change the mobility of certain metals, and thus, fully characterizing and understanding the site is of utmost importance. In addition, the wash water requires treatment before it can be discharged, as it is usually not completely free of smaller particles or organic particles. Also, measuring contaminant concentration in air particulates during excavation and treatment is an important step to ensure that a wider population is not being unduly exposed.

9.4.3 Phytoremediation

Phytoremediation, also called green remediation, botanoremediation, agroremediation, or vegetative remediation, can be defined as an in situ remediation strategy that uses vegetation and associated microbiota, soil amendments, and agronomic techniques to remove, contain, or render environmental contaminants harmless. The idea of using metal-accumulating plants to remove heavy metals and other compounds was first introduced in 1983, but the concept has been implemented for the past 300 years on wastewater discharges. Plants may break down or degrade organic pollutants or remove and stabilize metal contaminants. The methods used to phytoremediate metal contaminants are slightly different from those used to remediate sites polluted with organic contaminants. As it is a relatively new technology, phytoremediation is still mostly in its testing stages and as such has not been used in many places as a full-scale application. However, it has been tested successfully in many places around the world for many different contaminants. Phytoremediation is energy-efficient and esthetically pleasing method of remediating sites with low-to-moderate levels of contamination, and it can be used in conjunction with other more traditional remedial methods as a finishing step to the remedial process.

The advantages of phytoremediation compared with classical remediation are that (i) it is more economically viable using the same tools and supplies as agriculture; (ii) it is less disruptive to the environment and does not involve waiting for new plant communities to recolonize the site; (iii) disposal sites are not needed; (iv) it is more likely to be accepted by the public as it is more esthetically pleasing then traditional methods; (v) it avoids excavation and transport of polluted media, thus reducing the risk of spreading the contamination; and (vi) it has the potential to treat sites polluted with more than one type of pollutant. The disadvantages are as follows: (i) it is dependent on the growing conditions required by the plant (i.e., climate, geology, altitude, and temperature); (ii) large-scale operations require access to agricultural equipment and knowledge; (iii) success is dependent on the tolerance of the plant to the pollutant; (iv) contaminants collected in senescing tissues may be released back into the environment in autumn; (v) contaminants may be collected in woody tissues used as fuel; (vi) time taken to remediate sites far exceeds that of other technologies; and (vii) contaminant solubility may be increased leading to greater environmental damage and the possibility of leaching. Potentially useful phytoremediation technologies for remediation of heavy-metal-contaminated soils include phytoextraction (phytoaccumulation), phytostabilization, and phytofiltration.

9.4.3.1 Phytoextraction

Phytoextraction (phytoaccumulation) is the name given to the process where plant roots uptake metal contaminants from the soil and translocate them to their above soil tissues. A plant used for phytoremediation needs to be heavy-metal-tolerant, grows rapidly with a high biomass yield per hectare, has high metal-accumulating ability in the foliar parts, has a profuse root system, and has a high bioaccumulation factor. Phytoextraction is, no doubt, a publicly appealing (green) remediation technology. Two approaches have been proposed for phytoextraction of heavy metals, namely, continuous or natural phytoextraction and chemically enhanced phytoextraction.

Continuous phytoextraction is based on the use of natural hyperaccumulator plants with exceptional metal-accumulating capacity. Hyperaccumulators are species capable of accumulating metals at levels 100-fold greater than those typically measured in shoots of the common nonaccumulator plants. Once inside the plant, most metals are too insoluble to move freely in the vascular system so they usually form carbonate, sulfate, or phosphate precipitate immobilizing them in apoplastic (extracellular) and symplastic (intracellular) compartments. Hyperaccumulators have several beneficial characteristics but may tend to be slow growing and produce low biomass, and years or decades are needed to clean up contaminated sites. To overcome these shortfalls, chemically enhanced phytoextraction has been developed. The approach makes use of high biomass crops that are induced to take up large amounts of metals when their mobility in soil is enhanced by chemical treatment with chelating organic acids.

For more than 10 years, chelant-enhanced phytoextraction of metals from contaminated soils has received much attention as a cost-effective alternative to conventional techniques of enhanced soil remediation. When the chelating agent is applied to the soil, metal-chelant complexes are formed and taken up by the plant, mostly through a passive apoplastic pathway. Unless the metal ion is transported as a noncationic chelate, apoplastic transport is further limited by the high cation-exchange capacity of cell walls. Chelators have been isolated from plants that are strongly involved in the uptake of heavy metals and their detoxification. The chelating agent EDTA has become one of the most tested mobilizing amendments for less mobile/available metals such as lead. Chelating agents that have been isolated from plants are strongly involved in the uptake of heavy metals and their detoxification.

Depending on heavy metal concentration in the contaminated soil and the target values sought for in the remediated soil, phytoextraction may involve repeated cropping of the plant until the metal concentration drops to acceptable levels. The ability of the plant to account for the decrease in soil metal concentrations as a function of metal uptake and biomass production plays an important role in achieving regulatory acceptance. Theoretically, metal removal can be accounted for by determining metal concentration in the plant, multiplied by the reduction in soil metal concentrations. It should, however, be borne in mind that this approach may be challenged by several factors working together during field applications.

One of the key aspects of the acceptance of phytoextraction pertains to its performance, ultimate utilization of by-products, and its overall economic viability. Commercialization of phytoextraction has been challenged by the expectation that site remediation should be achieved in a time comparable with other cleanup technologies. Genetic engineering has a great role to play in supplementing the list of plants available for phytoremediation using engineering tools to insert into plants those genes that will enable the plant to metabolize a particular pollutant. A major goal of plant genetic engineering is to enhance the ability of plants to metabolize many of the compounds that are of environmental concern. Currently, some laboratories are using traditional breeding techniques, others are creating protoplast fusion hybrids, and still, others are looking at the direct insertion of novel genes to enhance the metabolic capabilities of plants. Overall, phytoextraction appears a very promising technology for the removal of metal pollutants from the environment and is at present approaching commercialization.

A serious challenge for the commercialization of phytoextraction has been the disposal of contaminated plant biomass especially in the case of repeated cropping where large tonnages of biomass may be produced. The biomass must be stored, disposed of, or utilized in an appropriate manner so as not to pose any environmental risk. The major constituents of biomass material are lignin, hemicellulose, cellulose, minerals, and ash. It possesses high moisture and volatile matter, low bulk density, and calorific value. Biomass is solar energy fixed in plants in form of carbon, hydrogen, and oxygen (oxygenated hydrocarbons) with a possible general chemical formula $CH_{1.44}O_{0.66}$. Controlled combustion and gasification of biomass can yield a mixture of producer gas (typically a mixture of carbon monoxide and nitrogen) and/or pyrolysis gas (pyrogas), which is any gas produced from a carbonaceous source by the application of high temperature (typically, >500°C, >930°F), which leads to the generation of thermal and electric energy (Speight, 2014b; Luque and Speight, 2015). Composting and compacting can be employed as volume reduction approaches to biomass reuse. Ashing of biomass can produce bio-ores especially after the phytomining of precious metals. Heavy metals such as cobalt, (Co), copper (Cu), iron (Fe), manganese (Mn), molybdenum (Mo), nickel (Ni), and zinc (Zn) are plant essential metals, and most plants have the ability to accumulate derivatives of these metals. The high concentrations of these metals in the harvested biomass can be *diluted* to accept a tolerable concentration by combining the biomass with clean biomass in formulations of fertilizer and fodder.

More generally, in terms of cite cleanup, *dilution* is the process in which a contaminant becomes less concentrated and reduces the risk because individual (floral and faunal) receptors are likely to be exposed to lower, less toxic concentrations of the hazard. In fact, the chronic effects of most dilute contaminants are extremely difficult to measure. By itself, however, dilution does not reduce the contaminant mass but does spread the area of potential exposure. This can be a distinct disadvantage since some contaminants are believed hazardous even at levels too dilute to be detected with standard field characterization techniques.

9.4.3.2 Phytostabilization

Phytostabilization, also referred to as in-place inactivation, is primarily concerned with the use of certain plants to immobilize soil sediment and sludge. Contaminant are absorbed and accumulated by roots, adsorbed onto the roots, or precipitated in the rhizosphere. This reduces or even prevents the mobility of the contaminants preventing migration into the groundwater or air and reduces the bioavailability of the contaminant, thus preventing spread through the food chain. Plants for use in phytostabilization should be able to (i) decrease the amount of water percolating through the soil matrix, which may result in the formation of a hazardous leachate; (ii) act as barrier to prevent direct contact with the contaminated soil; and (iii) prevent soil erosion and the distribution of the toxic metal to other areas. Phytostabilization can occur through the process of sorption, precipitation, complexation, or metal valence reduction. This technique is useful for the cleanup of lead (Pb), arsenic (As), cadmium (Cd), chromium (Cr), copper (Cu), and zinc (Zn). It can also be used to reestablish a plant community on sites that have been denuded due to the high levels of metal contamination. Once a community of tolerant species has been established, the potential for wind erosion (and thus spread of the pollutant) is reduced, and leaching of the soil contaminants is also reduced. Phytostabilization is advantageous because disposal of hazardous material/biomass is not required, and it is very effective when rapid immobilization is needed to preserve ground and surface waters.

9.4.3.3 Phytofiltration

Phytofiltration is the use of plant roots (rhizofiltration) or seedlings (blastofiltration), is similar in concept to phytoextraction, but is used to absorb or adsorb pollutants, mainly metals, from groundwater and aqueous waste streams rather than the remediation of polluted soils. Rhizosphere is the soil area immediately surrounding the plant root surface, typically up to a few millimeters from the root surface. The contaminants either are adsorbed onto the root surface or are absorbed by the plant roots. Plants used for rhizofiltration are not planted directly in situ but are acclimated to the pollutant first. Plants are hydroponically grown in clean water rather than soil, until a large root system has developed. Once a large root system is in place, the water supply is substituted for a polluted water supply to acclimatize the plant. After the plants become acclimatized, they are planted in the polluted area where the roots uptake the polluted water and the contaminants along with it. As the roots become saturated, they are harvested and disposed of safely. Repeated treatments of the site can reduce pollution to suitable levels as was exemplified in Chernobyl where sunflowers were grown in radioactively contaminated pools.

The range of application of inorganic compounds is extensive, and the fundamental inorganic chemicals (commodity inorganic chemicals) are a broad chemical category, and there are efforts being made to reduce the pollution before the pollutants enter the atmosphere.

9.5 POLLUTION PREVENTION

A major aspect of pollution prevention is to determine the inorganic chemicals that are toxic and/or dangerous to the flora and fauna, especially inorganic chemicals that can influence the flora and fauna of waterways. However, assessing the ability of inorganic chemicals to be toxic and dangerous to the flora and fauna is not always an easy task unless the properties of the inorganic chemicals are known and understood since different inorganic chemicals cause harm in different ways and to different organisms. In addition, the hazard posed by any specific inorganic chemical depends on the degree of toxicity of the chemical and the amount of the chemical enters the environment. Furthermore, some toxic inorganic chemicals do not break down easily in the environment, which allows these chemical to move up through the food chain. These persistent, bioaccumulative toxic inorganic chemicals can build up in the tissues of small organisms, which are then eaten by larger animals that are then, in turn, eaten by even larger animals, sometimes by humans.

To combat such effects, including bioaccumulation, knowledge of inorganic chemistry that leads to knowledge of the structure, properties, composition, reactions, and preparation of carbon-containing compounds becomes advantageous. This knowledge should not only include an understanding of the properties and behavior of inorganic chemicals (Chapters 4 and 7).

The inorganic chemical process industry plays an important role in the development of a country by providing a wide variety of products, which are being used in providing basic needs of rising demand. The inorganic chemical process industries use raw material derived from crude oil and natural gas, salt, oil and fats, biomass and energy from coal, natural gas, and a small percentage from renewable energy resources. Although initial manufacture of inorganic chemicals started with coal and alcohol from fermentation industry, due to availability of crude oil and natural gas, it dominated the scene, and now more than 90% of inorganic chemicals are produced from crude oil and natural gas routes. However, variable costs of crude oil and natural gas and continuous decrease in the reserves have spurred the chemical industry for alternative feedstock like coal, biomass, coalbed methane, shale gas, sand oil as an alternate source of fuel, and chemical feedstock.

Environmental awareness among the public and policymakers has been growing since the 1960s, when it became widely recognized that human activities were having harmful and large-scale effects on the environment (Chapters 1 and 8). Scientific and engineering research will continue to play an increasing role in both understanding the nature of chemical pollutants and the means by which the environment can be protected and cleaned from past spills and disposal of chemicals. Several themes for environmental protection are dominant and include (i) better assessment, management, and communication of risks; (ii) developing better scientific and technical understanding for cleanup; and (iii) setting appropriate goals and priorities for environmental cleanup initiatives. Further discussion of the potential of some of the cleanup methods is discussed below. A comprehensive discussion of the available technologies is beyond the scope of this book, but the discussion does touch upon technologies that could be the dominant technologies of the future.

In terms of preventing pollutants from entering the atmosphere, there has been development in the United States to develop clean coal technologies in which the pollutants entering the atmosphere from coal use are markedly reduced.

The term *clean coal technology* refers to a collection of technologies being developed to mitigate the environmental impact of energy generation from coal (Speight, 2013). When coal is used as a fuel source, the emissions of inorganic gases generated by the thermal decomposition of the coal include sulfur dioxide (SO_2), nitrogen oxides (NO_x), carbon dioxide (CO_2), mercury, and other chemical by-products that vary depending on the type of the coal being used. Thus, clean coal technologies are being developed to remove or reduce pollutant emissions to the atmosphere. Some of the techniques that would be used to accomplish this include (i) chemically washing minerals and impurities from the coal, (ii) gasification, (iii) improved technology for treating flue gases to remove pollutants to increasingly stringent levels and at higher efficiency, (iv) carbon capture and storage technologies to capture the carbon dioxide from the flue gas, and (v) dewatering lower rank coal (lignite, brown coal) to improve the calorific value and thus the efficiency of the conversion into electricity. Clean coal technology typically addresses atmospheric problems resulting from burning coal. Historically, the primary focus was on sulfur dioxide (SO_2) and on nitrogen oxides (NO_x), which are the most important gases in generation of acid rain, and particulate matter that causes visible air pollution. Because of the *clean coal technology program*, several different technological methods are available for the purpose of carbon capture as demanded by the clean coal concept: (i) precombustion capture, (ii) postcombustion capture, and (iii) oxyfuel combustion.

Precombustion capture involves gasification of a carbonaceous feedstock (such as coal) to form synthesis gas (Speight, 2014b; Luque and Speight, 2015)—a mixture of carbon monoxide and hydrogen—which may be sent to a *shift reactor* to produce a hydrogen and carbon-dioxide-rich gas mixture, from which the carbon dioxide can be efficiently captured and separated, transported, and ultimately sequestered (Chadeesingh, 2011). This technology is usually associated with the integrated gasification combined cycle (IGCC) process. *Postcombustion capture* refers to the capture of carbon dioxide from exhaust gases of combustion processes (Mokhatab et al., 2006; Speight, 2007, 2014a, 2017a; Kidnay et al., 2011; Bahadori, 2014). Oxyfuel combustion refers to a process in which a carbonaceous feedstock, such as coal, is burned in a mixture of recirculated flue gas and oxygen, rather than in air, which largely eliminates nitrogen from the flue gas enabling efficient and low-cost capture of the carbon dioxide.

Another option for environmental protection of the atmosphere is the various *gas processing* (also called *gas cleaning* or *gas refining*) that are used in plants that produce gases. Gas processing consists of separating all of the various hydrocarbons and fluids from the gas stream, (Fig. 9.1, using natural gas as the example) (Mokhatab et al., 2006; Speight, 2007, 2014a, 2017a; Kidnay et al., 2011).

The processes that have been developed to accomplish gas purification vary from a simple single-stage once-through washing-type operation to

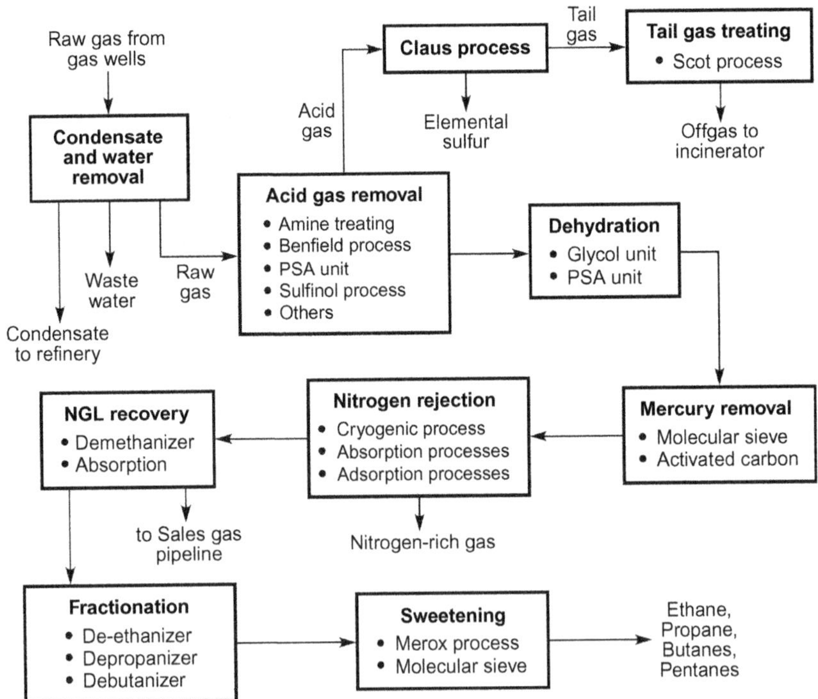

Fig. 9.1 Flow pattern for cleaning natural gas.

complex multistep recycling systems. In many cases, the process complexities arise because of the need for recovery of the materials used to remove the contaminants or even recovery of the contaminants in the original or altered form. In addition, the precise area of application of a given process is difficult to define, and several factors must be considered before process selection: (i) the types of contaminants in the gas stream, (ii) the concentrations of contaminants in the gas stream, (iii) the degree of contaminant removal desired, (iv) the selectivity of acid gas removal required, (v) the temperature of the gas to be processed, (vi) the pressure of the gas to be processed, (vii) the volume of the gas to be processed, (viii) the composition of the gas to be processed, (ix) the ratio of carbon dioxide to hydrogen sulfide ratio in the gas feedstock, and (x) the desirability of sulfur recovery due to process economics or environmental issues.

In terms of pollution of the land, *bioremediation* is generally considered to be an environmentally friendly technique of the future to restore soil and water to their respective original states by using indigenous microbes to break down and eliminate contaminants. Biological technologies are often used as a substitute to chemical or physical cleanup of chemical spills because biodegradation does not require as much equipment or labor as other methods; therefore, it is usually cheaper. It also allows cleanup workers to avoid contact with polluted soil and water.

These programs are helping companies to meet regulatory challenges by incorporating pollution control technologies into a portfolio of cost-effective regulatory compliance options for conventional and developmental coal-fired power plants.

REFERENCES

Bahadori, A., 2014. Natural Gas Processing: Technology and Engineering Design. Gulf Professional Publishing, Elsevier, Amsterdam.

Brar, S.K., Verma, M., Surampalli, R.Y., Misra, K., Tyagi, R.D., Meunier, N., Blais, J.F., 2006. Bioremediation of hazardous wastes: a review. Pract. Period. Hazard Toxic Radioact. Waste Manag. 10, 59–72.

Carson, R., 1962. Silent Spring. Houghton Mifflin Harcourt International, Geneva, IL.

Chadeesingh, R., 2011. The Fischer-Tropsch process. In: Speight, J.G. (Ed.), The Biofuels Handbook. The Royal Society of Chemistry, London, pp. 476–517 Part 3. (Chapter 5).

Habashi, F., 1994. Conversion reactions in inorganic chemistry. J. Chem. Educ. 71 (2), 130.

Kidnay, A.J., Parrish, W.R., McCartney, D.G., 2011. Fundamentals of Natural Gas Processing, second ed. CRC Press, Taylor & Francis Group, Boca Raton, FL.

Luque, R., Speight, J.G. (Eds.), 2015. Gasification for Synthetic Fuel Production: Fundamentals, Processes, and Applications. Woodhead Publishing, Elsevier, Cambridge.

Page, G.W., 1997. Contaminated Sites and Environmental Cleanup: International Approaches to Prevention, Remediation, and Reuse. Academic Press, New York, NY.

Singh, S., Kang, S.H., Mulchandani, A., Chen, W., 2008. Bioremediation: environmental clean-up through pathway engineering. Curr. Opin. Biotechnol. 19, 437–444.

Mokhatab, S., Poe, W.A., Speight, J.G., 2006. Handbook of Natural Gas Transmission and Processing. Elsevier, Amsterdam.

Speight, J.G., Arjoon, K.K., 2012. Bioremediation of Petroleum and Petroleum Products. Scrivener Publishing, Beverly, MA.

Speight, J.G., 2007. A Basic Handbook. GPC Books, Houston, TX.

Speight, J.G., 2013. The Chemistry and Technology of Coal, third ed. CRC Press, Taylor & Francis Group, Boca Raton, FL.

Speight, J.G., 2014a. The Chemistry and Technology of Petroleum, fifth ed. CRC Press, Taylor & Francis Group, Boca Raton, FL.

Speight, J.G., 2014b. Gasification of Unconventional Feedstocks. Gulf Professional Publishing, Elsevier, Oxford.

Speight, J.G., 2017a. Handbook of Petroleum Refining. CRC Press, Taylor & Francis Group, Boca Raton, FL.

Speight, J.G., 2017b. Environmental Organic Chemistry for Engineers. Butterworth-Heinemann, Elsevier, Cambridge, MA.

Tedder, D.W., Pohland, F.G., 2000. Emerging Technologies in Hazardous Waste Management. Kluwer Academic/Plenum Press, New York, NY.

Testa, S.M., Winegardner, D.L., 2000. Restoration of Contaminated Aquifers: Petroleum Hydrocarbons and Organic Compounds, second ed. Lewis Publishers, CRC Press, Taylor & Francis Group, Boca Raton, FL.

Wuana, R., Okieimen, F.E., 2011. Heavy metals in contaminated soils: a review of sources, chemistry, risks and best available strategies for remediation. ISRN Ecol. 2011, Article ID 402647, http://dx.doi.org/10.5402/2011/402647. https://www.hindawi.com/journals/isrn/2011/402647/.

APPENDIX

Table A1 The Chemical Elements Listed in Alphabetical Order

Element	Atomic Number	Symbol
Actinium	89	Ac
Aluminum	13	Al
Americium	95	Am
Antimony	51	Sb
Argon	18	Ar
Arsenic	33	As
Astatine	85	At
Barium	56	Ba
Berkelium	97	Bk
Beryllium	4	Be
Bismuth	83	Bi
Bohrium	107	Bh
Boron	5	B
Bromine	35	Br
Cadmium	48	Cd
Cesium	55	Cs
Calcium	20	Ca
Californium	98	Cf
Carbon	6	C
Cerium	58	Ce
Chlorine	17	Cl
Chromium	24	Cr
Cobalt	27	Co
Copernicium	112	Cn
Copper	29	Cu
Curium	96	Cm
Darmstadtium	110	Ds
Dubnium	105	Db
Dysprosium	66	Dy
Einsteinium	99	Es
Erbium	68	Er
Europium	63	Eu
Fermium	100	Fm
Flerovium	114	Fl
Fluorine	9	F
Francium	87	Fr
Gadolinium	64	Gd

Continued

Table A1 The Chemical Elements Listed in Alphabetical Order—cont'd

Element	Atomic Number	Symbol
Gallium	31	Ga
Germanium	32	Ge
Gold	79	Au
Hafnium	72	Hf
Hassium	108	Hs
Helium	2	He
Holmium	67	Ho
Hydrogen	1	H
Indium	49	In
Iodine	53	I
Iridium	77	Ir
Iron	26	Fe
Krypton	36	Kr
Lanthanum	57	La
Lawrencium	103	Lr
Lead	82	Pb
Lithium	3	Li
Livermorium	116	Lv
Lutetium	71	Lu
Magnesium	12	Mg
Manganese	25	Mn
Meitnerium	109	Mt
Mendelevium	101	Md
Mercury	80	Hg
Molybdenum	42	Mo
Neodymium	60	Nd
Neon	10	Ne
Neptunium	93	Np
Nickel	28	Ni
Niobium	41	Nb
Nitrogen	7	N
Nobelium	102	No
Osmium	76	Os
Oxygen	8	O
Palladium	46	Pd
Phosphorus	15	P
Platinum	78	Pt
Plutonium	94	Pu
Polonium	84	Po
Potassium	19	K
Praseodymium	59	Pr
Promethium	61	Pm

Table A1 The Chemical Elements Listed in Alphabetical Order—cont'd

Element	Atomic Number	Symbol
Protactinium	91	Pa
Radium	88	Ra
Radon	86	Rn
Rhenium	75	Re
Rhodium	45	Rh
Roentgenium	111	Rg
Rubidium	37	Rb
Ruthenium	44	Ru
Rutherfordium	104	Rf
Samarium	62	Sm
Scandium	21	Sc
Seaborgium	106	Sg
Selenium	34	Se
Silicon	14	Si
Silver	47	Ag
Sodium	11	Na
Strontium	38	Sr
Sulfur	16	S
Tantalum	73	Ta
Technetium	43	Tc
Tellurium	52	Te
Terbium	65	Tb
Thallium	81	Tl
Thorium	90	Th
Thulium	69	Tm
Tin	50	Sn
Titanium	22	Ti
Tungsten	74	W
Ununoctium	118	Uuo
Ununpentium	115	Uup
Ununseptium	117	Uus
Ununtrium	113	Uut
Uranium	92	U
Vanadium	23	V
Xenon	54	Xe
Ytterbium	70	Yb
Yttrium	39	Y
Zinc	30	Zn
Zirconium	40	Zr

Table A2 The Chemical Elements Listed in Order of the Atomic Number

Atomic Number	Symbol	Name
1	H	Hydrogen
2	He	Helium
3	Li	Lithium
4	Be	Beryllium
5	B	Boron
6	C	Carbon
7	N	Nitrogen
8	O	Oxygen
9	F	Fluorine
10	Ne	Neon
11	Na	Sodium
12	Mg	Magnesium
13	Al	Aluminum
14	Si	Silicon
15	P	Phosphorus
16	S	Sulfur
17	Cl	Chlorine
18	Ar	Argon
19	K	Potassium
20	Ca	Calcium
21	Sc	Scandium
22	Ti	Titanium
23	V	Vanadium
24	Cr	Chromium
25	Mn	Manganese
26	Fe	Iron
27	Co	Cobalt
28	Ni	Nickel
29	Cu	Copper
30	Zn	Zinc
31	Ga	Gallium
32	Ge	Germanium
33	As	Arsenic
34	Se	Selenium
35	Br	Bromine
36	Kr	Krypton
37	Rb	Rubidium
38	Sr	Strontium
39	Y	Yttrium
40	Zr	Zirconium
41	Nb	Niobium
42	Mo	Molybdenum

Table A2 The Chemical Elements Listed in Order of the Atomic Number—cont'd

Atomic Number	Symbol	Name
43	Tc	Technetium
44	Ru	Ruthenium
45	Rh	Rhodium
46	Pd	Palladium
47	Ag	Silver
48	Cd	Cadmium
49	In	Indium
50	Sn	Tin
51	Sb	Antimony
52	Te	Tellurium
53	I	Iodine
54	Xe	Xenon
55	Cs	Cesium
56	Ba	Barium
57	La	Lanthanum
58	Ce	Cerium
59	Pr	Praseodymium
60	Nd	Neodymium
61	Pm	Promethium
62	Sm	Samarium
63	Eu	Europium
64	Gd	Gadolinium
65	Tb	Terbium
66	Dy	Dysprosium
67	Ho	Holmium
68	Er	Erbium
69	Tm	Thulium
70	Yb	Ytterbium
71	Lu	Lutetium
72	Hf	Hafnium
73	Ta	Tantalum
74	W	Tungsten
75	Re	Rhenium
76	Os	Osmium
77	Ir	Iridium
78	Pt	Platinum
79	Au	Gold
80	Hg	Mercury
81	Tl	Thallium
82	Pb	Lead
83	Bi	Bismuth
84	Po	Polonium

Continued

Table A2 The Chemical Elements Listed in Order of the Atomic Number—cont'd

Atomic Number	Symbol	Name
85	At	Astatine
86	Rn	Radon
87	Fr	Francium
88	Ra	Radium
89	Ac	Actinium
90	Th	Thorium
91	Pa	Protactinium
92	U	Uranium
93	Np	Neptunium
94	Pu	Plutonium
95	Am	Americium
96	Cm	Curium
97	Bk	Berkelium
98	Cf	Californium
99	Es	Einsteinium
100	Fm	Fermium
101	Md	Mendelevium
102	No	Nobelium
103	Lr	Lawrencium
104	Rf	Rutherfordium
105	Db	Dubnium
106	Sg	Seaborgium
107	Bh	Bohrium
108	Hs	Hassium
109	Mt	Meitnerium
110	Ds	Darmstadtium
111	Rg	Roentgenium
112	Cn	Copernicium
113	Nh	Nihonium
114	Fl	Flerovium
115	Mc	Moscovium
116	Lv	Livermorium
117	Ts	Tennessine
118	Og	Oganesson

Table A3 The Chemical Elements Listed Alphabetically by Symbol

Symbol	Element	Atomic Number
Ac	Actinium	89
Ag	Silver	47
Al	Aluminum	13
Am	Americium	95
Ar	Argon	18
As	Arsenic	33
At	Astatine	85
Au	Gold	79
B	Boron	5
Ba	Barium	56
Be	Beryllium	4
Bh	Bohrium	107
Bi	Bismuth	83
Bk	Berkelium	97
Br	Bromine	35
C	Carbon	6
Ca	Calcium	20
Cd	Cadmium	48
Ce	Cerium	58
Cf	Californium	98
Cl	Chlorine	17
Cm	Curium	96
Cn	Copernicium	112
Co	Cobalt	27
Cr	Chromium	24
Cs	Cesium	55
Cu	Copper	29
Db	Dubnium	105
Ds	Darmstadtium	110
Dy	Dysprosium	66
Er	Erbium	68
Es	Einsteinium	99
Eu	Europium	63
F	Fluorine	9
Fe	Iron	26
Fl	Flerovium	114
Fm	Fermium	100
Fr	Francium	87
Ga	Gallium	31
Gd	Gadolinium	64
Ge	Germanium	32
H	Hydrogen	1

Continued

Table A3 The Chemical Elements Listed Alphabetically by Symbol—cont'd

Symbol	Element	Atomic Number
He	Helium	2
Hf	Hafnium	72
Hg	Mercury	80
Ho	Holmium	67
Hs	Hassium	108
I	Iodine	53
In	Indium	49
Ir	Iridium	77
K	Potassium	19
Kr	Krypton	36
La	Lanthanum	57
Li	Lithium	3
Lr	Lawrencium	103
Lu	Lutetium	71
Lv	Livermorium	116
Mc	Moscovium	115
Md	Mendelevium	101
Mg	Magnesium	12
Mn	Manganese	25
Mo	Molybdenum	42
Mt	Meitnerium	109
N	Nitrogen	7
Na	Sodium	11
Nb	Niobium	41
Nd	Neodymium	60
Ne	Neon	10
Nh	Nihonium	113
Ni	Nickel	28
No	Nobelium	102
Np	Neptunium	93
O	Oxygen	8
Og	Oganesson	118
Os	Osmium	76
P	Phosphorus	15
Pa	Protactinium	91
Pb	Lead	82
Pd	Palladium	46
Pm	Promethium	61
Po	Polonium	84
Pr	Praseodymium	59
Pt	Platinum	78
Pu	Plutonium	94

Table A3 The Chemical Elements Listed Alphabetically by Symbol—cont'd

Symbol	Element	Atomic Number
Ra	Radium	88
Rb	Rubidium	37
Re	Rhenium	75
Rf	Rutherfordium	104
Rg	Roentgenium	111
Rh	Rhodium	45
Rn	Radon	86
Ru	Ruthenium	44
S	Sulfur	16
Sb	Antimony	51
Sc	Scandium	21
Se	Selenium	34
Sg	Seaborgium	106
Si	Silicon	14
Sm	Samarium	62
Sn	Tin	50
Sr	Strontium	38
Ta	Tantalum	73
Tb	Terbium	65
Tc	Technetium	43
Te	Tellurium	52
Th	Thorium	90
Ti	Titanium	22
Tl	Thallium	81
Tm	Thulium	69
Ts	Tennessine	117
U	Uranium	92
V	Vanadium	23
W	Tungsten	74
Xe	Xenon	54
Yb	Ytterbium	70
Y	Yttrium	39
Zn	Zinc	30
Zr	Zirconium	40

Table A4 List and Formulas of Selected Inorganic Compounds

A

Aluminum bromide—$AlBr_3$
Aluminum iodide—AlI_3
Aluminum oxide—Al_2O_3
Aluminum chloride—$AlCl_3$
Aluminum fluoride—AlF_3
Aluminum hydroxide—$Al(OH)_3$
Aluminum nitrate—$Al(NO_3)_3$
Aluminum sulfide—Al_2S_3
Aluminum sulfate—$Al_2(SO_4)_3$
Ammonia—NH_3
Ammonium bicarbonate—NH_4HCO_3
Ammonium chromate—$(NH_4)_2CrO_4$
Ammonium chloride—NH_4Cl
Ammonium chlorate—NH_4ClO_3
Ammonium cyanide—NH_4CN
Ammonium dichromate—$(NH_4)_2Cr_2O_7$
Ammonium hydroxide—NH_4OH
Ammonium nitrate—NH_4NO_3
Ammonium sulfide—$(NH_4)_2S$
Ammonium sulfite—$(NH_4)_2SO_3$
Ammonium sulfate—$(NH_4)_2SO_4$
Ammonium persulfate—$(NH_4)_2S_2O_8$
Ammonium perchlorate—NH_4ClO_4
Ammonium thiocyanate—NH_4SCN
Antimony pentachloride—$SbCl_5$
Antimony pentafluoride—SbF_5
Antimony trioxide—Sb_2O_3
Arsine—AsH_3
Arsenic trioxide (arsenic(III) oxide)—As_2O_3
Arsenous acid—$As(OH)_3$

B

Barium carbonate—$BaCO_3$
Barium chlorate—$Ba(ClO_3)_2$
Barium chloride—$BaCl_2$
Barium chromate—$BaCrO_4$
Barium ferrate—$BaFeO_4$
Barium ferrite—$BaFe_2O_4$
Barium fluoride—BaF_2
Barium hydroxide—$Ba(OH)_2$
Barium iodide—BaI_2
Barium nitrate—$Ba(NO_3)_2$
Barium oxalate—$Ba(C_2O_4)$
Barium oxide—BaO

Table A4 List and Formulas of Selected Inorganic Compounds—cont'd

Barium peroxide—BaO_2
Barium sulfate—$BaSO_4$
Barium sulfide—BaS
Beryllium bromide—$BeBr_2$
Beryllium carbonate—$BeCO_3$
Beryllium chloride—$BeCl_2$
Beryllium fluoride—BeF_2
Beryllium hydride—BeH_2
Beryllium hydroxide—$Be(OH)_2$
Beryllium iodide—BeI_2
Beryllium nitrate—$Be(NO_3)_2$
Beryllium nitride—Be_3N_2
Beryllium oxide—BeO
Beryllium sulfate— $BeSO_4$
Beryllium sulfite—$BeSO_3$
Bismuth(III) oxide—Bi_2O_3
Bismuth oxychloride—BiOCl
Borane—diborane, B_2H_6; pentaborane, B_5H_9; decaborane, $B_{10}H_{14}$
Borax—$Na_2B_4O_7 \cdot 10H_2O$
Boric acid—H_3BO_3
Boron trichloride—BCl_3
Boron trifluoride—BF_3
Boron oxide—B_2O_3
Bromine monochloride—BrCl
Bromine pentafluoride—BrF_5
Bromine trifluoride—BrF_3

C

Cadmium bromide—$CdBr_2$
Cadmium chloride—$CdCl_2$
Cadmium fluoride—CdF_2
Cadmium iodide—CdI_2
Cadmium nitrate—$Cd(NO_3)_2$
Cadmium oxide—CdO
Cadmium sulfate—$CdSO_4$
Cadmium sulfide—CdS
Cesium bicarbonate—$CsHCO_3$
Cesium carbonate—Cs_2CO_3
Cesium chloride—CsCl
Cesium chromate—Cs_2CrO_4
Cesium fluoride—CsF
Calcium chlorate—$Ca(ClO_3)_2$
Calcium chloride—$CaCl_2$
Calcium chromate—$CaCrO_4$

Continued

Table A4 List and Formulas of Selected Inorganic Compounds—cont'd

Calcium fluoride—CaF_2
Calcium hydroxide—$Ca(OH)_2$
Calcium oxychloride—$CaOCl_2$
Calcium perchlorate—$Ca(ClO_4)_2$
Calcium sulfate (gypsum)—$CaSO_4$
Carbon dioxide—CO_2
Carbon disulfide—CS_2
Carbon monoxide—CO
Carbonic acid—H_2CO_3
Carbonyl chloride—$COCl_2$
Carbonyl fluoride—COF_2
Carbonyl sulfide—COS
Cerium(III) bromide—$CeBr_3$
Cerium(III) chloride—$CeCl_3$
Cerium(IV) sulfate—$Ce(SO_4)_2$
Chloric acid—$HClO_3$
Chlorine dioxide—ClO_2
Chlorine monoxide—ClO
Chlorine tetroxide (the peroxide)—$O_3ClOOClO_3$
Chlorine trifluoride—ClF_3
Chlorine trioxide—ClO_3
Chlorine—Cl_2
Chromic acid—CrO_3
Chromium trioxide (Chromic acid)—CrO_3
Chromium(II) chloride—$CrCl_2$ (also chromous chloride)
Chromium(II) sulfate—$CrSO_4$
Chromium(III) chloride—$CrCl_3$
Chromium(III) oxide—Cr_2O_3
Chromium(III) sulfate—$Cr_2(SO_4)_3$
Chromium(IV) oxide—CrO_2
Chromyl chloride—CrO_2Cl_2
Chromyl fluoride—CrO_2F_2
Cobalt(II) bromide—$CoBr_2$
Cobalt(II) carbonate—$CoCO_3$
Cobalt(II) chloride—$CoCl_2$
Cobalt(III) fluoride—CoF_3
Cobalt(II) nitrate—$Co(NO_3)_2$
Cobalt(II) sulfate—$CoSO_4$
Copper(I) chloride—$CuCl$
Copper(I) oxide—Cu_2O
Copper oxychloride—$H_3ClCu_2O_3$[1]
Copper(I) sulfide—Cu_2S
Copper(II) carbonate—$CuCO_3$
Copper(II) chloride—$CuCl_2$
Copper(I) fluoride— CuF
Copper(II) hydroxide—$Cu(OH)_2$

Table A4 List and Formulas of Selected Inorganic Compounds—cont'd

Copper(II) nitrate—$Cu(NO_3)_2$ Copper(II) oxide—CuO Copper(II) sulfate—$CuSO_4$ and the pentahydrate $CuSO_4 \cdot 5H_2O$ Copper(II) sulfide—CuS
D
Dichlorine monoxide—Cl_2O Dichlorine dioxide—Cl_2O_2 Dichlorine trioxide—Cl_2O_3 Dichlorine tetroxide (chlorine perchlorate)—$ClOClO_3$ Dichlorine hexoxide—Cl_2O_6 Dichlorine heptoxide—Cl_2O_7 Dichlorosilane—SiH_2Cl Dinitrogen pentoxide (nitronium nitrate)—N_2O_5 Dinitrogen tetroxide—N_2O_4 Disilane—Si_2H_6 Disulfur dichloride—S_2Cl_2
E
Erbium(III) chloride—$ErCl_3$ Europium(III) chloride—$EuCl_3$
G
Gold(I) chloride—$AuCl$ Gold(III) chloride—$(AuCl_3)_2$ Gold(I,III) chloride—Au_4Cl_8 Gold(III) fluoride—AuF_3 Gold(V) fluoride—AuF_5 Gold(I) bromide—$AuBr$ Gold(III) bromide—$(AuBr_3)_2$ Gold(I) iodide—AuI Gold(III) iodide—AuI_3 Gold(I) hydride—AuH Gold(III) oxide—Au_2O_3 Gold(I) sulfide—Au_2S Gold(III) sulfide—Au_2S_3
H
Hexafluorosilicic acid—H_2F_6Si Hydrazine—N_2H_4 Hydrazoic acid—HN_3 Hydroiodic acid—HI Hydrogen bromide—HBr Hydrogen chloride—HCl Hydrogen cyanide—HCN

Continued

Table A4 List and Formulas of Selected Inorganic Compounds—cont'd

Hydrogen fluoride—HF
Hydrogen peroxide—H_2O_2
Hydrogen selenide—H_2Se
Hydrogen sulfide—H_2S
Hydroxylamine—NH_2OH
Hypobromous acid—HOBr
Hypochlorous acid—HOCl
Hypophosphorous acid—H_3PO_2

I

Iodic acid—HIO_3
Iodine heptafluoride—IF_7
Iodine pentafluoride—IF_5
Iodine monochloride—ICl
Iodine trichloride—ICl_3
Iridium(IV) chloride—$IrCl_4$
Iron(II) chloride—$FeCl_2$ including hydrate
Iron(III) chloride—$FeCl_3$
Iron ferrocyanide—$Fe_7(CN)_{18}$
Iron(II) oxalate—FeC_2O_4
Iron(III) oxalate—$C_6Fe_2O_{12}$
Iron(II) oxide—FeO
Iron(III) nitrate—$Fe(NO_3)_3(H_2O)_9$
Iron(II,III) oxide—Fe_3O_4
Iron(III) oxide—Fe_2O_3
Iron(III) thiocyanate—$Fe(SCN)_3$
Iron(III) fluoride—FeF_3

L

Lead(II) carbonate—$Pb(CO_3)$
Lead(II) chloride—$PbCl_2$
Lead(II) iodide—PbI_2
Lead(II) nitrate—$Pb(NO_3)_2$
Lead hydrogen arsenate—$PbHAsO_4$
Lead(II) oxide—PbO
Lead(IV) oxide—PbO_2
Lead(II) phosphate—$Pb_3(PO_4)_2$
Lead(II) sulfate—$Pb(SO_4)$
Lead(II) sulfide—PbS
Lead tetroxide—Pb_3O_4[4]
Lithium aluminum hydride—$LiAlH_4$
Lithium bromide—LiBr
Lithium borohydride – $LiBH_4$
Lithium carbonate (lithium salt)—Li_2CO_3
Lithium chloride—LiCl

Table A4 List and Formulas of Selected Inorganic Compounds—cont'd

Lithium hypochlorite—$LiOCl$ Lithium chlorate—$LiClO_3$ Lithium perchlorate—$LiClO_4$ Lithium cobalt oxide—$LiCoO_2$ Lithium oxide—Li_2O Lithium peroxide—Li_2O_2 Lithium hydride—LiH Lithium hydroxide—$LiOH$ Lithium iodide—LiI Lithium nitrate—$LiNO_3$ Lithium sulfide—Li_2S Lithium sulfite—$HLiO_3S$ Lithium sulfate—Li_2SO_4 Lithium superoxide—LiO_2
M
Magnesium carbonate—$MgCO_3$ Magnesium chloride—$MgCl_2$ Magnesium oxide—MgO Magnesium perchlorate—$Mg(ClO_4)_2$ Magnesium phosphate—$Mg_3(PO_4)_2$ Magnesium sulfate—$MgSO_4$ Manganese(IV) oxide (manganese dioxide)—MnO_2 Manganese(II) sulfate monohydrate—$MnSO_4{\cdot}H_2O$ Manganese(II) chloride—$MnCl_2$ Manganese(III) chloride—$MnCl_3$ Manganese(IV) fluoride—MnF_4 Manganese(II) phosphate—$Mn_3(PO_4)_2$ Mercury(I) chloride—Hg_2Cl_2 Mercury(II) chloride—$HgCl_2$ Mercury fulminate—$Hg(ONC)_2$ Mercury(I) sulfate—Hg_2SO_4 Mercury(II) sulfate—$HgSO_4$ Mercury(II) sulfide—HgS Mercury(II) thiocyanate—$Hg(SCN)_2$ Metaphosphoric acid—HPO_3 Molybdenum(II) bromide—$MoBr_2$ Molybdenum(III) bromide—$MoBr_3$ Molybdenum(IV) carbide—MoC Molybdenum(II) chloride— Mo_6Cl_{12} Molybdenum(III) chloride—$MoCl_3$ Molybdenum(IV) chloride—$MoCl_4$ Molybdenum(V) chloride—Mo_2Cl_{10} Molybdenum trioxide—MoO_3

Continued

Table A4 List and Formulas of Selected Inorganic Compounds—cont'd

Molybdenum disulfide—MoS_2
Molybdenum hexacarbonyl—$Mo(CO)_6$
Molybdic acid—H_2MoO_4

N

Nickel(II) carbonate—$NiCO_3$
Nickel(II) chloride—$NiCl_2$ and hexahydrate
Nickel(II) fluoride—NiF_2
Nickel(II) hydroxide—$Ni(OH)_2$
Nickel(II) nitrate—$Ni(NO_3)_2$
Nickel(II) oxide—NiO
Nickel(II) sulfamate—$Ni(SO_3NH_2)_2$
Nickel(II) sulfide—NiS
Nitric acid—HNO_3
Nitrous acid—HNO_2
Nitrogen monoxide—NO
Nitrogen dioxide—NO_2
Nitrous oxide—N_2O (dinitrogen monoxide, laughing gas, NOS)
Nitrosylsulfuric acid—$NOHSO_4$

O

Oxygen difluoride—OF_2
Ozone—O_3

P

Perbromic acid—$HBrO_4$
Perchloric acid—$HClO_4$
Periodic acid—HIO_4
Perchloryl fluoride—$ClFO_3$
Persulfuric acid (Caro's acid)—H_2SO_5
Phosgene—$COCl_2$
Phosphine—PH_3
Phosphoric acid—H_3PO_4
Phosphorous acid (phosphoric(III) acid)—H_3PO_3
Phosphorus pentabromide—PBr_5
Phosphorus pentafluoride—PF_5
Phosphorus pentasulfide—P_4S_{10}
Phosphorus pentoxide—P_2O_5
Phosphorus sesquisulfide—P_4S_3
Phosphorus tribromide—PBr_3
Phosphorus trichloride—PCl_3
Phosphorus trifluoride—PF_3
Phosphorus triiodide—PI_3
Potash Alum—$K_2SO_4 \cdot Al_2(SO_4)_3 \cdot 24H_2O$
Potassium aluminum fluoride—$KAlF_4$
Potassium borate—$K_2B_4O_7 \cdot 4H_2O$

Table A4 List and Formulas of Selected Inorganic Compounds—cont'd

Potassium bromide—KBr
Potassium calcium chloride—$KCaCl_3$
Potassium bicarbonate—$KHCO_3$
Potassium bisulfite—$KHSO_3$
Potassium carbonate—K_2CO_3
Potassium chlorate—$KClO_3$
Potassium chlorite—$KClO_2$
Potassium chloride—KCl
Potassium cyanide—KCN
Potassium dichromate—$K_2Cr_2O_7$
Potassium dithionite—$K_2S_2O_4$
Potassium ferrioxalate—$K_3[Fe(C_2O_4)_3]$
Potassium ferricyanide—$K_3[Fe(CN)]_6$
Potassium ferrocyanide—$K_4[Fe(CN)]_6$
Potassium hydrogencarbonate—$KHCO_3$ (potassium bicarbonate)
Potassium hydrogen fluoride—HF_2K
Potassium hydroxide—KOH
Potassium iodide—KI
Potassium iodate—KIO_3
Potassium monopersulfate—$K_2SO_4{\cdot}KHSO_4{\cdot}2KHSO_5$
Potassium nitrate—KNO_3
Potassium perbromate—$KBrO_4$
Potassium perchlorate—$KClO_4$
Potassium periodate—KIO_4
Potassium permanganate—$KMnO_4$
Potassium sulfate—K_2SO_4
Potassium sulfite—K_2SO_3
Potassium sulfide—K_2S
Potassium thiocyanate—KSCN
Potassium titanyl phosphate—$KTiOPO_4$
Potassium vanadate—KVO_3
Praseodymium(III) chloride—$PrCl_3$
Protonated molecular hydrogen—$H_3{}^+$
Prussian blue (iron(III) hexacyanoferrate(II))—$Fe_4[Fe(CN)_6]_3$
Pyrosulfuric acid—$H_2S_2O_7$

S

Silane—SiH_4
Silica gel—$SiO_2{\cdot}nH_2O$
Silicic acid—$[SiO_x(OH)_{4-2x}]_n$
Silicofluoric acid—H_2SiF_6
Silicon boride—SiB_3
Silicon carbide—SiC
Silicon dioxide—SiO_2
Silicon monoxide—SiO

Continued

Table A4 List and Formulas of Selected Inorganic Compounds—cont'd

Silicon nitride—Si_3N_4
Silicon tetrabromide—$SiBr_4$
Silicon tetrachloride—$SiCl_4$
Silver bromate—$AgBrO_3$
Silver bromide—AgBr
Silver chlorate—$AgClO_3$
Silver chloride—AgCl
Silver chromate—Ag_2CrO_4
Silver fluoroborate—$AgBF_4$
Silver fulminate—AgCNO (silver cyanate)
Silver hydroxide—AgOH
Silver iodide—AgI
Silver nitrate—$AgNO_3$
Silver nitride—Ag_3N
Silver oxide—Ag_2O
Silver perchlorate—$AgClO_4$
Silver phosphate (silver orthophosphate)—Ag_3PO_4
Silver sulfate—Ag_2SO_4
Silver sulfide—Ag_2S
Silver(I) fluoride—AgF
Silver(II) fluoride—AgF_2
Sodium aluminate—$NaAlO_2$
Sodium arsenate—$H_{24}Na_3AsO_{16}$
Sodium azide—NaN_3
Sodium bicarbonate—$NaHCO_3$
Sodium bisulfite—$NaHSO_3$
Sodium borate—$Na_2B_4O_7$
Sodium borohydride—$NaBH_4$
Sodium bromate—$NaBrO_3$
Sodium bromide—NaBr
Sodium bromite—$NaBrO_2$
Sodium carbonate—Na_2CO_3
Sodium chlorate—$NaClO_3$
Sodium chloride—NaCl
Sodium chlorite—$NaClO_2$
Sodium cyanate—NaCNO
Sodium cyanide—NaCN
Sodium dichromate—$Na_2Cr_2O_7{\cdot}2H_2O$
Sodium dioxide—NaO_2
Sodium dithionite—$Na_2S_2O_4$
Sodium ferrocyanide—$Na_4Fe(CN)_6$
Sodium hydride—NaH
Sodium hydrogen carbonate (sodium bicarbonate)—$NaHCO_3$
Sodium hydrosulfide—NaSH
Sodium hydroxide—NaOH
Sodium hypobromite—NaOBr

Table A4 List and Formulas of Selected Inorganic Compounds—cont'd

Sodium hypochlorite—$NaOCl$
Sodium hypoiodite—$NaOI$
Sodium hypophosphite—$NaPO_2H_2$
Sodium iodate—$NaIO_3$
Sodium iodide—NaI
Sodium molybdate—Na_2MoO_4
Sodium monofluorophosphate (MFP)—Na_2PFO_3
Sodium nitrate—$NaNO_3$
Sodium nitrite—$NaNO_2$
Sodium oxide—Na_2O
Sodium perborate—$NaBO_3{\cdot}H_2O$
Sodium perbromate—$NaBrO_4$
Sodium percarbonate—$2Na_2CO_3{\cdot}3H_2O_2$
Sodium perchlorate—$NaClO_4$
Sodium periodate—$NaIO_4$
Sodium permanganate—$NaMnO_4$
Sodium peroxide—Na_2O_2
Sodium perrhenate—$NaReO_4$
Sodium persulfate—$Na_2S_2O_8$
Sodium phosphate, see trisodium phosphate—Na_3PO_4
Sodium selenate—Na_2O_4Se
Sodium selenide—Na_2Se
Sodium selenite—Na_2SeO_3
Sodium silicate—Na_2SiO_3
Sodium sulfate—Na_2SO_4
Sodium sulfide—Na_2S
Sodium sulfite—Na_2SO_3
Sodium tellurite—Na_2TeO_3
Sodium thiocyanate—$NaSCN$
Sodium thiosulfate—$Na_2S_2O_3$
Sodium tungstate—Na_2WO_4
Sodium uranate—$Na_2O_7U_2$
Sodium zincate—$H_4Na_2O_4Zn$[7]
Sulfamic acid—H_3NO_3S
Hydrogen sulfide (sulfane)—H_2S
Sulfur dioxide—SO_2
Sulfur hexafluoride—SF_6
Sulfur tetrafluoride—SF_4
Sulfuric acid—H_2SO_4
Sulfurous acid—H_2SO_3
Sulfuryl chloride—SO_2Cl_2

T

Thionyl chloride—$SOCl_2$
Tin(II) chloride (stannous chloride)—$SnCl_2$

Continued

Table A4 List and Formulas of Selected Inorganic Compounds—cont'd

Tin(II) fluoride—SnF_2
Tin(IV) chloride—$SnCl_4$
Titanium dioxide (titanium(IV) oxide)—TiO_2—Titanium dioxide (B)
Titanium nitride—TiN
Titanium(IV) bromide (titanium tetrabromide)—$TiBr_4$
Titanium(IV) carbide—TiC
Titanium(IV) chloride (titanium tetrachloride)—$TiCl_4$
Titanium(III) chloride—$TiCl_3$
Titanium(II) chloride—$TiCl_2$
Titanium(III) fluoride—TiF_3
Titanium(IV) iodide (titanium tetraiodide)—TiI_4
Tripotassium phosphate—K_3PO_4
Trisodium phosphate—Na_3PO_4
V
Vanadium oxytrichloride (vanadium(V) oxide trichloride)—$VOCl_3$
Vanadium(IV) chloride—VCl_4
Vanadium(II) chloride—VCl_2
Vanadium(II) oxide—VO
Vanadium(III) nitride—VN
Vanadium(III) bromide—VBr_3
Vanadium(III) chloride—VCl_3
Vanadium(III) fluoride—VF_3
Vanadium(IV) fluoride—VF_4
Vanadium(III) oxide—V_2O_3
Vanadium(IV) oxide—VO_2
Vanadium(IV) sulfate—$VOSO_4$
Vanadium(V) oxide—V_2O_5
W
Water—H_2O
Z
Zinc bromide—$ZnBr_2$
Zinc carbonate—$ZnCO_3$
Zinc chloride—$ZnCl_2$
Zinc cyanide—$Zn(CN)_2$
Zinc fluoride—ZnF_2
Zinc iodide—ZnI_2
Zinc oxide—ZnO
Zinc sulfate—$ZnSO_4$
Zinc sulfide—ZnS

Table A5 Melting Point, Boiling Point, Density, and Solubility of Common Inorganic Compounds

Compound	Melting point (°C)	Boiling point (°C)	Density (kg/m^3)	Solubility		
				Cold Water	Hot Water	Other Solvents
Aluminum						
Bromide, $AlBr_3$	97.5	263.3/ 747 mm	2640	d.	d.	s. al., CS_2, acet.
Chloride, $AlCl_3$	sb. 181		2440	d.	d.	s. al., eth., CCl_4; sl. s. bz.
Fluoride, AlF_3	sb. 1291	–	3070	0.5	1.71	i. a., alk., al., acet.
Iodide, AlI_3	191	360	3980	s.	s.	s. al., eth., CS_2, NH_3
Nitrate, $Al(NO_3)_3{\cdot}9H_2O$	73.5	d. 150	–	74	1.6	s. a., alk., al.
Nitride, AlN	>2200 (N_2)	d.	3260	d.	d.	d. a., alk.
Oxide, α-Al_2O_3	2015	2980	3970	i.	i.	v. sl. s. a., alk.
Sulfate, $Al_2(SO_4)_3$	d. 770	–	2710	36.4	98.1	s. dil. a.; sl. s. al.
Sulfide, Al_2S_3	1100	sb. 1500 (in N_2)	2020	d.	–	s. a., alk.; i. acet.
Ammonium						
Ammonia, NH_3	−77.7	−33.4	–	89.5/0	7.4	s. al., eth.
Acetate, $NH_4C_2H_3O_2$	114	d.	1170	148/4	d.	s. al.; sl. s. acet.
Azide, NH_4N_3	sb. 134 (exp.)		1346	25.3	37/40	s. al.; i. eth., bz.
Bromide, NH_4Br	sb. 452	–	2429	97/25	145	s. al., eth., acet, NH_3
Carbonate, $(NH_4)_2CO_3{\cdot}H_2O$	d. 58	–	–	84/15	d. 355	i. al., CS_2, NH_3
Chloride, NH_4Cl	sb. 340	–	1527	37.2	77.3	s. NH_3; sl. s. al.; i. acet., eth.
Chromate, $(NH_4)_2CrO_4$	d. 180	–	1910	34	d.	sl. s. NH_3, acet.; i.al.

Continued

Table A5 Melting Point, Boiling Point, Density, and Solubility of Common Inorganic Compounds—cont'd

Compound	Melting point (°C)	Boiling point (°C)	Density (kg/m^3)	Solubility		
				Cold Water	Hot Water	Other Solvents
Cyanide, NH_4CN	d. 36	–	1020/100	v. s.	d.	v. s. al.
Dichromate, $(NH_4)_2Cr_2O_7$	d. 170	–	2150	35.6	156	s. al.; i. acet.
Dihydrogen phosphate, $(NH_4)H_2PO_4$	190	–	1803	37.4	173	i. acet.
Fluoride, NH_4F	sb.	–	1009/25	100/0	d.	s. al.; i. NH_3
Formate, NH_4CHO_2	116	d. 180	1280	143	533/80	s. al., NH_3
Hydrogen phosphate, $(NH_4)_2HPO_4$	d. 155	–	1619	68.9	97.2/60	i. al., acet., NH_3
Hydrogen sulfide, NH_4HS	118/15 MPa	884/ 1.9 MPa	1170	128.1/0	d.	s. al., NH_3
Iodate, NH_4IO_3	d. 150	–	3309/21	2.06/15	–	–
Iodide, NH_4I	sb. 551	–	2514/25	172	250	v. s. al., acet., NH_3
Molybdate, $(NH_4)_2MoO_4$	d.	–	2276/25	s.d.	d.	s. a.; i. al., NH_3
Nitrate, NH_4NO_3	170	210/11	1725/25	192	871	s. al., acet., NH_3
Nitrite, NH_4NO_2	exp. 60–70	–	1690	v.s.	d.	s. al; i. eth.
Oxalate, $(NH_4)_2C_2O_4 \cdot H_2O$	d.	–	1500	4.45	34.64	i. NH_3
Sulfate, $(NH_4)_2SO_4$	d. 235	–	1769/50	75.4	103	i. al., acet., NH_3
Tartrate, $(NH_4)_2C_4H_4O$	d.	–	1601	63	86.9/60	sl. s. al.
Thiocyanate, NH_4CNS	149.6	d. 170	1305	170	346/60	s. al., acet., NH_3
Vanadate, meta-, NH_4VO_3	d. 200	–	2326	0.48	2.42/60	i. al., eth., NH_4Cl
Arsenic						
Bromide, $AsBr_3$	32.8	220	3540/25	d.	d.	s. HCl, CS_2
Chloride, $AsCl_3$	−8.5	130.2	2163	d.	d.	s. HCl, al., eth.

Fluoride, tri–, AsF_3	−6	−56	–	d.	d.	s. al., eth., bz.
Fluoride, penta–, AsF_5	−80	−52.8	–	s.	–	s. alk., al., eth., bz.
Hydride, AsH_3	−116.3	−55	1689/−85	0.2/0 (A)	–	sl. s. alk.; s. chl.
Iodide, AsI_3	146	403	4390/13	25-Jun	d.	s. al., eth., chl., bz
Oxide, tri, As_2O_3	315	457	3738	3.7	10.1	s. alk., HCl; i. eth.
Penta–, As_2O_5	d. 315	–	4320	150/16	76.4	s. alk., al., a.
Sulfide, tri–, As_2S_3	300	707	3430	5×10^{-5}/18	sl. s.	s. K_2S, $NaHCO_3$, al.
Penta–, As_2S_5	sb. d. 500	–	–	1.36×10^{-4}/0	d.	s. alk., HNO_3
Barium						
Acetate, $Ba(C_2H_3O_2)_2$	–	–	2468	72	74.8	i. al.
Bromate, $Ba(BrO_3)_2 \cdot H_2O$	$-H_2O$/170	d. 260	3990/18	0.66	5.7	s. acet.; i. al.
Bromide, $BaBr_2 \cdot 2H_2O$	$-H_2O$/75	$-2H_2O$/120	3584/24	104	149	s. al.
$BaBr_2$	857	d.	4781/24	104	149	v. s. al.; v. sl. s. acet.
Carbonate, $BaCO_3$	–	d.	4430	2×10^{-3}	6×10^{-3}	s. a., NH_4Cl; i. al.
Chlorate, $Ba(ClO_3)_2 \cdot H_2O$	$-H_2O$/120	−O/250	3180	34.4	105	sl. s. al., acet.
Chloride, $BaCl_2 \cdot 2H_2O$	$-2H_2O$/113	37.5/20	3097/24	35.7	58.8	sl. s. HCl, HNO_3
$BaCl_2$ (β)	961	1560	3917	35.7	58.8	v. sl. s. al.
Fluoride, BaF_2	1355	2137	4890	0.16	sl. s.	s. a., NH_4Cl
Hydride, BaH_2	d. 675	–	4210/0	d.	d.	d. a.
Hydroxide, $Ba(OH)_2 \cdot 8H_2O$	$-8H_2O$/78	–	2180/16	3.9	101/80	sl. s. al.; i. acet.
Iodate, $Ba(IO_3)_2$	d.		4998	0.035	197	s. HNO_3, HCl
Iodide, $BaI_2 \cdot 2H_2O$	$-H_2O$/100	$-2H_2O$/540	5150	220	276	s. al., acet.
BaI_2	740	200/15	5150/25	220	276	v. s. al., me., a.
Nitrate, $Ba(NO_3)_2$	592	d.	3240/23	9	34.4	sl. s. a.; i. al.
Nitrite, $Ba(NO_2)_2 \cdot H_2O$	d. 115	–	3173	67.5	30	v. s. a.; sl. s. al.
Oxide, BaO	2010	2000	5720	3.48	90.8/80	s. dil. a., al.; i. acet.
Perchlorate, $Ba(ClO_4)_2$	505	–	3200	198.5/25	562.3	v. s. al.

Continued

Table A5 Melting Point, Boiling Point, Density, and Solubility of Common Inorganic Compounds—cont'd

Compound	Melting point (°C)	Boiling point (°C)	Density (kg/m^3)	Solubility: Cold Water	Solubility: Hot Water	Solubility: Other Solvents
Peroxide, BaO_2	450	−O/800	4960	v. sl. s.	d.	s. dil. a.; i. acet
Sulfate, $BaSO_4$	1580	–	4500/15	2×10^{-4}	4×10^{-4}	sl. s. H_2SO_4
Sulfide, BaS	1200	–	4250/15	d.	d.	i. al., CS_2
Boron						
Bromide, BBr_3	−46	91.3	2643/18	d.	–	s. al., CCl_4
Chloride, BCl_3	−107	12.7	1349/11	d.	d.	d. al.
Fluoride, BF_3	−126.7	−99.9	–	1.06/0(A)	d.	s. conc. H_2SO_4; d. al.
Hydride, B_2H_6	−165.5	−92.5	447/−11	sl. s. d.	d.	s. NH_4OH, conc. H_2SO_4
B_4H_{10}	−120.8	16	560/−35	sl. s. d.	d.	d. al.; s. bz.
Iodide, BI_3	49.9	210	3350/50	d.	d.	d. al.;v. s. CS_2, bz., CCl_4
Nitride, BN	sb. 3000	–	2250	i.	sl. d.	sl. s. hot a.
Oxide, B_2O_3	450	1860	2460	1.1/0	16.4	s. a. al.
Boric acid, H_3BO_3	d. 169	–	1435/15	4.9	27.6	s. al.; s. acet.
Calcium						
Acetate, $Ca(C_2H_3O_2)_2$	d.	–	–	37.4/0	29.7	sl. s. al.
Bromide, $CaBr_2$	742	1810	3353/25	143	309	s. al., acet., a.
Carbide, CaC_2	–	2300	2200	d.	d.	–
Carbonate, $CaCO_3$ (aragonite)	d. 825	–	2930	0.00153/25	0.00190/75	s. a., NH_4Cl
Carbonate, $CaCO_3$ (calcite)	d. 899	–	2710/18	0.0014/25	0.0018/75	s. a., NH_4Cl
Chloride, $CaCl_2 \cdot 6H_2O$	$-4H_2O/30$	$-6H_2O/200$	1710/25	74.5	159	s. al.
$CaCl_2$	772	1940	2150/25	74.5	159	s. al., acet.

Fluoride, CaF_2	1420	2500	3180	0.0016/18	–	s. NH_4 slt.; sl. s. a.
Hydride, CaH_2	816 (H_2)	d. 600	1900	d.	d.	d. a.; i. bz.
Hydroxide, $Ca(OH)_2$	$-H_2O$/580	d.	2240	0.12	0.077	s. NH_4Cl aq., a.; i. al.
Iodide, CaI_2	778	1100	3956/25	209	426	s. al., acet., a.
Nitrate, $Ca(NO_3)_2 \cdot 4H_2O$	40	d. 132	1850	129	376	s. al., acet.
$(NO_3)_2$	561	–	2504/18	129	376	s. al., acet.
Nitrite, $Ca(NO_2)_2 \cdot 4H_2O$	$-2H_2O$/44	–	1674/0	48/0	68/42	s. al.
Oxide, CaO	2614	2850	3300	d.	d.	s. a.
Phosphate, ortho-$Ca_3(PO_4)_2$	1670	–	3140	0.002	d.	s. a.; i. al.
Silicate, meta-, $CaSiO_3$ (α)	1540	–	2905	0.0095/17	–	s. HCl
Sulfate, $CaSO_4 \cdot 2H_2O$	−1 5H_2O/130	$-2H_2O$/163	2320	0.274	0.189	s. a., NH_4 slt
Sulfate, $CaSO_4$	1450	–	2960	0.274	0.162	s. a., NH_4 slt
Sulfide, CaS	2525	d.	2500	sl. s. d.	–	d. a.
Tungstate, $CaWO_3$	–	–	6060	0.2	–	s. NH_4Cl; i. a., al.
Carbon						
Oxide, sub-, C_3O_2	−111.3	7	1114/0	d.	–	–
Monoxide, CO	−199	−191.50	–	3.5×10^{-2}/0 (A)	1.5×10^{-2}/ 60 (A)	s. al., bz., ac.
Dioxide, CO_2	sb. −78.48	–	1560/−79	1.68/0 (A)	0.36/60 (A)	s. a., alk., al., acet.
Oxysulfide, COS	−138.2	−50.2	1240/−87	0.54/20 (A)	–	v. s. CS_2; s. al.
Cyanogen, C_2N_2	−27.84	−20.7	958/−21	4.5/20 (A)	–	s. al., eth.
Chlorine						
Oxide, mono-, Cl_2O	−120	exp. 2.2	–	2/0 (A)	d.	s. alk., H_2SO_4
Dioxide, ClO_2	−60	exp. 11.1	–	20/4 (A)	d.	s. alk., H_2SO_4
Heptoxide, Cl_2O_7	−91.5	82	–	s. d.	–	s. bz.
Perchloric acid, $HClO_4$	−112	39/56	1764/22	39/51	∞	–

Continued

Table A5 Melting Point, Boiling Point, Density, and Solubility of Common Inorganic Compounds—cont'd

Compound	Melting point (°C)	Boiling point (°C)	Density (kg/m^3)	Solubility		
				Cold Water	Hot Water	Other Solvents
Chromium						
(III) Ammonium sulfate, $Cr_2(SO_4)_3$ $(NH_4)_2SO_4 \cdot 24H_2O$	94	$-9H_2O/100$	1720	15.8	32.7/40	ss. al., dil. a.
(II) Bromide, $CrBr_2$	842	–	4356	s.	s.	ss. al.
(III) $CrBr_3$	sb.	–	4250/25	i.	s.	v. s. al.; d. alk.
Carbonyl, $Cr(CO)_6$	d. 110	exp. 210	1770	i.	i.	sl. s. chl.; i. al., eth.
(II) Chloride, $CrCl_2$	824	–	2878/25	v. s.	v. s.	i. al., eth.
(III) Chloride, $CrCl_3$	1150	sb. 1300	2760/15	i	sl. s.	i. a., acet., al., eth.
(II) Fluoride, CrF_2	1100	>1300	4110	sl. s.	–	s. h. HCl; i. al.
(III) CrF_3	sb. 1100–1200	–	3800	i.	–	s. HF; sl. s. a.; i. al.
(III) Nitrate, $Cr(NO_3)_3 \cdot 9H_2O$	60	d. 100	–	77.4/25	s.	s. a., alk., al., acet.
(III) Oxide, Cr_2O_3	2330	4000	5210	i.	i.	i. a., alk., al.
(IV) CrO_3	196	d.	2700	161.7/0	167.4	s. conc. a., eth., al.
Oxychloride, CrO_2Cl_2	−96	117	1911	d.	d.	s. eth., ac.; d. al.
(III) Sulfide, Cr_2S_3	−S/1350	–	3770	i.	i.	s. HNO_3; d. al.
Cobalt						
(II) Acetate, $Co(C_2H_3O_2)_2$	$-4H_2O/140$	–	1705	s.	s.	s. a., al.
(II) Bromide, $CoBr_2$	678 (N_2)	–	4909/25	66.7/59	68.1/97	v. s. al.; s. eth., acet.
Carbonyl, $Co_2(CO)_8$	51	d. 52	1730/18	i.	i.	s. CS_2, eth.; sl. s. al.
(II) Chloride, $CoCl_2$	735	1049	3360/25	53	106	s. al. acet.; s. sl. eth.

Cyanide, $Co(CN)_2 \cdot 2H_2O$	$-2H_2O/280$	d. 300	1872/25 (anh.)	0.0042/18	–	s. KCN, HCl, NH_4OH
(II) Fluoride, CoF_2	1200	1400	4460/25	1.5/25	s.	sl. s. a; i. al., eth., bz.
(II) Hydroxide, $Co(OH)_2$	d.	–	3597/15	3.2×10^{-4}	–	s. a., NH_4 slt.
(II) Iodide, CoI_2	515 (vac.)	570 (vac.)	5680	159/0	420	v. s. al., acet.
(II) Nitrate, $Co(NO_3)_2 \cdot 6H_2O$	$-3H_2O/55$	d.	1870/25	96.8	339/91	v. s. al., acet.; sl. s. NH_3
(II) Oxide, CoO	1800	d.	6450	i.	i.	s. a. NH_4OH; i. al.
(III) Co_2O_3	d. 895	–	5180	i.	i.	s. a.; i. al.
(II) Sulfate, $CoSO_4 \cdot 7H_2O$	96.8	$-7H_2O/420$	1950/25	35.2	83	s. al.
(II) Sulfate, $CoSO_4$	d. 735	–	3710/25	36.2	83	sl. s. me.
Copper						
(II) Acetate, $Cu(C_2H_3O_2)_2 \cdot H_2O$	115	d. 240	1882	6.5	18	s. al.; eth.
(I) Bromide, CuBr	492	1345	4980	v. sl. s.	d.	s. a., NH_4OH; i. acet.
(II) $CuBr_2$	498	–	4770/25	127	132/50	s. al., acet., pyr.; i. bz.
(I) Chloride, CuCl	430	1490	4140	0.0062	–	s. HCl, NH_4OH, eth.; i. al.
(II) $CuCl_2$	620	d. 990	3386/25	73	107.9	v. s. al., h. a., acet.
(I) Cyanide, CuCN	473 (N_2)	d.	2920	i.	i.	s. HCl, KCN, NH_4OH
(II) Fluoride, $CuF_2 \cdot 2H_2O$	d.	–	2930/25	4.7	d.	s. al., a.; i. acet.
CuF_2	d. 950	–	4230	4.7	s.	s. dil. a.; i. al.
(I) Iodide, CuI	605	1290	5620	0.008/18	–	s. dil. HCl, KI, NH_3
(II) Nitrate, $Cu(NO_3)_2 \cdot 3H_2O$	114.5	d. 170	2320/25	125	985	v. s. al.
(I) Oxide, Cu_2O	1235	$-O/1800$	6000	i.	i.	s. HCl, NH_4Cl; sl. s. HNO_3; i. al.

Continued

Table A5 Melting Point, Boiling Point, Density, and Solubility of Common Inorganic Compounds—cont'd

Compound	Melting point (°C)	Boiling point (°C)	Density (kg/m^3)	Solubility		
				Cold Water	Hot Water	Other Solvents
(II) CuO	1326	–	6315	i.	i.	s. a., NH_4Cl, KCN
(II) Sulfate, $CuSO_4 \cdot 5H_2O$	$-4H_2O/110$	$-5H_2O/150$	2284	14.3	75.1	s. me.; i. al.
$CuSO_4$	d. 200	–	3603	14.3	75.1	sl. s. me.; i. al.
(II) Sulfide, CuS	d. 220	–	4600	$3.3 \times 10^{-5}/18$	–	s, a., KCN; i. al., alk.
Gold						
(I) Bromide, AuBr	d. 115	–	7900	i.	i.	s. NaCN; d. a.
(III) $AuBr_3$	97.5	–Br/160	–	sl. s.	–	s. eth., al.
(I) Chloride, AuCl	d. 170	–	7400	v. sl. s.	d.	s. HCl
(III), $AuCl_3$	d. 254	–	3900	68	v. s.	s. al., eth.; sl. s. NH_4OH; i. CS_2
(I) Iodide, AuI	d. 120	–	8250	v. sl. s.	d.	s. KI
Hydrogen						
Bromide, HBr	–86.82	–66.73	2770/–67	221/0	130	s. al.
Chloride, HCl	–114.22	–85.05	1187/–84.9	8.23/0	56.1/60	s. al., eth., bz.
Cyanide, HCN	–13.24	25.70	699/22	v. s.	v. s.	v. s. al.; s. eth.
Fluoride, HF	–83.1	19.9	987/19.5	v. s.	v. s.	–
Iodide, HI	–50.80	–35.36	2850/–47	0.43/0 (A)	v. s.	s. al.
Oxide-deuterium, D_2O	3.8	101.42	1105.6	∞	∞	–
Peroxide, H_2O_2	–0.41	150.2	1407/25	∞	–	s. al., eth.
Phosphide, H_3P	–133.78	–87.74	746/–90	0.26/17 (A)	i.	s. al., eth.
Selenide, H_2Se	–65.73	–41.3	–	3.77/4 (A)	2.7/ 22.5 (A)	s. CS_2, $COCl_2$
Sulfide, H_2S	–85.53	–60.34	–	4.53/0 (A)	1.18/60 (A)	s. CS_2; sl. s. al.
Telluride, H_2Te	–51	–2.3	2570/–20	s.	–	a. al., alk.

Iodine						
Bromide, mono-, IBr	sb. 50	–	4416/0	s. d.	–	s. al., eth., chl., CS_2
Chloride, mono-, ICl (α)	27.3	97.4	3182/0	d.	–	s. HCl, al., eth., CS_2
Chloride, tri-, ICl_3	d. 77	–	3177/15	s. d.	–	s. ac., al., bz., eth., CCl_4
Fluoride, penta-, IF_5	9.6	98	3750	d.	d.	d. a., alk.
Fluoride, hepta-, IF_7	sb. 4.5	–	2800/6	v. s.	d.	d. a., alk.
Iodic acid, HIO_3	d. 110	–	4630	309/16	574	v. s. al.; sl. s. HNO_3
Oxide, penta-, I_2O_5	d. 300	–	4799/25	187.4/13	v. s.	sl. s. dil. a.; i. al., eth., chl.
Iron						
(III) Ammonium sulfate, Fe $(NH_4)(SO_4)_2 \cdot 12H_2O$	40	$-12H_2O$/ 230	1710	44.2	v. s.	s. dil. a.; i. al.
(II) Ammonium sulfate, Fe $(NH_4)_2(SO_4)_2 \cdot 6H_2O$	d. 100	–	1860	20	53/80	i. al.
(II) Bromide, $FeBr_2$	d. 684	–	4636/25	109/10	170/96	s. al.; sl. s. bz.
(II) Carbonate, $FeCO_3$	d.	–	3800	0.0067/25	–	s. CO_2 aq.
Carbonyl, penta-, $Fe(CO)_5$	−21	103	1457	i.	–	s. al., eth., bz., alk.,
Carbonyl, nona, $Fe_2\ (CO)_9$	d. 80	–	2085/18	i.	i.	v. sl. s. al.; d. HNO_3
(II) Chloride, $FeCl_2$	677	1026	3160/25	62.6	95	v. s. al.; s. acet; i. eth.
(III) Chloride, $FeCl_3 \cdot 6H_2O$	37	280	–	55.1	536	s. al., eth.
(III) Chloride, $FeCl_3$	304	d. 319	2900/25	55.1	536	v. s. al., eth., acet.
Dicyclopentadiene, Fe $(C_5H_5)_2$ (ferrocene)	174	249	–	i.	–	s. organic solvents
(III) Fluoride, FeF_3	>1000	–	3520	sl. s.	s.	s. a.; i. al., eth.
(II) Iodide, $FeI_2 \cdot 4H_2O$	d. 94	–	2873	v. s.	d.	s. al., eth.
	50	d. 125	1684	150/0	v. s.	s. a., al., acet.

Continued

Table A5 Melting Point, Boiling Point, Density, and Solubility of Common Inorganic Compounds—cont'd

Compound	Melting point (°C)	Boiling point (°C)	Density (kg/m^3)	Solubility: Cold Water	Solubility: Hot Water	Solubility: Other Solvents
(III) Nitrate, $Fe(NO_3)_3 \cdot 9H_2O$						
(II) Oxide, $Fe_{0.95}O$	1377	–	5700	i.	i.	s. a.; i. al., alk.
(III) Oxide, Fe_2O_3	1565	–	5240	i.	i.	s. HCl, H_2SO_4
(II/III) Oxide, (magnetite), Fe_3O_4	1597	–	5180	i.	i.	s. conc. a.; i. alk., eth.
(II) Sulfate, $FeSO_4 \cdot 7H_2O$	64, $-6H_2O$/90	$-7H_2O$/300	1898	26.3	34.2/95	s. me.; sl. s. al.
(II) Sulfide, FeS	1190	d.	4740	6.2×10^{-4}/18	d.	s. d. a.
Lead						
Acetate, $Pb(C_2H_3O_2)_2$	280	–	3250	44.3	221/50	v. sl. s. al.
Bromide, $PbBr_2$	373	916	6660	0.84	4.71	s. a., KBr; i. al.
Carbonate, $PbCO_3$	d. 315	–	6600	1.1×10^{-4}	d.	s. a., alk.; i. al.
Chloride, $PbCl_2$	501	950	5850	1.1	3.34	s. NH_4 slt.; sl. s. dil. HCl, NH_3; i. al.
Chromate, $PbCrO_4$	844	d.	6120/15	5.8×10^{-6}/25	i.	s. a., alk.
Fluoride, PbF_2	855	1290	8240	0.064	–	s. HNO_3; i. acet.
Iodide, PbI_2	402	954	6160	0.063	0.41	s. alk., KI.; i. al.
Nitrate, $Pb(NO_3)_2$	d. 470	–	4530	56/0	127	s. aq. al., NH_3, alk.
Oxide, mono–, PbO (litharge)	890	–	9530	1.7×10^{-3}	–	s. HNO_3, alk., NH_4Cl
Oxide (red-lead), Pb_3O_4	d. 500	–	9100	i.	i.	s. h. HCl, ac.; i. al.
Oxide di–, PbO_2	d. 290	–	9375	i.	i.	s. dil. HCl; sl. s. ac.

Sulfate, $PbSO_4$	1170	–	6200	0.00425/25	0.0056/40	s. a., NH_4 slt.
Sulfide, PbS	1114	d.	7500	$8.6 \times 10^{-5}/13$	i.	s. a.; i. alk., al.
Tetraethyl, $Pb(C_2H_5)_4$	−136	d. 200	1659/11	i.	–	s. bz., al., eth.
Lithium						
Aluminum hydride, $LiAlH_4$	d. 125	–	917	d.	d.	s. eth.; exp. a.
Amide, $LiNH_2$	380–400	d. 750	1178/17	s.	d.	sl. s. NH_3, al; i. eth., bz.
Borohydride $LiBH_4$	d. 275	–	660	d.	d.	s. eth., sl. s. al. (d.)
Bromide, LiBr	547	1265	3464/25	168	266	s. al., eth.; sl. s. pyr.
Carbonate, Li_2CO_3	723	d. 1310	2110	1.33	0.72	s. a.; i. acet., al.
Chloride, LiCl	610	1325–1360	2068/25	84.7/25	128	s. al., acet.
Fluoride, LiF	848	1676	2635	0.14/25	0.14/35	s. HF.; i. al., acet.
Hydride, LiH	680	–	820	d.	d.	v. sl. s. a.; i. eth.
Iodide, LiI	450	1180	4076	165	481	v. s. al., acet., NH_4OH
Nitrate, $LiNO_3$	264	d. 600	2380	73.3	234	s. al., acet., NH_4OH
Nitrite, $LiNO_2 \cdot H_2O$	>100	d.	1615/0	93/0	340/50	v. s. al.
Oxide, Li_2O	>1700	–	2013/25	d.	d.	–
Perchlorate, $LiClO_4$	236	d. 430	2428	60/25	150/89	v. s. al., eth., acet.
Phosphate, Li_3PO_4	837	–	2537/18	0.039/18	–	s. a.; i. acet., NH_4OH
Sulfate, Li_2SO_4	845	–	2221	26.1/0	23	i. acet., al.
Sulfide, Li_2S	900–975	–	1660	v. s.	v. s.	v. s. al.
Magnesium						
Acetate, Mg $(C_2H_3O_2)_2 \cdot 4H_2O$	80	–	1454	v. s.	v. s.	v. s. al.
Bromide, $MgBr_2$	710	1280	3720/25	101/25	125	s. HBr., al., acet.
Carbonate, $MgCO_3$	d. 350	$-CO_2/900$	2958	0.006	sl. s.	s. a.
Chloride, $MgCl_2 \cdot 6H_2O$	d. 117	–	1569	55	73	v. s. al.

Continued

Table A5 Melting Point, Boiling Point, Density, and Solubility of Common Inorganic Compounds—cont'd

Compound	Melting point (°C)	Boiling point (°C)	Density (kg/m³)	Solubility: Cold Water	Solubility: Hot Water	Solubility: Other Solvents
Chloride, $MgCl_2$	708	1412	2320	55	73	s. al.
Fluoride, MgF_2	1260	2260	–	0.013/25	–	s. HNO_3; i. al.
Iodide, MgI_2	d. <637	–	4430/25	148/18	165	s. al., eth., NH_3
Nitrate, $Mg(NO_3)_2 \cdot 6H_2O$	89	d. 330	1636/25	72.7/25	253	s. al.
Nitride, Mg_3N_2	sb. 700 (vac.)	d. 800	2712/25	d.	d.	s. a.; i. al.
Oxide, MgO	2852	3600	3580/25	6×10^{-4}/25	3×10^{-5}	s. a., NH_4 slt.; i. al.
Perchlorate, $Mg(ClO_4)_2$	d. 251	–	2210/18	99/25	v. s.	s. al.; sl. s. eth.
Phosphate, $Mg_3(PO_4)_2 \cdot 22H_2O$	$-18H_2O$/ 100	d. 200	1640/15	0.0065	–	d. a.
Sulfate, $MgSO_4$	d. 1124	–	2660	33.7	73.8	s. al., eth.; i. acet.
Mercury						
(I) Bromide, Hg_2Br_2	sb. 345	–	7307	4×10^{-6}/26	v. sl. s.	s. a.; i. al., acet.
(II) $HgBr_2$	236	322	6109/25	0.55	4.0	s. al.; v. sl. s. eth.
(I) Chloride, Hg_2Cl_2	sb. 400	–	7150	2×10^{-4}	0.001/43	s. aq. reg.; sl. s. h. HNO_3, HCl; i. al.
(II) Chloride, $HgCl_2$	276	302	5440/25	6.9	48	v. s. al.; s. eth., ac.
(II) Cyanide, $Hg(CN)_2$	d.	–	3996	11.2/25	33	s. al., NH_3; i. bz.
(II) Fluoride, Hg_2F_2	570	d.	8730/15	d.	–	–
(I) HgF_2	d. 645	–	8950/15	d.	–	s. HF, dil. HNO_3
(I) Iodide, Hg_2I_2	sb. 140	–	7700	v. sl. s.	–	s. KI, NH_4OH; i. al., eth.
(II) HgI_2	259	354	6360/25	5.4×10^{-3}/22	sl. s.	s. al., eth.
(II) Nitrate, $Hg(NO_3)_2 \cdot 1/2\ H_2O$	79	d.	4390	v. s.	d.	s. HNO_3, acet.; i. al.

(I) Oxide, Hg_2O	d. 100	–	9800	i.	i.	s. HNO_3
(II) Oxide, HgO	d. 500	–	11100/4	$5.3 \times 10^{-3}/25$	0.039	s. a.; i. al., eth., alk.
(I) Sulfate, Hg_2SO_4	d.	–	7560	0.06/25	0.09	s. H_2SO_4, HNO_3
(II) $HgSO_4$	d.	–	6470	d.	–	s. a.; i. al., acet.
(II) Sulfide, HgS (α)	sb. 583	–	8100	$1 \times 10^{-6}/18$	–	s. aq. reg.; i. al., HNO_3
Molybdenum						
(V) Chloride, $MoCl_5$	194	268	2928	d.	d.	s. conc. a., chl.; s. d. al.
(VI) Fluoride, MoF_6	17	35	2550/18	s. d.	d.	s. NH_4OH, alk.
(VI) Oxide, MoO_3	795	sb. 1155	4692	0.18/23	2.05/70	s. HF, H_2SO_4, NH_4OH
(III) Sulfide, Mo_2S_3	d. 1100	–	5910/15	–	–	i. conc. HCl; d. h. HNO_3
(IV) Sulfide, MoS_2	1185	–	4800/14	i.	i.	s. aq. reg., h. H_2SO_4, HNO_3; i. dil. a.
Nickel						
Bromide, $NiBr_2$	963	–	5098/27	113/0	155	s. al., eth., NH_4OH
Carbonyl, $Ni(CO)_4$	−25	42.2	1320/17	0.018/9.8	–	s. HNO_3, aq. reg., al., eth.; i. dil. a./alk.
Chloride, $NiCl_2$	–	sb. 972	3550	62	88	s. al., NH_4OH
Fluoride, NiF_2	sb. 1000 (HF)	–	4630	2.56	2.58/90	s. a., alk., eth., NH_3
Iodide, NiI_2	797	–	5834	124/0	188	s. al.
Nitrate, $Ni(NO_3)_2 \cdot 6H_2O$	56.7	136.7	2050	94.2	238	s. al., NH_4OH
Oxide, NiO	1990	–	6670	i.	i.	s. a., NH_4OH
Sulfate, $NiSO_4 \cdot 6H_2O$	$-6H_2O/103$	–	2070	38	84	v. s. al., NH_4OH
Sulfide, NiS	797	–	5500	$3.6 \times 10^{-4}/18$	–	s. aq. reg., HNO_3

Continued

Table A5 Melting Point, Boiling Point, Density, and Solubility of Common Inorganic Compounds—cont'd

Compound	Melting point (°C)	Boiling point (°C)	Density (kg/m^3)	Solubility		
				Cold Water	Hot Water	Other Solvents
Nitrogen						
Chloride, tri-, NCl_3	−40	71, exp.	1653	i.	d.	s. chl., bz., CCl_4, CS_2
Fluoride, tri, NF_3	−207	−129	1537/−129	v. sl. s.	–	–
Hydrazine, N_2H_4	1.5	113.5	1.01/15	∞	∞	s. al.
Hydrazine hydrate, $N_2H_4 \cdot H_2O$	−40	118	1030	∞	∞	s. al.; i. eth., chl.
Hydrazine sulfate, $N_2H_4 \cdot H_2SO_4$	254	–	1370	2.9	14.4/80	i. al.
Hydroxylamine, NH_2OH	33.1	56	1200	s.	d.	s. a., al.; v. sl. s. eth., chl., bz.
Oxide, (ous), N_2O	−90.8	−88.5	–	1.3/0 (A)	0.57/25 (A)	s. al., eth., H_2SO_4
Oxide (ic), NO	−163.6	−151.8	1269/−150	7.1×10^{-2}/ 0 (A)	2.9×10^{-2}/ 60 (A)	s. aq. $FeSO_4$, CS_2; sl. s. H_2SO_4
Oxide, tri-, N_2O_3	−102	d. 3.5	1447/2	s.	d.	s. alk., a., eth.
Oxide, di-, NO_2	−11.2	21.1	1449	s. d.	–	s. alk., CS_2, chl.
Oxide, penta-, N_2O_5	30	d. 47	1642/18	s.	d.	s. chl.
Oxychloride, NOCl	−64.5	−6.4	1417/12	d.	d.	s. fuming H_2SO_4
Oxygen						
Fluoride, di-, OF_2	−223.8	−144.8	1900/−223.8	sl. s., d.	–	sl. s. a., alk.
Ozone, O_3	−192.7	−110.51	1614/−195.4	0.49/0 (A)	–	s. alk., oils

Palladium						
Chloride, $PdCl_2$	d. 500	–	4000/18	s.	s.	s. HCl, acet.
Nitrate, $Pd(NO_3)_2$	870	–	–	s. d.	—	s. HNO_3
Oxide, PdO	870	–	8700	i.	i.	i. aq. reg.
Phosphorus						
Bromide, tri–, PBr_3	−40	173	2852/15	d.	d.	s. eth., chl., CCl_4; d. al.
Chloride, tri–, PCl_3	−112	76	1574/21	d.	d.	s. eth., chl., CS_2, CCl_4
Chloride, penta–, PCl_5	sb. 162	–	1600	d.	d.	s. CCl_4, CS_2; d. a.
Fluoride, tri–, PF_3	−151.5	−101.5	–	d.	d.	s. al.; d. alk.
Fluoride, penta–, PF_5	−83	−75	–	d.	d.	d. alk.
Hydride, PH_3	−133	−87.7	–	0.26/20 (A)	–	–
Iodide, tri–, PI_3	61	d.	4180	d.	d.	v. s. CS_2
Oxide, tri–, P_2O_3	23.8	174 (N_2)	2135	d.	d.	s. CS_2, eth., chl., bz.
Oxide, penta–, P_2O_5	sb. 358	–	2390	d.	d.	s. H_2SO_4; i. acet.
Oxychloride, $POCl_3$	2	105	1675	d.	d.	d. al., a.
Sulfide, penta–, P_2S_5	288	514	2030	i.	d.	s. alk.; sl. s. CS_2
Phosphoric acid, ortho–, H_3PO_4	42.4	$-H_2O/213$	1834/18	548	v. s.	s. al.
Phosphorous acid, hypo, H_3PO_2	26.5	d. 130	1493	s.	v. s.	v. s. al., eth.
Platinum						
(II) Bromide, $PtBr_2$	d. 250	–	6650/25	i.	i.	s. HBr; i. al.
(IV) Bromide, $PtBr_4$	d. 180	–	5690/25	0.41	sl. s.	v. s. al., eth., HBr
(II) Chloride, $PtCl_2$	d. 581 (Cl_2)	–	6050	v. sl. s.	–	s. HCl; i. al., eth.
(IV) Chloride, $PtCl_4$	d. 370 (Cl_2)	–	4303/25	142/25	572/98	s. acet.; sl. s. al., NH_3; i, eth.

Continued

Table A5 Melting Point, Boiling Point, Density, and Solubility of Common Inorganic Compounds—cont'd

Compound	Melting point (°C)	Boiling point (°C)	Density (kg/m^3)	Solubility: Cold Water	Solubility: Hot Water	Solubility: Other Solvents
(IV) Oxide, PtO_2	450	–	10200	i.	i.	i. a., aq. reg.
Potassium						
Acetate, $KC_2H_3O_2$	292	–	1570/25	253	492/62	s. al.; i. eth., acet.
Aluminum sulfate, KAl $(SO_4)_2 \cdot 12H_2O$	$-9H_2O/65$	$-12H_2O/200$	1757	6.3	v. s.	s. dil. a.; i. al., acet.
Amide, KNH_2	335	–	–	d.	d.	s. NH_3; d. al.
Azide, KN_3	350 (vac.)	–	2040	49.6/17	106	s. al.; i. eth.
Borohydride, KBH_4	d. 500	–	1178	19.3	v. s.	sl. s. al.; i. eth.
Bromate, $KBrO_3$	d. 370	–	3270/18	6.91	49.9	sl. sl. al; i. acet.
Bromide, KBr	734	1435	2750/25	65	104	sl. s. al., eth.
Carbonate, K_2CO_3	891	d.	2428	111	156	i. al., acet.
Chlorate, $KClO_3$	356	d. 400	2320	7.3	56.3	s. aq. al., alk.
Chloride, KCl	770	1440	1984	34.2	56.3	s. alk., eth.; sl. s. al.
Chromate, K_2CrO_4	968	–	2732/18	62.7	80.1	i. al.
Citrate, $K_3C_6H_5O_7 \cdot H_2O$	d. 230	–	1980	148	–	sl. s. al.
Cyanide, KCN	634	–	1520/16	50	100	s. me.; sl. s. al.
Dichromate, $K_2Cr_2O_7$	398	d. 500	2676/25	12.2	102	i. al.
Dihydrogen phosphate, KH_2PO_4	252.6	–	2338	22.4	83.4/90	i. al.
Ferricyanide, $K_3Fe(CN)_6$	d.	–	1850/25	48.8/25	77.5	s. acet.; i. al.
Ferrocyanide, $K_4Fe(CN)_6 \cdot 3H_2O$	$-3H_2O/70$	d.	1850/17	28.2	74.9	s. acet.; i. al., eth.
Fluoride, KF	860	1505	2480	95	150/80	s. HF; i. al.
Formate, $KCHO_2$	167	d.	1910	331/18	657/80	s. al.; i. eth.

Hydride, KH	d.	–	1470	d.	d.	i., eth., bz.; d. a.
Hydrogen carbonate, $KHCO_3$	d. 100–200	–	2170	22.4	60/60	i. al.
Hydrogen phthalate, $KHC_8H_4O_4$	–	–	1636	11.4/25	33	–
Hydrogen sulfate, $KHSO_4$	214	d.	2322	48.6	121.7	i. al.
Hydroxide, KOH	360	1320	2044	112	182	v. s. al.; i. eth.
Iodate, KIO_3	d. >100	–	3930/32	8.1	32.3	s. KI.; i. al.
Iodide, KI	680	1340	3130	144	208	s. al., acet., NH_3
Metabisulfite, $K_2S_2O_5$	d. 190	–	2340	44.5	125/94	sl. s. al.
Nitrate, KNO_3	334	d. 400	2109/16	31.6	245	i. al., eth.
Nitrite, KNO_2	440	d.	1915	307	413	v. s. NH_3; sl. s. al.
Oxalate, $K_2C_2O_4{\cdot}H_2O$	$-H_2O$/100	–	2127/4	35.7	80.2	–
Oxide, mono-, K_2O	d. 350	–	2320/0	v. s., d.	v. s., d.	s. al., eth.
Perchlorate, $KClO_4$	d. 400	–	2520/10	0.75/0	21.8	v. sl. s. al.; i. eth.
Periodate, KIO_4	−O/300	–	3618/15	0.5/25	6.8/97	v. sl. s. KOH
Permanganate, $KMnO_4$	d. <240	–	2703	6.3	25.3/65	v. s. me., acet.; s. a.
Persulfate, $K_2S_2O_8$	d. <100	–	2477	4.7	–	i. al.
Sulfate, K_2SO_4	1069	1689	2662	11.1	24.1	i. al., acet.
Sulfide, K_2S	840	–	1805/14	s.	v. s.	s. al.; i. eth.
Tetrafluoroborate, KBF_4	d. 350	–	2498	0.44	6.27	sl. s. al., eth.; i. alk.
Thiocyanate, KCNS	173	d. 500	1886/14	217	687	s. al., acet.
Selenium						
Bromide, mono-, Se_2Br_2	d. 227	–	3604/15	d.	d.	s. chl., CS_2; d. al.
Chloride, mono-, Se_2Cl_2	−85	d. 130	2770/25	d.	d.	s. CS_2, chl, bz.; d. al.
Chloride, tetra-, $SeCl_4$	170–196	d. 288	2600	d.	d.	s. $POCl_3$; i. al., CS_2
Hydride, SeH_2	−60.4	−41.5	2004/−41.5	0.038/4 (A)	2.7/22 (A)	s. CS_2

Continued

Table A5 Melting Point, Boiling Point, Density, and Solubility of Common Inorganic Compounds—cont'd

Compound	Melting point (°C)	Boiling point (°C)	Density (kg/m^3)	Solubility: Cold Water	Solubility: Hot Water	Solubility: Other Solvents
Oxide, di-, SeO_2	sb. 316	—	3950/15	39.4/16	82.5/65	s. al., acet., bz., ac.
Oxychloride, $SeOCl_2$	8.5	176.4	2420/22	d.	d.	s. CS_2, CCl_4, bz., chl.
Selenic acid, H_2SeO_4	60	d. 260	2950	1300/30	∞	s. H_2SO_4; i. NH_3; d. al.
Selenous acid, H_2SeO_3	d. 70/−H_2O	–	3004/15	167	385/90	v. s. al.; i. NH_3
Silicon						
Bromide, $SiBr_4$	5.4	154	2771/25	d.	d.	d. H_2SO_4
Chloride, $SiCl_4$	−69	57.6	1483	d.	d.	d. al.
Fluoride, SiF_4	−90.2	−86	1660/−95	d.	d.	s. al., HF
Hydride, SiH_4	−185	−112	680/−85	i.	–	i. al.; d. KOH
Oxide (quartz), SiO_2	1610	2230	2650	i.	i.	s. HF; v. sl. s. alk.
Silver						
Bromide, AgBr	432	d. >1300	6473/25	8.4×10^{-6}	3.7×10^{-4}	s. KCN, $Na_2S_2O_3$; i. al.
Carbonate, Ag_2CO_3	d. 218	–	6077	0.0032	0.05	s. NH_4OH, Na_2SO_3; i. al.
Chloride, AgCl	455	1550	5560	8.9×10^{-5}/10	0.0021	s. NH_4OH, KCN, $Na_2S_2O_3$
Chromate, Ag_2CrO_4	–	–	5625	0.0026	0.064	s. NH_4OH, KCN
Cyanide, AgCN	d. 320	–	3950	2.3×10^{-5}	–	s. NH_4OH, KCN, HNO_3, $Na_2S_2O_3$
Fluoride, AgF	435	1160	5852/16	172	205	sl. s. NH_4OH
Iodide, AgI (β)	558	1506	6010/15	2.6×10^{-7}/25	2.5×10^{-6}/60	s. KCN; sl. s. NH_4OH
Nitrate, $AgNO_3$	212	d. 444	4352/19	218	733	s. eth.; v. sl. s. al.

Nitrite, $AgNO_2$	d. 140	–	4453/26	0.155/0	1.36/60	s. ac., NH_4OH; i. al.
Oxide, Ag_2O	d. 230	–	7143/17	0.0013	0.0055	s. a., KCN, NH_4OH, al.
Sulfate, Ag_2SO_4	652	d. 1085	5450/29	0.80	1.41	s. a., NH_4OH; i. al.
Sulfide, Ag_2S (argentite)	825	d.	7317	8.4×10^{-15}	–	s. KCN, a.
Sodium						
Acetate, $NaC_2H_3O_2$	324	–	1528	119/0	170	sl. s. al.
Aluminum fluoride, Na_3AlF_6 (cryolite)	1000	–	2900	0.042/25	0.135	d. alk.; i. HCl
Arsenate, $Na_3AsO_4{\cdot}12H_2O$	86.3	–	1760	19/15	–	s. al.
Azide, NaN_3	d.	–	1846	41.7/17	–	sl. s. al.; i. eth.
Borate, tetra-, $Na_2B_4O_7{\cdot}10H_2O$	$-8H_2O/60$	$-10H_2O/$ 320	1730	2.64	39.3	v. sl. s. al.; i. a.
Borate, tetra-, $Na_2B_4O_7$	741	d. 1575	2367	2.64	39.3	i. al.
Borohydride, $NaBH_4$	d. 400	–	1074	55/25	v. s.	s. al., pyr.; i. eth.
Bromide, NaBr	755	1390	3203/25	90.8	118.8	sl. s. al.
Carbonate, $Na_2CO_3{\cdot}10H_2O$	$-H_2O/34$	–	1440/15	22.2	44.7	i. al.
Carbonate, Na_2CO_3	851	d.	2532	22.2	44.7	sl. s. al.; i. acet.
Chlorate, $NaClO_3$	248–261	d.	2409/15	79/0	230	s. al., NH_3
Chloride, NaCl	801	1413	2165/25	35.8	39.4	sl. s. al., NH_3; i. a.
Chromate, $Na_2CrO_4{\cdot}10H_2O$	19.9	–	1483	87.3/30	60	sl. s. a.; i. ac.
Citrate, $Na_3C_6H_5O_7{\cdot}2H_2O$	$-2H_2O/150$	–	–	30.7/25	–	sl. s. al.
Cyanide, NaCN	564	1496	–	48/10	82/35	s. NH_3; sl. s. al.
Dichromate, $Na_2Cr_2O_7{\cdot}2H_2O$	$-2H_2O/100$	d. 400 (anh.)	2520/13	180	433	i. al.
Dihydrogen phosphate, $NaH_2PO_4{\cdot}H_2O$	$-H_2O/100$	d. 204	2040	52/0	370	v. sl. s. eth., chl., tol.

Continued

Table A5 Melting Point, Boiling Point, Density, and Solubility of Common Inorganic Compounds—cont'd

Compound	Melting point (°C)	Boiling point (°C)	Density (kg/m^3)	Solubility: Cold Water	Solubility: Hot Water	Solubility: Other Solvents
Fluoride, NaF	993	1695	2558/41	4.06	5.08	s. HF; v. sl. s. al.
Hydrogen carbonate, $NaHCO_3$	$-CO_2$/270	–	2159	9.5	14.2	sl. s. al.
Hydrogen phosphate, $Na_2HPO_4 \cdot 12H_2O$	$-5H_2O$/35	$-12H_2O$/100	1520	1.6	35/34	i. al.
Hydrogen sulfate, $NaHSO_4$	>315	d.	2435/13	28.7/25	104	sl. s. al.
Hydroxide, NaOH	318	1390	2130	109	347	v. s. al.; i. acet., eth.
Iodide, NaI	651	1304	3667/25	179	302	v. s. al., acet.
Metabisulfite, $Na_2S_2O_5$	d. >150	–	1400	54	81.7	sl. s. al.
Molybdate, Na_2MoO_4	687	–	3280/18	65	84	–
Nitrate, $NaNO_3$	307	d. 380	2261	87	177	v. s. NH_3; sl. s. al.
Nitrite, $NaNO_2$	271	d. 320	2168/0	81.5/15	160	v. s. NH_3; s. al.; sl. s. eth.
Oxalate, $Na_2C_2O_4$	d. 250–270	–	2340	3.4	6.5	i. al., eth.
Oxide, mono-, Na_2O	sb. 1275	–	2270	d.	d.	d. al.
Oxide, per, Na_2O_2	d. 460	–	2805	s.	d.	d. al.; s. dil. a.
Perchlorate, $NaClO_4 \cdot H_2O$	130	d. 482	2020	182/15	248/50	s. al.
Permanganate, $NaMnO_4 \cdot 3H_2O$	d. 170	–	2470	v. s.	v. s.	s. NH_3, d. alk.
Phosphate, $Na_3PO_4 \cdot 12H_2O$	d. 73	–	1620	8.8	68/70	i. CS_2, al.
Sulfate, $Na_2SO_4 \cdot 10H_2O$	32.4	$-10H_2O$/100	1464	19.1	42.2	i. al.
Sulfate, Na_2SO_4	897	–	2680	19.1	42.2	s. H_2SO_4; i. al.
Sulfide, mono-, Na_2S	1170	–	1856/14	15.4/10	57.2/90	sl. s. al.; i. eth.
Sulfite, $Na_2SO_3 \cdot 7H_2O$	$-7H_2O$/150	–	1539/15	27.1	128.3/80	sl. s. al.
Tetrafluoroborate, $NaBF_4$	d. 384	–	2470	108/26	210	sl. s. al., d. H_2SO_4

Thiosulfate, $Na_2S_2O_3 \cdot 5H_2O$	d. 48	$-5H_2O/100$	1729/17	70.0	191.3	s. NH_3; v. sl. s. al.
Tungstate, Na_2WO_4	698	–	4179	73	97	i. al., eth.
Sulfur						
Chloride, mono-, S_2Cl_2	−80	137	1678	d.	d.	s. CS_2, eth., bz.
Chloride, di-, SCl_2	−78	57	1621/15	d.	d.	s. bz., CCl_4; d. al.
Fluoride, hexa-, SF_6	−50.2	–	1880/−50.5	v. sl. s.	sl. s.	s. al., KOH
Hydride, H_2S	−85.49	−60.33	–	4.37/0 (A)	1.86/40 (A)	s. al., eth., CS_2
Oxide, di-, SO_2	−73.2	−10.0	1434/−10	0.23/0 (A)	0.0058/90 (A)	s. al., ac., H_2SO_4
Oxide, tri-, SO_3	16.85	43.3	1970	d.	d.	s. al., H_2SO_4, ac.
Sulfuric acid, H_2SO_4	10.3	330	1841	∞	∞	d. al.
Sulfonyl chloride, SO_2Cl_2	−54.1	69.1	1667	d.	d.	s. ac., bz.
Sulfonyl fluoride, SO_2F_2	−136.7	−55.4	1700	0.05 (A)	–	s. al., CCl_4; sl. s. alk.
Thionyl bromide, $SOBr_2$	−52	138	2680/18	d.	d.	s. bz., chl., CS_2
Thionyl chloride, $SOCl_2$	−104.5	75.7	1679	d.	d.	s. bz., chl.; d. a., al.
Thionyl fluoride, SOF_2	−110.5	−43.8	1780/−100	d.	d.	s. eth., bz., chl., acet.
Tin						
(II) Bromide, $SnBr_2$	215.5	620	5117/17	85.2/0	222.5	s. al., eth., acet.
(IV) Bromide, $SnBr_4$	31	202/734 Torr	3340/35	s. d.	d.	s. acet.
(II) Chloride, $SnCl_2$	246	652	3950/25	270/15	–	s. al., eth., acet., pyr.
(IV) Chloride, $SnCl_4$	−33	114.1	2226	s.	d.	s.CS_2, eth.
(IV) Fluoride, SnF_4	sb. 705	–	4780	v. s.	d.	–
(II) Iodide, SnI_2	320	717	5285	0.98/30	4.03	v. s. NH_4OH, HI aq.
(IV) Iodide, SnI_4	144.5	364.5	4473/0	s.	d.	v. s. CS_2; s. bz., CCl_4
(II) Oxide, SnO	d. 1080	–	6446/0	i.	i.	s. a., alk.; sl. s. NH_4Cl

Continued

Table A5 Melting Point, Boiling Point, Density, and Solubility of Common Inorganic Compounds—cont'd

Compound	Melting point (°C)	Boiling point (°C)	Density (kg/m^3)	Solubility		
				Cold Water	Hot Water	Other Solvents
(IV) Oxide, SnO_2	1630	sb.1800–1900	6950	i.	i.	s. fus. alk.; i. aq. reg.
(II) Sulfate, $SnSO_4$	d. >360	–	–	18.8/19	18.1	s. H_2SO_4
(II) Sulfide, SnS	882	1230	5220/25	$2 \times 10^{-6}/18$	–	d. alk., HCl
(IV) Sulfide, SnS_2	d. 600	–	4500	$2 \times 10^{-4}/18$	–	d. alk.; i. a.
Titanium						
(IV) Bromide, $TiBr_4$	39	230	2600	d.	–	s. al., eth.
(III) Chloride, $TiCl_3$	d. 440	–	2640	s.	s.	v. s. al.; s. HCl; i. eth.
(IV) Chloride, $TiCl_4$	−25	136.4	1726	d.	d.	s. dil. HCl, al.
(IV) Fluoride, TiF_4	sb. 284	–	2798	s. d.	–	s. H_2SO_4, al., pyr.
(II) Oxide, TiO	1750	>3000	4930	–	–	s. dil. H_2SO_4; i. HNO_3
(IV) Oxide, TiO_2 (rutile)	1843	2500–3000	4260	i.	i.	s. H_2SO_4, alk.
Vanadium						
(IV) Chloride, VCl_4	−28	149	1816/30	s. d.	–	s. al., eth., chl.
(IV) Fluoride, VF_4	d. 325	–	2975/23	–	–	s. acet.; sl. s. al., chl.
(IV) Oxide, VO_2	1967	–	4339	i.	i.	s. a., alk.
(V) Oxide, V_2O_5	690	d. 1750	3357/18	0.8	sl. s.	s. a., alk.; i. al.
Vanadyl (V) chloride, $VOCl_3$	−77	126.7	1829	s. d.	–	s. al., eth., ac.
Zinc						
Acetate, Zn $(C_2H_3O_2)_2 \cdot 2H_2O$	$-2H_2O/100$	–	1735	30	44.6	s. al.
Bromide, $ZnBr_2$	394	650	420/25	447	675	v. s. al., eth., acet.

Carbonate, $ZnCO_3$	$-CO_2/300$	–	4398	0.001/15	–	s. a., alk.; i. acet.
Chloride, $ZnCl_2$	283	732	2910/25	368	615	v. s. al., eth.
Cyanide, $Zn(CN)_2$	d. 800	–	1852	5×10^{-4}	–	s. alk., KCN; i. al.
Fluoride, ZnF_2	872	1500	4950/25	1.62	s.	s. h. a.; i. al.
Iodide, ZnI_2	446	d. 624	4736/25	432/18	511	s. a., al., eth.
Nitrate, $Zn(NO_3)_2 \cdot 6H_2O$	36.4	$6H_2O$/105–131	2065/14	139/30	900/70	v. s. al.
Oxide, ZnO	1975	–	5606	$1.6 \times 10^{-4}/29$	–	s. a., alk; i. al.
Sulfate, $ZnSO_4 \cdot 7H_2O$	100	$-7H_2O/280$	1957/25	59	60.5	sl. s. al.
Sulfide, ZnS (wurtzite)	sb. 1185	–	3980	$6.9 \times 10^{-4}/18$	v. sl. s.	v. s. a.; i. ac.

a., acid; ac., acetic acid; acet., acetone; anh., anhydrous; aq., aqueous; al., alcohol (ethanol); alk., alkali; bz., benzene; c., cold; chl., chloroform; conc., concentrated; d., decomposes; dil., dilute; eth., ether; exp., explodes; fus., fused; h., hot; i., insoluble; me., methanol; pyr., pyridine; s., soluble; sb., sublimes; sl., slightly; slt., salts; tol., toluene; v., very; vac., vacuum; ∞, miscible in all proportions.

Solubility is the solubility in cold water (20°C) or hot water (100°C) unless stated otherwise.

Table A6 Common Techniques for the Characterization of Inorganic Compounds

Physical characteristics	
Melting/boiling point	Inorganic compounds are often ionic and so have very high melting points. While some inorganic compounds are solids with accessible melting points and some are liquids with reasonable boiling points, there are not the exhaustive tabulations of melting/boiling point data for inorganic compounds that exist for organics. In general, melting and boiling points are not useful in identifying an inorganic compound, but they can be used to assess its purity, if they are accessible
Color	Inorganic compounds, in contrast to many organic compounds, are very colorful. Unfortunately, color alone is not reliable indicator of a compound's identity, but it is useful when one is following a separation on a column
Crystal shapes	Information about the arrangement of the particles in the solid from the shape of a well-formed crystal can be obtained when observing the visual changes in the crystal when it is rotated under a polarizing microscope
Elemental analysis	Elemental analysis is one of the most useful methods available to characterize a compound
Mass spectroscopy	In inorganic chemistry, this is most often used to determine the molar mass of a chemical. When mass spectroscopy data are combined with an elemental analysis, the chemical formula of the substance can be determined. Analysis of a compound's fragmentation pattern can be used to gain structural information
Chromatography	This technique is most often used to separate a product from a complex reaction mixture. If a known sample of the compound is available, it can be identified in a reaction mixture by spiking the analyte with the known. Most chromatography in inorganic chemistry is on solutions (e.g., column chromatography and HPLC, both normal and reverse phase), because the high boiling point of many inorganic compounds precludes GC analysis

Table A6 Common Techniques for the Characterization of Inorganic Compounds—cont'd

Spectroscopic/structural methods	
UV-Vis spectroscopy	The absorption bands in the UV-Vis and near IR can be used to determine the general type of atom bound to a metal and the geometry about the metal
Circular dichroism (CD)	CD spectroscopy measures the degree to which a sample rotates circularly polarized light. CD is similar to absorption spectroscopy, but because of different selection rules for the electronic transitions, peak intensities and widths may be different between CD and absorption (the energy of a peak must still, however, be the same)
IR spectroscopy	In simple compounds, the number, energy, and intensity of the IR transitions are directly related to the geometry of compound and to which atoms are bound to which other atoms. For complex compounds involving large organic moieties, the IR becomes more difficult to interpret. IR is useful to determine the presence of complex counter ions like PF_6^-, ClO_4^-, and BF_4^-, because they have distinctive absorptions in the IR
Raman spectroscopy	Raman spectroscopy is complementary to IR absorption spectroscopy. By considering what peaks are present or absent, in the two spectra of a compound, the molecular geometry can be determined. Raman is also useful because it can, depending on instrument design, scan to very low frequencies ($\sim$100 cm^{-1}) and thus observe transitions too low for IR absorption
NMR spectroscopy	The nuclear magnetic resonance (NMR) spectra of inorganic chemicals are often more complicated than organic chemicals because other nuclei also have nuclear magnetic moments. So, in addition to the familiar ^{1}H ($I=1/2$) and ^{13}C ($I=1/2$), there are $\sim$90 other elements that have at least one NMR-active nucleus

Continued

Table A6 Common Techniques for the Characterization of Inorganic Compounds—cont'd

Electron paramagnetic resonance (EPR or ESR)	EPR is for paramagnetic compounds with an odd number of unpaired electrons (EPR can be done when there are an even number of unpaired electrons, but it is a much harder experiment). In many ways, EPR and NMR are the same, and there is even a technique that combines them (electron-nuclear double resonance, ENDOR, spectroscopy)
X-ray diffraction	Single-crystal X-ray diffraction is the most powerful X-ray technique for inorganic chemists. From precise measurement of the intensity and angles at which an X-ray beam diffracts off a crystal, the arrangement of the atoms can be reconstructed. Some inorganic compounds (e.g., rocks and minerals) cannot be obtained as single crystals. In these cases, the X-ray powder diffraction method can be used to obtain the dimensions of the unit cell for use in identification (there is a large, indexed catalog of lattice constants for many minerals)

Table A7 Common Physical Constants Used in Inorganic Chemistry[a]

Acceleration due to gravity, g
$9.80665\ m/s^2$
Atomic mass unit, amu
1.66054×10^{-27} kg
Avogadro number, N_A
6.0221367×10^{23} particles/mol
Bohr magneton, μ_B
$\mu_B = eh/(4\pi m_e)$
$9.2740154 \times 10^{-24}$ J/T
Bohr radius, a_0
5.29177×10^{-11} m
Boltzmann constant, k_B
1.380658×10^{-23} J/K ($=8.617385 \times 10^{-5}$ eV/K)
Electron radius, r_e
2.81792×10^{-15} m
Faraday constant, F
$F = N_A e$
9.64846×10^4 C/mol
Gas constant, R
$R = N_A k_B$
$8.31451\ m^2 kg/s^2$ K mol
Molar volume, V_{mol}
$22.41383\ m^3/kmol$
Permeability of a vacuum, μ_0
$\mu_0 = 4\pi \times 10^{-7}\ T^2\ m^3/J$
$12.566370614 \times 10^{-7}\ T^2\ m^3/J$
Planck constant, h
$6.6260755 \times 10^{-34}$ J s
$h/(2\pi) = 1.05457266 \times 10^{-34}$ J s
Speed of light, c
2.99792458×10^8 m/s

[a]Listed alphabetically.

CONVERSION FACTORS

1 AREA

1 square centimeter (1 cm^2) = 0.1550 square inches
1 square meter 1 (m^2) = 1.1960 square yards
1 hectare = 2.4711 acres
1 square kilometer (1 km^2) = 0.3861 square miles
1 square inch (1 $inch^2$) = 6.4516 square centimeters
1 square foot (1 ft^2) = 0.0929 square meters
1 square yard (1 yd^2) = 0.8361 square meters
1 acre = 4046.9 square meters
1 square mile (1 mi^2) = 2.59 square kilometers

2 CONCENTRATION CONVERSIONS

1 part per million (1 ppm) = 1 microgram per liter (1 μg/L)
1 microgram per liter (1 μg/L) = 1 milligram per kilogram (1 mg/kg)
1 microgram per liter (μg/L) $\times 6.243 \times 10^8$ = 1 lb per cubic foot (1 lb/ft^3)
1 microgram per liter (1 μg/L) $\times 10^{-3}$ = 1 milligram per liter (1 mg/L)
1 milligram per liter (1 mg/L) $\times 6.243 \times 10^5$ = 1 pound per cubic foot (1 lb/ft^3)
1 gram mole per cubic meter (1 g mol/m^3) $\times 6.243 \times 10^5$ = 1 pound per cubic foot (1 lb/ft^3)
10,000 ppm = 1%, w/w
1 ppm hydrocarbon in soil × 0.002 = 1 lb of hydrocarbons per ton of contaminated soil

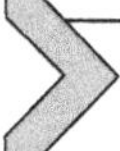

3 NUTRIENT CONVERSION FACTOR

1 pound, phosphorus × 2.3 (1 lb P × 2.3) = 1 pound, phosphorous pentoxide (1 lb P_2O_5)
1 pound, potassium × 1.2 (1 lb K × 1.2 = 1 pound, potassium oxide (1 lb K_2O)

4 TEMPERATURE CONVERSIONS

°F = (°C × 1.8) + 32
°C = (°F − 32)/1.8
(°F − 32) × 0.555 = °C
Absolute zero = −273.15°C
Absolute zero = −459.67°F

5 SLUDGE CONVERSIONS

1700 lb wet sludge = 1 yd^3 wet sludge
1 yd^3 sludge = wet tons/0.85
Wet tons sludge × 240 = gallons sludge
1 wet ton sludge × % dry solids/100 = 1 dry ton of sludge

6 VARIOUS CONSTANTS

Atomic mass	(mu) = $1.6605402 \times 10^{-27}$
Avogadro's number	(N) = 6.0221367×10^{23} mol^{-1}
Boltzmann's constant	(k) = 1.380658×10^{-23} J K^{-1}
Elementary charge	(e) = $1.60217733 \times 10^{-19}$C
Faraday's constant	(F) = 9.6485309 × 104C mol^{-1}
Gas (molar) constant	(R) = $k\ N\tilde{}8.314510$ J mol^{-1} K-1 = 0.08205783 L atm mol^{-1} K^{-1}
Gravitational acceleration	(g) = 9.80665 m s^{-2}
Molar volume of an ideal gas at 1 atm and 25°C	($V_{\text{ideal gas}}$) = 24.465 L mol^{-1}
Planck's constant	(h) = $6.6260755 \times 10^{-34}$ J s
Zero, Celsius scale	(0°C) = 273.15°K

7 VOLUME CONVERSION

Barrels (petroleum, US) to Cu feet multiply by 5.6146
Barrels (petroleum, US) to Gallons (US) multiply by 42
Barrels (petroleum, US) to Liters multiply by 158.98
Barrels (US, liq.) to Cu feet multiply by 4.2109

Barrels (US, liq.) to Cu inches multiply by 7.2765 × 103
Barrels (US, liq.) to Cu meters multiply by 0.1192
Barrels (US, liq.) to Gallons multiply by (US, liq.) 31.5
Barrels (US, liq.) to Liters multiply by 119.24
Cubic centimeters to Cu feet multiply by 3.5315×10^{-5}
Cubic centimeters to Cu inches multiply by 0.06102
Cubic centimeters to Cu meters multiply by 1.0×10^{-6}
Cubic centimeters to Cu yards multiply by 1.308×10^{-6}
Cubic centimeters to Gallons (US liq.) multiply by 2.642×10^{-4}
Cubic centimeters to Quarts (US liq.) multiply by 1.0567×10^{-3}
Cubic feet to Cu centimeters multiply by 2.8317×10^{4}
Cubic feet to Cu meters multiply by 0.028317
Cubic feet to Gallons (US liq.) multiply by 7.4805
Cubic feet to Liters multiply by 28.317
Cubic inches to Cu cm multiply by 16.387
Cubic inches to Cu feet multiply by 5.787×10^{-4}
Cubic inches to Cu meters multiply by 1.6387×10^{-5}
Cubic inches to Cu yards multiply by 2.1433×10^{-5}
Cubic inches to Gallons (US liq.) multiply by 4.329×10^{-3}
Cubic inches to Liters multiply by 0.01639
Cubic inches to Quarts (US liq.) multiply by 0.01732
Cubic meters to Barrels (US liq.) multiply by 8.3864
Cubic meters to Cu cm multiply by 1.0×10^{6}
Cubic meters to Cu feet multiply by 35.315
Cubic meters to Cu inches multiply by 6.1024×10^{4}
Cubic meters to Cu yards multiply by 1.308
Cubic meters to Gallons (US liq.) multiply by 264.17
Cubic meters to Liters multiply by 1000
Cubic yards to Bushels (Brit.) multiply by 21.022
Cubic yards to Bushels (US) multiply by 21.696
Cubic yards to Cu cm multiply by 7.6455 × 105
Cubic yards to Cu feet multiply by 27
Cubic yards to Cu inches multiply by 4.6656×10^{4}
Cubic yards to Cu meters multiply by 0.76455
Cubic yards to Gallons multiply by 168.18
Cubic yards to Gallons multiply by 173.57
Cubic yards to Gallons multiply by 201.97
Cubic yards to Liters multiply by 764.55
Cubic yards to Quarts multiply by 672.71

Cubic yards to Quarts multiply by 694.28
Cubic yards to Quarts multiply by 807.90
Gallons (US liq.) to Barrels (US liq.) multiply by 0.03175
Gallons (US liq.) to Barrels (petroleum, US) multiply by 0.02381
Gallons (US liq.) to Bushels (US) multiply by 0.10742
Gallons (US liq.) to Cu centimeters multiply by 3.7854×10^3
Gallons (US liq.) to Cu feet multiply by 0.13368
Gallons (US liq.) to Cu inches multiply by 231
Gallons (US liq.) to Cu meters multiply by 3.7854×10^{-3}
Gallons (US liq.) to Cu yards multiply by 4.951×10^{-3}
Gallons (US liq.) to Gallons (wine) multiply by 1.0
Gallons (US liq.) to Liters multiply by 3.7854
Gallons (US liq.) to Ounces (US fluid) multiply by 128.0
Gallons (US liq.) to Pints (US liq.) multiply by 8.0
Gallons (US liq.) to Quarts (US liq.) multiply by 4.0
Liters to Cu centimeters multiply by 1000
Liters to Cu feet multiply by 0.035315
Liters to Cu inches multiply by 61.024
Liters to Cu meters multiply by 0.001
Liters to Gallons (US liq.) multiply by 0.2642
Liters to Ounces (US fluid) multiply by 33.814

8 WEIGHT CONVERSION

1 ounce (1 ounce) = 28.3495 grams (18.2495 g)
1 pound (1 lb) = 0.454 kilogram
1 pound (1 lb) = 454 grams (454 g)
1 kilogram (1 kg) = 2.20462 pounds (2.20462 lb)
1 stone (English) = 14 pounds (14 lb)
1 ton (US; 1 short ton) = 2000 lb
1 ton (English; 1 long ton) = 2240 lb
1 metric ton = 2204.62262 pounds
1 tonne = 2204.62262 pounds

9 OTHER APPROXIMATIONS

14.7 pounds per square inch (14.7 psi) = 1 atmosphere (1 atm)

1 kilopascal (kPa) $\times 9.8692 \times 10^{-3} = 14.7$ pounds per square inch (14.7 psi)
1 $yd^3 = 27$ ft^3
1 US gallon of water = 8.34 lb
1 imperial gallon of water = 10 lb
1 $ft^3 = 7.5$ gallon = 1728 cubic inches = 62.5 lb
1 $yd^3 = 0.765$ m^3
1 acre-inch of liquid = 27150 gallons = 3.630 ft^3
1-foot depth in 1 acre (in-situ) = 1613 × (20%–25% excavation factor) = ~2000 yd^3
1 yd^3 (clayey soils-excavated) = 1.1–1.2 tons (US)
1 yd^3 (sandy soils-excavated) = 1.2–1.3 tons (US)
Pressure of a column of water in psi = height of the column in feet by 0.434.

GLOSSARY

A

Abiotic Not associated with living organisms; synonymous with *abiological.*

Abiotic transformation The process in which a substance in the environment is modified by nonbiological mechanisms.

Absorption The penetration of atoms, ions, or molecules into the bulk mass of a substance.

Abyssal zone The portion of the ocean floor below 3281–6561 ft where light does not penetrate and where temperatures are cold and pressures are intense; this zone lies seaward of the continental slope and covers approximately 75% of the ocean floor; the temperature does not rise above 4°C (39°F); since oxygen is present, a diverse community of invertebrates and fishes do exist, and some have adapted to harsh environments such as hydrothermal vents of volcanic creation.

Acid A chemical capable of donating a positively charged hydrogen atom (proton, $H+$) or capable of forming a covalent bond with an electron pair; an acid increases the hydrogen ion concentration in a solution, and it can react with certain metals.

Acid anhydride An organic compound that reacts with water to form an acid.

Acidity The capacity of the water to neutralize OH^-.

Acidophiles Metabolically active in highly acidic environments and often have a high heavy metal resistance.

Acids, bases, and salts Many inorganic compounds are available as acids, bases, or salts.

Acyclic A compound with straight or branched carbon-carbon linkages but without cyclic (ring) structures.

Adhesion The degree to which oil will coat a surface, expressed as the mass of oil adhering per unit area. A test has been developed for a standard surface that gives a semiquantitative measure of this property.

Adsorption The retention of atoms, ions, or molecules on to the surface of another substance.

Aerobe An organism that needs oxygen for respiration and hence for growth.

Aerobic In the presence of or requiring oxygen; an environment or process that sustains biological life and growth or occurs only when free (molecular) oxygen is present.

Aerobic bacteria Any bacteria requiring free oxygen for growth and cell division.

Aerobic conditions Conditions for growth or metabolism in which the organism is sufficiently supplied with oxygen.

Aerobic respiration The process whereby microorganisms use oxygen as an electron acceptor.

Aerosol A colloidal-sized atmospheric particle.

Alcohol An organic compound with a carbon bound to a hydroxyl (—OH) group; a hydroxyl group attached to an aromatic ring is called a phenol rather than an alcohol; a compound in which a hydroxy group (—OH) is attached to a saturated carbon atom (e.g., ethyl alcohol, C_2H_5OH).

Aldehyde An organic compound with a carbon bound to a —(C=O)—H group; a compound in which a carbonyl group is bonded to one hydrogen atom and to one alkyl group (RC(=O)H).

Algae Microscopic organisms that subsist on inorganic nutrients and produce organic matter from carbon dioxide by photosynthesis.

Aliphatic compound Any organic compound of hydrogen and carbon characterized by a linear chain or branched chain of carbon atoms; three subgroups of such compounds are alkanes, alkenes, and alkynes.

Alkalinity The capacity of water to accept H^+ ions (protons).

Alkaliphiles Organisms that have their optimum growth rate at least 2 pH units above neutrality.

Alkalitolerants Organisms that are able to grow or survive at pH values above 9 but their optimum growth rate is around neutrality or less.

Alkane (paraffin) A group of *hydrocarbons* composed of only carbon and hydrogen with no double bonds or aromaticity. They are said to be "saturated" with hydrogen. They may by straight chain (normal), branched, or cyclic. The smallest alkane is methane (CH_4); the next, ethane (CH_3CH_3); then propane ($CH_3CH_2CH_3$); and so on.

Alkanes The homologous group of linear (acyclic) aliphatic hydrocarbons having the general formula C_nH_{2n+2}; alkanes can be straight chains (linear), branched chains, or ring structures; often referred to as paraffins.

Alkene (olefin) An unsaturated *hydrocarbon*, containing only hydrogen and carbon with one or more double bonds but having no aromaticity. *Alkenes* are not typically found in crude oils but can occur as a result of heating.

Alkenes Acyclic branched or unbranched hydrocarbons having one carbon-carbon double bond (—C=C—) and the general formula C_nH_{2n}; often referred to as olefins.

Aliphatic compounds A broad category of hydrocarbon compounds distinguished by a straight or branched open-chain arrangement of the constituent carbon atoms, excluding aromatic compounds; the carbon-carbon bonds may be either single or multiple bonds—alkanes, alkenes, and alkynes are aliphatic hydrocarbons.

Alkoxide An ionic compound formed by removal of hydrogen ions from the hydroxyl group in an alcohol using a reactive metal such as sodium or potassium.

Alkyl A molecular fragment derived from an alkane by dropping a hydrogen atom from the formula; examples are methyl (CH_3) and ethyl (CH_2CH_3).

Alkyne A compound that consists of only carbon and hydrogen that contains at least one carbon-carbon triple bond; alkyne names end with *yne*.

Alkyl groups A *hydrocarbon* functional group (C_nH_{2n+1}) obtained by dropping one hydrogen from a fully saturated compound, for example, methyl (—CH_3), ethyl (—CH_2CH_3), propyl (—$CH_2CH_2CH_3$), or isopropyl (($CH_3)_2CH$—).

Alkyl radicals Carbon-centered radicals derived formally by removal of one hydrogen atom from an alkane, for example, the ethyl radical (CH_3CH_2).

Alkynes The group of acyclic branched or unbranched hydrocarbons having a carbon-carbon triple bond (—C≡C—).

Ambient The surrounding environment and prevailing conditions.

Amide An organic compound that contains a carbonyl group bound to nitrogen; the simplest amides are formamide ($HCONH_2$) and acetamide (CH_3CONH_2).

Amine An organic compound that contains a nitrogen atom bound only to carbon and possibly hydrogen atoms; examples are methylamine, CH_3NH_2; dimethylamine, CH_3NHCH_3; and trimethylamine, $(CH_3)_3N$.

Amino acid A molecule that contains at least one amine group (—NH_2) and at least one carboxylic acid group (—COOH); when these groups are both attached to the same carbon, the acid is an α-amino acid—α-amino acids are the basic building blocks of proteins.

Amorphous solid A noncrystalline solid having no well-defined ordered structure.

Amphoteric molecule A molecule that behaves both as an acid and as a base, such as hydroxy pyridine.

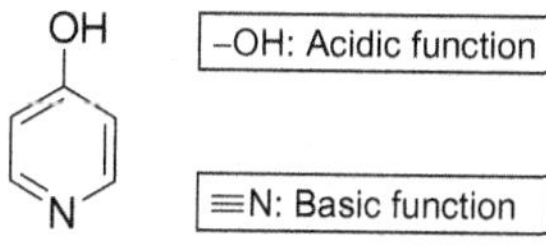

Anaerobe An organism that does not need free-form oxygen for growth. Many anaerobes are even sensitive to free oxygen.

Anaerobic A biologically mediated process or condition not requiring molecular or free oxygen; relating to a process that occurs with little or no oxygen present.

Anaerobic bacteria Any bacteria that can grow and divide in the partial or complete absence of oxygen.

Anaerobic respiration The process whereby microorganisms use a chemical other than oxygen as an electron acceptor; common substitutes for oxygen are nitrate, sulfate, and iron.

Analyte The component of a system to be analyzed—for example, chemical elements or ions in groundwater sample.

Anion An atom or molecule that has a negative charge; a negatively charged ion.

Anoxic An environment without oxygen.

Aphotic zone The deeper part of the ocean beneath the photic zone, where light does not penetrate sufficiently for photosynthesis to occur.

API gravity An American Petroleum Institute measure of *density* for petroleum:

$$\text{API gravity} = [141.5/(\text{specific gravity at } 15.6\,°\text{C}) - 131.5]$$

Freshwater has a gravity of 10 degrees API. The scale is commercially important for ranking oil quality; heavy oils are typically <20 degrees API; medium oils are 20–35 degrees API; light oils are 35–45 degrees API.

Aquasphere The water areas of the Earth, also called the hydrosphere.

Aquatic chemistry The branch of environmental chemistry that deals with chemical phenomena in water.

Aquifer A water-bearing layer of soil, sand, gravel, rock, or other geologic formations that will yield usable quantities of water to a well under normal hydraulic gradients or by pumping.

Arene A hydrocarbon that contains at least one aromatic ring.

Aromatic Organic cyclic compounds that contain one or more benzene rings; these can be monocyclic, bicyclic, or polycyclic hydrocarbons and their substituted derivatives. In aromatic ring structures, every ring carbon atom possesses one double bond.

Aromatic ring An exceptionally stable planar ring of atoms with resonance structures that consist of alternating double and single bonds, such as benzene.

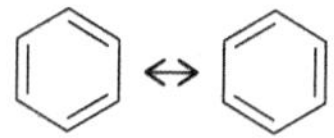

Aromatic compound A compound containing an aromatic ring; aromatic compounds have strong, characteristic odors.

Aryl A molecular fragment or group attached to a molecule by an atom that is on an aromatic ring.

Asphaltene fraction A complex mixture of heavy organic compounds precipitated from crude oil and *bitumen* by natural processes or in laboratory by addition of excess *n*-pentane or *n*-heptane; after precipitation of the *asphaltene fraction*, the remaining oil or *bitumen* consists of *saturates*, *aromatics*, and *resins*.

Assay Qualitative or (more usually) quantitative determination of the components of a material or system.

Association colloids Colloids that consist of special aggregates of ions and molecules (micelles).

Asymmetrical carbon A carbon atom covalently bonded to four different atoms or groups of atoms.

Atmosphere The thin layer of gases that covers the surface of the Earth; composed of two major components, nitrogen 78.08% and oxygen 20.955% with smaller amounts of argon 0.934%, carbon dioxide 0.035%, neon 1.818×10^{-3}%, krypton 1.14×10^{-4}%, helium 5.24×10^{-4}%, and xenon 8.7×10^{-6}%; and may also contain 0.1%–5% water by volume, with a normal range of 1%–3%; the reservoir of gases moderates the temperature of the Earth, absorbs energy and damaging ultraviolet radiation from the sun, transports energy away from equatorial regions, and serves as a pathway for vapor-phase movement of water in the hydrologic cycle.

Atomic number The atomic number is equal to the number of positively charged protons in the nucleus of an atom that determines the identity of the element.

Atomic radius The relative size of an atom; among the main group of elements, atomic radii mostly decrease from left to right across rows in the periodic table; metal ions are smaller than their neutral atoms, and nonmetallic anions are larger than the atoms from which they are formed; atomic radii are expressed in angstrom units of length (Å).

ATSDR Agency for Toxic Substances and Disease Registry.

Attenuation The set of human-made or natural processes that either reduce or appear to reduce the amount of a chemical compound as it migrates away or is disposed from one specific point toward another point in space or time, for example, the apparent reduction in the amount of a chemical in a groundwater plume as it migrates away from its source; degradation, dilution, dispersion, sorption, or volatilization are common processes of attenuation.

Autoignition temperature (AIT) A fixed temperature above which a flammable mixture is capable of extracting sufficient energy from the environment to self-ignite.

Autotrophs Organisms or chemicals that use carbon dioxide and ionic carbonates for the C that they require.

Avogadro's number The number of molecules (6.023×10^{23}) in 1 g mol of a substance.

B

Bacteria Single-celled prokaryotic microorganisms that may be shaped as rods (bacillus), spheres (coccus), or spirals (vibrios, spirilla, and spirochetes).

Benthic zone The ecological region at the lowest level of a body of water such as an ocean or a lake, including the sediment surface and some subsurface layers; organisms living in this zone (benthos or benthic organisms) generally live in close relationship with the substrate bottom; many such organisms are permanently attached to the bottom; because light does not penetrate very deep ocean water, the energy source for the benthic ecosystem is often organic matter from higher up in the water column that sinks to the depths.

Benzene A colorless liquid formed from both anthropogenic activities and natural processes; widely used in the United States and ranks in the top 20 chemicals used; a natural part of crude oil, gasoline, and cigarette smoke; one of the major components of JP-8 fuel.

Bimolecular reaction The collision and combination of two reactants involved in the rate-limiting step.

Bioaccumulation The accumulation of substances, such as pesticides, or other chemicals in an organism; occurs when an organism absorbs a chemical—possibly a toxic chemical—at a rate faster than that at which the substance is lost by catabolism and excretion; the longer the biological half-life of a toxic substance, the greater the risk of chronic poisoning, even if environmental levels of the toxin are not very high; see Biomagnification.

Bioaugmentation A process in which acclimated microorganisms are added to soil and groundwater to increase biological activity. Spray irrigation is typically used for shallow contaminated soils, and injection wells are used for deeper contaminated soils.

Biochemical oxygen demand (BOD) An important water quality parameter; refers to the amount of oxygen utilized when the organic matter in a given volume of water is degraded biologically.

Biodegradation The natural process whereby bacteria or other microorganisms chemically alter and break down organic molecules; the breakdown or transformation of a chemical substance or substances by microorganisms using the substance as a carbon and/or energy source.

Bioinorganic compounds Natural and synthetic compounds that include metallic elements bonded to proteins and other biological chemistries.

Biological marker (biomarker) Complex organic compounds composed of carbon, hydrogen, and other elements that are found in oil, *bitumen*, rocks, and sediments and which have undergone little or no change in structure from their parent organic molecules in living organisms; typically, biomarkers are isoprenoids, composed of isoprene subunits; biomarkers include compounds such as pristane, phytane, triterpane derivatives, steranes derivatives, and porphyrin derivatives.

Biomagnification The increase in the concentration of heavy metals (i.e., mercury) or organic contaminants such as chlorinated hydrocarbons in organisms as a result of their consumption within a food chain/web; an example is the process by which contaminants such as polychlorobiphenyl derivatives (PCBs) accumulate or magnify as they move up the food chain—PCBs concentrate in tissue and internal organs—and as big fish eat little fish, they accumulate all the PCBs that have been eaten by everyone below them in the food chain; can occur as a result of (1) persistence, in which the chemical cannot be broken down by environmental processes; (2) food chain energetics, in which the concentration of the chemical increases progressively as it moves up a food chain; and (3) a low or nonexistent rate of internal degradation or excretion of the substance that is often due to water insolubility.

Bioremediation A treatment technology that uses biological activity to reduce the concentration or toxicity of contaminants; materials are added to contaminated environments to accelerate natural biodegradation.

Biosphere A term representing all of the living entities on the Earth.

Biota Living organisms.

Bitumen A complex mixture of *hydrocarbonaceous constituents* of natural or pyrogenous origin or a combination of both.

Boiling liquid expanding vapor explosion (BLEVE) An event that occurs when a vessel ruptures, which contains a liquid at a temperature above its atmospheric pressure boiling point; the explosive vaporization of a large fraction of the vessel contents, possibly followed by the combustion or explosion of the vaporized cloud if it is combustible (similar to a rocket).

Boiling point The temperature at which a liquid begins to boil—that is, it is the temperature at which the vapor pressure of a liquid is equal to the atmospheric or external pressure. The boiling point distributions of crude oils and petroleum products may be in a range from 30°C to in excess of 700°C (86–1290°F).

Breakdown product A compound derived by chemical, biological, or physical action on a chemical compound; the breakdown is a process that may result in a more toxic or a less toxic compound and a more persistent or less persistent compound than the original compound.

BTEX The collective name given to benzene, toluene, ethylbenzene, and the xylene isomers (*p*-, *m*-, and *o*-xylene); a group of volatile organic compounds (VOCs) found in petroleum hydrocarbons, such as gasoline, and other common environmental contaminants.

BTX The collective name given to benzene, toluene, and the xylene isomers (*p*-, *m*-, and *o*-xylene); a group of volatile organic compounds (VOCs) found in petroleum hydrocarbons, such as gasoline, and other common environmental contaminants.

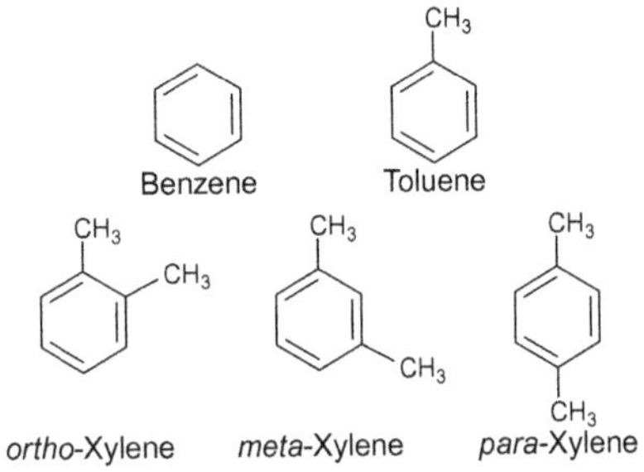

C

Carbenium ion A generic name for carbocation that has at least one important contributing structure containing a tervalent carbon atom with a vacant *p* orbital.

Carbanion The generic name for anions containing an even number of electrons and having an unshared pair of electrons on a carbon atom (e.g., Cl_3C^-).

Carbon Element number 6 in the periodic table of elements.

Carbon preference index (CPI) The ratio of odd to even *n*-alkanes; odd/even CPI *alkanes* are equally abundant in petroleum but not in biological material—a CPI near 1 is an indication of petroleum.

Carbon tetrachloride A manufactured compound that does not occur naturally; produced in large quantities to make refrigeration fluid and propellants for aerosol cans; in the past, carbon tetrachloride was widely used as a cleaning fluid, in industry and dry cleaning businesses, and in the household; also used in fire extinguishers and as a fumigant to kill insects in grain—these uses were stopped in the mid-1960s.

Carbonyl group A divalent group consisting of a carbon atom with a double bond to oxygen; for example, acetone (CH_3—(C=O)—CH_3) is a carbonyl group linking two methyl groups.

Carboxylic acid An organic molecule with a —CO_2H group; hydrogen atom on the —CO_2H group ionizes in water; the simplest carboxylic acids are formic acid (H-COOH) and acetic acid (CH_3-COOH).

Catabolism The breakdown of complex molecules into simpler ones through the oxidation of organic substrates to *provide* biologically available energy—adenosine triphosphate (ATP) is an example of such a molecule.

Catalysis The process where a catalyst increases the rate of a chemical reaction without modifying the overall standard Gibbs energy change in the reaction.

Catalyst A substance that alters the rate of a chemical reaction and may be recovered essentially unaltered in form or amount at the end of the reaction.

Cation exchange The interchange between a cation in solution and another cation in the boundary layer between the solution and the surface of negatively charged material such as clay or organic matter.

Cation-exchange capacity (CEC) The sum of the exchangeable bases plus total soil acidity at a specific pH, usually 7.0 or 8.0. When acidity is expressed as salt extractable acidity, the cation-exchange capacity is called the effective cation-exchange capacity (ECEC), because this is considered to be the CEC of the exchanger at the native pH value; usually expressed in centimoles of charge per kilogram of exchanger (cmol/kg) or millimoles of charge per kilogram of exchanger.

Cellulose A polysaccharide, polymer of glucose, that is found in the cell walls of plants; a fiber that is used in many commercial products, notably paper.

CERCLA Comprehensive Environmental Response, Compensation, and Liability Act. This law created a tax on the chemical and petroleum industries and provided broad federal authority to respond directly to releases or threatened releases of hazardous substances that may endanger public health or the environment.

Chain reaction A reaction in which one or more reactive reaction intermediates (frequently radicals) are continuously regenerated, usually through a repetitive cycle of elementary steps (the *propagation step*); for example, in the chlorination of methane by a radical mechanism, Cl is continuously regenerated in the chain propagation steps:

$$Cl\cdot + CH_4 \rightarrow HCl + H_3C\cdot$$
$$H_3C\cdot + Cl_2 \rightarrow CH_3Cl + Cl\cdot$$

In chain polymerization reactions, reactive intermediates of the same types, generated in successive steps or cycles of steps, differ in relative molecular mass.

Check standard An analyte with a well-characterized property of interest, for example, concentration, density, and other properties that are used to verify method, instrument, and operator performance during regular operation; *check standards* may be obtained from a certified supplier, may be a pure substance with properties obtained from the literature or may be developed in-house.

Chemical bond The forces acting among two atoms or groups of atoms that lead to the formation of an aggregate with sufficient stability to be considered as an independent molecular species.

Chemical dispersion In relation to oil spills, this term refers to the creation of oil-in-water *emulsions* by the use of chemical dispersants made for this purpose.

Chemical induction (coupling) When one reaction accelerates another in a chemical system, there is said to be a chemical induction or coupling. Coupling is caused by an intermediate or a by-product of the inducing reaction that participates in a second reaction; chemical induction is often *observed* in oxidation-reduction reactions.

Chemical reaction A process that results in the interconversion of chemical species.

Chemical species An ensemble of chemically identical molecular entities that can explore the same set of molecular energy levels on the timescale of the experiment; the term is applied equally to a set of chemically identical atomic or molecular structural units in a solid array.

Chemical waste Any solid, liquid, or gaseous waste material that, if improperly managed or disposed of, may pose substantial hazards to human health and the environment.

Chemical weight The weight of a molar sample as determined by the weight of the molecules (the molecular weight); calculated from the weights of the atoms in the molecule.

Chemistry The science that studies matter and all of the possible transformations of matter.

Chemotrophs Organisms or chemicals that use chemical energy derived from oxidation-reduction reactions for their energy needs.

Chlorinated solvent A volatile organic compound containing chlorine; common solvents are trichloroethylene, tetrachloroethylene, and carbon tetrachloride.

Chlorofluorocarbon Gases formed of chlorine, fluorine, and carbon whose molecules normally do not react with other substances; formerly used as spray-can propellants, they are known to destroy the protective ozone layer of the Earth.

Chromatography A method of chemical analysis where compounds are separated by passing a mixture in a suitable carrier over an absorbent material; compounds with different absorption coefficients move at different rates and are separated.

***Cis-trans* isomers** The difference in the positions of atoms (or groups of atoms) relative to a reference plane in an organic molecule; in a *cis* isomer, the atoms are on the same side of the molecule, but are on opposite sides in the *trans* isomer; sometimes called stereoisomers; these arrangements are common in alkenes and cycloalkanes.

Clay A very fine-grained soil that is plastic when wet but hard when fired; typical clay minerals consist of silicate and aluminosilicate minerals that are the products of weathering reactions of other minerals; the term is also used to refer to any mineral of very small particle size.

Clean Water Act The Clean Water Act establishes the basic structure for regulating discharges of pollutants into the waters of the United States. It gives EPA the authority to implement pollution control programs such as setting wastewater standards for industry; also continued requirements to set water quality standards for all contaminants in surface waters and makes it unlawful for any person to discharge any pollutant from a point source into navigable waters, unless a permit was obtained under its provisions.

Cluster compounds Ensembles of bound atoms; typically larger than a molecule yet more defined than a bulk solid.

Coefficient of linear thermal expansion The ratio of the change in length per degree Celcius to the length at 0°C.

Cofferdam (also called a *coffer*) A temporary enclosure built within or in pairs across a body of water and constructed to allow the enclosed area to be pumped out.

Coke A hard, dry substance containing carbon that is produced by heating bituminous coal or other carbonaceous materials to a very high temperature in the absence of air; used as a fuel.

Colloidal particles Particles that have some characteristics of both species in solution and larger particles in suspension, which range in diameter form about 0.001 μm to approximately 1 μm, and which scatter white light as a light blue hue observed at right angles to the incident light.

Combination reactions Reactions where two substances combine to form a third substance; an example is two elements reacting to form a compound of the elements and is shown in the general form $A + B \rightarrow AB$; examples include

$$2Na(s) + Cl_2(g) \rightarrow 2NaCl(s) \text{ and } 8Fe + S_8 \rightarrow 8FeS$$

Cometabolism The process by which compounds in petroleum may be enzymatically attacked by microorganisms without furnishing carbon for cell growth and division; a variation on biodegradation in which microbes transform a contaminant even though the contaminant cannot serve as the primary energy source for the organisms. To degrade the contaminant, the microbes require the presence of other compounds (primary substrates) that can support their growth.

Complex modulus A measure of the overall resistance of a material to flow under an applied stress, in units of force per unit area. It combines *viscosity* and elasticity elements to provide a measure of "stiffness," or resistance to flow. The *complex modulus* is more useful than *viscosity* for assessing the physical behavior of very non-Newtonian materials such as *emulsions*.

Complex inorganic chemicals Molecules that consist of different types of atoms (atoms of different chemical elements) which, in chemical reactions, are decomposed with the formation of several other chemicals.

Compound The combination of two or more different elements, held together by chemical bonds; the elements in each compound are always combined in the same proportion by mass (law of definite proportion).

Concentration Composition of a mixture characterized in terms of mass, amount, volume, or number concentration with respect to the volume of the mixture.

Condensation aerosol Formed by condensation of vapors or reactions of gases.

Conservative constituent or compound One that does not degrade and is unreactive and its *movement* is not retarded within a given environment (aquifer, stream, and contaminant plume).

Constituent An essential part or component of a system or group (i.e., an ingredient of a chemical mixture); for example, benzene is one constituent of gasoline.

Contaminant A pollutant unless it has some detrimental effect, can cause deviation from the normal composition of an environment; a pollutant that causes deviations from the normal composition of an environment. Are not classified as pollutants unless they have some detrimental effect.

Coordination compounds Compounds where the central ion, typically a transition metal, is surrounded by a group of anions or molecules.

Corrosion Oxidation of a metal in the presence of air and moisture.

Covalent bond A region of *relatively* high electron density between atomic nuclei that results from sharing of electrons and that *gives* rise to an attractive force and a characteristic internuclear distance; carbon-hydrogen bonds are covalent bonds.

Cracking The process in which large molecules are broken down (thermally decomposed) into smaller molecules; used especially in the petroleum refining industry.

Critical point The combination of critical temperature and critical pressure; the temperature and pressure at which two phases of a substance in equilibrium become identical and form a single phase.

Critical pressure The pressure required to liquefy a gas at its critical temperature; the minimum pressure required to condense gas to liquid at the critical temperature; a substance is still a fluid above the critical point, neither a gas nor a liquid, and is referred to as a supercritical fluid; expressed in atmosphere or psi.

Critical temperature The temperature above which a gas cannot be liquefied, regardless of the amount of pressure applied; the temperature at the critical point (end of the vapor pressure curve in phase diagram); at temperatures above critical temperature, a substance cannot be liquefied, no matter how great the pressure; expressed in °C.

Culture The growth of cells or microorganisms in a controlled artificial environment.

Cycloalkanes (naphthene and cycloparaffin) A saturated, cyclic compound containing only carbon and hydrogen. One of the simplest *cycloalkanes* is cyclohexane (C_6H_{12}); steranes derivatives and triterpane derivatives are branched naphthene derivatives consisting of multiple condensed five- or six-carbon rings.

D

Daughter product A compound that results directly from the degradation of another chemical.

Decomposition reactions Reactions in which a single compound reacts to give two or more products; an example of a decomposition reaction is the decomposition of mercury (II) oxide into mercury and oxygen when the compound is heated; a compound can also decompose into a compound and an element, or two compounds.

Deflagration An explosion with a flame front moving in the unburned gas at a speed below the speed of sound (1250 ft/s).

Degradation The breakdown or transformation of a compound into by-products and/or end products.

Dehydration reaction (condensation reaction) A chemical reaction in which two organic molecules become linked to each other via covalent bonds with the removal of a molecule of water; common in synthesis reactions of organic chemicals.

Dehydrohalogenation Removal of hydrogen and halide ions from an alkane resulting in the formation of an alkene.

Denitrification Bacterial reduction of nitrate to nitrite to gaseous nitrogen or nitrous oxides under anaerobic conditions.

Density The mass per unit volume of a substance. *Density* is temperature-dependent, generally decreasing with temperature. The density of oil relative to water, its specific gravity, governs whether a particular oil will float on water. Most fresh crude oils and fuels will float on water. Bitumen and certain residual fuel oils, however, may have densities greater than water at some temperature ranges and may submerge in water. The density of a spilled oil will also increase with time as components are lost due to weathering.

Detection limit (in analysis) The minimum single result that, with a stated probability, can be distinguished from a representative blank value during the laboratory analysis of substances such as water, soil, air, rock, and biota.

Detonation An explosion with a shock wave moving at a speed greater than the speed of sound in the unreacted medium.

1,4-Dichlorobenzene A chemical used to control moths, molds, and mildew and to deodorize restrooms and waste containers; does not occur naturally but is produced by chemical companies to make products for home use and other chemicals such as resins; most of the 1,4-dichlorobenzene enters the environment as a result of its use in moth repellent products and in toilet deodorizer blocks. Because it changes from a solid to a gas easily (sublimes), almost all 1,4-dichlorobenzene produced is released into the air.

Dichloroelimination Removal of two chlorine atoms from an alkane compound and the formation of an alkene compound within a reducing environment.

Dichloromethane (CH_2Cl_2) An organic solvent often used to extract organic substances from samples; toxic, but much less so than chloroform or carbon tetrachloride, which were previously used for this purpose.

Differential thermal analysis (DTA) and thermogravimetric analysis Techniques that may be used to measure the water of crystallization of a salt and the thermal decomposition of hydrates.

Dihaloelimination Removal of two halide atoms from an alkane compound and the formation of an alkene compound within a reducing environment.

Diols Chemical compounds that contain two hydroxy (—OH) groups, generally assumed to be, but not necessarily, alcoholic; aliphatic diols are also called glycols.

Direct emissions Emissions from sources that are owned or controlled by the reporting entity.

Dispersant (chemical dispersant) A chemical that reduces the surface tension between water and a hydrophobic substance such as oil. In the case of an oil spill, dispersants facilitate the breakup and dispersal of an oil slick throughout the water column in the form of an oil-in-water emulsion; chemical dispersants can only be used in areas where biological damage will not occur and must be approved for use by government regulatory agencies.

Dispersion aerosol Formed by grinding of solids, atomization of liquids, or dispersion of dusts; a colloidal-sized particle in the atmosphere formed.

Dissolved oxygen (DO) The key substance in determining the extent and kinds of life in a body of water.

Double bond A covalent bond resulting from the sharing of two pairs of electrons (four electrons) between two atoms.

Double-displacement reactions Reactions where the anions and cations of two different molecules switch places to form two entirely different compounds. These reactions are in the general form

$$AB + CD \rightarrow AD + CB$$

An example is the reaction of lead (II) nitrate with potassium iodide to form lead (II) iodide and potassium nitrate:

$$Pb(NO_3)_2 + 2KI \rightarrow PbI_2 + 2KNO_3$$

A special kind of double-displacement reaction takes place when an acid and base react with each other; the hydrogen ion in the acid reacts with the hydroxyl ion in the base causing the formation of water. Generally, the product of this reaction is some ionic salt and water:

$$HA + BOH \rightarrow H_2O + BA$$

An example is the reaction of hydrobromic acid (HBr) with sodium hydroxide:

$$HBr + NaOH \rightarrow NaBr + H_2O$$

Downgradient In the direction of decreasing static hydraulic head.

Drug Any substance presented for treating, curing, or preventing disease in human beings or in animals; a drug may also be used for making a medical diagnosis, managing pain, or for restoring, correcting, or modifying physiological functions.

E

Ecology The scientific study of the relationships between organisms and their environments.

Ecological chemistry The study of the interactions between organisms and their environment that are mediated by naturally occurring chemicals.

Ecology The study of environmental factors that affect organisms and how organisms interact with these factors and with each other.

Ecosystem A community of organisms together with their physical environment, which can be viewed as a system of interacting and interdependent relationships; this can also include processes such as the flow of energy through trophic levels and the cycling of chemical elements and compounds through living and nonliving components of the system; the trophic level of an organism is the position it occupies in a food chain; ecosystem is a term representing an assembly of mutually interacting organisms and their environment in which materials are interchanged in a largely cyclical manner.

Electron acceptor The atom, molecule, or compound that receives electrons (and therefore is reduced) in the energy-producing oxidation-reduction reactions that are essential for the growth of microorganisms and bioremediation—common electron acceptors in bioremediation are oxygen, nitrate, sulfate, and iron.

Electron affinity The electron affinity of an atom or molecule is the amount of energy released or spent when an electron is added to a neutral atom or molecule in the gaseous state to form a negative ion.

Electron configuration of an atom The extranuclear structure; the arrangement of electrons in shells and subshells; chemical properties of elements (their valence states and reactivity) can be predicted from the electron configuration.

Electron donor The atom, molecule, or compound that donates electrons (and therefore is oxidized); in bioremediation, the organic contaminant often serves as an electron donor.

Electronegativity The tendency of an atom to attract electrons in a chemical bond; nonmetals have high electronegativity, fluorine being the most electronegative, while alkali metals possess least electronegativity; the electronegativity difference indicates polarity in the molecule.

Elimination A reaction where two groups such as chlorine and hydrogen are lost from adjacent carbon atoms and a double bond is formed in their place.

Empirical formula The simplest whole-number ratio of atoms in a compound.

Emulsan Is a polyanionic heteropolysaccharide bioemulsifier produced by *Acinetobacter calcoaceticus* RAG-1; used to stabilize oil-in-water emulsions.

Emulsion A stable mixture of two immiscible liquids, consisting of a continuous phase and a dispersed phase. Oil and water can form both oil-in-water and water-in-oil emulsions. The former is termed as dispersion, while *emulsion* implies the latter. Water-in-oil emulsions formed from petroleum and brine can be grouped into four stability classes: stable, a formal emulsion that will persist indefinitely; mesostable, which gradually degrade over time due to a lack of one or more stabilizing factors; entrained water, a mechanical mixture characterized by high viscosity of the petroleum component that impedes separation of the two phases; and unstable, which are mixtures that rapidly separate into immiscible layers.

Emulsion stability Generally accompanied by a marked increase in *viscosity* and elasticity, over that of the parent oil that significantly changes behavior. Coupled with the increased volume due to the introduction of brine, emulsion formation has a large effect on the choice of countermeasures employed to combat a spill.

Emulsification The process of *emulsion* formation, typically by mechanical mixing. In the environment, *emulsions* are most often formed as a result of wave action. Chemical agents can be used to prevent the formation of *emulsions* or to "break" the *emulsions* to their component oil and water phases.

Endergonic reaction A chemical reaction that requires energy to proceed. A chemical reaction is endergonic when the change in free energy is positive.

Endothermic reaction A chemical reaction in which heat is absorbed.

Engineered bioremediation A type of remediation that increases the growth and degradative activity of microorganisms by using engineered systems that supply nutrients, electron acceptors, and/or other growth-stimulating materials.

Enhanced bioremediation A process that involves the addition of microorganisms (e.g., fungi, bacteria, and other microbes) or nutrients (e.g., oxygen and nitrates) to the subsurface environment to accelerate the natural biodegradation process.

Entering group An atom or group that forms a bond to what is considered to be the main part of the substrate during a reaction, for example, the attacking nucleophile in a bimolecular nucleophilic substitution reaction.

Enthalpy of formation (ΔH_f) The energy change or the heat of reaction in which a compound is formed from its elements; energy cannot be created or destroyed but is converted from one form to another; the enthalpy change (or heat of reaction) is

$$\Delta H = H_2 - H_1$$

H_1 is the enthalpy of reactants and H_2 the enthalpy of the products (or heat of reaction); when H_2 is less than H_1, the reaction is exothermic, and ΔH is negative, that is, temperature increases; when H_2 is greater than H_1, the reaction is endothermic, and the temperature falls.

Entropy A thermodynamic quantity that is a measure of disorder or randomness in a system; the total entropy of a system and its surroundings always increases for a spontaneous process; the standard entropies, S°, are entropy values for the standard states of substances.

Environment The total living and nonliving conditions of an organism's internal and external surroundings that affect an organism's complete life span; the conditions that surround someone or something; the conditions and influences that affect the growth, health, progress, etc. of someone or something; the total living and nonliving conditions (internal and external surroundings) that are an influence on the existence and complete life span of the organism.

Environmental analytical chemistry The application of analytic chemical techniques to the analysis of environmental samples—in a regulatory setting.

Environmental biochemistry The discipline that deals specifically with the effects of environmental chemical species on life.

Environmental chemistry The study of the sources, reactions, transport, effects, and fates of chemical species in water, soil, and air environments, and the effects of technology thereon.

Environmentalist A person working to solve environmental problems, such as air and water pollution, the exhaustion of natural resources, and uncontrolled population growth.

Environmental pollution The contamination of the physical and biological components of the Earth system (atmosphere, aquasphere, and geosphere) to such an extent that normal environmental processes are adversely affected.

Environmental science The study of the environment, its living and nonliving components, and the interactions of these components.

Environmental studies The discipline dealing with the social, political, philosophical, and ethical issues concerning man's interactions with the environment.

Enzyme A macromolecule, mostly proteins or conjugated proteins produced by living organisms, that facilitate the degradation of a chemical compound (catalyst); in general, an enzyme catalyzes only one reaction type (reaction specificity) and operates on only one type of substrate (substrate specificity); any of a group of catalytic proteins that are produced by cells and that mediate or promote the chemical processes of life without themselves being altered or destroyed.

Epoxidation A reaction wherein an oxygen molecule is inserted in a carbon-carbon double bond, and an epoxide is formed.

Epoxides A subclass of epoxy compounds containing a saturated three-membered cyclic ether. See *Epoxy compounds*.

Epoxy compounds Compounds in which an oxygen atom is directly attached to two adjacent or nonadjacent carbon atoms in a carbon chain or ring system, thus cyclic ethers.

Equipment blank A sample of analyte-free media that has been used to rinse the sampling equipment. It is collected after completion of decontamination and prior to sampling. This blank is useful in documenting and controlling the preparation of the sampling and laboratory equipment.

Ester A compound formed from an acid and an alcohol; in esters of carboxylic acids, the —COOH group and the —OH group lose a molecule of water and form a —COO— bond (R_1 and R_2 represent organic groups):

$$R_1COOH + R_2OH \rightarrow R_1COOR_2 + H_2O$$

Ether A compound with an oxygen atom attached to two hydrocarbon groups. Any carbon compound containing the functional group C–O–C, such as diethyl ether ($C_2H_5O{\cdot}C_2H_5$).

Ethylbenzene A colorless, flammable liquid found in natural products such as coal tar and crude oil; it is also found in manufactured products such as inks, insecticides, and paints; a minor component of JP-8 fuel.

Eurkaryotes Microorganisms that have well-defined cell nuclei enclosed by a nuclear membrane.

Eutrophication The growth of algae may become quite high in very productive water, with the result that the concurrent decomposition of dead algae reduces oxygen levels in the water to very low values.

Exothermic reaction A reaction that produces heat and absorbs heat from the surroundings.

***Ex situ* bioremediation** A process that involves removing the contaminated soil or water to another location before treatment.

F

Facultative anaerobes Microorganisms that use (and prefer) oxygen when it is available but can also use alternate electron acceptors such as nitrate under anaerobic conditions when necessary.

Fate The ultimate disposition of the inorganic chemical in the ecosystem, either by chemical or biological transformation to a new form that (hopefully) is nontoxic (degradation) or, in the case of an ultimately persistent inorganic pollutants, by conversion to a less offensive chemicals or even by sequestration in a sediment or other location that is expected to remain undisturbed.

Fatty acids Carboxylic acids with long hydrocarbon side chains; most natural fatty acids have hydrocarbon chains that do not branch; any double bonds occurring in the chain are *cis* isomers—the side chains are attached on the same side of the double bond.

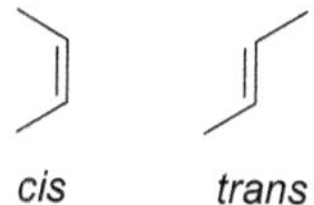

Fauna All of the animal life of any particular region, ecosystem, or environment; generally, the naturally occurring or indigenous animal life (native animal life).

Fermentation The process whereby microorganisms use an organic compound as both electron donor and electron acceptor, converting the compound to fermentation products such as organic acids, alcohols, hydrogen, and carbon dioxide; microbial metabolism in which a particular compound is used both as an electron donor and an electron acceptor resulting in the production of oxidized and reduced daughter products.

Field capacity or in situ (field water capacity) The water content on a mass or volume basis, remaining in soil two or three days after having been wet with water and after free drainage is negligible.

Fingerprint A chromatographic signature of relative intensities used in oil-oil or oil-source rock correlations; mass chromatograms of steranes derivatives or terpane derivatives are examples of fingerprints that can be used for qualitative or quantitative comparison of crude oil.

Flammability limits A gas mixture will not burn when the composition is lower than the lower flammable limit (LFL); the mixture is also not combustible when the composition is above the upper flammability limit (UFL).

Flammable chemical (flammable substance) A chemical or substance is usually termed flammable if the flash point of the chemical or substance is below 38°C (100°F).

Flash point The temperature at which the vapor over a liquid will ignite when exposed to an ignition source. A liquid is considered to be flammable if its *flash point* is less than 60°C. *Flash point* is an extremely important factor in relation to the safety of spill cleanup operations. Gasoline and other light fuels can ignite under most ambient conditions and therefore are a serious hazard when spilled. Many freshly spilled crude oils also have low *flash points* until the lighter components have evaporated or dispersed.

Flora The plant life occurring in a particular region or time; generally, the naturally occurring or indigenous plant life (native plant life).

Fluids Liquids; also a generic term applied to all substances that flow freely, such as gases and liquids.

Foam A colloidal suspension of a gas in a liquid.

Fog A term denoting high level of water droplets.

Fraction One of the portions of a chemical mixture separated by chemical or physical means from the remainder.

Free radical A molecule with an odd number of electrons—they do not have a completed octet and often undergo vigorous redox reactions.

Fugitive emissions Emissions that include losses from equipment leaks or evaporative losses from impoundments, spills, or leaks.

Functional group An atom or a group of atoms attached to the base structure of a compound that has similar chemical properties irrespective of the compound to which it is a part; a means of defining the characteristic physical and chemical properties of families of organic compounds.

Fungi Nonphotosynthetic organisms, larger than bacteria, aerobic, and can thrive in more acidic media than bacteria. Important function is the breakdown of cellulose in wood and other plant materials.

G

Gas Matter that has no definite volume or definite shape and always fills any space given in which it exists.

Gas chromatography (GC) A separation technique involving passage of a gaseous moving phase through a column containing a fixed liquid phase; it is used principally as a quantitative analytical technique for compounds that are volatile or can be converted to volatile forms.

Gaseous nutrient injection A process in which nutrients are fed to contaminated groundwater and soil via wells to encourage and feed naturally occurring microorganisms—the most common added gas is air in the presence of sufficient oxygen, microorganisms convert many organic contaminants to carbon dioxide, water, and microbial cell mass. In the absence of oxygen, organic contaminants are metabolized to methane, limited amounts of carbon dioxide, and trace amounts of hydrogen gas. Another gas that is added is methane. It enhances degradation by cometabolism in which as bacteria consume the methane, they produce enzymes that react with the organic contaminant and degrade it to harmless minerals.

GC-MS Gas chromatography-mass spectrometry.

GC-TPH GC detectable total petroleum hydrocarbons, that is the sum of all GC-resolved and unresolved hydrocarbons. The resolvable hydrocarbons appear as peaks, and the unresolvable hydrocarbons appear as the area between the lower baseline and the curve defining the base of the resolvable peaks.

Geologic time The span of time that has passed since the creation of the Earth and its components; a scale use to measure geologic events millions of years ago.

Geosphere A term representing the solid earth, including soil, which supports most plant life.

Glycerol A small molecule with three alcohol groups ($HOCH_2CHOHCH_2OH$); basic building block of fats and oils.

$$\begin{array}{c} HOCH_2 \\ | \\ HOCH \\ | \\ HOCH_2 \end{array}$$

Gram-equivalent weight (nonredox reaction) The mass in grams of a substance equivalent to 1 g-atom of hydrogen, 0.5 g-atom of oxygen, or 1 g-ion of the hydroxyl ion; can be determined by dividing the molecular weight by the number of hydrogen atoms or hydroxyl ions (or their equivalent) supplied or required by the molecule in a reaction.

Gram-equivalent weight (redox reaction) The molecular weight in grams divided by the change in oxidation state.

Gravimetric analysis A technique of quantitative analytical chemistry in which a desired constituent is efficiently recovered and weighed.

Greenhouse effect The warming of an atmosphere by its absorption of infrared radiation while shortwave radiation is allowed to pass through.

Greenhouse gases Any of the gases whose absorption of solar radiation is responsible for the greenhouse effect, including carbon dioxide, ozone, methane, and the fluorocarbons.

Guest molecule (or ion) An organic or inorganic ion or molecule that occupies a cavity, cleft, or pocket within the molecular structure of a host molecular entity and forms a complex with the host entity or that is trapped in a cavity within the crystal structure of a host.

H

Half-life (abbreviated to $t_{1/2}$) The time required to reduce the concentration of a chemical to 50% of its initial concentration; units are typically in hours or days; the term is commonly used in nuclear physics to describe how quickly (radioactive decay) unstable atoms undergo radioactive decay, or how long stable atoms survive the potential for radioactive decay.

Halide An element from the halogen group, which includes fluorine, chlorine, bromine, iodine, and astatine.

Halogen Group 17 in the periodic table of the elements; these elements are the reactive nonmetals and are electronegative.

Hardness scale (Mohs' scale) A measure of the ability of a substance to abrade or indent one another; the Mohs hardness is based on a scale from 1 to 10 units in which diamond, the hardest substance, is given a value of 10 Mohs and talc given a value of 0.5.

Hazardous waste A potentially dangerous chemical substance that has been discarded, abandoned, neglected, released, or designated as a waste material, or one that may interact with other substances to pose a threat.

Haze A term denoting decreased visibility due to the presence of particles.

Heat capacity ($C\rho$) The quantity of thermal energy needed to raise the temperature of an object by 1°C; the heat capacity is the product of mass of the object and its specific heat, $C\rho = \text{mass} \times \text{specific heat}$.

Heat of fusion (ΔH_{fus}) The amount of thermal energy required to melt 1 mol of the substance at the melting point; also termed as latent heat of fusion and expressed in kcal/mol or kJ/mol.

Heat of vaporization (ΔH_{vap}) The amount of thermal energy needed to convert 1 mol of a substance to vapor at boiling point; also known as latent heat of vaporization and expressed kcal/mol or kJ/mol.

Henry's law The relation between the partial pressure of a compound and the equilibrium concentration in the liquid through a proportionality constant known as Henry's law constant.

Henry's law constant The concentration ratio between a compound in air (or vapor) and the concentration of the compound in water under equilibrium conditions.

Herbicide A chemical that controls or destroys unwanted plants, weeds, or grasses.

Heterocyclic An organic group or molecule containing rings with at least one noncarbon atom in the ring.

Heterogeneous Varying in structure or composition at different locations in space.

Heterotroph An organism that cannot synthesize its own food and is dependent on complex organic substances for nutrition.

Heterotrophic bacteria Bacteria that utilize organic carbon as a source of energy; organisms that derive carbon from organic matter for cell growth.

Heterotrophs Organisms or chemicals that obtain their carbon from other organisms.

Hopane A pentacyclic *hydrocarbon* of the *triterpane* group believed to be derived primarily from bacteriohopanoids in bacterial membranes.

Homogeneous Having uniform structure or composition at all locations in space.

Homologue A compound belonging to a series of compounds that differ by a repeating group; for example, propanol ($CH_3CH_2CH_2OH$), *n*-butanol ($CH_3CH_2CH_2CH_2OH$), and *n*-pentanol ($CH_3CH_2CH_2CH_2CH_2OH$) are homologues; they belong to the homologous series of alcohols, $CH_3(CH_2)_nOH$.

Hydration The addition of a water molecule to a compound within an aerobic degradation pathway.

Hydrocarbon One of a very large and diverse group of chemical compounds composed only of carbon and hydrogen; the largest source of hydrocarbons is petroleum crude oil; the principal constituents of crude oils and refined petroleum products.

Hydrogen bond A form of association between an electronegative atom and a hydrogen atom attached to a second, relatively electronegative atom; best considered as an electrostatic interaction, heightened by the small size of hydrogen, which permits close proximity of the interacting dipoles or charges.

Hydrogenation A reaction where hydrogen is added across a double or triple bond, usually with the assistance of a catalyst; a process whereby an enzyme in certain microorganisms catalyzes the hydrolysis or reduction of a substrate by molecular hydrogen.

Hydrogenolysis A reductive reaction in which a carbon-halogen bond is broken, and hydrogen replaces the halogen substituent.

Hydrology The scientific study of water.

Hydrolysis A chemical transformation process in which a chemical reacts with water. In the process, a new carbon-oxygen bond is formed with oxygen derived from the water molecule, and a bond is cleaved within the chemical between carbon and some functional group.

Hydrophilic Water-loving; the capacity of a molecular entity or of a substituent to interact with polar solvents, in particular with water, or with other polar groups; hydrophilic molecules dissolve easily in water but not in fats or oils.

Hydrophilic colloids Generally, macromolecules, such as proteins and synthetic polymers, that are characterized by strong interaction with water resulting in spontaneous formation of colloids when they are placed in water.

Hydrophilicity The tendency of a molecule to be solvated by water.

Hydrophobic Fear of water; the tendency to repel water.

Hydrophobic colloids Colloids that interact to a lesser extent with water and are stable because of their positive or negative electric charges.

Hydrophobic interaction The tendency of hydrocarbons (or of lipophilic hydrocarbon-like groups in solutes) to form intermolecular aggregates in an aqueous medium, and analogous intramolecular interactions.

Hydrosphere The water areas of the Earth; also called the aquasphere.

Hydroxylation Addition of a hydroxyl group to a chlorinated aliphatic hydrocarbon.

Hydroxyl group A functional group that has a hydrogen atom joined to an oxygen atom by a polar covalent bond (—OH).

Hydroxyl ion One atom each of oxygen and hydrogen bonded into an ion (OH^-) that carries a negative charge.

Hydroxyl radical A radical consisting of one hydrogen atom and one oxygen atom; normally does not exist in a stable form.

I

Indirect emissions Emissions that are a consequence of the activities of the reporting entity but occur at sources owned or controlled by another entity.

Infiltration rate The time required for water at a given depth to soak into the ground.

Inhibition The decrease in rate of reaction brought about by the addition of a substance (inhibitor), by virtue of its effect on the concentration of a reactant, catalyst, or reaction intermediate.

Inoculum A small amount of material (either liquid or solid) containing bacteria removed from a culture in order to start a new culture.

Inorganic Pertaining to, or composed of, chemical compounds that are not organic, that is, contain no carbon-hydrogen bonds; examples include chemicals with no carbon and those with carbon in nonhydrogen-linked forms.

Inorganic acid An inorganic compound that elevates the hydrogen concentration in an aqueous solution; alphabetically, examples are carbonic acid (HCO_3), a weak inorganic acid. Hydrochloric acid (HCl), a highly corrosive, strong inorganic acid with many uses. Hydrofluoric acid (HF), a weak inorganic acid that is highly reactive with silicate, glass, metals, and semimetals. Nitric acid (HNO_3), a highly corrosive and toxic strong inorganic acid. Phosphoric acid, not considered a strong inorganic acid; found in solid form as a mineral and has many industrial uses. Sulfuric acid, a highly corrosive inorganic acid. It is soluble in water and widely used.

Inorganic base an inorganic compound that elevates the hydroxide concentration in an aqueous solution; alphabetically, examples are ammonium hydroxide (ammonia water), a solution of ammonia in water; calcium hydroxide (lime water), a weak base with many industrial uses; magnesium hydroxide, referred to as brucite when found in its solid mineral form; sodium bicarbonate (baking soda), a mild alkali; sodium hydroxide (caustic soda), a strong inorganic base and used widely in industrial and laboratory environments.

Inorganic chemistry The study of inorganic compounds, specifically the structure, reactions, catalysis, and mechanism of action.

Inorganic compound A compound that consists of an ionic component (an element from the periodic table) and an anionic component; a compound that does not contain carbon chemically bound to hydrogen; carbonates, bicarbonates, carbides, and carbon oxides are considered inorganic compounds, even though they contain carbon; a large number of compounds occur naturally, while others may be synthesized; in all cases, charge neutrality of the compound is key to the structure and properties of the compound.

Inorganic reaction chemistry Inorganic chemical reactions fall into four broad categories: combination reactions, decomposition reactions, single-displacement reactions, and double-displacement reactions.

Inorganic salts Inorganic salts are neutral, ionically bound molecules and do not affect the concentration of hydrogen in an aqueous solution.

Inorganic synthesis The process of synthesizing inorganic chemical compounds; is used to produce many basic inorganic chemical compounds.

In situ In its original place; unmoved; unexcavated; remaining in the subsurface.

In situ bioremediation A process that treats the contaminated water or soil where it was found.

Interfacial tension The net energy per unit area at the interface of two substances, such as oil and water or oil and air. The air/liquid interfacial tension is often referred to as surface tension; the SI units for *interfacial tension* are millinewtons per meter (mN/m). The higher the *interfacial tension*, the less attractive the two surfaces are to each other, and the more size of the interface will be minimized. Low surface tensions can drive the spreading of one fluid on another. The surface tension of an oil, together its viscosity, affects the rate at which spilled oil will spread over a water surface or into the ground.

Intermolecular forces Forces of attraction that exist between particles (atoms, molecules, and ions) in a compound.

Internal standard (IS) A pure analyte added to a sample extract in a known amount, which is used to measure the relative responses of other analytes and surrogates that are components of the same solution. The *internal standard* must be an analyte that is not a sample component.

Intramolecular (1) Descriptive of any process that involves a transfer (of atoms, groups, electrons, etc.) or interactions (such as forces) between different parts of the same molecular entity; (2) relating to a comparison between atoms or groups within the same molecular entity.

Intrinsic bioremediation A type of bioremediation that manages the innate capabilities of naturally occurring microbes to degrade contaminants without taking any engineering steps to enhance the process.

Inversions Conditions characterized by high atmospheric stability that limits the vertical circulation of air, resulting in air stagnation and the trapping of air pollutants in localized areas.

Ionic bond A chemical bond or link between two atoms due to an attraction between oppositely charged (positive-negative) ions.

Ionic bonding Chemical bonding that results when one or more electrons from one atom or a group of atoms is transferred to another. Ionic bonding occurs between charged particles.

Ionic compounds Compounds where two or more ions are held next to each other by electric attraction.

Ionic liquids An ionic liquid is a salt in the liquid state or a salt with a melting point lower than 100°C (212°F); variously called liquid electrolytes, ionic melts, ionic fluids, fused salts, liquid salts, or ionic glasses; powerful solvents and electrically conducting fluids (electrolytes).

Ionic radius A measure of ion size in a crystal lattice for a given coordination number (CN); metal ions are smaller than their neutral atoms, and nonmetallic anions are larger than the atoms from which they are formed; ionic radii depend on the element, its charge, and its coordination number in the crystal lattice; ionic radii are expressed in angstrom units of length (Å).

Ionization energy The ionization energy is the energy required to remove an electron completely from its atom, molecule, or radical.

Ionization potential The energy required to remove a given electron from its atomic orbital; the values are given in electron volts (eV).

Isomers Compounds that have the same number and types of atoms—the same molecular formula—but differ in the structural formula, that is, the manner in which the atoms are combined with each other.

Isotope A variant of a chemical element which differs in the number of neutrons in the atom of the element; all isotopes of a given element have the same number of protons in each atom and different isotopes of a single element occupy the same position on the periodic table of the elements.

IUPAC International Union of Pure and Applied Chemistry.

K

Ketone An organic compound that contains a carbonyl group (R_1COR_2).

L

Lag phase The growth interval (adaption phase) between microbial inoculation and the start of the exponential growth phase during which there is little or no microbial growth.

Latex A polymer of *cis*-1-4 isoprene; milky sap from the rubber tree *Hevea brasiliensis*.

Law A system of rules that are enforced through social institutions to govern behavior; can be made by a collective legislature or by a single legislator, resulting in statutes, by the executive through decrees and regulations, or by judges through binding precedent; the formation of laws themselves may be influenced by a constitution (written or tacit) and the rights encoded therein; the law shapes politics, economics, history, and society in various ways and serves as a mediator of relations between people. See also Regulation.

Leaving group An atom or group (charged or uncharged) that becomes detached from an atom in what is considered to be the residual or main part of the substrate in a specified reaction.

Lignin A complex amorphous polymer in the secondary cell wall (middle lamella) of woody plant cells that cements or naturally binds cell walls to help make them rigid; highly resistant to decomposition by chemical or enzymatic action; also acts as support for cellulose fibers.

Limnology The branch of science dealing with the characteristics of freshwater, including biological properties and chemical and physical properties.

Lipophilic F-loving; applied to molecular entities (or parts of molecular entities) having a tendency to dissolve in fatlike (e.g., hydrocarbon) solvents.

Lipophilicity The affinity of a molecule or a moiety (portion of a molecular structure) for a lipophilic (fat soluble) environment. It is commonly measured by its distribution behavior in a biphasic system, either liquid-liquid (e.g., partition coefficient in octanol/water).

Lithosphere The part of the geosphere that is directly involved with environmental processes through contact with the atmosphere, the hydrosphere, and living things. Varies from 50 to 100 km in thickness. Consists of outer mantle and crust.

Loading rate The amount of material that can be absorbed per volume of soil.

LTU Land treatment unit; a physically delimited area where contaminated land is treated to remove/minimize contaminants and where parameters such as moisture, pH, salinity, temperature, and nutrient content can be controlled.

M

Macromolecule A large molecule of high molecular mass composed of more than 100 repeated monomers (single chemical units of lower relative mass); a large complex molecule formed from many simpler molecules.

Mass number The number of protons plus the number of neutrons in the nucleus of an atom.

Matter Any substance that has inertia and occupies physical space; can exist as solid, liquid, gas, plasma, or foam.

Measurement A description of a property of a system by means of a set of specified rules that maps the property on to a scale of specified values by direct or mathematical comparison with specified references.

Mechanical explosion An explosion due to the sudden failure of a vessel containing a non-reactive gas at a high pressure.

Melting point The temperature when matter is converted from solid to liquid.

Mesosphere The portion of the atmosphere of the Earth where molecules exist as charged ions caused by interaction of gas molecules with intense ultraviolet (UV) light.

Metabolic by-product A product of the reaction between an electron donor and an electron acceptor; metabolic by-products include volatile fatty acids, daughter products of chlorinated aliphatic hydrocarbons, methane, and chloride.

Metabolism The physical and chemical processes by which foodstuffs are synthesized into complex elements; complex substances are transformed into simple ones, and energy is made available for use by an organism; thus, all biochemical reactions of a cell or tissue, both synthetic and degradative, are included; the sum of all of the enzyme-catalyzed reactions in living cells that transform organic molecules into simpler compounds used in biosynthesis of cellular components or in extraction of energy used in cellular processes.

Metabolize A product of metabolism.

Methanogens Strictly anaerobic archaebacteria; able to use only a very limited spectrum of substrates (e.g., molecular hydrogen, formate, methanol, methylamine, carbon monoxide, or acetate) as electron donors for the reduction of carbon dioxide to methane.

Methanogenic The formation of methane by certain anaerobic bacteria (methanogens) during the process of anaerobic fermentation.

Methyl a group (—CH_3) derived from methane; for example, CH_3Cl is methyl chloride (systematic name, chloromethane) and CH_3OH is methyl alcohol (systematic name, methanol).

Micelles a spherical cluster formed by the aggregation of soap molecules in water.

Microclimate a highly localized climatic conditions; the climate that organisms and objects on the surface are exposed to close to ground, under rocks, and surrounded by vegetation and often quite different from the surrounding macroclimate.

Microcosm a diminutive, representative system analogous to a larger system in composition, development, or configuration.

Microorganism an organism of microscopic size that is capable of growth and reproduction through biodegradation of food sources, which can include hazardous contaminants; microscopic organisms including bacteria, yeasts, filamentous fungi, algae, and protozoa; a living organism too small to be seen with the naked eye; includes bacteria, fungi, protozoans, microscopic algae, and viruses.

Microbe the shortened term for microorganism.

Mineralization the biological process of complete breakdown of organic compounds, whereby organic materials are converted to inorganic products (e.g., the conversion of hydrocarbons to carbon dioxide and water); the release of inorganic chemicals from organic matter in the process of aerobic or anaerobic decay.

Mist liquid particles.

Mixed waste any combination of waste types with different properties or any waste that contains both hazardous waste and source, special nuclear, or by-product material; as defined by the US EPA, mixed waste contains both hazardous waste (as defined by RCRA and its amendments) and radioactive waste (as defined by AEA and its amendments).

Modulus of elasticity the stress required to produce unit strain to cause a change of length (Young's modulus), or a twist or shear (shear modulus), or a change of volume (bulk modulus); expressed as dynes/cm^2.

Moiety A term generally used to signify part of a molecule; for example, in an ester R^1COOR^2, the alcohol moiety is R^2O.

Molality (m) The gram moles of solute divided by the kilograms of solvent.

Molarity (M) The gram moles of solute divided by the liters of solution.

Molecular weight The mass of 1 mol of molecules of a substance.

Molecule The smallest unit in a chemical element or compound that contains the chemical properties of the element or compound.

Mole fraction The number of moles of a component of a mixture divided by the total number of moles in the mixture.

Monoaromatic Aromatic hydrocarbons containing a single benzene ring.

Monosaccharide A simple sugar such as fructose or glucose that cannot be decomposed by hydrolysis; colorless crystalline substances with a sweet taste that have the same general formula, $C_nH_{2n}O_n$.

MTBE Methyl tertiary-butyl ether; is a fuel additive that has been used in the United States since 1979. Its use began as a replacement for lead in gasoline because of health hazards associated with lead. MTBE has distinctive physical properties that result in it being highly soluble, persistent in the environment, and able to migrate through the ground. Environmental regulations have required the monitoring and cleanup of MTBE at petroleum contaminated sites since February 1990; the program continues to monitor studies focusing on the potential health effects of MTBE and other fuel additives.

N

Native fauna The native and indigenous animal of an area.

Native flora The native and indigenous plant life of an area.

Natural organic matter (NOM) An inherently complex mixture of polyfunctional organic molecules that occurs naturally in the environment and is typically derived from the decay of floral and faunal remains; although they do occur naturally, the fossil fuels (coal, crude oil, and natal gas) are usually not included in the term *natural organic matter*.

NCP National Contingency Plan—also called the National Oil and Hazardous Substances Pollution Contingency Plan; provides a comprehensive system of accident reporting, spill containment, and cleanup, and established response headquarters (National Response Team and Regional Response Teams).

Nernst equation An equation that is used to account for the effect of different activities upon electrode potential:

$$E = E^0 + \frac{2.303RT}{nF}\log\frac{\text{Reactants}}{\text{Products}} = E^0 + \frac{0.0591}{n}\log\frac{\text{Reactants}}{\text{Products}}$$

Nitrate enhancement A process in which a solution of nitrate is sometimes added to groundwater to enhance anaerobic biodegradation.

Nonpoint source pollution Pollution that does not originate from a specific source. Examples of nonpoint sources of pollution include the following: (1) sediments from construction, forestry operations and agricultural lands; (2) bacteria and microorganisms from failing septic systems and pet wastes; (3) nutrients from fertilizers and yard debris; (4) pesticides from agricultural areas, golf courses, athletic fields and residential yards, oil, grease, antifreeze, and metals washed from roads, parking lots, and driveways; (5) toxic chemicals and cleaners that were not disposed of correctly; and (6) litter thrown onto streets, sidewalks, and beaches or directly into the water by individuals. See Point source pollution.

Normality (N) The gram equivalents of solute divided by the liters of solution.

Nucleophile A chemical reagent that reacts by forming covalent bonds with electronegative atoms and compounds.

Nuclide A nucleus rather than to an atom—isotope (the older term); it is better known than the term nuclide and is still sometimes used in contexts where the use of the term nuclide might be more appropriate; identical nuclei belong to one nuclide, for example, each nucleus of the carbon-13 nuclide is composed of six protons and seven neutrons.

Nutrients Major elements (e.g., nitrogen and phosphorus) and trace elements (including sulfur, potassium, calcium, and magnesium) that are essential for the growth of organisms.

O

Oceanography The science of the ocean and its physical and chemical characteristics.

Octane A flammable liquid (C_8H_{18}) found in petroleum and natural gas; there are 18 different octane isomers that have different structural formulas but share the molecular formula C_8H_{18}; used as a fuel and as a raw material for building more complex organic molecules.

Octanol-water partition coefficient (K_{ow}) The equilibrium ratio of a chemical's concentration in octanol (an alcoholic compound) to its concentration in the aqueous phase of a two-phase octanol-water system, typically expressed in log units (log K_{ow}); K_{ow} provides an indication of a chemical's solubility in fats (lipophilicity), its tendency to bioconcentrate in aquatic organisms, or sorb to soil or sediment.

Oleophilic Oil-seeking or oil-loving (e.g., nutrients that stick to or dissolve in oil).

Order of reaction A chemical rate process occurring in systems for which concentration changes (and hence the rate of reaction) are not themselves measurable, provided it is possible to measure a chemical flux.

Organic Compounds that contain carbon chemically bound to hydrogen; often contain other elements (particularly O, N, halogens, or S); chemical compounds based on carbon that also contain hydrogen, with or without oxygen, nitrogen, and other elements.

Organic carbon (soil) partition coefficient (K_{oc}) The proportion of a chemical sorbed to the solid phase at equilibrium in a two-phase water/soil or water/sediment system expressed on an organic carbon basis; chemicals with higher K_{oc} values are more strongly sorbed to organic carbon and, therefore, tend to be less mobile in the environment.

Organic chemistry The study of compounds that contain carbon chemically bound to hydrogen, including synthesis, identification, modeling, and reactions of those compounds.

Organic liquid nutrient injection An enhanced bioremediation process in which an organic liquid, which can be naturally degraded and fermented in the subsurface to result in the generation of hydrogen. The most commonly added for enhanced anaerobic bioremediation include lactate, molasses, hydrogen release compounds (HRCs), and vegetable oils.

Organochlorine compounds (chlorinated hydrocarbons) Organic pesticides that contain chlorine, carbon, and hydrogen (such as DDT); these pesticides affect the central nervous system.

Organometallic compounds Compounds that include carbon atoms directly bonded to a metal ion.

Organophosphorus compound A compound containing phosphorus and carbon; many pesticides and most nerve agents are organophosphorus compounds, such as malathion.

Osmotic potential Expressed as a negative value (or zero), indicates the ability of the soil to dissolve salts and organic molecules; the reduction of soil water osmotic potential is caused by the presence of dissolved solutes.

OPA Oil Pollution Act of 1990; an act that addresses oil pollution and establishes liability for the discharge and substantial threat of a discharge of oil to navigable waters and shorelines of the United States.

Oven dry The weight of a soil after all water has been removed by heating in an oven at a specified temperature (usually in excess of 100°C, 212°F) for water; temperatures will vary if other solvents have been used.

Oxidation The transfer of electrons away from a compound, such as an organic contaminant; the coupling of oxidation to reduction (see below) usually supplies energy that microorganisms use for growth and reproduction. Often (but not always), oxidation results in the addition of an oxygen atom and/or the loss of a hydrogen atom.

Oxygen enhancement with hydrogen peroxide An alternative process to pumping oxygen gas into groundwater involves injecting a dilute solution of hydrogen peroxide. Its chemical formula is H_2O_2, and it easily releases the extra oxygen atom to form water and free oxygen. This circulates through the contaminated groundwater zone to enhance the rate of aerobic biodegradation of organic contaminants by naturally occurring microbes. A solid peroxide product (e.g., oxygen-releasing compound (ORC¨)) can also be used to increase the rate of biodegradation.

Oxidation-reduction reactions (redox reactions) Reactions that involve oxidation of one reactant and reduction of another.

Ozone (O_3) A form of oxygen containing three atoms instead of the common two (O_2); formed by high-energy ultraviolet radiation reacting with oxygen.

P

PAHs Polycyclic aromatic hydrocarbons. Alkylated *PAHs* are *alkyl group* derivatives of the parent *PAHs*. The five target alkylated *PAHs* referred to in this report are the alkylated naphthalene, phenanthrene, dibenzothiophene, fluorene, and chrysene series.

Paraffin An alkane.

Pathogen An organism that causes disease (e.g., some bacteria or viruses).

Perfluorocarbon (PFC) A derivative of hydrocarbons in which all of the hydrogens have been replaced by fluorine.

Permeability The capability of the soil to allow water or air movement through it. The quality of the soil that enables water to move downward through the profile, measured as the number of inches per hour that water moves downward through the saturated soil.

Permeable reactive barrier (PRB) A subsurface emplacement of reactive materials through which a dissolved contaminant plume must move as it flows, typically under natural gradient and treated water exits the other side of the permeable reactive barrier.

pH A measure of the acidity or basicity of a solution; the negative logarithm (base 10) of the hydrogen ion concentration in gram ions per liter; a number between 0 and 14 that describes the acidity of an aqueous solution; mathematically, the pH is equal to the negative logarithm of a the concentration of H_3O^+ in solution.

Phenol A molecule containing a benzene ring that has a hydroxyl group substituted for a ring hydrogen.

Phenyl A molecular group or fragment formed by abstracting or substituting one of the hydrogen atoms attached to a benzene ring.

Photic zone The upper layer within bodies of water reaching down to about 200 m, where sunlight penetrates and promotes the production of photosynthesis; the richest and most diverse area of the ocean.

Phototrophs Organisms or chemicals that utilize light energy from photosynthesis.

Physical change Refers to the change that occurs when a material changes from one physical state to another without the formation of intermediate substances of different composition in the process, such as the change from gas to liquid.

Phytodegradation The process in which some plant species can metabolize VOC contaminants. The resulting metabolic products include trichloroethanol, trichloroacetic acid, and dichloracetic acid; mineralization products are probably incorporated into insoluble products such as components of plant cell walls.

Phytovolatilization The process in which VOCs are taken up by plants and discharged into the atmosphere during transpiration.

PM_{10} Particulate matter below 10 μ in diameter; this corresponds to the particles inhalable into the human respiratory system, and its measurement uses a size selective inlet.

$PM_{2.5}$ Particulate matter below 2.5 μ in diameter; this is closer to but slightly finer than the definitions of respirable dust that have been used for many years in industrial hygiene to identify dusts that will penetrate the lungs.

Point emissions Emissions that occur through confined air streams as found in stacks, ducts, or pipes.

Point source pollution Any single identifiable source of pollution from which pollutants are discharged, such as a pipe. Examples of point sources include (1) discharges from wastewater treatment plants, (2) operational wastes from industries, and (3) combined sewer outfalls. See Nonpoint source pollution.

Polar compound An organic compound with distinct regions of positive and negative charge. *Polar compounds* include alcohols, such as sterols, and some *aromatics*, such as monoaromatic steroids. Because of their polarity, these compounds are more soluble in polar solvents, including water, compared with nonpolar compounds of similar molecular structure.

Pollutant A substance present in greater than natural concentration as a result of human activity that has a net detrimental effect upon its environment or upon something of value in that environment.

Polymer A large molecule made by linking smaller molecules (monomers) together.

Pour point The lowest temperature at which an oil will appear to flow under ambient pressure over a period of 5 s. The *pour point* of crude oils generally varies from −60°C to 30°C. Lighter oils with low *viscosities* generally have lower *pour points*.

Primary substrates The electron donor and electron acceptor that are essential to ensure the growth of microorganisms. These compounds can be viewed as analogous to the food and oxygen that are required for human growth and reproduction.

Producers Organisms or chemicals that utilize light energy and store it as chemical energy.

Prokaryotes Microorganisms that lack a nuclear membrane so that their nuclear genetic material is more diffuse in the cell.

Propagule Any part of a plant (e.g., bud) that facilitates dispersal of the species and from which a new plant may form.

Propane A colorless, odorless, flammable gas (C_3H_8) found in petroleum and natural gas; used as a fuel and as a raw material for building more complex organic molecules.

Protozoa Microscopic animals consisting of single eukaryotic cells.

R

Radical (free radical) A molecular entity such as CH_3 and Cl possessing an unpaired electron.

Radioactive decay The process by which the nucleus of an unstable atom loses energy by emitting radiation.

Rate A derived quantity in which time is a denominator quantity so that the progress of a reaction is measured with time.

Rate constant, *k* See *Order of reaction.*

Rate-controlling step (rate-limiting step and rate-determining step) The elementary reaction having the largest control factor exerts the strongest influence on the rate; a step having a control factor much larger than any other step is said to be rate-controlling.

Recalcitrant Unreactive, nondegradable and refractory.

Receptor An object (animal, vegetable, or mineral) or a locale that is affected by the pollutant.

Redox (reduction-oxidation reactions) Oxidation and reduction occur simultaneously; in general, the oxidizing agent gains electrons in the process (and is reduced) while the reducing agent donates electrons (and is oxidized).

Reducers Organisms or chemicals that break down chemical compounds to more simple species and thereby extract the energy needed for their growth and metabolism

Reduction The transfer of electrons to a compound, such as oxygen, that occurs when another compound is oxidized.

Reductive dehalogenation A variation on biodegradation in which microbially catalyzed reactions cause the replacement of a halogen atom on an organic compound with a hydrogen atom. The reactions result in the net addition of two electrons to the organic compound.

Refractive index (index of refraction) The ratio of wavelength or phase velocity of an electromagnetic wave in a vacuum to that in the substance; a measure of the amount of refraction a ray of light undergoes as it passes through a refraction interface: a useful physical property to identify a pure compound.

Regulation A concept of management of complex systems according to a set of rules (laws) and trends; can take many forms: legal restrictions promulgated by a government authority, contractual obligations (such as contracts between insurers and their insureds), social regulation, coregulation, third-party regulation, certification, accreditation, or market regulation. See Law.

Releases On-site discharge of a toxic chemical to the surrounding environment; includes emissions to the air, discharges to bodies of water, releases at the facility to land, and contained disposal into underground injection wells.

Releases (to air, point and fugitive air emissions) All air emissions from industry activity; point emissions occur through confined air streams as found in stacks, ducts, or pipes; fugitive emissions include losses from equipment leaks, or evaporative losses from impoundments, spills, or leaks.

Releases (to land) Disposal of toxic chemicals in waste to on-site landfills, land-treated or incorporation into soil, surface impoundments, spills, leaks, or waste piles. These activities must occur within the boundaries of the facility for inclusion in this category.

Release (to underground injection) A contained release of a fluid into a subsurface well for the purpose of waste disposal.

Releases (to water and surface water discharges) Any releases going directly to streams, rivers, lakes, oceans, or other bodies of water: any estimates for storm water runoff and nonpoint losses must also be included.

Resins The name given to a large group of *polar compounds* in oil. These include hetero-substituted *aromatics*, acids, ketones, alcohols, and monoaromatic steroids. Because of their polarity, these compounds are more soluble in *polar* solvents, including water, than the nonpolar compounds, such as *waxes* and *aromatics*, of similar molecular weight. They are largely responsible for oil *adhesion*.

Respiration The process of coupling oxidation of organic compounds with the reduction of inorganic compounds. such as oxygen, nitrate, iron (III), manganese (IV), and sulfate.

Rhizodegradation The process whereby plants modify the environment of the root zone soil by releasing root exudates and secondary plant metabolites. Root exudates are typically photosynthetic carbon, low-molecular-weight molecules, and high-molecular-weight organic acids. This complex mixture modifies and promotes the development of a microbial community in the rhizosphere. These secondary metabolites have a potential role in the development of naturally occurring contaminant-degrading enzymes.

Rhizosphere The soil environment encompassing the root zone of the plant.

RRF Relative response factor.

S

Saturated hydrocarbon A saturated carbon-hydrogen compound with all carbon bonds filled; that is, there are no double or triple bonds, as in olefins or acetylenes.

Saturated solution A solution in which no more solute will dissolve; a solution in equilibrium with the dissolved material.

Saturation The maximum amount of solute that can be dissolved or absorbed under prescribed conditions.

SIM (Selecting Ion Monitoring) Mass spectrometric monitoring of a specific mass/charge (m/z) ratio. The *SIM* mode offers better sensitivity than can be obtained using the full scan mode.

Simple inorganic chemicals Molecules that consist of one-type atoms (atoms of one element) that, in chemical reactions, cannot be decomposed to form other chemicals.

Single-displacement reactions Reactions where one element trades places with another element in a compound. These reactions come in the general form of

$$A + BC \rightarrow AC + B.$$

Examples include (i) magnesium replacing hydrogen in water to make magnesium hydroxide and hydrogen gas

$$Mg + 2H_2O \rightarrow Mg(OH)_2 + H_2$$

and (ii) the production of silver crystals when a copper metal strip is dipped into silver nitrate

$$Cu(s) + 2AgNO_3(aq) \rightarrow 2Ag(s) + Cu(NO_3)_2(aq)$$

Smoke The particulate material assessed in terms of its blackness or reflectance when collected on a filter, as opposed to its mass; this is the historical method of measurement of particulate pollution; particles formed by incomplete combustion of fuel.

Solid-state compounds A diverse class of compounds that are solid at standard temperature and pressure and exhibit unique properties as semiconductors.

Solubility The amount of a substance (solute) that dissolves in a given amount of another substance (solvent); a measure of the solubility of an inorganic chemical in a solvent, such as water; generally, ionic substances are soluble in water and other polar solvents while the nonpolar, covalent compounds are more soluble in the nonpolar solvents; in sparingly soluble, slightly soluble, or practically insoluble salts, degree of solubility in water and occurrence of any precipitation process may be determined from the solubility product, K_{sp}, of the salt—the smaller the K_{sp} value, the lower the solubility of the salt in water.

Soluble Capable of being dissolved in a solvent.

Solute Any dissolved substance in a solution.

Solution Any liquid mixture of two or more substances that is homogeneous.

Solvolysis Generally, a reaction with a solvent involving the rupture of one or more bonds in the reacting solute: more specifically, the term is used for substitution, elimination, or fragmentation reactions in which a solvent species is the nucleophile; hydrolysis, if the solvent is water or alcoholysis if the solvent is an alcohol.

Specific heat The amount of heat required to raise the temperature of one gram of a substance by 1°C; the specific heat of water is 1 cal or 4.184 J.

Stable As applied to chemical species, the term expresses a thermodynamic property, which is quantitatively measured by relative molar standard Gibbs energies; a chemical species A is more stable than its isomer B under the same standard conditions.

Standard potential Used to predict if a species will be oxidized or reduced in solution (under acidic or basic conditions) and whether any oxidation-reduction reaction will take place.

Starch A polysaccharide containing glucose (long-chain polymer of amylose and amylopectin) that is the energy storage reserve in plants.

Stoichiometry The calculation of the quantities of reactants and products (among elements and compounds) involved in a chemical reaction.

Stoke's Law

$$v = \frac{gd^2(\rho_1 - \rho_2)}{18\eta}$$

Stratosphere The portion of the atmosphere of the Earth where ozone is formed by the reaction of ultraviolet light on dioxygen molecules.

Strong acid An acid that releases H^+ ions easily—examples are hydrochloric acid and sulfuric acid.

Strong base A basic chemical that accepts and holds proton tightly—an example is the hydroxide ion.

Styrene A human-made chemical used mostly to make rubber and plastics; present in combustion products, such as cigarette smoke and automobile exhaust.

Sublimation The direct vaporization or transition of a solid directly to a vapor without passing through the liquid state.

Substrate A chemical species of particular interest, of which the reaction with some other chemical reagent is under observation (e.g., a compound that is transformed under the influence of a catalyst); also the component in a nutrient medium, supplying microorganisms with carbon (C-substrate) and nitrogen (N-substrate) as food needed to grow.

Surface-active agent A compound that reduces the surface tension of liquids, or reduces interfacial tension between two liquids or a liquid and a solid; also known as surfactant, wetting agent, or detergent.

Surface tension Caused by molecular attractions between the molecules of two liquids at the surface of separation.

Sustainable development Development and economic growth that meets the requirements of the present generation without compromising the ability of future generations to meet their needs; a strategy seeking a balance between development and conservation of natural resources.

Sustainable enhancement An intervention action that continues until such time that the enhancement is no longer required to reduce contaminant concentrations or fluxes.

Steranes A class of tetracyclic, saturated biomarkers constructed from six isoprene subunits ($\sim C_{30}$). *Steranes* are derived from sterols, which are important membrane and hormone components in eukaryotic organisms. Most commonly used *steranes* are in the range of C_{26}–C_{30} and are detected using m/z 217 mass chromatograms.

Surrogate analyte A pure analyte that is extremely unlikely to be found in any sample, which is added to a sample aliquot in a known amount and is measured with the same procedures used to measure other components. The purpose of a *surrogate analyte* is to monitor the method performance with each sample.

T

Terminal electron acceptor (TEA) A compound or molecule that accepts an electron (is reduced) during metabolism (oxidation) of a carbon source; under aerobic conditions, molecular oxygen is the terminal electron acceptor; under anaerobic conditions, a variety of terminal electron acceptors may be used. In order of decreasing redox potential, these terminal electron acceptors include nitrate, manganese (Mn^{3+} and Mn^{6+}), iron (Fe^{3+}), sulfate, and carbon dioxide; microorganisms preferentially utilize electron acceptors that provide the maximum free energy during respiration; of the common terminal electron acceptors listed above, oxygen has the highest redox potential and provides the most free energy during electron transfer.

Terpanes A class of branched, cyclic alkane biomarkers including *hopanes* and tricyclic compounds.

Terpenes Hydrocarbon solvents, compounds composed of molecules of hydrogen and carbon; they form the primary constituents in the aromatic fractions of scented plants, for example, pine oil and turpentine and camphor oil.

Tetrachloroethylene (perchloroethylene) A human-made chemical that is widely used for dry cleaning of fabrics and for metal-degreasing operations; also used as a starting material (building block) for making other chemicals and is used in some consumer products such as water repellents, silicone lubricants, fabric finishers, spot removers, adhesives, and wood cleaners; can stay in the air for a long time before breaking down into other chemicals or coming back to the soil and water in rain; much of the tetrachloroethylene that gets into water and soil will evaporate; because tetrachloroethylene can travel easily through soils, it can get into underground drinking water supplies.

Thermal conductivity A measure of the rate of transfer of heat by conduction through unit thickness, across unit area for unit difference of temperature; measured as calories per second per square centimeter for a thickness of 1 cm and a temperature difference of 1°C; units are cal/cm s°K or W/cm°K.

Thermodynamics The study of the energy transfers or conversion of energy in physical and chemical processes defines the energy required to start a reaction or the energy given out during the process.

Toluene A clear, colorless liquid that occurs naturally in crude oil and in the tolu tree; produced in the process of making gasoline and other fuels from crude oil; used in making paints, paint thinners, fingernail polish, lacquers, adhesives, and rubber, and in some printing and leather tanning processes; a major component of JP-8 fuel.

Total n-alkanes The sum of all resolved *n-alkanes* (from C_8 to C_{40} plus pristane and phytane).

Total five alkylated PAH homologues the sum of the five target PAHs (naphthalene, phenanthrene, dibenzothiophene, fluorene, and chrysene) and their alkylated (C_1 to C_4) homologues, as determined by GCMS. These five target alkylated PAH homologous series are oil-characteristic aromatic compounds.

Total aromatics The sum of all resolved and unresolved aromatic hydrocarbons including the total of BTEX and other alkyl benzene compounds, total five target alkylated PAH homologues, and other EPA priority PAHs.

Total saturates The sum of all resolved and unresolved aliphatic hydrocarbons including the total n-alkanes, branched alkanes, and cyclic saturates.

Total suspended particulate matter The mass concentration determined by filter weighing, usually using a specified sampler which collects all particles up to approximately 20 μ depending on wind speed.

Toxicity A measure of the toxic nature of a chemical; usually expressed quantitatively as LD_{50} (median lethal dose) or LC_{50} (median lethal concentration in air)—the latter refers to inhalation toxicity of gaseous substances in air; both terms refer to the calculated concentration of a chemical that can kill 50% of test animals when administered.

Toxicological chemistry The chemistry of toxic substances with emphasis upon their interactions with biological tissue and living organisms.

TPH Total petroleum hydrocarbons; the total measurable amount of petroleum-based hydrocarbons present in a medium as determined by gravimetric or chromatographic means.

Transfers A transfer of toxic (organic) chemicals in wastes to a facility that is geographically or physically separate from the facility reporting under the toxic release inventory; the quantities reported represent a movement of the chemical away from the reporting facility; except for off-site transfers for disposal, these quantities do not necessarily represent entry of the chemical into the environment.

Transfers (POTWs) Waste waters transferred through pipes or sewers to a publicly owned treatment works (POTW); treatment and chemical removal depend on the chemical's nature and treatment methods used; chemicals not treated or destroyed by the POTW are generally released to surface waters or land filled within the sludge.

Transfers (to disposal) Wastes that are taken to another facility for disposal generally as a release to land or as an injection underground.

Transfers (to energy recovery) Wastes combusted off-site in industrial furnaces for energy recovery; treatment of an organic chemical by incineration is not considered to be energy recovery.

Transfers (to recycling) Wastes that are sent off-site for the purposes of regenerating or recovering still valuable materials; once these chemicals have been recycled, they may be returned to the originating facility or sold commercially.

Transfers (to treatment) Wastes moved off-site for either neutralization, incineration, biological destruction, or physical separation; in some cases, the chemicals are not destroyed but prepared for further waste management.

1,1,1-Trichloroethane Does not occur naturally in the environment; used in commercial products, mostly to dissolve other chemicals; beginning in 1996, 1,1,1-trichloroethane was no longer made in the United States because of its effects on the ozone layer; because of its tendency to evaporate easily, the vapor form is usually found in the environment; 1,1,1-trichloroethane also can be found in soil and water, particularly at hazardous waste sites.

Trichloroethylene A colorless liquid that does not occur naturally; mainly used as a solvent to remove grease from metal parts and is found in some household products, including typewriter correction fluid, paint removers, adhesives, and spot removers.

Triglyceride An ester of glycerol and three fatty acids; the fatty acids represented by "R" can be the same or different.

$$\begin{array}{c} RCOOCH_2 \\ | \\ RCOOCH_2 \\ | \\ RCOOCH_2 \end{array}$$

Triterpanes A class of cyclic saturated *biomarkers* constructed from six isoprene subunits; cyclic terpane compounds containing two, four, and six isoprene subunits are called monoterpane (C_{10}), diterpane (C_{20}), and *triterpane* (C_{30}), respectively.

Trophic The trophic level of an organism is the position it occupies in a food chain.

Troposphere The portion of the atmosphere of the Earth that is closest to the surface.

Tyndall effect The characteristic light scattering phenomenon of colloids results from those being the same order of size as the wavelength of light.

U

UCM Unresolved complex mixture of hydrocarbons on, for example, a gas chromatographic tracing; the UCM appear as the *envelope* or *hump area* between the solvent baseline and the curve defining the base of resolvable peaks.

Underground storage tank A storage tank that is partially or completely buried in the earth.

Unsaturated compound An organic compound with molecules containing one or more double bonds.

Unsaturated zone The zone between land surface and the capillary fringe within which the moisture content is less than saturation, and pressure is less than atmospheric; soil pore spaces also typically contain air or other gases; the capillary fringe is not included in the unsaturated zone (See *Vadose zone*).

Upgradient In the direction of increasing potentiometric (piezometric) head. See also *Downgradient.*

US EPA US Environmental Protection Agency.

USGS US Geological Survey.

V

Vadose zone The zone between land surface and the water table within which the moisture content is less than saturation (except in the capillary fringe), and pressure is less than atmospheric; soil pore spaces also typically contain air or other gases; the capillary fringe is included in the vadose zone.

Valence state of an atom The power of an atom to combine to form compounds; determines the chemical properties.

Vapor pressure The pressure exerted by a solid or liquid in equilibrium with its own vapor; depends on temperature and is characteristic of each substance; the higher the vapor pressure at ambient temperature, the more volatile the substance.

Viscosity The resistance of a fluid to shear, movement or flow; a function of the composition of a fluid; the viscosity of an ideal, noninteracting fluid does not change with shear rate—such fluids are called Newtonian; expressed as g/cm s or poise; 1 P = 100 cP.

Volatile Readily dissipating by evaporation.

Volatile organic compounds (VOC) Organic compounds with high vapor pressures at normal temperatures. VOCs include light saturates and aromatics, such as pentane, hexane, BTEX and other lighter substituted benzene compounds, which can make up to a few percent of the total mass of some crude oils.

W

Water solubility The maximum amount of a chemical that can be dissolved in a given amount of pure water at standard conditions of temperature and pressure; typical units are milligrams per liter (mg/L), gallons per liter (gal/L), or pounds per gallon (lbs/gal).

Waxes Waxes are predominately straight-chain *saturates* with melting points above 20°C (generally, the *n*-alkanes C_{18} and higher molecular weight).

Weak Acid An acid that does not release H^+ ions easily—an example is acetic acid.

Weak base A basic chemical that has little affinity for a proton—an example is the chloride ion.

Weathering Processes related to the physical and chemical actions of air, water, and organisms after oil spill. The major weathering processes include evaporation, dissolution, dispersion, photochemical oxidation, water-in-oil *emulsification*, microbial degradation, adsorption onto suspended particulate materials, interaction with mineral fines, sinking, sedimentation, and formation of tar balls.

Wet deposition The term used to describe pollutants brought to ground either by rainfall or by snow; this mechanism can be further subdivided depending on the point at which the pollutant was absorbed into the water droplets.

Wilting point The largest water content of a soil at which indicator plants, growing in that soil, wilt and fail to recover when placed in a humid chamber.

X

Xylenes The term that refers to all three types of xylene isomers (meta-xylene, ortho-xylene, and para-xylene); produced from crude oil; used as a solvent and in the printing, rubber, and leather industries, as well as a cleaning agent and a thinner for paint and varnishes; a major component of JP-8 fuel.

Z

Zwitterion A particle that contains both positively charged and negatively charged groups; for example, amino acids ($H_2NHCHRCO_2H$) can form zwitterions ($^+H_3NCHRCOO^-$).

INDEX

Note: Page numbers followed by *f* indicate figures, and *t* indicate tables.

A

Absorption, 359*t*, 362–363
Absorption spectroscopy
  infrared, 182–183*t*, 277–278
  UV-Vis, 182–183*t*, 277
Acid-base chemistry, 98–108, 99*t*
  acid-base reaction, 104
  Brønsted-Lowry acids and bases, 104–105
  Lewis acids and bases, 105–106
  oxides and salts, 106–108
Acid-base reaction, 104, 449
Acidic precipitation, 12
Acid rain, 262*t*, 369, 375, 381
Acids
  corrosive, 428*t*
  uses, 216*t*
Actinide elements, 83
Adsorption, 359–361, 359*t*
  inorganic chemicals, 316–317
Aerobic biodegradation, intrinsic, 435
Aerosol particle, 43
Aggregation, 39–42*t*, 374–375
Agriculture
  pesticides in, 25
  soil pollution, 381
Agroremediation. *See* Phytoremediation
Airborne particles
  in aerosols, 374–375
  surface catalysis of, 355–356
Air contaminants, inorganic, 232*t*
Air pollutants, 239*t*
  and effects, 241*t*, 249–251*t*
  primary, 248
  secondary, 248
  sources, 241*t*
  toxic, 248
Air pollution, 391, 400, 402
  sources, 242–253, 244*t*
Alkali earths, 65*t*
Alkali metals, 65*t*, 217
Alkalis reactions, 104
Aluminum, 67–74*t*, 257–258*t*, 297–304*t*, 327–331*t*, 499–521*t*
Ammonia, 122–124, 249–251*t*
Ammonium, 499–521*t*
Ammonium nitrate, 125–126
Ammonium phosphate, 126–129
Ammonium sulfate, 129–130
Anhydrous ammonia, 123
Anions, 58*t*
Anthropogenic pollutants, 233
Anthropogenic waste, 381
Anthroposphere, 23–24
Antimony, 67–74*t*, 255*t*, 257–258*t*, 297–304*t*, 327–331*t*
Antoine equation, 199
Aquasphere, 19–23
  chemistry in, 370–377
  freshwater region, 21–22
  marine region, 22–23
Aquatic biome, 19
Aquatic sediments, 445
  contamination in, 445
Aquatic toxicity, 372*t*
Aqueous-phase chemicals, 445–446
Arsenic, 174–177*t*, 255*t*, 257–258*t*, 327–331*t*, 499–521*t*
Asbestos, 255*t*
  fibers, 172–178
Ashing, effect of, 17
Asphalt, 461
Atmosphere
  chemistry in, 18–19, 367–370
  composition, 15–17, 16*t*
  pollutants in, 324
  structure, 13–19
Atomic Energy Act, 413
Atom radius of element, 184
Avogadro's law, 189

B

Bacteria (microbes), on chemicals, 433
BAFs. *See* Bioaccumulation factors (BAFs)
Barium, 67–74*t*, 174–177*t*, 255*t*, 257–258*t*, 297–304*t*, 327–331*t*, 499–521*t*
Basalt, 39–42*t*
Bases
  corrosive, 428*t*
  properties of, 211*t*
Bauxite, 67–74*t*, 297–304*t*
BCFs. *See* Bioconcentration factors (BCFs)
Beryllium, 67–74*t*, 255*t*, 257–258*t*, 297–304*t*, 327–331*t*
Best available technology (BAT), 405
Best conventional technology (BCT), 405
Bioaccumulation factors (BAFs), 320–321
Bioaccumulative chemicals, 286
Bioaugmentation, 454–455
Bioconcentration, chemical, 321–322
Bioconcentration factors (BCFs), 284–285, 291, 320–322
Biodegradation, 364–366, 365*t*, 433–435, 453–455
  intrinsic, 435
  as natural process, 434
  temperature influences rate of, 434–435
Bioinorganic chemistry, 53
Bioinorganic compounds, 75–76
Biological contamination, drinking water, 417
Biological system, elements in, 4*t*
Biomass, 241–242
Bioremediation. *See* Biodegradation
Biosedimentation, 317, 373–374
Biosphere
  chemistry in, 377–381
  terrestrial, 23–30
Biostimulation, 454–455
Biota-sediment accumulation factor (BSAF), 321–322
Bitumen. *See* Asphalt
Black pigments, 165–167*t*
Blue pigments, 165–167*t*
Boiler slag, 268, 268*t*, 273–274
Boiling point, inorganic compounds, 191–192, 275–276, 499–521*t*
Bond, 83
  dissociation energy, 186
  lengths, 181–187
  order, 183
  strengths, 181–187
  types, 84*t*, 85–88, 85–87*f*
Boron, 499–521*t*
Botanoremediation. *See* Phytoremediation
Bottom ash, 267–268, 268*t*, 271–273
Boyle's law, 189
Brine
  properties, 131*t*
  purification, 130–134
Brønsted-Lowry acids and bases, 104–105
Brown pigments, 165–167*t*

C

Cadmium, 174–177*t*, 255*t*, 257–258*t*, 327–331*t*
Calcination, 137
Calcium, 499–521*t*
Carbon, 499–521*t*
Carbon dioxide
  properties, 179*t*
  source, 249–251*t*
Carbon monoxide, 246–247, 251
  environmental effects, 241*t*
  source, 249–251*t*
Carbon steel, corrosion, 228*t*
Catalysis
  heterogeneous, 225–227
  homogeneous, 225
Catalysts, 223–227, 355–357
  characteristics, 223*t*
  role, 224
Cations, 58*t*
  in inorganic chemistry, 78*t*
  transition metal and metal, 59*t*
Caustic soda process, 134–135
CFCs. *See* Chlorofluorocarbons (CFCs)
CFR. *See* Code of Federal Regulations (CFR)
Charles' law, 189
Chelant-enhanced phytoextraction, of metals, 470–471
Chelating agents, chemistry of cleaning, 439–440
Chelating organic acids, 467–468

Chelators, 470–471
Chemical elements
alphabetically by symbol, 485–487*t*
in alphabetical order, 479–481*t*
in order of atomic number, 482–484*t*
Chemicals
contamination, 283–284
drinking water, 417
effects in environment, 422–423
extractant for soil washing, 467–469
mixtures, 420
oxidation, 337
PBT criteria for, 320
physical and chemical properties of, 173*t*
pollutant
classification, 1
inorganic, 1–2
safety guidelines, for working with chemicals, 305*t*
substance, 420
transformation process, inorganic chemicals, 333–334, 337–344
types, 319–323
waste, 238
Chemical Substances Inventory of the Toxic Substances Control Act, 421
Chemisorption, 362
Chemistry and engineering, 44–46
Chlor-alkali chemicals, 135–139
Chloride, 257–258*t*, 327–331*t*
Chlorine, 55–56, 139–140, 252, 499–521*t*
Chlorofluorocarbons (CFCs), 224, 253
Chromatography, 276–277, 522–524*t*
Chromite, 67–74*t*, 297–304*t*
Chromium, 174–177*t*, 255*t*, 257–258*t*, 327–331*t*, 436, 499–521*t*
Circular dichroism (CD) spectroscopy, 522–524*t*
Citric acid, 468*f*
Clausius-Clapeyron equation, 192, 199
Claus process, 157–160
Clay, 39–42*t*, 67–74*t*, 297–304*t*
minerals, 360, 361*t*
soil, 438
Clean Air Act (CAA), 397–402
Clean Air Act and Amendments, 114, 400–402
Clean coal technology, 475
Cleanup, environmental, 435–453
chemistry of cleaning, 437–443
builders, 440–441
chelating agents, 439–440
preservatives, 442–443
soil types, 438–439
solvents, 441–442
surfactants, 439
containment, 450–451
conversion, 446–450
passive remediation technologies, 451–453
physical removal process, 443–446
Clean Water Act (CWA), 393–394, 403–406
Climate change, 240
Cluster compounds, 75–76
Coagulation, 374
Coal
ash, 267–268, 267*t*
pulverized, 271, 273
Coal-fired power plants, 261, 269
Coarse soil, 26
Cobalt, 67–74*t*, 297–304*t*, 499–521*t*
Code of Federal Regulations (CFR), 37, 386, 414–416
Cofferdam, 450–451
Colloidal stability, 374–375
Color
groundwater, 326*t*
inorganic compounds, 522–524*t*
Combination reactions, 93, 340–341, 447
Commodity chemicals, 112–113, 236–237
Complexation process, 343
Comprehensive Environmental Response Compensation and Liability Act (CERCLA), 406–409, 424
Conductance, 210
Contaminants, 238, 283, 318, 454*t* *See also specific types of contaminant*
anthropogenic sources of, 286
in environment, 323–331
inorganic, 284
radionuclide, 454*t*
Contamination, 305
air, land, and water, 432
in aquatic sediments, 445
chemical, 283–284

Conversion factors
area, 527
concentration conversions, 527
nutrient, 527
sludge conversions, 528
temperature conversions, 528
various constants, 528
volume conversion, 528–530
weight conversion, 530
Coordination compounds, 75–76
Copper, 67–74*t*, 174–177*t*, 257–258*t*, 297–304*t*, 327–331*t*, 499–521*t*
Corrosion, 227–228, 228*t*
Covalent bond, 84*t*, 183–184
Cradle-to-grave analysis, 427–428
Creaming process, 374–375
Critical compressibility factor, 215
Crude chemicals, 112–113
Crystal, 172, 188
shapes, inorganic compounds, 522–524*t*
Crystallography
X-ray, 182–183*t*
diffraction, 280
Cyanide, 257–258*t*, 327–331*t*

D

d-block electrons, 171–172
DDT. *See* Dichlorodiphenyltrichloroethane (DDT)
Decomposition reactions, 93, 340, 447
Degradation, 11–12
Demulsification, 313
Density, inorganic compounds, 193–195, 194*t*, 499–521*t*
Desorption, 360–361
Detergents, synthetic, 439, 442
Diamond, 39–42*t*
Diatomite, 39–42*t*
Dichlorodiphenyltrichloroethane (DDT), 25, 265–266, 427
Dielectric constant, 209
Dielectric loss, 209
Diffraction
X-ray, 522–524*t*
crystallography, 280
Dilution, inorganic chemicals, 363
Dipole-dipole interaction, 84*t*
Dipole moment, 186
Dispersion of chemicals, 308–310
Disproportionation, 354
Dissolution, inorganic chemicals, 310–312
Dissolved inorganic chemicals, 311–312
Dissolved solids, 327–331*t*
Dolomite, 39–42*t*
Domestic hazardous waste. *See* Household hazardous waste (HHW)
Double-displacement reaction, 94, 340, 448
Drinking water, contamination, 392, 417–419

E

Ecobalance. *See* Cradle-to-grave analysis
EDTA. *See* Ethylenediaminetetraacetic acid (EDTA)
E-factor, 292–294, 339
Electrical properties, inorganic compounds, 208–212, 211*t*
Electrochemical series, of metals, 183–184, 184*t*
Electrokinetics, 456*t*
Electrolyte, 210–211
Electromotive force, 209–210
Electron affinity, 92–93
Electron paramagnetic resonance (EPR) spectroscopy, 182–183*t*, 279–280, 522–524*t*
Electron spin resonance (ESR) spectroscopy, 279–280
Elemental analysis, inorganic compounds, 276, 522–524*t*
Elements, 67–74*t*, 109*t*, 296–304*t*
atom radius of, 184
in biological system, 4*t*
equivalent mass of, 97
periodic table, 3*f*, 52*f*, 185*f*, 243*f*
Emergency Planning and Community Right-to-Know Act (EPCRA), 114, 408
Emission factor, 12
Emulsification, inorganic chemicals, 312–313
Emulsion, 313
Encapsulation, 456*t*
Endangered Species Act (ESA), 383–386

Endocrine disrupters, 28
Engineering, chemistry and, 44–46
Entrainment, 337
Environment, 8–30
  chemical effects in, 54
  contaminants in, 323–331
  contamination in, 259–260
  inorganic chemicals in, 37–44
  inorganic chemistry and, 30–33
Environmental chemistry, 1, 46, 284*f*
Environmental law, 386–388, 394. *See also* Environmental regulations
Environmental pollutants, 238–239, 436–437
Environmental pollution, 231–234, 283, 453–454
Environmental Protection Agency (EPA), 43–44, 389, 397, 403–404, 413–416, 420, 425
  category of, 390
  CERCLA, 407–408
  Clean Air Act, 401–402
  Clean Water Act, 404–405
  hazardous wastes, 415
  RCRA, 416–417
  SDWA, 417–418
  TSCA, 420–421
Environmental Quality Improvement Act, 397
Environmental regulations, 383–389, 384–387*t*
  CERCLA, 406–409
  Clean Air Act, 397–402
  Clean Water Act, 403–406
  Hazardous Materials Transportation Act, 409–410
  hazardous waste regulations, 423–424, 425*f*
  OSHA, 410–412
  RCRA, 412–417
  requirements, 424–426
  SDWA, 417–419
  TSCA, 419–421, 419*t*
Environmental remediation, 435–437
EPA. *See* Environmental Protection Agency (EPA)
Equilibrium constant, 80, 217
Equilibrium vapor pressure, 199
Ethylenediaminetetraacetic acid (EDTA), 440*f*
Evaporation
  inorganic chemicals, 313–314
  and sublimation, 337
Expression environmental cleanup, 435–453
Ex situ immobilization techniques, 458
Ex situ soil washing, 465
Ex situ solidification-stabilization, 461–462

F

F-block elements, 81–82
Federal Motor Carrier Safety Regulations, 44
Federal Water Pollution Control Act, 403–404, 413
Feldspar, 39–42*t*, 67–74*t*, 297–304*t*
Fertilization, 435
Fibers, asbestos, 172–178
Fine chemicals, 112–113, 113*t*
  inorganic, 113*t*
Flocculation, 374
Fluidized bed reactors, 287
Fluoride, 255*t*, 257–258*t*, 327–331*t*
Fluorite (fluorspar), 39–42*t*, 67–74*t*, 297–304*t*
Fly ash, 267–271
  Class C, 270–271
  Class F, 270
Forbidden explosives, 220
Forest, 29–30
Fossil fuels, 240, 242, 247–248
  combustion, 368–369
Fourier transform infrared (FTIR) spectrometer, 278
Freezing point, inorganic compounds, 195–196
Furnace bottom ash (FBA), 272

G

Gallium, 67–74*t*, 297–304*t*
Gamma decay, 345
Garnet, 39–42*t*
Gas
  cleaning technologies, 399–400
  physical properties, 11*t*, 76*t*, 190*t*
Gaseous inorganic substances, 245

Gasoline, reformulated, 402
Geosphere, 23–30
Gold, 39–42*t*, 67–74*t*, 297–304*t*, 499–521*t*
Granite, 39–42*t*
Granular triple superphosphate (GTSP), 127
Graphite, 39–42*t*
Great Lakes Water Quality Agreement of 1978, 404
Green chemistry, 45
Green environment, 44–45
Greenhouse effect, 248
Greenhouse gas, 14
Green pigments, 165–167*t*
Green remediation. *See* Phytoremediation
Groundwater
  color, 326*t*
  contaminant sources to, 257–258*t*
  inorganic contaminants in, 327–331*t*
  odor, 326*t*
  pH, 326*t*
  physical properties, 326*t*
  pollution, 258
  taste, 326*t*
  turbidity, 326*t*
Gypsum, 39–42*t*, 67–74*t*, 297–304*t*

H

Haber-Bosch process, 119–120, 147
Half-life, 352–353
Half reaction, 354
Halite (sodium chloride, salt), 39–42*t*, 67–74*t*, 297–304*t*
Halogens, 66*t*
Hardness, 327–331*t*
Harmful inorganic chemicals, 6
Harmless chemicals, 5, 38
Hazardous air pollutants (HAPs), 400
Hazardous and Solid Waste Amendments (HSWA), 412, 418, 423–424
Hazardous Materials Transportation Act, 409–410
Hazardous waste, 37, 238
  characteristics, 415
  ignitable, 413
  reactive, 414
  regulations, 423–424
Heavy chemicals, 112–113, 113*t*
Heavy metals, 20, 28–29, 259–260, 262–264, 266–267, 450
  contaminated
    sites, remediation, 455–473, 456*t*
    soils, 455–457
  immobilization, inorganic amendments, 459*t*
  in land, 265*t*
  in plant, 265*t*
  properties, 29*t*
  in soil, 265*t*, 322
  source, 249–251*t*
Heptane, 246
Heterochain polymers, 162
Heterogeneous catalysis, 225–227, 356–357
  vs. homogeneous catalysis, 226*t*
Heterosphere, 15
HHW. *See* Household hazardous waste (HHW)
Highest occupied molecular orbital (HOMO), 106
Home generated special materials. *See* Household hazardous waste (HHW)
Homochain polymers, 162
Homogeneous catalysis, 225, 357
  vs. heterogeneous catalysis, 226*t*
Homosphere, surface-based, 14–15
Household cleaning agents, 440
Household hazardous waste (HHW), 43
Hydrides, melting/boiling point temperatures, 376*t*
Hydrocarbon
  fuels, 428–432
  petroleum-derived, 433
Hydrochloric acid, 140–141
Hydroelectric power, 255
Hydrofluoric acid, 141–143
Hydrogen, 499–521*t*
  bond, 84*t*
  production, 143
  radioactive isotope of, 352
Hydrolysis reactions, 346–348
Hydrosphere. *See* Aquasphere
Hyperaccumulators, 470
Hypothetical chemical reaction, 217

I

Ideal fluid, 207
Ideal gas law, 189–191
Immobilization technique, 456*t*, 458–465
  assessment of efficiency and capacity of, 464–465
  phytoremediation, 469–473
  solidification/stabilization, 459–462
  vitrification, 462–464
Incinerator bottom ash (IBA), 272
Indigenous chemicals, 5, 38–43
Indium, 67–74*t*, 297–304*t*
Industrial catalysts, 356
Industrial inorganic chemical manufacturers, 287, 288*t*
Industrial inorganic chemistry, 111–113
Industrial pollution, 389–390
Industrial revolution, 381
Infrared (IR) absorption spectroscopy, 182–183*t*, 277–278
Infrared (IR) spectroscopy, 522–524*t*
Inhibitors, 356
Inorganic acids, 102*t*
Inorganic air contaminants, 232*t*
Inorganic bases, 102*t*
Inorganic chemical contaminants (IOCs), 255*t*, 283–284, 317–318
Inorganic chemicals
  industry, 115–122
    historical aspects, 118–120
    modern industry, 120–122
    products of, 111
  production, 122–161
    ammonia, 122–124
    ammonium nitrate, 125–126
    ammonium phosphate, 126–129
    ammonium sulfate, 129–130
    brine purification, 130–134
    caustic soda processing, 134–135
    chlor-alkali chemicals, 135–139
    chlorine processing, 139–140
    feedstocks and effluents for, 116–118*t*
    hydrochloric acid, 140–141
    hydrofluoric acid, 141–143
    hydrogen production, 143
    nitric acid, 143–147
    nitrogen compounds, 147–149
    phosphoric acid, 149–151
    phosphorus compounds, 151–152
    sodium carbonate, 152–154
    sulfuric acid, 154–157
    sulfur recovery, 157–160
    titanium dioxide, 161
Inorganic chemistry
  and environment, 30–33
  metal cations in, 78*t*
Inorganic compounds, 2, 32*t*, 319–320, 333
  adsorption, 316–317
  biodegradation, 364–366, 365*t*
  boiling point, 191–192, 275–276, 499–524*t*
  characterization, 274–280
  chemical properties, 335*t*
  chemical transformation, 333–334, 337–344
  chromatography, 276–277
  classes of, 55
  classification, 63–80
  cleanup methods for spills of, 455*t*
  colligative properties, 187–213
  color, 522–524*t*
    and crystal shape, 276
  corrosion, 334–335, 336*t*
  critical properties, 213–216
  crystal shapes, 522–524*t*
  density, 193–195, 194*t*, 499–521*t*
  dilution, 363
  dispersion, 308–310
  dissolution, 310–312
  E-factors, 339
  electrical properties, 208–212, 211*t*
  elemental analysis, 276, 522–524*t*
  emulsification, 312–313
  in environment, 37–44
  EPR spectroscopy, 279–280
  evaporation, 313–314
  formula weight, 192–193
  freezing point, 195–196
  ideal gas law, 189–191
  infrared absorption spectroscopy, 277–278
  issues, 336–337
  leaching, 314–316
  list and formulas, 488–498*t*

Inorganic compounds *(Continued)*
low volatility of, 188
magnetic properties, 212
melting point, 195, 275–276, 499–524*t*
NMR spectroscopy, 279
nomenclature of, 56–63
optical properties, 212–213
partition coefficient, 198–199
phase transition, 213–214, 213–214*t*, 214*f*
physical constants in, 525*t*
physical properties, 11*t*, 76*t*, 187–213, 275–277, 335*t*, 522–524*t*
physical transformation, 344–345
pollutant, 1–2, 287–289, 429–430*t*
classification, 235*t*
pollution by, 236–237, 323
product, 427–432
qualitative analysis, 274
quantitative analysis, 274
Raman spectroscopy, 278–279
reaction, 93–98
refractive index, 206–207, 206*t*
released into environment, 305–319
sedimentation, 316–317
solubility, 197–198, 499–521*t*
sorption process, 357–359, 359*t*
absorption, 362–363
adsorption, 359–361
specific heat, 213
spectroscopic methods for identification of, 182–183*t*
spectroscopic/structural methods, 277–280, 522–524*t*
spreading, 317–318
sublimation, 318–319, 318*t*, 319*f*
thermal conductivity, 207
triple point, 196, 197*t*
use and misuse of, 33–37
UV-Vis absorption spectroscopy, 277
vapor pressure, 199–205, 200–205*t*
viscosity, 207–208
Inorganic contaminants, 284
in groundwater, 327–331*t*
Inorganic-contaminated soil, 438
Inorganic gases, 77
Inorganic pigments, 164–168, 165–167*t*
Inorganic pollutants, 28, 234–236, 234*t*, 285*t*, 306*t*, 333, 379–380
persistent, 334–336
sources, 238–267
air pollution, 242–253, 244*t*
sources of, 430–432*t*
Inorganic pollution, 231–233
Inorganic polymers, 161–164
Inorganic reactions, 345–346
hydrolysis, 346–348
photolysis, 348–351
radioactive decay, 351–353, 351*t*
rearrangement reactions, 353
redox, 354–355
Inorganic salts, 102*t*
Inorganic water pollutants, 254*t*
In-place inactivation. *See* Phytostabilization
In situ immobilization techniques, 458
Intermolecular bond, 84*t*, 85
International Union of Pure and Applied Chemistry (IUPAC), 7, 56–58
Intramolecular bond, 84*t*
Intrinsic biodegradation, 435
aerobic, 435
IOCs. *See* Inorganic chemical contaminants (IOCs)
Iodine, 499–521*t*
Ion exchange, 359, 379–380
Ionic bond, 84*t*, 85, 171–172, 187
Ionic compound, 4, 58–60
Ionic liquids
low-temperature, 62
properties, 62*t*
Ionization energy, 88–92
Ionization potential, 88–91, 89–91*t*
Ions, polyatomic, 60–61, 61*t*
Iron, 257–258*t*, 327–331*t*, 499–521*t*
Iron ore, 39–42*t*, 67–74*t*, 297–304*t*
Isotope, 62–63
IUPAC. *See* International Union of Pure and Applied Chemistry (IUPAC)

K

Kelvin temperature, 189–191, 190*t*
Kinematic viscosity, 208
Kyoto Protocol, 36

L

Land pollution, 261–267
Lanthanide metals, 83
Leaching, 314–316, 337
Lead, 67–74*t*, 257–258*t*, 297–304*t*, 327–331*t*, 499–521*t*
  properties and behavior, 174–177*t*
  source, 249–251*t*
LED. *See* Light-emitting diodes (LED)
Lewis acids and bases, 105–106
Life-cycle analysis/assessment. *See* Cradle-to-grave analysis
Light-emitting diodes (LED), 208–209
Light metal, 29*t*
Limestone, 39–42*t*
Limited tonnage chemicals, 112–113
Liquid, physical properties, 11*t*, 76*t*, 190*t*
Lithium, 67–74*t*, 297–304*t*, 499–521*t*
LMWOAs. *See* Low-molecular-weight organic acids (LMWOAs)
Loss on ignition (LOI), 270
Loss tangent, 209
Lowest unoccupied molecular orbital (LUMO), 106
Low-molecular-weight organic acids (LMWOAs), 467–468

M

Magnesium, 499–521*t*
Magnetic properties, inorganic compounds, 212
Manganese, 67–74*t*, 257–258*t*, 297–304*t*, 327–331*t*
Marine environment, 372–373
Mass spectrometry, 280
Mass spectroscopy, 522–524*t*
Maximum achievable control technology (MACT), 401
Maximum contaminant level goals (MCLGs), 418
Melting point, inorganic compounds, 195, 275–276, 499–521*t*
Mercury, 255*t*, 257–258*t*, 327–331*t*, 499–521*t*
  environmental effects, 241*t*
  properties and behavior, 174–177*t*
Metal-bearing solids, 263
Metal cations, in inorganic chemistry, 78*t*
Metal-contaminated soil, remediation of, 457
Metal ions, 439–440
Metalloids, 65–66*t*
Metals, 64–75
  chelant-enhanced phytoextraction of, 470–471
  electrochemical series of, 183–184, 184*t*
  ion, nomenclature, 59*t*
  lanthanide, 83
  nonradioactive, 436
  properties and behavior, 174–177*t*
Mica, 67–74*t*, 297–304*t*
  minerals, 39–42*t*
Microbes
  on chemicals, 433
  transform, 433–434
Microbial bioremediation, factors for, 365*t*
Microbial degradation, 337, 374
Microbiological contaminants, 417
Microorganism, 433
Mineralization, 433
Minerals, 39–42*t*, 67–75*t*, 108–109, 294–304, 296*t*
Mitigation, 454–455
Mobility reduction, 456*t*
Molecular forces, 191
Molecular geometry, of molecule, 181
Molecular symmetry, 181
Molecule, molecular geometry of, 181
Molybdenum, 67–74*t*, 297–304*t*, 499–521*t*
Monitored natural attenuation (MNA), 451–452
Muriatic acid. *See* Hydrochloric acid

N

National Environmental Education Act, 397
National Environmental Policy Act (NEPA), 397
National Pollutant Discharge Elimination System (NPDES), 405–406
Nation primary drinking water standard (NPDWS), 418
Natural attenuation, 454–455
Natural gas
  cleaning, 476*f*
  steam reforming of, 123

Natural products chemicals, 5
Neutralization reaction, 448
Newtonian fluid, 207–208
Nickel, 67–74*t*, 255*t*, 257–258*t*, 297–304*t*, 327–331*t*, 499–521*t*
Nitrate, 255*t*, 257–258*t*, 327–331*t*
Nitric acid, 143–147
Nitrite, 255*t*, 257–258*t*, 327–331*t*
Nitrogen, 499–521*t*
  photolysis, 350–351
  production, 147–149
Nitrogen dioxide, 249–251*t*, 251
Nitrogen oxide, 241*t*, 247–248
Nitrous oxide, 179*t*
Nomenclature
  of inorganic compounds, 56–63
  of metals ion, 59*t*
Nonaqueous phase chemicals, 445–446
Nonindigenous chemicals, 43–44
Nonmetals, 65–66*t*, 66–75
Nonpoint source (NPS) pollution, 292
Nonpolar bonds, 187
Nonradioactive metals, 436
Nuclear magnetic resonance (NMR) spectroscopy, 182–183*t*, 279, 522–524*t*
Nuclide, 63
Nutrient enrichment, 435

O

Occupational Safety and Health (OSHA), 410–412
Odor, groundwater, 326*t*
Oil-in-water emulsion, 312–313
Optical properties, inorganic compounds, 212–213
Ores, 297–304*t*
Organic carbon, 43
Organic chemicals, 2
  solubility of, 9–11
  volatility of, 180
Organic-contaminated soils, 438
Organometallic chemistry, 53
Organometallic compounds, 75–76
Orthophosphoric acid. *See* Phosphoric acid
Oxidation, 59*t*
Oxidation-reduction reactions. *See* Redox reactions
Oxides of nitrogen, 249–251*t*
Oxides of sulfur, 249–251*t*
Oxidizers, 428*t*
Oxidizing chemicals, 218
Oxoanions, 66*t*
Oxyfuel combustion, 475–476
Oxygen, 499–521*t*
Ozone, 247–248, 251–252
  environmental effects, 241*t*
  source, 249–251*t*

P

Palladium, 499–521*t*
Particle aggregation, 374–375
Particulate matter, 252
  environmental effects, 241*t*, 245
  source, 249–251*t*
  types, 252
Partition coefficient, inorganic compounds, 198–199
Pathway-engineering techniques, 453–454
Peaty soil, 438
Peptization, 374–375
Periodic table, 81–83
  of elements, 3*f*, 52*f*, 89–91*t*, 185*f*, 243*f*
Perlite, 67–74*t*, 297–304*t*
Permeable reactive barrier, 450–451
Peroxyacetyl nitrate, 252*f*
Persistent, bioaccumulative, and toxic (PBT) criteria for chemicals, 320
Persistent inorganic pollutants (PIPs), 34, 289–291, 307, 334–336
Pesticides, 324–325, 427
  in agriculture, 25
Petroleum-derived hydrocarbons, 433
PGM. *See* Platinum group metals (PGM)
pH
  groundwater, 326*t*
  scale, 104*f*
  soil, 380
Phase transition, inorganic compounds, 213–214, 213–214*t*, 214*f*
Phosphate, 441
  rock, 39–42*t*, 67–74*t*, 297–304*t*
Phosphoric acid, 149–151

Phosphorus, 151–152, 499–521*t*
Photochemical smog, 368–369
Photodecomposition. *See* Photolysis
Photodissociation. *See* Photolysis
Photolysis, 348–351
  indirect/sensitized, 349–350
  photochemical mechanism, 349
  in troposphere, 350
Physical contamination, drinking water, 417
Physical separation, 456*t*
Physical transformation process, inorganic chemicals, 344–345
Phytoaccumulation, 470–472
Phytoextraction, 470–472
Phytofiltration, 473
Phytoremediation, 469–473
Phytostabilization, 472–473
Pigments, inorganic, 164–168, 165–167*t*
PIPs. *See* Persistent inorganic pollutants (PIPs)
Planetary boundary layer, 15
Platinum, 499–521*t*
Platinum group metals (PGM), 67–74*t*, 297–304*t*
Polar bonds, 187
Pollutants, 437 *See also specific types of pollutants*
  air, 239*t*, 241*t*, 248, 249–251*t*
  anthropogenic, 233
  classification, 1
  environmental, 436–437
  inorganic chemical, 1–2, 429–430*t*
  inorganic, sources of, 430–432*t*
  soil, 239*t*
  water, 239*t*
Pollution. *See also specific types of pollution*
  air, 391, 400, 402
  defined, 283
  industrial, 389–390
  inorganic chemicals, 236–237, 323
  prevention, 473–477
  soil, 27–30, 381, 394
  water, 371, 391–394, 403–406
Polyatomic ions, 60–61, 61*t*
Polyelectrolytes, 210
Polymer
  heterochain, 162
  homochain, 162
  inorganic, 161–164
Polyphosphazene, 161–162*f*
Polysilane, 161–162, 161–162*f*
Polysiloxane, 161–162, 161–162*f*
Postcombustion capture, 475–476
Potash, 39–42*t*, 67–74*t*, 297–304*t*
Potassium, 499–521*t*
Potentially responsible parties (PRPs), 407
Precipitation, 343, 380–381, 448
Precombustion capture, 475–476
Premanufacture notification (PMN) requirement, 421
Preservatives, chemistry of cleaning, 442–443
Primary air pollutants, 248
Proton transfer reaction, 215–216
Publicly owned treatment works (POTW), 405
Pulverized coal, 271, 273
Pulverized fuel ash. *See* Fly ash
Purple pigments, 165–167*t*
Pyrite, 39–42*t*, 67–74*t*, 297–304*t*
Pyrophoric chemical, 218, 222

Q

Qualitative analysis, inorganic compounds, 274
Quantitative analysis, inorganic compounds, 274
Quartz, 67–74*t*, 297–304*t*
  crystals, 39–42*t*

R

Radiation, ultraviolet, 253
Radioactive decay, 345, 351–353, 351*t*
Radioactive isotope, of hydrogen, 352
Radiological contamination, drinking water, 417
Radionuclide, 351–352, 451
  contaminants, 177*t*, 454*t*
  properties and treatment methods for, 177*t*
Raman spectroscopy, 182–183*t*, 278–279, 522–524*t*
Rare-earth elements, 67–74*t*, 81–82, 297–304*t*

RCRA. *See* Resource Conservation and Recovery Act (RCRA)
Reaction
 inorganic chemical, 93–98
 terminology, 95*t*
Reactive chemicals, 216–223
 Class I chemicals, 218
 Class II chemicals, 218
 Class III chemicals, 218–219
 Class IV chemicals, 219
 Class V chemicals, 219
 Class VI chemicals, 219
 Class VII chemicals, 220
 Class VIII chemicals, 220
 general aspects, 217–220
 water-sensitive chemicals, 220–223, 221*t*
Rearrangement reaction, 353
Redox reactions, 354–355, 449
Red pigments, 165–167*t*
Reformulated gasoline, 402
Refractive index, inorganic compounds, 206–207, 206*t*
Regulatory law, 114
Relative density (specific gravity), 193–194
Relative permittivity, 209
Remediation
 environmental, 435–436
  chemistry of, 437
 of heavy metal-contaminated sites, 455–473, 456*t*
 of metal-contaminated soil, 457
Renewable energy, 240
Resource Conservation and Recovery Act (RCRA), 37, 43–44, 412–417, 423–425
Rock, phosphate, 67–74*t*
Routes of exposure, 417, 419*t*
Run-of-the-pile triple superphosphate (ROP-TSP), 127–128

S

Safe Drinking Water Act (SDWA), 114, 393–394, 417–419
Saline soil, 438
Sandstone, 39–42*t*
Sandy soil, 438
Seawater, inorganic constituents of, 452*t*
Secondary air pollutants, 248
Sedimentation, 374–375
 inorganic chemicals, 316–317
Sediments, 260, 445
 aquatic, 445
Seeding, 435
Selenium, 174–177*t*, 255*t*, 257–258*t*, 327–331*t*, 499–521*t*
Self-purification, 375
Semiconductor, 208
 lasers, 208–209
 silicon, 208–209
Semivolatile organic gases (SVOCs), 369
Sewage sludge, 262–263
Significant new use rules (SNURs), 408
Silica, 39–42*t*, 67–74*t*, 297–304*t*
 fibers, 212
Silicon, 499–521*t*
 semiconductors, 208–209
Silicone, 162–163
 polymers, 163*f*
Silt, 260, 438
Silver, 67–74*t*, 174–177*t*, 257–258*t*, 297–304*t*, 327–331*t*, 499–521*t*
Single-crystal X-ray diffraction, 280, 522–524*t*
Single-displacement reaction, 94, 340, 448
Smog, 249–251*t*
 photochemical, 368–369
 source, 249–251*t*
Soap, 439
Soda ash, 67–74*t*, 297–304*t*
Sodium, 257–258*t*, 327–331*t*, 499–521*t*
Sodium carbonate, 67–74*t*, 152–154, 297–304*t*, 441
Sodium silicate, 441
Sodium tripolyphosphate (STPP), 441
Soil
 characteristics, 377–378, 378*t*
 chemistry of cleaning, 438–439
 clay, 438
 combination, 438
 composition, 26–27
 contamination, 247–248
 flushing, 456*t*
 heavy metals in, 322, 455–457
 organic-contaminated, 438
 peaty, 438
 pH, 380
 pollutants, 239*t*
 saline, 438

sandy, 438
silty, 438
vapor extraction, 456*t*
Soil pollution, 27–30, 261, 381, 394
acid rain, 262*t*
agricultural activity, 262*t*
industrial activity, 262*t*
sources, 262*t*
Soil washing, 456*t*, 465–469
chemical extractants for, 467–469
ex situ, 465
principles, 466–467
Solidification, 456*t*
Solids, physical properties, 11*t*, 76*t*, 190*t*
Solid state, 75–76
Solid-state catalysis. *See* Heterogeneous catalysis
Solid waste, 124, 412–414, 423
Solubility, inorganic compounds, 197–198, 499–521*t*
Solvay process, 119, 137
Solvents, chemistry of cleaning, 441–442
Sorption process, 357–358, 380–381
absorption, 362–363
adsorption, 359–361
factors, 358–359
irreversible, 359
types, 359
Specific gravity (relative density), 193–194
Specific heat property, inorganic compounds, 213
Spectroscopy. *See also* Absorption spectroscopy
circular dichroism, 522–524*t*
electron paramagnetic resonance, 182–183*t*
EPR, 279–280, 522–524*t*
ESR, 522–524*t*
infrared, 522–524*t*
absorption, 182–183*t*
mass, 522–524*t*
NMR, 182–183*t*, 522–524*t*
Raman, 278–279, 522–524*t*
UV-Vis, 522–524*t*
absorption, 182–183*t*
Spilled chemicals, guidelines for segregation of, 428*t*
Spreading, inorganic chemicals, 317–318
Stabilization/immobilization, 456*t*
Stabilizing agents, 374–375
State implementation plans (SIP), 401–402
State Water Pollution Control Revolving Fund, 404
Steam methane reforming, 123
Stoichiometric coefficients, 96
Stoichiometry, 95
Sublimation
evaporation and, 337
inorganic chemicals, 318–319, 318*t*, 319*f*
Sulfate, 257–258*t*, 327–331*t*
Sulfur, 39–42*t*, 67–74*t*, 297–304*t*, 499–521*t*
recovery, 157–160
Sulfur dioxide, 247, 252
environmental effects, 241*t*
source, 249–251*t*
Sulfuric acid, 154–157
Superfund. *See* Comprehensive Environmental Response, Compensation, and Liability Act (CERCLA)
Superfund Amendments and Reauthorization Act (SARA), 407–408, 418
Surface active agent. *See* Surfactants
Surface-based homosphere, 14–15
Surface emission, 16
Surface organometallic compounds, 226–227
Surfactants, chemistry of cleaning, 439
Synthetic ammonia, 123
Synthetic detergents, 439, 442

T

Talc, 39–42*t*
Tantalum, 67–74*t*, 297–304*t*
Tartaric acid, 468*f*
Taste, groundwater, 326*t*
Terrestrial biosphere, 23–30
chemistry in, 377–381
Thallium, 255*t*, 257–258*t*, 327–331*t*
Thermal conductivity, inorganic compounds, 207
Thermal desorption, 456*t*
Thermal pollution, 253–254
Thermophilic bacteria, 364–365
Tin, 499–521*t*
Titanium, 67–74*t*, 297–304*t*, 499–521*t*
Titanium dioxide, 161

Toxic air pollutant, 248
Toxic chemicals, 395, 408, 419–420, 419*t*, 428*t*
 inorganic, 453
Toxic metals, 27
Toxic soil pollutants, 27–28
Toxic Substances Control Act (TSCA), 409, 419–421, 419*t*
Transition metal, 65*t*
 and metal cations, 59*t*
Triple point, inorganic compounds, 196, 197*t*
Trona, 39–42*t*, 67–74*t*, 297–304*t*
Troposphere, photolysis reactions in, 350
TSCA. *See* Toxic Substances Control Act (TSCA)
Tungsten, 67–74*t*, 297–304*t*
Turbidity, groundwater, 326*t*

U

Ultraviolet radiation, 253
Ultraviolet-visible (UV-Vis) spectroscopy, 277, 522–524*t*
 absorption, 182–183*t*
Underground injection control (UIC) systems, 418
United Nations Framework Convention on Climate Change (UNFCCC), 36
United States Environmental Protection Agency (US EPA), 44
Uranium, 67–74*t*, 297–304*t*, 352
Used Oil Recycling Act, 412, 423–424
US Environmental Protection Agency (US EPA), 345–346, 433

V

Vanadium, 67–74*t*, 297–304*t*, 499–521*t*
Van der Waals interaction, 84*t*
Vapor density, 193–194
Vapor pressure, inorganic compounds, 199–205, 200–205*t*
Vegetative remediation. *See* Phytoremediation
Very large-scale integrated circuits (VLSIs), 208
Vibrational spectroscopy. *See* Infrared (IR) absorption spectroscopy
Viscosity, inorganic compounds, 207–208
Vitrification, 456*t*
Volatile organic compounds (VOCs), 248, 253, 357, 369

W

Waste
 corrosive, 414
 disposal, 381
 management, 388–389, 396
Water
 primary inorganic chemical contaminants in, 255*t*
 properties, 179*t*
Water impounded hopper (WIH) system, 272
Water-in-oil emulsion, 312–313
Water pollutants, 20, 239*t*, 254*t*
Water pollution, 253–261, 371, 391–394, 403–406
Water quality laws, 392, 395–396
Water-reactive chemicals. *See* Water-sensitive chemicals
Water reservoir, 255
Water-sensitive chemicals, 220–223, 221*t*, 428*t*
Water vapor, 14, 16
Wellman-Lord process, 159–160
Wet-bottom boilers, 273
Wet deposition, 16
White pigments, 165–167*t*

X

X-ray crystallography, 182–183*t*
X-ray diffraction, 522–524*t*
 crystallography, 280

Y

Yellow pigments, 165–167*t*

Z

Zeolites, 39–42*t*, 67–74*t*, 297–304*t*
Zinc, 67–74*t*, 174–177*t*, 257–258*t*, 297–304*t*, 327–331*t*, 499–521*t*

Printed in the United States
By Bookmasters